Introduction to Radar Remote Sensing

IEEE Press

445 Hoes Lane

Piscataway, NJ 08854

Introduction to Radar Remote Sensing

Howard Zebker

Stanford University
Stanford, CA, USA

IEEE Press
WILEY

Published by John Wiley & Sons, Inc., Hoboken, New Jersey.

Library of Congress Cataloging-in-Publication Data has been applied for:

Hardback ISBN: 9781394426447
ePDF ISBN: 9781394426461
ePub ISBN: 9781394426454
oBook ISBN: 9781394426478

Cover Design: Wiley
Cover Image: 3DJustincase/stock.adobe.com

This textbook is dedicated to my mentor, colleague, and friend Len Tyler, without whose early interest and help this book would not have been possible. Len and I co-taught an introductory radar course for graduate and advanced undergraduate students for many years, and this work is derived primarily from the notes and problems developed from that work. His insights and subsequent approach to presenting this material in a new perspective well-suited for students interested in radar as a remote sensing tool were seminal in what became this book. I am most grateful for this and regretful that he was not able to see this through to completion, but his spirit has guided the development of the work since the beginning.

Contents

About the Author

Howard Zebker is the Kwoh-Ting Li Professor of Engineering and Professor of Geophysics at Stanford University. His research group specializes in interferometric radar remote sensing applications and technique development, as applied to studies of Earth and solar system. Originally a microwave engineer, he built support equipment for the SEASAT satellite synthetic aperture radar and designed airborne radar systems. He later developed imaging radar polarimetry, a technique for measurement of the radar scattering matrix of a surface. He is best known for the development of radar interferometry, or InSAR, leading to spaceborne and airborne sensors capable of measuring topography to meter-scale accuracy and surface deformation to mm scale. He has served as science team member for a number of NASA space missions, including the Spaceborne Imaging Radar series, Cassini Mission to Saturn, and several missions leading up to the present-day NISAR radar remote sensing satellite. His work has primarily concentrated on analysis of data acquired by the radar/radiometer instrument and the development of new application areas. He is a Member of the National Academy of Engineering and a Fellow of the AGU, IEEE, and Electromagnetics Academy. He has served on a number of National Academy panels, mostly for the Space Studies Board, plus the Naval Studies Board Advanced Radar Technology Panel. He holds a joint appointment with the Department of Geophysics and the Department of Electrical Engineering at Stanford, where he also serves as Associate Chair for admissions.

Preface

The origins of radar are arguably scientific. The vast, overwhelming impetus for its development, however, came from military necessity in WWII, continued throughout the "Cold War" era, and still drives many of the newer technologies that are applied to modern systems. So important had radar become in military operations that the dramatic and sudden deflation of the old Soviet Union's massive investment in radar systems for defense by the revelation of "stealth" technology in the West is sometimes stated as a major factor in the collapse of the Eastern Bloc. From these military roots have sprung countless civil and environmental applications that today foster the development of ever more complex systems built on the foundations of the earlier defense emphasis.

The period at the end of WWII found the inventors of radar in a strong technical position with the intense, recent experience of wartime, ready to carry on in new areas. This resulted in the rapid development of entirely new fields, as well as the application of radars to problems of peacetime. Radar astronomy and the imaging techniques of radio astronomy grew out of this period. Methods applied to the study of the moon and planets were also applied in the form of new mapping systems to Earth, with closely parallel developments in military and civilian sectors. Technical developments leading to extremely powerful systems for ballistic missile defense proved useful for the scientific study of Earth's atmosphere and ionosphere. Air defense systems and techniques provided the basis of air traffic control; airborne radar for fighter interceptors became weather avoidance radars in modern airliners. At present, the "plowshares" of radar are numerous and pervasive in society. How many know that the airliner in which they are riding is not permitted to take off without an operating radar if the flight path will bring it into the vicinity of thunderstorms?

In this setting, the literature of "radar" comprises well over one million papers and archived reports, plus a large number of books, conference reports, and uncounted internal documents held by industrial concerns. Despite the enormous number of works and its central role in the operation of technical society, there appears to be no introductory work suitable for a broad presentation in a general context of the principles and applications of modern radars. There are some excellent works to be sure, but these tend to be specialized, perhaps in an attempt to manage the scope.

As the authors of these notes, we are interested in providing an introduction to the subject that sets out the basic characteristics of radars from first principles, in a manner suitable for readers with technical and scientific training who are interested in a broad range of applications. At the same time, we wish to promote thinking about new applications, particularly in remote sensing and scientific uses of radars (one of us is an engineer-geophysicist, the other an engineer-astronomer). To this end, we draw on a wide range of techniques and examples from the past, as well as from developing systems likely to be important in the future.

This volume is intended as an introduction to the several most fundamental and significant of these topic areas and to show how each fits into or affects the performance of a radar system. As indicated, the authors are particularly interested in the use of radar for environmental remote sensing in the broad meaning of this term. For this reason, the applications touch not only sensing of environmental parameters such as wave height and direction at sea, or tracking of hazardous effluents in the atmosphere, but also local and global aspects of solid Earth geophysics. Consequently, our approach to the subject is intended to provide a unified overview that can be applied across all radar systems. From these fundamentals, we hope that readers can build systems best adapted to meet the challenges of the future.

About the Companion Website

This book is accompanied by a companion website.

www.wiley.com/go/zebker

To access the website materials, click on the link under Related Resources on the book's page on wiley.com. You can also find this by searching http://www.wiley.com for the ISBN 9781394426447.

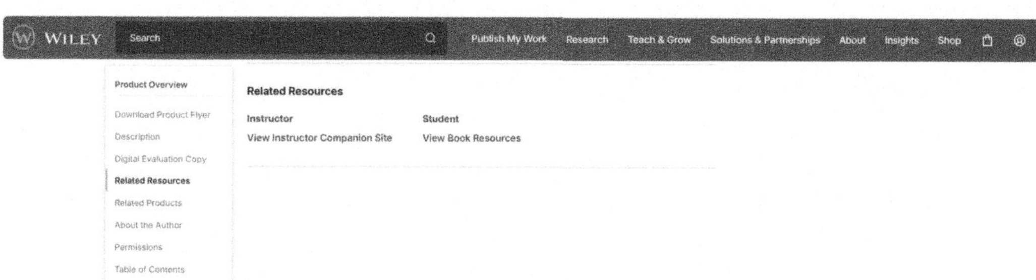

The website includes:
Problems with solutions

1

Introduction and Overview

This book is a broad introduction to the theory and practice of modern remote sensing radars. Radars are one of the earliest engineering systems developed yet one that is relevant today, as new technologies enable the application of fundamental principles to modern needs. We begin with a brief historical background that provides an interesting example of improvised, creative design, as well as an illustrative example of two fundamental modes of radar operation.

The well-known acronym stands for radio detection and ranging, derived from its original purpose, which was to sense the presence of remote or obscure objects and to simultaneously estimate the distance to those objects. The source of the term appears to be the US Navy, which first coined it in the early years of World War II, an intense era of radar development, but the US Army and the British also claim it. While maintaining its original technical meaning, today the term "radar" is used in common speech to describe a range of subjective inferences as well. Today, in addition to its technical "systems" aspect, the word has become an ordinary noun in common speech where it refers to a physiological state of awareness. In many ways, though, the two uses are very similar as the original radar referred to a means of "seeing" with radio waves while subjective "sensing" refers to the use of human intelligence and learning beyond vision.

In popular conception, radar devices operate by emitting a "searching" beamed, burst of electromagnetic energy into space, and then "listening out" for an echo. When an echo is received, a target is said to be "detected." Simultaneously, the position of the target is determined from the combination of the direction of the beam and the time required for the pulse to reach the target and for the echo to return, which allows calculation of the distance, or "range," to the target.[1] Most radars operate somewhat in this fashion. Furthermore, there are many subtleties not revealed by the simple picture just described. As one example, targets are typically in motion with respect to the radar, and the relative motion of the target has an important impact on the optimal design of radar systems. Beyond such subtleties, an increasing number of important systems rely on configurations that are very different from a simple "listen for the echo" system.

Our goal in this work is to set forth the underlying principles governing the performance of all radars, with an emphasis on those of importance of broad interest for civil, geophysical, and environmental applications of remote sensing. At the same time, we hope to convey some of the practical lore of radar that current and future radar users, designers, analysts, and managers of radar systems will find helpful in understanding and addressing broad aspects of radar design related to remote sensing. Radar grew from an amalgam of skills drawn from several fields, including such

1 Readers will note that much of the jargon associated with radar systems arises from historical military roots, which drove early development. These terms remain today as they describe specific traits even if they are no longer used in their original context.

esoteric areas as electromagnetic wave propagation and scattering, optics, antennas, and theories of information, signal detection, and games, to name some of the more prominent ones. Currently, work on modern radar systems continues both to depend on and motivate these fields. For this reason, it is necessary to draw on sometimes disparate sources in order to construct a reasonably complete picture of how radar systems work and the manner in which various components of radar systems are optimized to best address a particular design problem.

Successful application of radar to scientific studies and exploration requires a thorough understanding of the subject of investigation as well as the intended application area. This is the *sine qua non* of the scientific use of radar. As an example, one who knows radar cannot become an oceanographer solely by studying the characteristics of radar echoes from the sea. But individuals or teams of individuals possessing detailed knowledge of physical oceanography and radar have produced marvelous systems for remotely sensing, e.g., from ships or satellites, properties of the sea based on observations of radar signals reflected by ocean waves. Among the several skills needed, successful ventures of this type inherently require an understanding of the interaction of electromagnetic waves with the rough surface of the sea. Consequently, in this work, we seek to provide an understanding of how radar echoes are formed. Stated differently and more generally, we address the interactions of radar wave interactions with objects and surfaces inherent in remote sensing of the environment, which is more specifically known as "scattering."

1.1 Concepts and Origins

The fundamental properties of electromagnetic waves necessary for the functioning of a radar, including their reflection by metallic and dielectric objects, were demonstrated in 1886 by Heinrich Hertz, who carried out a series of laboratory experiments and demonstrations of wave generation and reflection phenomena at a wavelength of only 0.66 m. Subsequently, the earliest system developments required half a century to emerge. The earliest practical demonstration of reflections of radio waves for detection of remote objects followed at the turn of the century, as documented by patents for a land-based system awarded in 1904 to Christian Hülsmeyer on the basis of a *detection* of ships at ranges up to 3 km.[2] Marconi, who founded and built a thriving business on long-wave radio communications, suggested in 1922 the use of self-contained electromagnetic detection systems for use onboard ships. The first known demonstration of a practical US ship detection system in the United States also occurred in 1922 when Taylor & Young at the US Naval Research Laboratory used a 5-m wavelength, continuous-wave interference system to detect a wooden ship steaming on the Potomac River near Washington, DC. Three years later, in 1925, Breit & Tuve of the Carnegie Institution of Washington used a pulsed system to determine the height of the ionosphere, thereby providing an example of a time-domain implementation of a ranging system operating at a roughly 75 km range. By then, the fact of remote sensing of objects by electromagnetic systems was well established, although it would be several more years before a compact system with collocated transmitter and receiver and improved precision was achieved, and 15 more before the acronym RADAR was created. Progress in the 1930s was extremely rapid following these early steps, with many of the earliest and continuing developments in the United States at the Naval Research Laboratory carried out in secrecy. Similar development work was in progress in England and Germany, with the British making critical contributions to efforts in the United States.

2 That is, Hülsmeyer's device could remotely sense the arrival or departure of a ship with the range of his apparatus, but could not accurately determine the position.

Research and development related to military radars provided the technical underpinnings and skills central to later developments in electronic computers as well as the machines of high-energy physics. For example, design and operation of the large particle accelerators rely on the availability of very high-power klystron microwave amplifiers developed for radar to provide the energizing waves for the electron and positron beams. Picosecond resolution timing circuits employed originally in nuclear reaction analyzers and now routinely in semiconductor laboratories are the descendants of trigger circuits developed for switching early radars. At the same time, radar benefited from developments in electronic capabilities. For example, data modems and the earliest computer networks were created as links connecting multiple radar observation sites with a central data processing facility for the purpose of tracking aircraft.

Figure 1.1 illustrates the principle of the interference system employed by Taylor & Young in 1922 to detect the presence of vessels on the Potomac River. The experimenters placed a continuous wave, or "CW," transmitter and receiver on opposite shores. Signals radiated by the transmitter reached the receiver by the direct path between the two, and also by a reflection path from boats on the river. Both signals appear at the receiving antenna, where the output signal is the linear sum of the two inputs. The receiver was a simple amplifier with a rectifier serving as the "detector" for which the receiver output was the signal envelope, i.e., a voltage proportional to the magnitude of the summed input signal. The phase and amplitude differences between the direct and reflected signals depended on the difference in the two path lengths and on the "disturbance" of the illuminating wave, which in turn depended on the electromagnetic wave *scattering* properties of the boat. As a result of the boat's motion, the characteristics of the reflected signal, in particular its *time delay* with respect to the signal received directly from the transmitter, varied at the receiver. Each change of 2π in the phase difference between the two signals indicated a change in the path difference of one wavelength, which was apparent at the output of the detector. Through this mechanism, the motion of the boat resulted in a regular pattern of variations in the receiver output, thus revealing the boat's presence and some information about its progress on the river (see Figure 1.1, inset).

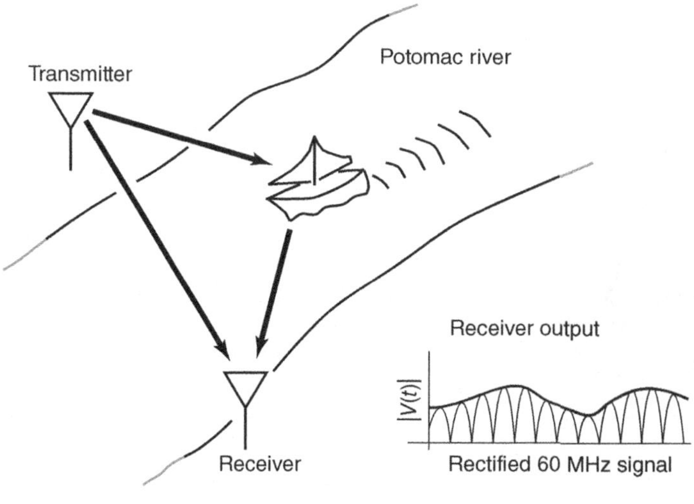

Figure 1.1 Geometry of original Taylor & Young interference radar system, 1922. Signals reflected from boat on the river and those propagating directly from the transmitter "interfere" at the receiver. Beats between direct and reflected signals are "detected" by rectifying the receiver output.

This experiment is widely accepted as the first demonstration of a radar system, although Hülsmeyer's patent precedes it, while others reject it as incomplete.

A little thought and some simple calculations will show that the rate of signal variation can be related to the combination of the position and velocity of the boat relative to the transmitter and receiver. On the other hand, the strength of the modulation is related to the size, shape, orientation, and position of the boat, in that these affect the strength and phase of the scattering. So we immediately see that the performance of the radar depends on the combination of the hardware system, the geometry, and the electric wave scattering characteristics of the "target." The geometry and arrangement of the Taylor & Young experiment are referred to as a *bistatic* configuration, denoting the well-separated transmitter and receiver, as well as aspects of the system design that depend on the separation. Users of Wi-Fi and cellular systems will recognize the parallels between the phenomena described above and the multipath communication problem in which interference between directly propagating and reflected digital signals can have deleterious effects on the performance of the data link. The underlying interference phenomenon is physically the same as that observed directly by Taylor & Young, with the primary difference being use of an unmodulated waveform to detect the boat.

Figure 1.2 illustrates a second type of system fielded by Breit & Tuve in 1925, just three years later. In this instance, the experimenters for the first time, employed a *pulsed* radar to measure the time required for an electromagnetic wave to travel to and from the radar located on the ground and Earth's ionosphere, which is the shell of ionized, conducting gas found in Earth's outer atmosphere. Breit & Tuve found it necessary to place the transmitter and receiver on opposite sides of a small hill in order to prevent the transmitted signal from overloading the receiver amplifier. Signals emitted by the transmitter reached the receiver along an near vertical, transmitter-ionosphere-receiver path

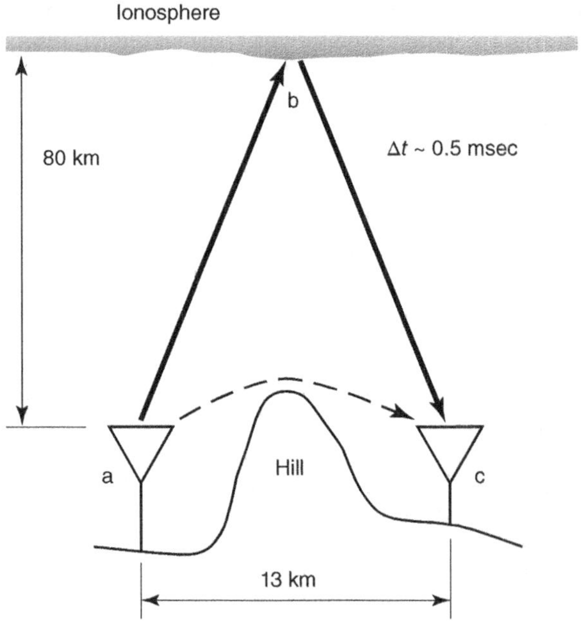

Figure 1.2 Breit & Tuve setup for Ionospheric height measurement. Height of the ionosphere was first measured by Breit & Tuve using a pulsed-signal reflection system. Folklore has it that the hill in this experiment was Capitol Hill in Washington, D.C., although we have not been able to firmly validate this claim.

$\overline{abc}$. In addition, there was considerable leakage between the transmitter and receiver by diffraction, scattering, and ground wave propagation via the over-the-hill path $\overline{ac}$. Because the shortest pulses that could be generated with the technology of the period were of the order of milliseconds in duration, signals traveling to the receiver along the two paths overlapped somewhat in time (see Figure 1.1). In the absence of the hill acting as a screen, the ionospheric echo would have been undetectable in the presence of the strong pulse from the nearby transmitter.

The arrangement of Breit & Tuve's measurement is fundamentally different from that of Taylor & Young's in that the transmitter and receiver are geometrically close to each other as viewed from the reflection point in the ionosphere. Except for the need to use the hill to improve the dynamic range of the system, the transmitter and receiver could share a common location. A configuration in which the transmitter and receiver are collocated is referred to as *monostatic*, implying that only *backscattered* echoes are observed (Figure 1.3).

Early British work was closely related to that in the United States and led the field in important ways. By the beginning of 1940, a number of radar stations operating at 25 MHz ($\lambda = 12$ m) were in place along the eastern and southern shores of England and Scotland for the detection of enemy aircraft approaching the coast from mainland Europe. This system was known by the code name "Chain Home." Chain Home could operate in either a cooperative interference mode employing two or more stations, similar to the Taylor & Young experiment on the Potomac, or in a single station mode with the same site transmitting and receiving, similar to the original Breit & Tuve configuration albeit with the use of antennas directed toward the horizon. Thus, the Chain Home system could operate in either a monostatic or bistatic mode. These radars provided data on the location and speed of approaching aircraft that were critical to the efficient use of the numerically weaker British fighter force.

A British invention critical to the future course of radar design was made by Boot and Randall in 1940. The *magnetron* oscillator/amplifier was the first high-power microwave tube capable of producing very large pulse powers at frequencies above about 300 MHz, corresponding to wavelengths much shorter than 1 m. This development made possible the construction of pulsed transmitters capable of producing transmissions of only 1–10 μsec duration, or 100–1000 times shorter than those employed by Breit & Tuve. The result was a practical means of achieving high time-of-flight range resolution in combination with good angular accuracy with conveniently sized antennas. These characteristics made possible efficient, compact designs suitable for use as detection and localization systems onboard aircraft and ships, as well as for early aircraft remote sensing systems. Subsequently, the vast majority of radars are refined extensions of early monostatic

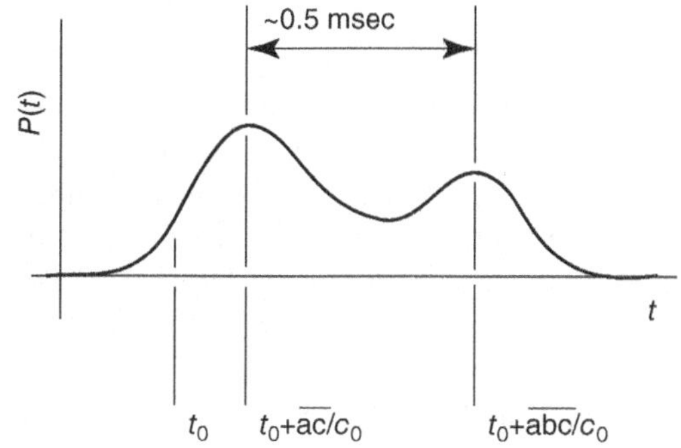

Figure 1.3 Breit & Tuve ionospheric echo: direct and echo signals received by Breit & Tuve overlapped as a result of length of transmitter pulse. Transmission time is t_0, speed of light is c_0.

designs, although accumulated progress in transmitter, receiver, antenna, and signal processing technology renders a modern radar virtually unrecognizable at first glance as a successor to the earlier forms.

The systems employed in the two experiments by Taylor & Young and Breit & Tuve are the forerunners of all modern radar designs for which the four corners of the design space are (i) monostatic, (ii) bistatic geometries, (iii) pulse-modulation, and (iv) continuous-wave transmission. Modern radar systems provide examples of all four basic types, although bistatic systems and geometry are relatively unusual. The primary conceptual extensions beyond the early stages are the broad development of modulation and signal processing techniques and the use of *multistatic geometries* in which the number of transmitters plus receivers is more than two, requiring the addition of a fifth vertex to our bounding box.

The literature of radar contains literally millions of entries, and no one can be familiar with the full breadth of this work. Furthermore, an individual can devote an entire career to one of the subspecialities related to radar, such as transmitter or antenna development, signal processor and detector design, or analysis of radio wave scattering. But not to worry, the fundamentals or radar systems are relatively few, and we introduce and explain the important concepts here.

The appendices contain a bibliography providing a selection of works and critical comments, which the reader can use to guide further exploration of the field.

1.2 Modern Uses and Related Applications

What remained at the end of 1925, and is ongoing to the present, is the task of understanding fundamental constraints on the performance of radar systems and then in technological development, with the goal of finding refined implementations that approach known limits of optimum performance. While the frontiers of the last century were, for example, the limits of resolution, optimal design of waveforms, and digital realization of signal processing, current problems in remote sensing deal with issues such as optimal methods to distinguish between cultural and natural features of Earth's surface, estimation of the global biomass, the thickness of arctic permafrost, and the depth of the ice layer on Jupiter's moon Europa, detection of minute changes in the topography of Earth, and estimation of surface winds over the ocean. While the genesis and primary development of radar has its roots in military and navigation systems, the techniques and technology of radar have migrated to many other fields, as have the applications and the questions addressed.

Today, many radars are designed specifically for scientific and civil purposes, while others are adaptations of military radar systems to civilian ends. Consider that there are at least four broad categories of radar, as listed in Table 1.1.

While the concept of radar and the ideas discussed in this volume consider primarily electromagnetic systems, we note that there are also a number of systems based on the use of mechanical, rather than electromagnetic waves. Indeed, the waves employed in radar-like systems need not be electromagnetic. Table 1.2 lists some examples of sensing techniques that are either near or direct analogs of radar. In these instances, radar methods are adapted for the use of acoustic or seismic waves, or electromagnetic waves are used in an unusual manner or in an unconventional setting. Predating these are the evolved biological adaptations evident in nature, but which exhibit striking parallels to modern systems.

As indicated above, the topic of radar draws from many disciplines. The mélange of ideas and concepts underpinning the development of modern radar touches or incorporates a wide range of topics presently considered part of electrical engineering systems and principles, encompassing coding, information, detection, and game theory, signal processing, very high-speed switching, communications, computation, and data networking, all resting on basic physical principles,

Table 1.1 Some major categories of radar systems. Lists such as these are necessarily incomplete. Note the breadth of applications stretching across a wide range of cultural and economic segments of society.

Category	Function	Subjects of interest
Astronomy	Scientific study	Planets, asteroids, comets, Sun
	Space exploration	Navigation, planetary sensing
Civil	Surveillance and control	Ships, aircraft, motor traffic
	Detection	Door openers, burglar alarms, collision avoidance
	Sports	Ball speed
	Weather	Rain tonight?
Defense	Surveillance and detection	Vehicles, personnel
	Tracking, guidance	Terrain avoidance, landing systems
Environment	Scientific	Atmosphere, ocean, solid Earth
	Geophysical	Earth's shape, continental drift
	Observation	Nocturnal birds and insects in flight
	Monitoring	Seismic shift, sea state, ice flows
	Policy and management	Crop yields, forest health

Table 1.2 Radar-related technologies. With the exception of lidar, which is the optical analog of radar, the types and examples listed exploit the properties of mechanical, rather than electromagnetic waves. Seismic probing can be based on the use of natural sources, as in the example of earthquakes, or intentional man-made sources such as controlled explosions. An understanding of the principles of radar can be applied directly to the mechanisms of many of these examples.

Probe type	Functions	Examples
Acoustic probes	Imaging	Medical diagnosis, nondestructive testing
	Material science	Determination of physical properties
	Acoustic microscope	Physical science experiments
Lidar	Remote sensing	Determination of effluent species
	Ranging	Surveying, Earth motions, space science
Seismology	Earth sensing	Imaging Earth's interior
	Exploration	Location and evaluation of minerals
Sonar	Oceanography	Mapping of ocean bottom, navigation
	Military	Underwater detection and communication
Biological	Prey detection	Bats, moths, porpoises
	Navigation	Dolphins, whales

including electromagnetics, wave propagation and scattering, and aspects of atmospheric science, oceanography, and geophysics, etc. Each of these topics, and many others, comes into play in some important aspect of the radar problem. Every system designer needs to be aware of these issues and, in the technical specialties, some designs require collective expertise in one or more areas.

In this volume, we show how the several most significant of the topics above fit into or affect the performance of a radar system. To begin, Chapters 2 through 10 address fundamental issues of radar design and performance applicable to all radar systems, largely independent of intended application. The remaining chapters build on the fundamentals to develop a foundation for the study of advanced systems.

There are many histories of the early years of radar and its development, each interesting in its own right. We note that some of these works are replete with interesting examples of what was done and why, but also reveal the thinking behind how things are done. Two of the more popular are those by R.M. Page *Origins of Radar*, and A. Price, *Instruments of Darkness*, (*v.* Bibliography) which provide a comprehensive study of developments in the United States during WWII. A well-researched, recent history is *The Invention That Changed the World...*, by R. Buderi. Buderi's discussion nicely ties together the prominent American threads leading to important developments, primarily at MIT Radiation Laboratory, but also including the early British contributions.

Buderi's work is likely to become a standard reference for US contributions in WWII through the height of the Cold War. A more personal account of early developments from a British point of view is given by E. G. Bowen in *Radar Days*, which describes contributions by the author to early research and development in radar, and his leadership of the development of airborne systems; the climate of the initial work in the field closely resembles that in the competitive startup environment of today, and is instructive reading. The work by H. von Kroge at GEMA is also a personal account reflecting the development of radar in Germany. Such works are valuable for their contribution to the history of our subject and its technology, and for the insights they give into the early applications, the ways in which technical innovation occurs, and the evolution of countermeasures that often drive military radar development. Some of the technical thrusts/reports presage the tactics and strategies employed by modern computer manufacturers in jockeying for position in the modern marketplace.

In the rest of this work, as you will see, based on the large number of ways in which the basic physical phenomena can be exploited, we take a very broad view as to the definition of radar and include a wide range of configurations and unconventional systems of operation within the rubric. We endeavor to show that the basic principles of radar design, operation, and analysis have expanded to unanticipated applications in many topical fields. In one example, a modern manifestation of the Taylor & Young geometry, unmodulated signals transmitted by the Voyager and Cassini spacecraft and received on Earth have been used to study properties of the material making up the Rings of Saturn. "Scattering" of the radio signals by particles within the rings can be detected on the ground at times when the spacecraft passes near or behind the ring system. In such instances as these and for the Taylor & Young experiments, we think in terms of a disturbance of the sensing wave rather than an "echo" because the disturbed wave does not return to the source the way an "echo" does. While these two experiments at first glance seem quite different, the analysis and interpretation of the data are nearly the same, albeit accomplished with very different technologies. This is the trend of the future, in which developments in communication and signal processing make practical ideas and implementations that would astound the workers of "early days."

2

Basic Radar Measurements

As seen in Chapter 1, radar systems are based on the detection and measurement of disturbances or reflections—the general term is *scattering*—of electromagnetic waves by an object of interest. While there are a great many ways in which this can happen, as we shall see in later chapters, from the perspective of an observer at the location of the radar receiver, an electromagnetic signal possesses only a limited number of discernible characteristics or parameters. Radars operate on the basis of these physical properties—features of electromagnetic waves that we can isolate and measure—rather than some arbitrary, desirable characteristic that we might wish upon them. Here we summarize the basic properties of electromagnetic waves and identify associated observables, i.e., measurable characteristics, that can serve as the basis for the design and operation of real systems, and give examples of basic system architectures.

2.1 Wave Observables

The simplest representation for an electromagnetic signal in free space wave is a linearly polarized plane wave with cosinusoidal form, which we write as:

$$\mathbf{E}(t, \mathbf{r}) = \mathbf{A}\cos[2\pi ft - \mathbf{k} \cdot \mathbf{r} + \phi] \tag{2.1}$$

where $\mathbf{E}(t, \mathbf{r})$ represents the time, t, and space variation, $\mathbf{r}$, of an electric-field plane wave propagating in the direction of the *wave vector* $\mathbf{k}$. While the individual variables on the right-hand side of Eq. (2.1) characterize four basic physical parameters of the wave: (i) the temporal rate of oscillation, or *frequency* f, of the wave, (ii) the spatial variation, $\mathbf{k}$, with rate of spatial variation $|\mathbf{k}|$ and *direction of propagation* $\mathbf{k}/|\mathbf{k}|$, (iii) the angular offset of the source relative to the origin of our coordinate system, or *phase* ϕ, and (iv) the vector magnitude of the wave force, or *wave polarization*, $\mathbf{A}$, with *amplitude* $|\mathbf{A}|$ and *direction* of the field $\mathbf{A}/|\mathbf{A}|$. In addition, the total value of the argument of the cosine in Eq. (2.1) is the *instantaneous phase* which determines the value of the cosine at a point in space-time. A similar expression for the accompanying magnetic field, $\mathbf{H}$, would have the same form but with the orientation as prescribed by Maxwell's equations.

In addition to the four wave parameters above, there is an implicit, fifth observable, which we refer to as *modulation*, that depends on the manner in which the wave is manipulated at its source. Modulation imposes a *time variation* on the wave parameters. Most common, perhaps, is switching of the wave On and Off by forcing the magnitude $|\mathbf{A}|$ to vary between zero and some finite value. Modulation of the waveform allows measurement of the *time delay* or *flight time* between the instant in which a radar signal is emitted by the transmitter and that when it arrives at the receiver. This is accomplished by measuring the time that lapses between the imposition of a particular modulation pattern on the signal emitted by the radar transmitter and that time when the same pattern of time variation is observed in the received signal waveform. The ability to measure time delay between signal transmission and reception on the basis of time variation of the wave parameters is perhaps the most fundamental—certainly the most exploited—of the five observables in that it directly constrains any solutions for the transmitter-target-receiver path length in an accurate manner. Modulation can be applied to any of the physical parameters of the wave, so that in general we can have $\mathbf{A}(t)$, $f(t)$, $\mathbf{k}(t)$, and $\phi(t)$. It should be recognized, though, that these four are not all independent, as we shall see below.

Taking the four physical parameters in turn, the temporal frequency of a wave denotes the number of oscillations per unit time at a fixed point in space. As frequency is perhaps the best known and understood property of the wave, we do not discuss it further here.

The spatial rate of variation of the wave is proportional to temporal frequency and inversely proportional to wavelength. This is inherent in the magnitude of the wave vector,

$$|\mathbf{k}| = \frac{2\pi}{\lambda} = \frac{2\pi f}{c} \tag{2.2}$$

where λ is the wavelength and c is the speed of light in the medium. From this, $(\mathbf{k} \cdot \mathbf{r})/(2\pi f)$ is the time delay for a signal to reach position $\mathbf{r}$ at the speed of light.

It is easy to show that the expression,

$$\mathbf{k} \cdot \mathbf{r} = \text{constant} \tag{2.3}$$

is the equation of a plane perpendicular to $\mathbf{k}$, and that the maximum rate of spatial change is in the direction $\mathbf{k}$.[1] Increasing the value of the constant on the right-hand side of Eq. (2.3) by 2π changes the position of the plane by the one wavelength in the $\mathbf{k}$ direction. Viewed from another perspective, changing the position vector, $\mathbf{r}$, by one wavelength parallel to $\mathbf{k}$ changes the value of the dot product in Eq. (2.3) by 2π radians. From this, points of fixed instantaneous phase on the curve of the cosine line up along planes perpendicular to $\mathbf{k}$, and repeat, modulo(2π), regularly spaced by one wavelength in direction $\mathbf{k}$. Examination of the argument of the cosine, Eq. (2.1), as time progresses, shows that the cosinusoidal waveform moves in the direction of $\mathbf{k}$ at the speed of light. Thus, Eq. (2.1) describes a wave propagating in the direction $\mathbf{k}$ while $\mathbf{k} \cdot \mathbf{r}$ describes the total phase retardation at position $\mathbf{r}$ with respect to the origin of our spatial coordinate system.

The phase offset, ϕ, is a measure of the timing offset of the oscillations of our wave relative to the phase measured at the origin. Setting $\mathbf{r} = 0$ in Eq. (2.1) shows that the oscillations of the reference wave have a positive zero crossing of the cosine at time $t = -\phi/2\pi f$. Positive values of ϕ indicate that the field strength in Eq. (2.1) crosses zero going positive earlier than, i.e., before, a reference wave with no phase offset, and vice versa for negative values of ϕ. Because of the cyclic nature of the cosine, the behavior of the wave is repetitive for each change in ϕ by an integer multiple of 2π. For this reason, the use of ϕ as a measure of time is ambiguous by integer multiples of $1/f$.

The polarization vector $\mathbf{A}$ describes both the strength of the wave and the direction in which the electric force acts. When we are concerned primarily with the wave strength, we use the term amplitude to refer to $|\mathbf{A}|$. We reserve the term polarization for instances in which the direction of the electromagnetic field, i.e., the orientation of $\mathbf{A}$ is to be emphasized. Polarization of a radiated wave is determined by the structure of the transmitting antenna, but usually is modified by scattering interactions with the radar target, with objects along the propagation path, and in certain cases by propagation through anisotropic media such as Earth's ionosphere. At the receiver, the component of a polarized wave sensed is determined by the structure of the receiving antenna.

---

1 Write out the vector form of Eq. (2.3), as a scalar equation by expanding the dot product, and then compare the result with the standard form for the equation of a plane. See the problem at the end of this chapter.

As indicated above, the basic electromagnetic wave characteristics of amplitude, frequency, phase, and polarization are often functions of time resulting from a combination of intentional modulation of the radiated signal imposed at the transmitter, propagation through a time-varying transmission medium, and from scattering by moving or fluctuating objects. Variations of $\mathbf{k}$ also occur, but are almost always a consequence of a change in the frequency at the transmitter or as a result of the Doppler effect.

Apart from temporal variations of $\mathbf{k}$, there are situations in which the engineering and scientific analysis is simplified by considering $\mathbf{k}$ as the independent variables. Rewriting Eq. (2.1) to reflect the possibility of time variation of these parameters, we can express the right-hand side as:

$$\mathbf{A}(t)\cos[2\pi f(t)t - \mathbf{k}(t)\cdot\mathbf{r}(t) + \phi(t)] \tag{2.4}$$

This more general form is used often throughout the remainder of this book. As an example, modulation of the transmitted signal can be thought of as a means for "labeling" segments of the transmitted waveform by introducing known time variations into the amplitude, frequency, phase, or polarization at the transmitter. Unlike phase, which is always ambiguous on the scale of the wavelength, the use of simple waveforms for the modulation can result in deterministic relationships between the signal time of flight and the distances of transmitter and receiver along direct and/or scattered paths.

2.2 Application of Wave Observables

From the above, it is apparent that the number of variables available for manipulation in radar signal design is limited. Similarly, the number of wave parameters that can be measured and interpreted is also small. Nonetheless, these may be exploited to infer many properties of the objects observed by radar, with broad applications as follows:

Amplitude and intensity describe the strength of the wave, which is often an important factor in the interpretation of the physical characteristics of the echoing object. The polarization vector $\mathbf{A}$ represents the electric field strength $\mathbf{E}$ in V/m. The wave intensity is given by $I = |\mathbf{E}|^2/2\eta$ W/m^2, where η is the wave impedance of the medium. In some applications, it is more convenient to use the scalar amplitude $|\mathbf{E}|$ rather than the intensity; however, both are important. Obviously, the intensity and amplitude are equivalent scalar measures of the wave strength.

Frequency, or more precisely, the change in frequency associated with motion of a scattering object, allows precise determination of the motion of scattering objects based on the Doppler effect, and is commonly used for this purpose. In addition, time manipulation of the transmitted frequency represents an important, fundamental method of modulation.

Phase provides for precise measurements of relative signal arrival times at two or more receivers, which can be used to determine direction of arrival, as well as to determine small changes in radar signal time of flight. Such information is useful in determining the position, i.e., *localization*, of points within a scene or of individual scattering features relative to the radar. Given accurate observations, changes in position can be determined on the scale of the radar wavelength.

Time of flight is a measure of the radar propagation path length between transmitter, T, and receiver, R, followed by the signal. This quantity is given by $\Delta t = \overline{\mathrm{TR}}/c$, and also is referred to as the *flight time* and the *propagation time delay* of the signal.

Polarization is useful in interpreting characteristics of the echo source objects. Echoes from smooth objects and/or regularly structured objects tend to be more systematically polarized than those from irregular ones.

Wave vector orientation of a radar signal at the receiver is synonymous with the wave direction of arrival, which is related to the location of a scattering object. Some systems, those with multiple antenna configurations, are able to extract the three-dimensional direction of the wave vector. Simpler systems with a single measurement aperture providing two components of the direction are more common.

Every radar measurement and application is based on the use of one or more of the six observables above. Some systems quantitatively extend the observational set, for example, by resorting to simultaneous independent measurements at multiple frequencies, or by measurements of the variation in observable parameters as functions of frequency and time. Although such extensions can lead to very complex systems, the complexity should not be allowed to obscure the basic constraint that all inferences from radar must be based on observations of some or all of the six parameters described above. In working through the remainder of this book, it is helpful to keep in mind the opportunities opened up when combinations of these observables are available, as well as fundamental limitations imposed by the basic properties of electromagnetic waves. Yet there are many design possibilities: infinite choices of modulation, configurations predicated on statistical measures, and variations in geometry.

The form of temporal modulation employed strongly affects the accuracy and precision with which various of these characteristics can be measured while providing the means to sort out the time-history of the transmissions and echo signals. The principal ways in which this can be accomplished and the basis for the selection of particular temporal waveforms have been the subject of many innovative papers in the literature and is a major topic in the latter sections of this book.

2.3 Use of Time and Frequency Measurements in Simple Radars

Improved instrumentation later led to the use of the Doppler effect to improve the precision of relative velocity estimates, and this type of measurement forms the basis, for example, of precision tracking of spacecraft, as well as more prosaic applications, such as the use of radar "guns" to measure the speed of a pitcher's fastball. In this section, we outline the basic measurement of time-delay and Doppler shift and relate these to basic properties of the radar signal. We illustrate how measurements of range, supplemented by those of direction, can be used to monitor the landing of aircraft. Similarly, measurements of Doppler shift can be used to determine the speed of a baseball.

2.3.1 Timing Measurements

In modern radar, the time delay of the echo signal was the first, and continues to be the most widely used of the basic wave properties. That historically radar development was driven by a military need to measure the distance, or range. Subsequent realization that time-differenced values of range measurements normalized by time difference between observations yield the relative velocity between the radar instrument and reflecting objects led to additional applications.

Consider a thought experiment in which an electromagnetic pulse of energy is radiated from an antenna, propagates to a reflecting object and returns to its site of origin where it is recognized as an echo of the original, transmitted pulse. Here the observable is the time delay between the switching of $|\mathbf{A}(t)|$ from zero to some finite value, thereby initiating radiation of a signal toward the target, and the time that the beginning of the returning echo waveform is observed at the receiver. The measured time delay between transmission and reception of the pulse corresponds to the time that the electromagnetic energy was in flight. This time depends on the path traveled by the energy pulse and its speed along that path, as suggested by the observing geometry as illustrated for two targets in Figure 2.1.

Because the speed of light in the atmosphere is known and nearly constant, it is straight-forward to infer the distance from the radar to the reflecting object from a measurement of the time delay. Usually, the receiver will be nearly co-located with the transmitter, although in bistatic geometries the transmitter and receiver subsystems can be widely separated. The time delay from transmission of a radar pulse to its reception is the ratio of the round-trip distance to the speed of light, and is easily discerned on a plot of signal intensity versus time provided that the signal levels are adequate relative to the system noise. Such a plot is illustrated in Figure 2.2. In this simple example, the total round trip distance is 15,000 m for Object 1 located at a distance of 7500 m from the radar, and is 18,000 m for Object 2 at a distance of 9000 m. At the speed of light, for the echo from Object 1, the time delay is,

$$\tau_1 = \frac{\text{distance}}{\text{speed of light}} = \frac{15 \times 10^3\ \text{m}}{3 \times 10^8\ \text{m/s}} = 50\ \mu\text{s} \tag{2.5}$$

while for Object 2 the result is,

$$\tau_2 = \frac{18 \times 10^3\ \text{m}}{3 \times 10^8\ \text{m/s}} = 60\ \mu\text{s} \tag{2.6}$$

It is usually more convenient to think of the distance between the measurement site and the scattering object, or R, than it is the round-trip distance.

In monostatic observing geometries, the expected time delay will be the distance from the instrument to the target divided by the propagation velocity, multiplied by two to account for the

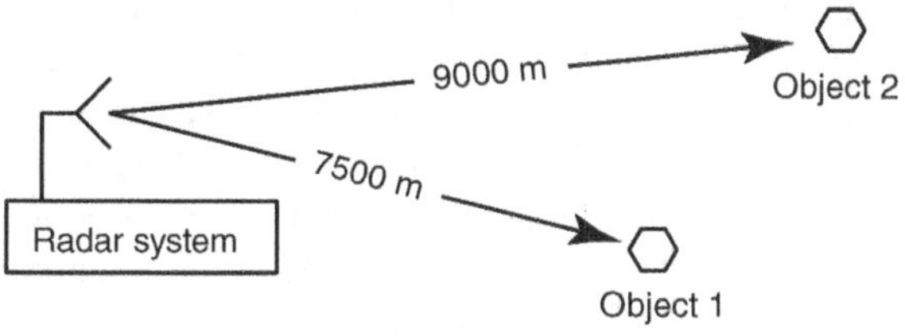

Figure 2.1 Radar system with two reflecting objects. Schematic defining range relationship between two objects sensed by a single radar. For distances as shown the round trip path followed by the radar signal is twice the distances shown.

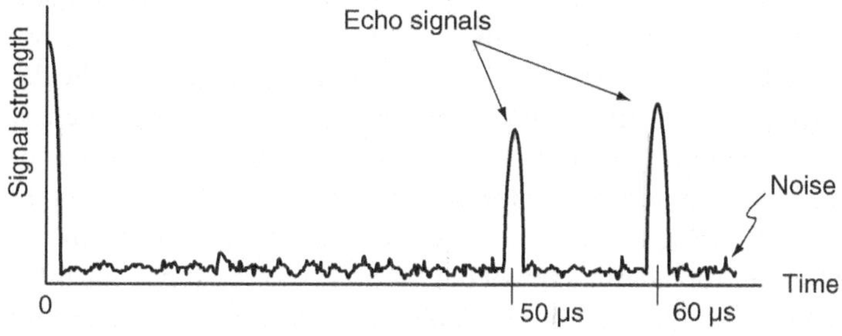

Figure 2.2 Echoes from two objects. Plot shows variation of received signal power as a function of time with echoes from objects a ranges of 7500 and 9000 m, as in Figure 2.1. Plot is typical of echo signals in the presence of noise.

round-trip distance in which the signal traverses the same path twice. Expressing range R as a function of time delay, τ for this case,

$$R = \frac{c\tau}{2} \tag{2.7}$$

where c is the speed of light in the medium. Equivalently, we can think of the ratio of the distance to time delay as one half that of the speed of light,

$$\frac{R}{\tau} = \frac{c}{2} = \begin{array}{l} 150\,\text{Mm/s} \\ 150\,\text{km/ms} \\ 150\,\text{m}/\mu\text{s} \\ 150\,\text{mm/ns} \end{array} \tag{2.8}$$

Thus, the relationship between time delay and distance may be considered simply as a distance of 150 m for each microsecond of time delay, a useful conversion factor to keep in mind for many common radar problems. When considering longer distances, such as determining the distance to the moon or other planetary bodies, a more appropriate conversion factor might be 150 km per millisecond delay, or even 150,000 km per second of delay.[2]

Often, it is more convenient to think directly in terms of the *flight time* of the radar pulse than range. In many instances, we refer to this as the *light time*, in order to emphasize that this is the time required at the speed of light for the pulse to make the round trip from transmitter to scattering object and back. We observe the time interval between the transmission of the signal and the reception of its echo.[3]

Armed with only the above analysis, we can easily understand one common radar application—a system that can track the relative positions of aircraft approaching an airport in order to allow air traffic controllers to sequence landings safely. For this purpose, suppose that in Figure 2.1 the radar system is located near the runways and that the reflectors represent not fixed targets but incoming aircraft instead. In this scenario, each aircraft will produce an echo signal at a time delay proportional to its distance from the airport. If, in addition, the radar antenna creates a narrow rotating beam to scan the horizon, then the direction of arrival of the echo, as determined from the antenna's pointing direction at the time the echo is observed, reveals the angular position, or azimuth compass direction, of each aircraft. It is straightforward to present the results in polar coordinates using the observed azimuth angle and range coordinates as the plotting variables. The resulting display device, called a plan position indicator, or PPI, originally comprised a large circular cathode ray tube displaying a line from the center to the perimeter sweeping out a narrow area indicating the instantaneous pointing direction of the radar antenna, Figure 2.3. The angle of reference line was synchronized with a rotating radar antenna, while the line itself was actually swept from the display center outward in synchronization with the time delay corresponding to echoes from increasing ranges. An air traffic controller, by examining a series of plotted radar positions as displayed on the PPI, can direct each plane to adjust its approach speed so that only one will require use of the runway at any particular arrival time. Analog radars such as these were closely coupled to the physics of the observations. Air traffic control radars augment our natural vision by supplying information needed to operate safely at night, in other conditions of obscured vision, and during rough weather.

---

2 Yet another value is approximately one astronomical unit, or $\approx 3 \times 10^{11}$ m per 1000 s of round-trip flight time delay!
3 In planetary radar astronomy and in tracking deep space vehicles, the term *round-trip light time* is used.

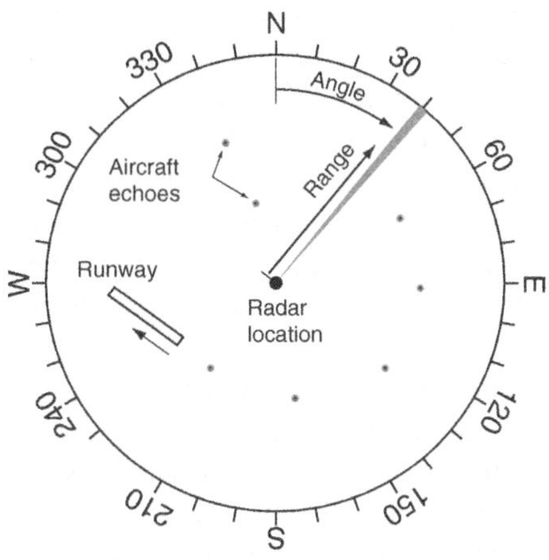

Figure 2.3 A plan position indicator display, or PPI. The radar is located at the center of the map display; scanning of the area within the range circle is accomplished by rotation of a narrow beam transmitting/receiving antenna represented by narrow wedge emanating from the radar location. Any backscattering objects appear as small dots which are mapped onto a polar plot labeled in distance (range) and compass direction (azimuth angle). Arriving aircraft are sequenced to approach the runway in an orderly fashion. Here, the approach path is CW around the radar location, as suggested by the dots, each of which represents a single observation.

2.3.2 Doppler Measurements

Another fundamental property of a radar target, the radial component of its velocity, can be inferred directly from its radar echo. Specifically, the radial motion of an object toward or away can be inferred from *range-rate* information inherent in the echo. One method for obtaining $\dot{R}$, mentioned above, is to estimate the value from the change in range between observations made at two different times. For this,

$$v_{\text{radial}} = \dot{R} \simeq \frac{R_2 - R_1}{t_2 - t_1} = \frac{\Delta R}{\Delta t} \quad \text{(m/s)} \tag{2.9}$$

where, v_{radial}, measured as expected with positive sign for increasing range, is the radial component of the target velocity relative to the radar, and the differences in positions over the time interval approximates the derivative.

A second method is to exploit the Doppler shift. In classical terms, the Doppler shift observed by radar is the result of reflections from a moving object. For backscatter, to first order, the Doppler shift is,

$$\Delta f_{\text{Dop}} = f_{\text{rec}} - f_{\text{trans}} = -\frac{2\dot{R}}{\lambda} \quad \text{(Hz)} \tag{2.10}$$

where f_{rec} and f_{trans} are the received and transmitted frequencies, respectively, λ is the wavelength of the transmitted signal, and $\dot{R} = dR/dt$. Note that in this formulation we can think of the round-trip Doppler effect as just twice the radial velocity measured in units of radar wavelength, together with an appropriate sign, Eq. (2.10). One can also use the more conventional formula,

$$\Delta f_{\text{Dop}} = -2\frac{f}{c}\dot{R} \quad \text{(Hz)} \tag{2.11}$$

or

$$v_{\text{radial}} = -\lambda\frac{\Delta f_{\text{Dop}}}{2} \quad \text{(m/s)} \tag{2.12}$$

As is the case of the range formulas, the factor of two in the expressions above results from two-way propagation of the signal along the path because the round trip path length changes at twice the rate of change of range. Physically, ignoring relativistic effects, the round-trip Doppler effect can be thought of in terms of a signal that undergoes two intermediate frequency shifts. The first of these is associated with interaction of the transmitted signal with the moving target in which the frequency "seen" by the target is Doppler shifted by $\dot{R}/\lambda$; scattered energy reradiated back toward the receiver of a monostatic radar undergoes a second frequency shift of the same sign and essentially the same value as that when the illumination signal first arrived at the target. In the second part of this calculation, we can often ignore the small difference in frequency due to the first Doppler shift in the calculation of the second. Taken together, the result is to double the one-way effect.

Equations (2.10) and (2.11) are accurate for measurements of objects moving at speeds that are very much less than the speed of light, i.e., less than about $c/10$. For most applications, this accommodates all terrestrial flying objects and space vehicles in Earth orbit.

Since we are usually interested in the motion of the radar "target," and recognizing that $\dot{R}$ is the component of velocity along the line-of-sight from the radar, we can write,

$$\upsilon_{\text{radial}} = -\lambda \frac{\Delta f_{\text{Dop}}}{2} \quad (\text{m/s}) \tag{2.13}$$

where υ_{radial} is explicitly the radial velocity.

Doppler shift can be observed in a number of ways. Conceptually, the simplest and most convenient, perhaps, is to view the frequency of the echo signal on a spectrum analyzer, in which case the frequency change appears as a shift in the position of the spectral line or lines associated with the echo waveform. An example in abstract form is suggested by Figure 2.4.

Realistic systems that are designed to measure the velocity of objects often use a single frequency, or continuous wave, "CW," waveform.

An example of a radar system using Doppler information is that of a pitching coach employing a radar "gun" to determine the speed of a baseball passing over home plate. Speed guns operate by (i) radiating a CW signal and (ii) using a small fraction of the transmitter output as a comparison signal for determining the Doppler shift of echoes from an approaching pitched ball. The second step typically is accomplished by taking the algebraic product of the received signal with the sample of the transmitted signal, a process known as "mixing" of the two. The Doppler shift is given by

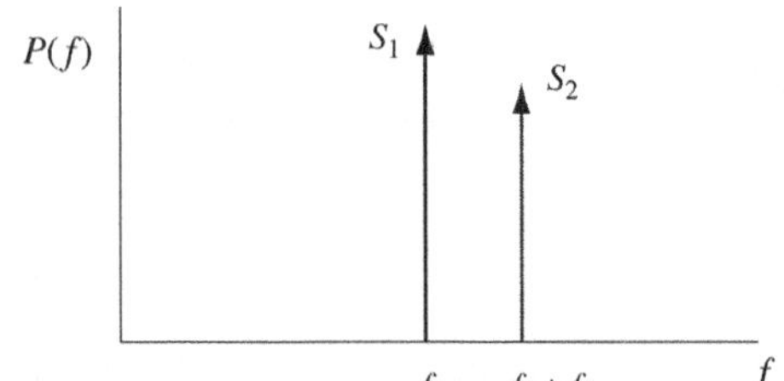

Figure 2.4 Illustrating a Doppler shifted waveform. Power spectrum of a transmitted narrow bandwidth signal is shifted in frequency by interaction with a moving object. Received echo appears at a frequency that is different that transmitted by f_{Dop}, Eq. (2.10). For monostatic radar, frequency is increased if the scattering object is moving toward the radar, is unchanged if motion is transverse to radar line-of-sight, and is negative if object is moving away from radar. More complex waveforms with broad spectral features behave in the same manner.

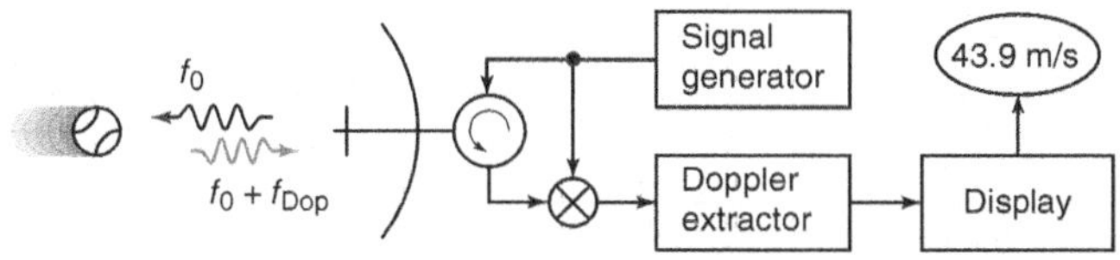

Figure 2.5 Block diagram of radar speed gun. A radar "gun" measures the Doppler shift induced by a moving object. Speed of approach can be calculated from Eq. (2.13). Connection among signal generator, antenna, and mixer, ⊗, is accomplished with a "circulator," a nonlinear device structured so that a signal entering one port is connected to the adjacent port located CCW from the entry port, ↺. Output from the signal generator at frequency f_0 flows to the antenna, and is radiated; input signals received by the antenna at frequency $f_0 + f_{\mathrm{Dop}}$ flow to the mixer. Output of the mixer is the product of the echo signal and a sample of the transmitted signal.

low-pass filtered output of the mixer. Figure 2.5 indicates the basic functional elements of a speed gun. In this application, distance from the radar to the target is not of concern, but the relative velocity of the two is. By counting the frequency of the extracted, Doppler shifted, waveform and by knowing the wavelength of the transmitted signal, we can calculate the speed of the ball with respect to the radar by applying Eq. (2.13) above.

Measurements of echo signal intensity and polarization are also useful in specialized radar systems. As we shall see, in addition to its range, the intensity of scattering from any object is related to a combination of the object's shape, structure, and material properties; echo polarization depends on the target structure primarily, and less so on the materials.

Sometimes both range and Doppler measurements are combined in a more complex system. Suppose that the air traffic control radar described previously measures not only the time delay of the radar pulses, but also the Doppler shift. Then the flight controllers could know the radar-to-aircraft line-of-sight velocity of each approaching aircraft as well as its distance; this is effective if the radar is positioned roughly along the flight path. Alternatively, the total velocity can be estimated by converting the range-azimuth PPI display, Figure 2.3, to rectangular coordinates and differencing the sequential two-dimensional position measurements. In practice, both types of measurement are employed, but the choice between the two depends on considerations related to specific applications. Air traffic control radars use the latter method as the total velocity is needed, independently of an individual aircraft's direction of flight. Doppler measurements, for example, are used in modern weather radars to determine wind speeds within thunderstorms and tornados, since the motion of the airmass is such that the important aspects of the motion can be inferred from line-of-sight observations alone.

2.3.3 System Elements

The example measurements above and in previous discussions suggest a wide variety of applications for similar measurements of time-delay and frequency. In fact, all share a few common functional elements. And all radars comprise subsystems that include, in some form, the system elements of Table 2.1.

Using these elements, we can draw a generic block diagram applicable to all conventional radar systems. This is given in Figure 2.6. Similar elements will be found in derivative systems, such as *sonars*, i.e., echoing systems based on sound waves, and lidars, i.e., echoing systems based on optical technology.

Table 2.1 Elements and functions common to all radars. Cf. Figure 2.6.

System element	Function
Oscillator	Provide clock signal for other system elements
Modulator	Generate and conditions signal waveforms
Transmitter	Amplify signals
Antenna(s)	Transmit and receive signals
Receiver	Amplify and filter echo signals, convert to digital form
Data processor	Apply various algorithms to extract information from echo signals

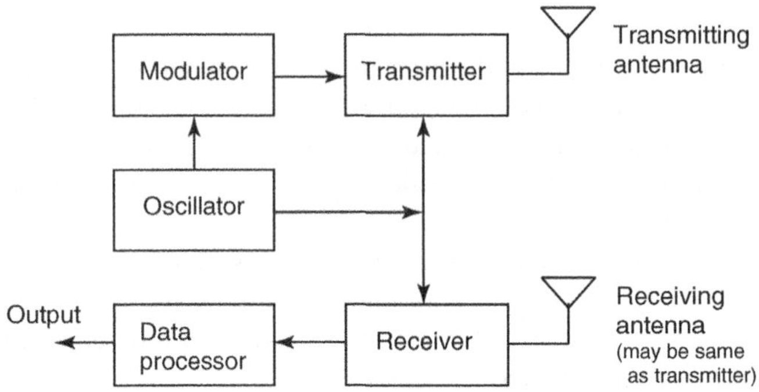

Figure 2.6 Generic radar block diagram. Functional elements common to all radars (see text).

2.4 Side-looking Radar Concept: An Imaging Radar System

In this section, we introduce the concept of *side-looking radar*, or SLR, designed for the purpose of creating two-dimensional images of surface reflectivity. This type of radar is a powerful tool for remote sensing, and very different from the systems with rotating dish antennas typically seen in films and introduced conceptually above. Rather, SLR systems employ a radar in translational motion, with a narrow-beam antenna directed to one side of the radar path, together with measurements of signal intensity vs. range to produce high-resolution images of Earth's surface.

The purpose of this type of imaging radar is to sort out the locations of all the scatterers on the surface, and display the brightness from each scatterer in a systematic way. The product of these systems is an "image strip" representing the scattering properties of the terrain parallel to the flight path of the aircraft or spacecraft that serves as the radar platform. Instead of a rotating antenna to scan an area surrounding a fixed point, an SLR system transmits a series of pulses to illuminate a *swath* of ground located to the side of the path defined by the motion of the platform carrying the radar. Scattering from each point on the ground within the illumination pattern of the antenna returns some of the incident energy to the radar. Figure 2.7 provides a simplified view of the geometry in which the coordinates are defined with respect to an idealized linear flight path followed by the radar platform.

In contrast with the systems intended to measure the range and velocity of individual objects, an SLR characterizes an entire "scene" made up of myriad discrete resolution elements, each of which characterizes the composite scattering from a small, well-defined area within the imaged swath. While we think of the scene as being scanned as a result of the motion of the radar platform, sensing of the scene remains within the province of radar scattering from each resolution element.

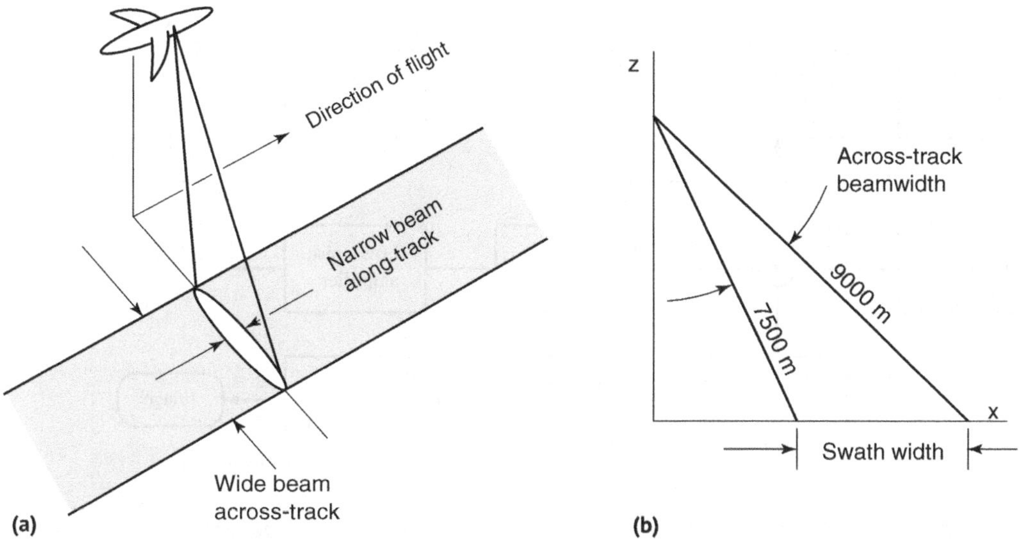

Figure 2.7 Side-looking radar geometry. (a) Radar carried on aircraft has antenna pattern directed to the right-hand side of flight path. Highly elliptical beam pattern progressively illuminates a swath parallel to aircraft track. At each position the antenna beam pattern defines a narrow strip in the across-track direction; range measurements define locations within the illuminated strip. (b) Example range triangle viewed along the direction of flight. Minimum and maximum ranges to edges of illuminated swath are determined by the across-track antenna beamwidth and the altitude of the vehicle. Ranges here correspond to example in Figures 2.1 and 2.2, for illustration only.

Obviously, scanning of the scene requires a very different approach from that intended to monitor a fixed, localized region of space, as described to this point. Nevertheless, SLR and the range and Doppler systems described above have much in common.

Here, we describe a *real-aperture* SLR, in which the radar discriminates among echo signals from multiple locations abeam on side of a well-defined flight path by recording the time that points on the surface are illuminated by the antenna beam. As we shall see, a measurement of time of flight, together with the known geometry is sufficient to determine the locations of all scattering elements in the scene. Identification and discrimination among surface features in the direction perpendicular to the flight path are achieved by the range to the surface, while resolution parallel to the path is provided by the transverse resolution of the antenna. Figure 2.8 provides a simplified example of a real-aperture system showing that the radar need only perform a few simple functions to capture the observational data required to form the image. For improved performance, the Doppler frequency of each echo can be incorporated into the radar processor to obtain finer along-track resolution, as described in detail in Chapters 11 and 12.

Returning to Figure 2.7, the term *across track* refers to the direction on the mapped surface perpendicular to the flight path; similarly, the term *along track* refers to the direction on the surface parallel to the flight path. Assume, for example, that the geometry and radar parameters are such that the range at the near edge of the illuminated swath is 7500 m, and 9000 m at the far edge, so that the swath width corresponds to a 1500 m change in the range direction as measured along the slant path between the aircraft and the ground, perpendicular to the flight path. With a judicious choice of antenna, a narrow cross segment of the swath in the along-track direction can be selected by the antenna beam, as this is essentially the only area illuminated by the transmitted waveform and from which echoes are sensed. If a suitable radar waveform is employed to define particular

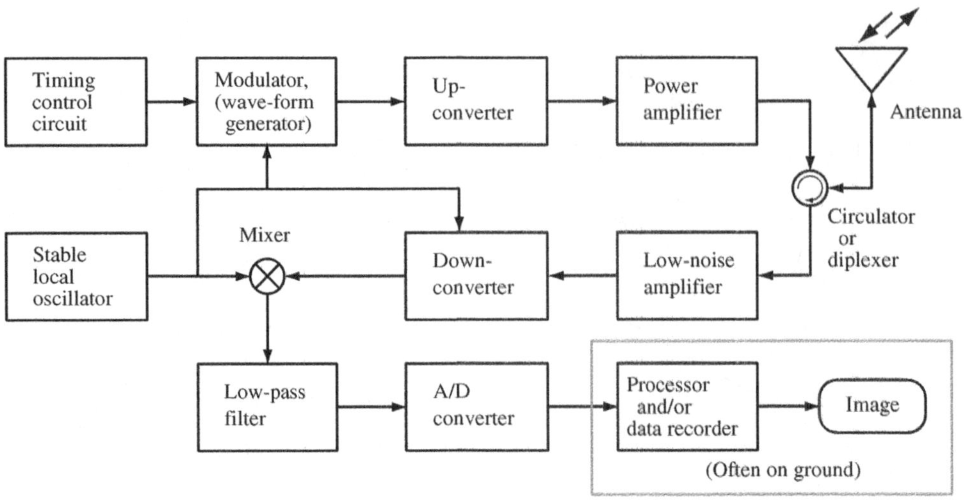

Figure 2.8 Simplified block diagram for side-looking imaging radar. System requires same elements and functions identified in Table 2.1 and Figure 2.6. Upper path comprises modulator and transmitter; middle path includes reference oscillator used to derive transmitter and provide receiver reference signals, elements of a super-heterodyne receiver, and mixer to prepare for signal processing; lower path contains signal-processing elements.

segments in the range direction, then a two-dimensional map of the radar reflectivity within the swath can be formed as the radar moves along the flight track.

In its simplest realization, the transmitted waveform is a short pulse of energy radiated periodically as the radar moves along the flight path, sequentially illuminating strips of the surface perpendicular to the long dimension of the swath. The areal resolution of this map depends on the combination of the resolutions in range for the cross-track direction and in antenna beamwidth for the along-track direction. Modern imaging radar systems operate in a manner similar to that just described, but make use of complex signal waveforms and apply precise signal processing to the problem.

In visualizing the operation of an SLR system, keep in mind the very large difference between the speed of the aircraft and the speed at which the radar range changes. In our example, the radar pulse scans the 1500 m cross-track distance in only 10 μs (cf. Figures 2.1 and 2.2). Even a very high-speed aircraft, moving in the along-track direction at, say, 300 m/s, travels only 3 mm in the 10 μs interval between the time of the echo from the nearest point on the swath and that when the echo is received from the far side of the swath. Similarly, an Earth-orbiting satellite with velocity 7 km/s moves much less than 1 m in this same time interval. Consequently, for the purpose of visualizing the geometry, an aircraft can be considered as stationary during the time of each observation, while changing observation points along the path between observations of echoes. Later, when considering more advanced systems, we shall see that even the relatively slow motion of the radar platform can be put to very good use in refined designs, however. Even then, it is helpful to maintain the perspective here of the large difference between the speed of platform and that of the radar signals in space.

In practice, the duration of the transmitted pulse might be very brief, or it might be a modulated transmission, lasting longer than a simple pulse, in which fine range resolution is obtained by use of time-waveform compression. In either instance, the pulse waveform is transmitted repetitively so that the transmitted signal might have the appearance of that in Figure 2.9. The received waveform,

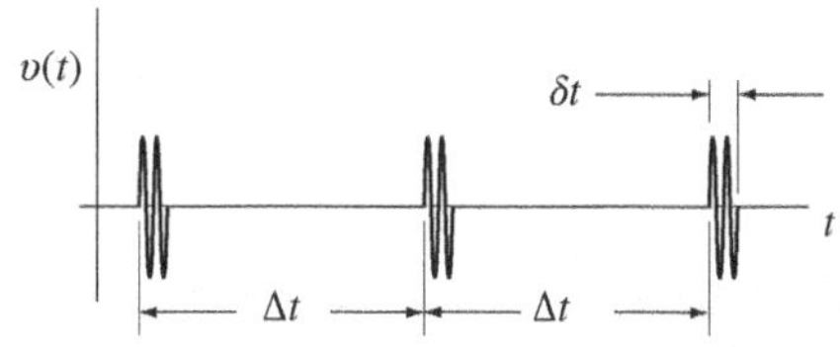

Figure 2.9 Transmitted waveform of a simple imaging radar. Plot depicts the radiated waveform of the transmitted signal. Transmitted signal comprises a brief oscillatory pulse. In a conventional radar system employing simple pulses the difference between the time separation of the pulses, Δt, and the pulse duration, δt, typically is orders of magnitude. For example, ranges of $\Delta t \sim 0.001$ to 0.1 s and $\delta t \sim 0.000{,}000{,}1$ to $0.000{,}01$ s are common.

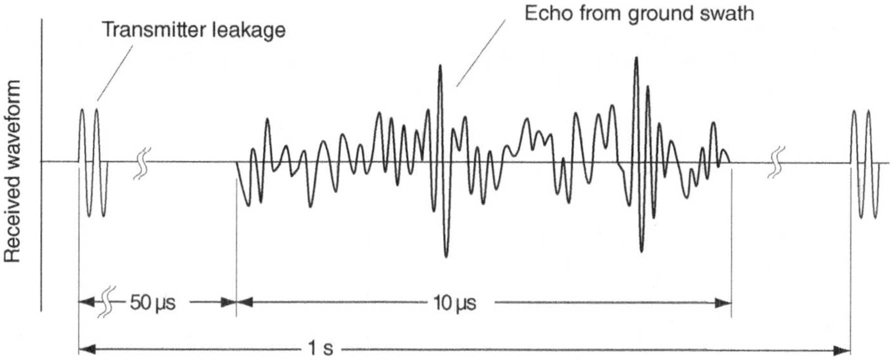

Figure 2.10 Received waveform from within a side-looking swath. Figure illustrates the relationship of transmitted pulse and echo from an extended swath. Trace represents the signal at the terminals of the receiving antenna of our simple SLR. Some portion of the transmitter pulse is always present in the receiver output as the result of finite isolation between the transmitting and receiving elements of the system. Transmitter leakage signal can be used as a fiducial mark in the receiver data. Echo signal from the image mapping strip on the ground is spread out in time as a result of the range variation across the ground swath. In our example, the first ground echo appears 50 μs after start of the transmitter pulse radiation and continues for 10 μs. Here the interval between transmitted pulses is 1 s. Note breaks in scale.

on the other hand, is always made up of a sequence of echoes that are longer than the transmitted signal as a result of interaction with the extended area of the surface in the cross track direction of the swath. The stretched echo begins at the time of first reception from points located at the near edge of the swath and continues until the return at greatest range is reached at the most distant point across the swath. The intensity of the received echo as a function of time is controlled by the combination of varying range and the strength of the radar backscattering across the illuminated surface. Thus, the strength of the received waveform is controlled by the degree to which the surface returns radar energy to the receiver as a function of time delay, with the overall echo duration controlled by the difference in echo times from the near- and far-swath locations. An SLR echo from a fairly uniform surface might resemble the sketch in Figure 2.10. On the other hand, if the scene within the swath were to contain a single strong reflector point amidst many weaker ones, then the response corresponding to a single transmitted pulse would more closely resemble that in Figure 2.11.

As suggested above, assume that the antenna beamwidth is quite narrow in the along-track direction. There are many techniques to generate such a narrow antenna patterns, ranging from use of a very long structure parallel to the flight path, which is impractical on most aircraft, to the use of signal processing techniques to create a virtual, but functionally effective, "synthetic aperture."

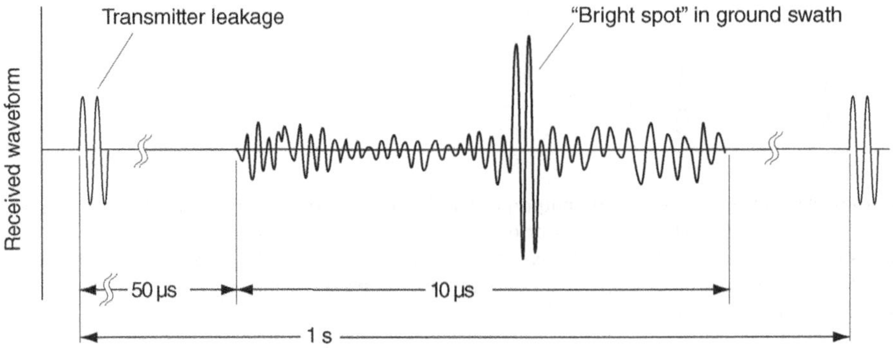

Figure 2.11 Received waveform with echo containing a strongly reflecting feature. "Bright spot" within the swath occurs when a point within the antenna beam includes an unusually strong, localized scattering feature or area. This appears in the echo waveform as a brief interval of markedly increased amplitude of the received waveform. Cf. Figure 2.10.

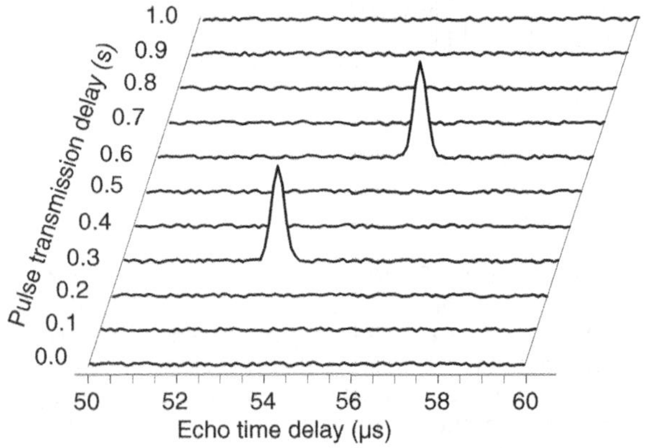

Figure 2.12 Received waveform with echoes from two bright points. Echoes are identified by strong signal for which echo strength clearly exceeds small fluctuations in baseline due to system noise. Echo from transmission at 0.3 s is received 53.5 μs later; echo from transmission at 0.6 s is received after 56 μs. Plot depicts a system in which the along-track width of the antenna beam on the surface is sufficient to isolate echoes from points separated in the along-track direction by the distance the radar moves in 0.1 s.

Regardless of the processing approach, we build an image of the ground by observing a series of echoes from repeating transmitted pulses emitted sequentially along the flight path and displaying these in a raster format, in which the trace of each echo is written directly adjacent to its predecessor. If we were to observe a region containing two bright point scattering points located on the surface, and then lay down a series of echo lines corresponding to the narrow antenna beam next to each other, we might obtain a result similar to that in Figure 2.12. The figure is idealized in that it shows two echoes of equal strength registered with the transmission time coordinates.

Displaying the same data as above in image format, we obtain Figure 2.13, where the representation is in terms of signal intensity within the transmission-time, echo-delay space. A full image requires more closely spaced pulse transmissions in the along-track dimension, which can be achieved by decreasing the intervals between pulse transmissions or reducing the speed of the radar platform. The size of the bright points represents the impulse response of the radar to a point target in echo delay, as controlled primarily by system bandwidth, and in the along-track direction, as controlled primarily by the width of the antenna beam in the along-track direction.

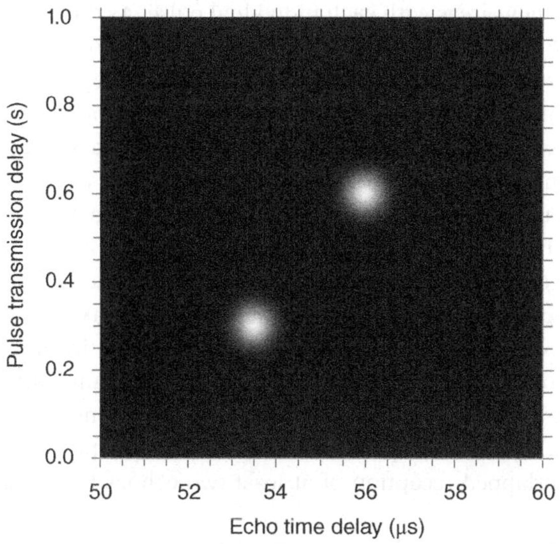

Figure 2.13 Image containing two bright points. Radar and scattering point geometry is same as for Figure 2.12. An image such as this requires either much higher sampling rates or interpolation to display the features of the response. In a simple radar, this is achieved by decreasing the along-track pulse separation to the point where the antenna response from adjacent scans begin to overlap, which accounts for the spreading of the point response in the vertical dimension of the image.

We can convert the time delays to actual distances to form more useful geometric maps of the surface. The across-track range direction separation, ΔR, as determined from the difference of the time delay Δt to the two points is,

$$\Delta\tau = 2.5\ \mu\text{s} \Rightarrow \Delta R = \frac{3 \times 10^{8} \times 2.5 \times 10^{-6}}{2} = 375\ \text{m} \tag{2.14}$$

Keep in mind that in Eq. (2.14), the range coordinate is based on the flight time of the radar pulse that travels on a straight line path between the radar and a given location on the surface. Consequently, e.g., the abscissa in Figure 2.12 gives the slant distance from the radar to points on the mapped surface; this distance is always greater than the distance along the surface within the swath (cf. Figure 2.7b). Further geometric transformations are required to relate the time-delay of SLR observations to locations in true surface coordinates.

In the along-track direction, the radar sample lines are in turn separated by 0.3 s. If the aircraft velocity is 200 m/s, then,

$$\Delta a_z = 0.3\ \text{s} \times 200\ \text{m/s} = 60\ \text{m} \tag{2.15}$$

In this instance, the distance between the two positions in the ground is the same as the distance moved by the radar between two successive pulses.

The minimum detectable separation of two objects at the same along-track range, or the *along-track resolution*, of our simple radar is set by the projection of the antenna pattern onto the surface. While faster pulsing of the transmitter leads to more frequent measurements, and hence observations which are more closely spaced in the along-track direction, the ability to discriminate among closely-spaced targets in the along-track dimension is limited by the effectiveness of the antenna response in restricting the observed echo energy to a thin swath on the ground. To see how this works, consider that at sufficiently low rates of pulse transmission, the motion of the platform causes the antenna beam to illuminate entirely new areas with each individual pulse. As the interval between pulse transmissions decreases, at some point, the areas illuminated by two successive pulses must overlap in the along-track direction. This occurs when the along-track separation between pulses is less than the along-track dimension of the antenna footprint in the swath—the point at which our ability to discriminate between distinct scattering features in two successive pulse transmission/reception cycles begins to degrade. From this, we see that the width of the antenna beam in the along-track direction ultimately limits the along-track resolution. On the other hand, the scene is undersampled when the rate of pulsing is too low.

Similarly, our ability to discriminate among distinct scattering features in the cross-track direction can be lost as a result of ambiguity if the pulse repetition rate becomes too high, making the interpretation of the image difficult or impossible. In our example, the swath is 10 μs in time, corresponding to 1500 m in range, which is consistent with the cross-track antenna beamwidth, as suggested in Figure 2.1. It follows that transmission of pulses more frequently than once each 10 μs—as might be desirable in connection with a more complex waveform than the simple pulse used here—would result in simultaneous, overlapped reception of at least two echoes from the same across-track range swath.

For the reasons above, the interval between pulses, the choice of the swath width, and the design of the antenna beam are all interrelated. Obviously, the most straightforward resolution of this conflict is to transmit one pulse each time the antenna is moved one beamwidth. As we shall see, there are elegant, well-developed, design approaches for managing such conflicts while optimizing radar system performance. But a firm understanding of the fundamental geometric considerations we have explored here is key to understanding this field.

2.5 Basis for More Complex Systems

Most, but not all, radars determine range by a rather direct measurement of time delay, which is then related to distance by $R = (c\tau)/2$. A myriad of applications follows from this simple relationship. Advanced radars also employ measurements of Doppler frequency shift to determine range rate by $\Delta f = 2\dot{R}/\lambda$, again an extremely useful relationship which permits additional inferences regarding target location and motion. All radars make use of the wave observables introduced at the beginning of this chapter, and all are constructed of basic building blocks that include the elements listed in Table 2.1.

Many ingenious methods for construction, combination, and configuration of the basic elements exist, supporting the range of applications discussed in Chapter 1. Likewise, the theory describing these elements is both complex and highly developed, much more so than one might imagine from the simple pictures just presented. At the same time, the basic physical quantities displayed in image form, such as the two simple examples, above should be kept in mind when evolving complexity begins to obscure the underlying principles.

3

The Radar Equation

Armed with basic concepts from the first two chapters, we now begin a quantitative analysis of radar systems. At several points previously, we have referred to conceptual transmission of "illuminating" signals controlled in part by antenna "beams," and the reception of "echoes" from a "target" object, all without the need to be particularly precise. To go further, we need quantitative tools describing transmission, redirection, and reception of radar signals. Central to this process is the interaction of the illuminating waves with a scattering object in such a way that energy from the radar transmitter is redirected to the radar receiver. Previously, we used the concept of a radar target to specify the object of interest, but we must also realize—especially in the context of radar remote sensing—that in a given instance this process can involve a large number of scattering objects that are often spread over a large area of interest. Under many circumstances, however, the interaction of radar signals with these objects can be identified and dealt with one by one. So we begin by quantifying the strength of the received signal in terms of radar system parameters and the characteristics of the scattering by a single target object. In this chapter we develop the dependence of received signal power on radar system parameters and measurement geometry.

3.1 Radar Equation and Radar Cross Section

3.1.1 Radar Equation

The *radar equation* quantifies the relationship among the several parameters of importance to radar system performance. The needed expression is derived easily from first principles by making use of a standard method for describing the effectiveness of a *scattering object*, our notional *target*, in redirecting a portion of the incident power from the radar transmitter to the radar receiver. This returned power is known as the *radar echo*.

To visualize the way in which the radar equation is structured, consider the situation in Figure 3.1, which illustrates the geometric relationships among the prominent system elements. These are the transmitter and transmitting antenna, the scattering object, and the receiving antenna and the receiver. Distances between the various elements are needed, as well as the geometric configuration, which is implicit at this point. The needed variables and their corresponding physical meanings and units are as follows:

P_t—*Power* delivered to the transmitting antenna (watts).
P_r—*Power* collected from scattered waves and guided to receiver input by receiving antenna (watts).
R_t—*Range*, i.e., distance, separating transmitting point and scattering object (m).
R_r—*Range*, i.e., distance, separating scattering object and receiving point (m).
G_t—*Gain* of transmitting antenna relative to an isotropic radiator (dimensionless).

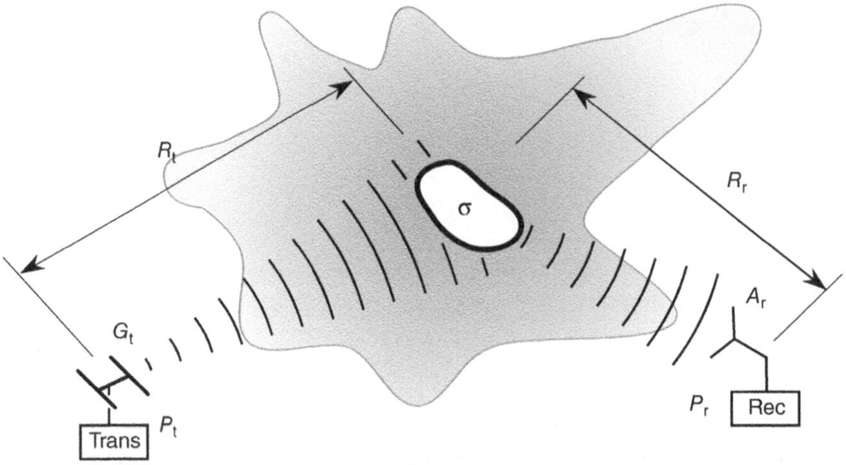

Figure 3.1 Scattering geometry for the generalized, or bistatic radar equation. Transmitter (Trans) illuminates an object having radar scattering cross section σ. Interaction of the incident spherical wave with the object scatters the incident illumination in all directions. A small part of the scattered energy is captured by the receiving antenna and fed to the receiver (Rec).

A_r—*Aperture*, effectively, of the receiving antenna (m^2).
σ—*Radar cross section* (RCS) of a scattering object (m^2).

The last of these, σ, is the least familiar. It represents a characteristic area or effective scattering surface of the scattering object. The use of σ for this purpose is traditional. With these definitions, the radar equation can be written as:

$$\mathrm{Pr} = \left(\frac{P_t}{4\pi R_t^2}\right) \times (G_t) \times (\sigma) \times \left(\frac{1}{4\pi R_r^2}\right) \times (A_r) \tag{3.1}$$

The five factors identified by parentheses on the right-hand side, as mentioned earlier, can be interpreted directly in physical terms, as follows:

$\frac{P_t}{4\pi R_t^2}$ is the power density, at distance R_t from the transmitter, of a spherical wave carrying total power of P_t. Stated more precisely, this is the power supplied by the transmitter, normalized by the area of a sphere at the distance of the scattering object, i.e., by the area through which it would flow if radiated isotropically, or uniformly in all directions. The units are watts per unit area of a sphere of radius R_t. Alternatively, the factor $P_t \times 1/4\pi R_t^2$ is sometimes described as a "spatial attenuation," although there is no actual dissipation of energy.[1] The mechanism of the reduction in power density (W/m^2) is the spherical spreading of the energy with distance from the source.

G_t is a dimensionless ratio called the *antenna gain*, which characterizes the degree to which the transmitting antenna concentrates the radiated power in the direction of interest relative to a hypothetical isotropic device. Thus, the product of the first two factors in Eq. (3.1) gives the power density incident at the scattering object. Note though, that the degree of concentration is a function of direction with respect to the frame of the antenna, which is implicit here. Any dissipative power loss in the antenna is also incorporated into this factor.

1 The term *spatial dispersion* is a more accurate description. In systems analysis, the effect is treated as a loss of energy, though.

σ is the RCS, of the object of interest. This quantity represents the ability of a scattering object to redirect power in the transmitted incident wave toward the receiver. By definition, RCS *is scaled so that if all the power intercepted by the scatterer were radiated isotropically*, then it would match the power density actually present at the receiver. Thus, σ incorporates aspects both of a receiving antenna (refer to the following), by extracting power from an incident illuminating wave, and of a transmitting antenna, by launching a new, so-called scattered wave. The product of σ and the first two factors, $P_t G_t / 4\pi R_t^2$, is the power scattered in the direction of the receiver, scaled as though the scattering were isotropic.[2]

$\frac{1}{4\pi R_t^2}$ scales the scattered power by the area of a sphere through which it flows at a distance R_r of the receiver from the scattering object. This convenient usage is consistent with the of isotropic scaling in the definition of σ. Multiplying by the previous three factors, the inverse area renormalizes the scattered power to give the power density at the receiver.

A_r is the effective aperture or the *capture area* of the receiving antenna. That is, the antenna—regardless of physical size or shape—extracts the power in A_r m^2 flowing in the direction from the location of the scatterer with cross section σ toward the location of A_r. The product of all five factors is the power input to a receiver whose input impedance is matched for maximum power transfer to that of the antenna. As noted above for G_t and σ, A_r is a function of direction with respect to its reference frame.

In the above, note that transmitter and receiver locations are arbitrary and do not have any particular relationship to one another. Systems in which the transmitter and receiver are well-separated are referred to as *bistatic radars*, in recognition of two, usually fixed operating locations. This is the general case.[3] As mentioned on p. 4, the first radar of Taylor & Young fits this description. Bringing the transmitting and receiving locations close together in Figure 3.1 reduces the above to the essential backscatter geometry effectively exploited by Breit & Tuve. Collocating the transmitter and receiver results in the common geometry of modern monostatic systems.

3.1.2 Radar Cross Section

For purposes of describing the scattering properties of the radar target, it is useful to express σ in terms of the ratio of the strength of the scattered wave at the receiving location to that of the wave incident upon the target. For this purpose, note that the intensity of the incident wave, i.e., the power per unit area of the wavefront, can be expressed in terms of the incident electric field at the scattering object,

$$\frac{P_t\, G_t}{4\pi R_t^2} = \frac{|E_i|^2}{2\eta_0} \tag{3.2}$$

and similarly, at the receiver,

$$\frac{P_r}{A_r} = \frac{|E_s|^2}{2\eta_0} \tag{3.3}$$

---

2 This normalization of σ does not imply that the scattering is actually isotropic. Typically, radar wave scattering is highly variable with direction from the scattering object. Isotropic equivalent scaling is applied in *all directions*; however, regardless of the actual variation of scattering with aspect. Details are in Chapter 4.

3 A moving transmitter can be used in conjunction with a moving receiver. Logically, this might be called a "bi-dynamic radar," with similar variations for a moving transmitter and stationary receiver, etc. Such terminology is unusual.

where E_i represents the incident electric field strength at the location of the scatterer and E_s the scattered field strength at the location of the receiver; the factor η_0 in the denominator is the impedance of free space. With this observation, we can describe the RCS in terms of the incident and scattered power densities, or equivalently, in terms of the fields,

$$\sigma = \frac{|E_s|^2}{|E_i|^2} 4\pi R_r^2 \tag{3.4}$$

From the form in Eq. (3.4), we see that σ is the ratio of the scattered to the incident power densities scaled by the factor $4\pi R_r^2$. As before, this last factor is included because, at any significant distance from the scattering object, the scattered waves appear to be emerging from what is effectively a point source. Use of the $4\pi R_r^2$ normalization in Eq. (3.4) compensates for the spherical spreading or *spatial dispersion* of the scattered energy, thereby producing a range-independent definition of RCS. The embedded factor of 4π arises from the isotropic normalization of the RCS.[4]

There are several hidden parameters in the radar equation as written in Eq. (3.1). Accurate calculations of the radar echo strength require that we know the polarization of both the transmitting and receiving antennas, the scattering properties of the target in response to varying polarization, frequency of operation, orientation of the target, and the conditions of propagation, as well as other, less obvious, factors.[5] For the most part, such effects can be treated as independent variables that modify the values of the several factors that we have already identified and discussed, most notably the antenna gains and the RCS.

In practical applications, it is implicit in Eq. (3.4) that the transmitter and receiver are located at a considerable distance from the target. Therefore, while overall we recognize that these fields are spherical waves, the radius of curvature of the incident wavefronts is so large compared with the dimensions of the scatterer *and* the receiving antenna as to appear *locally* flat. Otherwise, we could not simultaneously treat the local fields incident on the target and on the receiving antenna as a plane wave. Expanding on this point, accurate application of the radar equation requires that we keep two points in mind. First, the transmitter, receiver, and target must be sufficiently well-separated so that the field strengths can be calculated on the basis of spherical expansion of the wavefront. Second, although we know the radiated and scattered wavefronts are spherical, the distances R_t and R_r also must be sufficiently large that we can neglect the curvature on local scales roughly comparable with the sizes of the scattering object and the receiving antenna. These two conditions are implicit in moving from Eqs. (3.2) and (3.3) to Eq. (3.4), in which σ is expressed in terms of a ratio of the incident and scattered power carried by plane waves, independent of range.

In theoretical work, Eq. (3.4) is taken as the definition of σ, where the incident and scattered fields are assumed to be plane waves.

When the conditions above are violated σ becomes a function of higher-order terms in the field curvature—the value of σ then depends on R_t and R_r. Failure of the incident plane wave assumption often occurs in practice, especially when using large antennas or observing large targets at short wavelengths, and on radar test ranges. It is well to be aware of this consideration as it can cause trouble if not recognized and compensated.

---

4 For comparison, in optics and astronomy, the factor of 4π is dropped from the range normalization, which is expressed in solid angle. This provides certain conveniences for telescopic observations. This leads to a difference of a factor of 4π between the scaling used in the definition of RCS (Eq. 3.4) and that of the *differential scattering cross section* commonly used in the optical literature, where the 4π is omitted.

5 For example, the scattering properties of some materials are nonlinear. For such, the σ is a function of the strength of the illuminating signal.

In the above manipulations we have moved from a measurement-based, operational definition of σ, Eq. (3.1), to a formal expression in terms of a ratio of field strengths, Eq. (3.4). The two are equivalent, however, and both are useful. It is convenient to use the measurement-based approach when reducing or interpreting observations. In such situations, the parameters typically correspond to variables in Eq. (3.1), except that one of these, often σ, is unknown. In theoretical studies of scattering, one posits the incident field and seeks the scattered field at great distances from a specified object of interest. In the latter instance, the more compact definition of Eq. (3.4) is the convenient form.

3.1.3 Antenna Gain, Aperture, and Far-field

Antenna gain and effective aperture depend on the orientation of the physical devices with respect to the wave direction of interest. We indicate this by explicit reference to standard polar angles used to indicate the direction of transmission and reception,

$$G_t = G_t(\theta, \phi) \tag{3.5}$$

$$A_r = A_r(\theta', \phi') \tag{3.6}$$

Clearly, $\theta, \phi, \theta', \phi'$ must be defined with respect to some well-established reference system such as a geographic frame, but often that of a carrier vehicle.

The *effective aperture* of an antenna is linked directly to its gain through the relationship,

$$G = 4\pi \frac{A_{\text{eff}}}{\lambda^2} \tag{3.7}$$

where A_{eff} is the effective aperture or "capture area" of the antenna.[6] This relationship is useful in finding modified forms of the radar equation, in estimating the gain of an antenna for which the physical aperture is well-defined, such as a large-aperture dish antenna or a flat panel array, and for estimating the effective aperture of antennas for which the aperture is not evident, such as a simple dipole. In the latter case, the gain can be calculated or measured to provide a basis for finding the effective aperture. Unless indicated otherwise, here A_t and A_r always refer to the effective apertures of the transmitting and receiving antennas under discussion, respectively.

The gain and aperture functions of a typical radar antenna are tightly concentrated in a small angular region about the maximum, perhaps in a conical beam of only a few degrees or less in angular width. Such behavior is associated with large values of gain and effective aperture. Figure 3.2 illustrates the structure of a realistic pattern expressed as a function of polar angle, θ, measured from the beam center. The concentration of transmitted energy—and reciprocally the enhanced sensitivity upon reception—associated with this type of pattern is required for practical operation of many systems. This is not always the case, though. For some systems, where observation over a large angular spread is desired, an extremely broad beam may be needed. An example is a satellite fence designed to monitor space vehicles, orbital debris, or perhaps to study the influx of meteors striking Earth.

For reasons similar to those given for the RCS, unqualified values for G_t and A_r apply only for large values of R_t and R_r. But how far is far enough? As before, it is the distance from a given source beyond which the fields can be approximated by those of a plane wave, dubbed the *far field*. In practice, this approximation is formulated in terms for a simple criterion guaranteeing that the

---

6 Equation 3.7 holds for all antennas. The proof is based on reciprocity, and is valid in linear, time-invariant media; it can be found in any standard text on antenna theory.

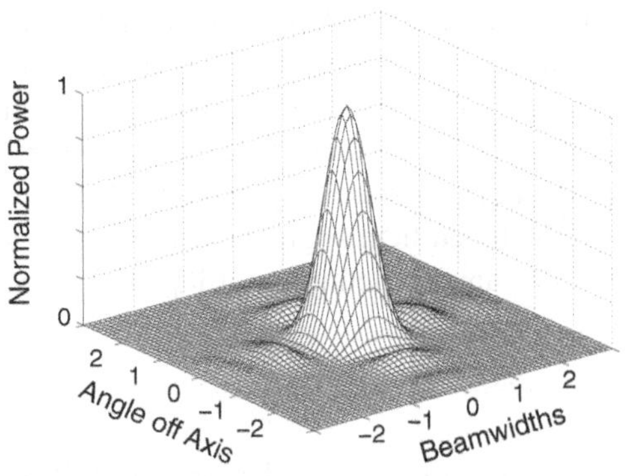

Figure 3.2 Typical narrow beam antenna response pattern. Plot indicates the response, A_r, of antenna to signals arriving from angles θ_x, θ_y from the x, y axes, respectively. Angles are measured in "beamwidths," here taken as the distance from the maximum response to the first zeros measured parallel to horizontal axes. Owing to reciprocity, the same plot represents the degree of concentration of energy or power obtained when transmitting, G_t. Many radar systems employ directional beams so that response is concentrated in particular directions, as suggested by the plot; but this is not always the case. The beam is often represented in standard polar coordinates, i.e., θ, ϕ, with $\theta = 0$ aligned with the direction of maximum response, and ϕ the angle of rotation about the maximum response.

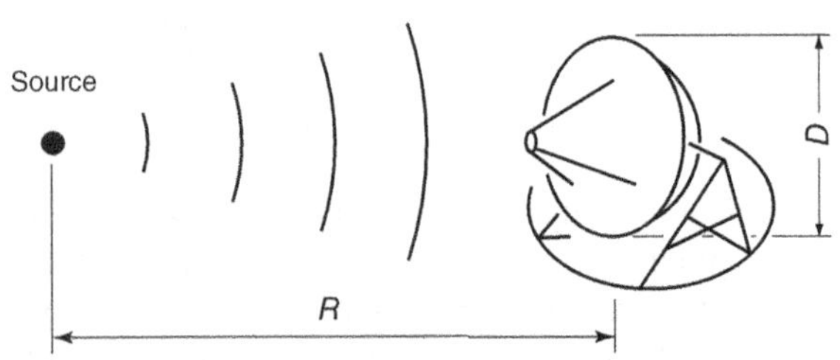

Figure 3.3 Illustrating source at distance R for antenna of size D. Radiation from a source at a finite distance R results in spherical waves across a receiving aperture. Far-field criterion indicates when incident fields can be approximated by plane waves.

variation in phase across the antenna aperture, or phase error, associated with the source-antenna geometry will be less than a specified maximum. That is for $R \geq R_{ff}$, where R_{ff} is the minimum distance to the far field, the phase difference between the actual arriving spherical wavefront and that of a plane wave arriving traveling in the same direction, must not exceed some agreed value. By consensus, the tolerable phase error is often taken to be $\Delta\phi = \pi/8$, although larger or smaller values are used if required for a given application.

Figure 3.3 provides an illustration. Considering a spherical wave generated at the "Source," and calculating the minimum range from the tolerable phase error, it is easy to show that,

$$R > R_{ff} = \frac{2D^2}{\lambda} \Rightarrow \Delta\phi\bigg|_{\text{aperture}} < \frac{\pi}{8}\text{ rad or }\frac{\lambda}{16}\text{wavelengths or }22.5^\circ \tag{3.8}$$

For R greater than R_{ff} above, the variation in phase of a spherical wavefront across an aperture of diameter D is always less than $\pi/8$ rad, and decreases with increasing distance from the source.

In describing the far-field limit, we used an example of reception by a large-aperture antenna with a well-defined diameter. The same limit applies to the definition of the boundary of the far field for the target scattering object. In this instance, R is the distance from the illuminating antenna and our D is the maximum diameter of the scattering object.

Considering now the transmitting case, for reasons that are not intuitive, R_{ff}, as calculated above, is also the boundary for the far field of a large transmitting antenna. Inside this boundary, the field structures become increasingly complex as R is reduced and the R^{-2} variation of power density with distance breaks down, and with it the accuracy of the product of the first two factors in Eq. 3.1.

The three parameters of the radar equation, G_t, A_r, and σ are all defined in the geometric far field as defined above. For ranges beyond the calculated R_{ff}, these parameters are independent of range, i.e., R_t and R_r.

Placing a source or receiver at a distance R greater than the minimum range specified guarantees that incident wavefronts are locally "flat" to at least the reference phase accuracy of $\pi/8$. Violation of the range minimum criterion is an indication that use of the simple radar equation, Eq. (3.1) with its R^{-2} variation with range, should be reconsidered.[7]

In order to obtain an idea of the quantities involved, consider some famous and some typical antennas, Table 3.1. Obviously, antennas in the size range tens of meters are not uncommon, but the largest can be an order of magnitude larger than that! Wavelengths of a few centimeters are also in common use.

The minimum distance to the far field can be considerable and is often a problem for instrumentation systems designed to measure the properties of large antennas, as illustrated by few examples in Table 3.2. When the required distance is too large to be practical, experimental characterization of antenna performance cannot be accomplished without resorting to use of remote satellite sources or radio stars for calibration. Alternatively, on a laboratory scale, the use of complex

Table 3.1 Size and wavelengths of operation for selected antennas. Wavelength ranges are approximate.

Facility	Diameter (m)	λ (m)
Big Dish, Stanford, CA[a]	46	0.10–10
Deep Space Station, Goldstone, CA[b]	70	0.01–1
Radio Telescope, Effelsberg, FRG[c]	100	0.001–1
Radio Telescope, Green Bank, WV[d]	105	0.07–2(?)
Arecibo Observatory, Puerto Rico[e]	305	0.03–1
One Kilometer Aperture, var.[f]	1000	0.001–1
Very large array, Soccoro, NM[g]	27,000	0.001–1
Typical radar	$0.3 \sim 20$	0.03–2
Air surveillance radar	15	0.30–2
Typical spacecraft	5–15	0.01–0.20

a. ca. 1960, mesh surface, Newtonian/Cassegrain.
b. ca. 1970, solid with mesh extension, Cassegrain.
c. ca. 1972, solid with mesh extension.
d. ca. 2002, elliptical 110 × 100 m (value 105 is geometric mean), offset Newtonian.
e. ca. 1963, fitted with Gregorian feed 1997.
f. ca. 2015, array with 1 km^2 collecting area, goal of the International Astronomical Union.
g. ca. 1980, 27 element "Y"-shaped array of 25 m diameter. Cassegrain reflectors, max. baseline 35 km.

7 See e.g., Stutzman and Thiele, *Antenna Theory and Design* or Balanis, *Antenna Theory, Analysis, and Design*, where there are several entries in the index under "field regions." Most standard antenna texts discuss the relationships between near field and far field. Note that the minimum range given in Eq. (3.8) pertains strictly to the geometric situation of Figure 3.3. A similar but stronger restriction pertains to the geometry of the scattering object; see Chapter 4.

Table 3.2 Far-field minimum ranges. Minimum distance to the far field from example apertures and wavelengths in Table 3.1. Example R_{ff}s are calculated from Eq. (3.8). Note that the first three facilities operate as radars, in addition to other modes.

Facility	λ (m)	R_{ff} (km)
Big dish	1	4.2
Big dish	0.1	42
Goldstone deep space station	0.03	300
Arecibo planetary radar	0.12	1550
One kilometer aperture	0.03	67,000
Air surveillance	0.30	1.5

optics to create local regions of uniform phase for test purposes, and/or the application of analytical procedures to computationally transform measurements of the antenna "near-field" allow prediction of far-field performance. Such procedures are often necessary.

In some situations, the far-field limit is an issue for the design of operating radars. As Table 3.2 illustrates, such instances arise when short wavelengths are combined with the use of large apertures. This, and other, similar, physical limits are always a consideration when applying the radar equation in a new situation, at least until the limitations are well-understood. In addition, beginning to think in terms of the physical picture involving a combination of spherical and plane, waves is useful in developing a broad understanding for some of the wave propagation effects associated with advanced systems.

3.2 Monostatic Radar Equation

3.2.1 Definition and Dependence on Wavelength

Thus far, our discussion of the radar equation has referenced the general case in which the transmitter and receiver occupy distinctly different locations. Now consider the more usual arrangement in which the transmitter and receiver share a common antenna, or employ two antennas that are sufficiently close together in terms of other geometric aspects of the system that they can be treated as colocated. As indicated previously, such radars are referred to as *monostatic radars*, or by default, simply *radars,* as this is by far the most frequent practical configuration. In such instances, as seen from the target object, the directions of incidence and scattering are effectively antiparallel. Additionally, the distance from the transmitter and the receiver to the target is the same, or essentially so. With these constraints, Eq. (3.1) reduces to the *monostatic radar equation*. With $R_t = R_r = R$, we have,

$$P_r = \frac{P_t \, G \, A \, \sigma}{(4\pi)^2 R^4} \tag{3.9}$$

which is a common form of radar equation applicable only to monostatic systems. The two antennas may occupy the exact same location, as when a single antenna serves both as transmitter and receiver, or they may be separate units physically located very close to one another.

When the same or identical antennas transmit and receive, we can use the general relationship between antenna gain and aperture Eq. (3.7) to rewrite Eq. (3.9) in two alternative forms. We obtain these from the radar equation by eliminating first the effective aperture and then the gain by substitution of Eq. (3.7) in Eq. (3.9). This gives,

$$P_{\mathrm{r}} = \frac{P_{\mathrm{t}} A_{\mathrm{eff}}^2 \sigma}{4\pi R^4 \lambda^2} \propto \lambda^{-2} \tag{3.10}$$

$$P_{\mathrm{r}} = \frac{P_{\mathrm{t}} G^2 \lambda^2 \sigma}{(4\pi)^3 R^4} \propto \lambda^2 \tag{3.11}$$

Thus, we see that the received power is either inversely proportional to or proportional to the square of wavelength, when either the effective aperture or gain is independent of frequency, and the other parameters are held constant. For example, "dish" type antennas have effective apertures that approximate their physical cross section. In this instance, Eq. (3.10) tells us that the received power in a system comprising two such antennas increases as $f^2 \propto \lambda^{-2}$. Similarly, resonant "half-wavelength" dipoles have the same gain, independently of the design frequency—although the physical length of such antennas varies with frequency. Equation (3.11) tells us that the received power in a system comprising a pair of resonant dipoles increases as $f^{-2} \propto \lambda^2$. While Eqs. (3.10) and (3.11) are useful for many purposes, they are confusing as to the underlying dimensionality of the system.

The expressions above can be rearranged to show that it is the wavelength, in fact, that forms the relevant scale in the radar equation. Using the expression for received power in terms of the aperture size, all dimensions can be expressed entirely in units of λ. Multiplying and dividing by λ^4, Eq. (3.10) gives,

$$P_{\mathrm{r}} = \frac{P_{\mathrm{t}}}{4\pi} \frac{1}{(R/\lambda)^4} \cdot \frac{A^2}{\lambda^4} \cdot \frac{\sigma}{\lambda^2} \tag{3.12}$$

where R is now scaled by λ while A and σ are scaled by λ^2, with the result that the scaled quantities are dimensionless. This tells us that for any system with constant radiated power, the received power remains constant with any change in frequency, provided that *all* physical dimensions of the system are also rescaled by the corresponding change in wavelength.

3.2.2 Maximum Operating Range

Designers of traditional radar systems intended exclusively for purposes of surveillance and detection are concerned with sensitivity at a given range and with the maximum range of operation. Let $P_{\mathrm{r}_{\min}}$ be the minimum useful received power in a given system. Rearranging Eq. (3.1) once again,

$$R_{\max} = \left[\frac{1}{P_{\mathrm{r}_{\min}}} \frac{P_{\mathrm{t}} G_{\mathrm{t}} A_{\mathrm{t}} \sigma}{(4\pi)^2} \right]^{\frac{1}{4}} \tag{3.13}$$

Many system engineers consider this expression the fundamental form of the radar equation because it represents—in typical, but not all situations—the most extreme design condition for specified system power, antenna characteristics, and target scattering properties. Note that the maximum range for a specified power increases only as the fourth root of the other parameters, so range is not easily extended beyond a current state-of-the-art value. Obviously, Eq. (3.13) also can be rewritten in forms corresponding to those in Eqs. (3.11) and (3.10).

Modern systems can be much more complex. Consider the simple side-looking imaging system of Figure 2.7. In that example, we are concerned primarily with the quality of the image generated from the side-looking radar (SLR) observations. Since the range to different across-track locations varies between 7500 and 9000 m, we are concerned with performance over an interval of ranges, rather than just the maximum range. And there are other complications as well, as there are trade-offs between, e.g., resolution and image quality.

3.3 Comparison of Radar Signals with Direct Path Signals

It is useful to compare the strength of bistatically scattered radar signals with that of normal communication links. Consider the geometry in Figure 3.4, which is the same as that of Figure 3.1 but with the direct path with distance R_d has been added. Letting P_{radar} represent the received signal along the radar path and P_{direct} represent the received signal along the direct path, we can write,

$$\frac{P_{radar}}{P_{direct}} = \frac{\left[\frac{R_t G_t A_r \sigma}{(4\pi)^2 R_t^2 R_r^2}\right]}{\left[\frac{P_t G_t' A_r'}{4\pi R_d^2}\right]} = \frac{1}{4\pi}\frac{G_t A_r}{G_t' A_r'} \cdot \frac{\sigma R_d^2}{R_t^2 R_r^2} \tag{3.14}$$

where the new quantities G_t' and A_r' are the antenna parameters for signals transmitted and received, respectively, along the direction of the direct path. Usually, and unless the target lies near the line connecting transmitter and receiver, $\sigma << R_i^2$, where i refers to the three paths: d, t, and r, while $R_d \leq R_t + R_r$ and $G_t A_r / G_t' A_r' > 1$. The ratio of RCS to the range factors, $\sigma/(R_t^2 R_r^2)$, typically is much less than all other factors, though, including the ratio of the antenna parameters. That is, the power received over the radar path is almost always very much less than that along the direct path.

The conclusion, as mentioned earlier, can be understood in physical terms as a consequence of the reradiation of the scattered signals by the radar target. The initial radiation of signal by the transmitter undergoes a spherical spreading, which leads to an R^{-2} propagation loss, the first factor in the list following Eq. (3.1), p. 26. As a consequence, the transmitted signal becomes progressively weaker with increasing values of R, losing one fourth of its intensity for each doubling of the range for both the direct and radar paths. In addition, there is a second spherical spreading associated with reradiation of the scattered energy from the target object along the radar-to-receiver path, the fourth factor in the list beneath Eq. (3.1). The effect of this second spreading is multiplicative with the first, very roughly doubling the *magnitude* of the loss along the scattered path for most geometries.

Exceptions occur for instances in which one of the scattering paths is very short as compared with the other, so that a sufficiently large RCS can compensate for the doubled spherical-spreading effect. The numbers can be interesting—some test calculations will provide a feel for the behavior of the radar equation as compared with the Friis equation for point-to-point communication. On the other hand, the signal along the direct path—although it may suffer from less favorable values for the equipment parameters—is weakened only once due to spherical spreading of the area of the wave.

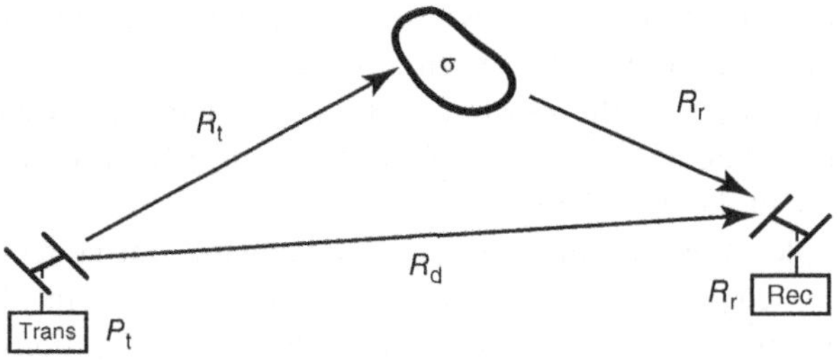

Figure 3.4 Comparing geometry of radar scattering and direct paths. Although the direct path length, R_d, typically is comparable to the total of the transmitter-plus-target-plus-receiver path, $R_t + R_r$ the net loss along the direct path is almost always substantially less than along the scattered signal path (see text).

3.4 Typical Values of RCS

We need some familiarity with the value of the RCS for various objects. The analysis of electromagnetic scattering is difficult, depending on the complexity of the objects considered—sufficiently so that accurate analytical methods of analysis are not available for most objects. Consequently, the study of RCS relies to a great extent on measurement and calibration. Exact values are known for a small number of canonical examples from classical solution of Maxwell's Equations, and for a large number of interesting shapes and generic configurations derived from detailed analysis, approximation, and numerical computation. Results can be found in the vast literature of electromagnetic scattering and in treatises on the subject.

Here we note a few useful rules of estimation that provide an indication of what to expect from a variety of possible target objects. The first of these pertains to conducting objects comparable with or significantly larger than a wavelength; viewed from aspects exclusive of forward scattering, the RCS of these objects is very roughly that of the projected area of the object. Second, the RCS of dielectric objects is usually lower than that of metal objects. Third, there are special geometric configurations for which neither of the two previous statements is accurate. These exceptions usually involve: (i) some sort of focusing phenomena leading to an RCS that is exceptionally larger than would be otherwise anticipated on the basis of size alone; or (ii) some sort of unusual shape that renders the RCS exceptionally less than anticipated. Values in Table 3.3 of RCS observed at centimeter wavelengths show the range for some common objects.

For purposes of testing and calibration, the use of reference targets for which the scattering behavior is well understood is standard practice. The sphere is among the most commonly used reference objects. For homogeneous spheres and near-spheres that are large with respect to the wavelength, i.e., $a > \lambda$, where a is the radius, we have the simple formula,

$$\left.\sigma\right|_{\text{sphere}} \simeq \rho\pi a^2 = \text{reflection coefficient} \times \text{projected area} \tag{3.15}$$

where ρ is the specular power reflection coefficient at normal incidence to a slab of the same material as the sphere. This formula applies only to external reflections from the surface of the sphere. It fails for scattering from highly transparent spheres, e.g., raindrops, in which internal reflections play important roles in backscattering, for near-forward scatter, and for certain other angles. Thus,

Table 3.3 Typical values of RCS for representative objects.

Object	RCS (m^2)[a]
Insect	10^{-5}
Bird	0.01
Bicycle with rider	1
Person	0.1–2
Rowboat, motorboat	0.2–2
Small aircraft[b]	1
Boeing 747[b]	100–400
Automobile[b]	100
Ship (10,000 tons)	$\approx$ tonnage

a. Values are for cm-λ.

b. Strongly dependent on viewing aspect.

we can use the same formula to calculate the monostatic RCS of a small object such as a baseball, and of a very large object such as our moon!

$$\sigma\Big|_{\text{baseball}} \simeq 0.1 \times \pi(0.03)^2 \simeq 2.7 \times 10^{-4}\ \text{m}^2 \tag{3.16}$$

$$\sigma\Big|_{\text{Moon}} \simeq 0.07 \times \pi(1.736 \times 10^6)^2 \simeq 6.63 \times 10^{11}\ \text{m}^2 \tag{3.17}$$

We note the RCS of a conducting sphere with radius $\pi^{-\frac{1}{2}}$ m is 1 m^2.

In the examples, the restriction that the radius be greater than the wavelength is not a problem for the moon(!), but it restricts us to wavelengths less than a few centimeters for the baseball, and to less than about 10 μm, say, for scattering from a "BB." Equation (3.15) is satisfactory for spheres of "lossy," or opaque, materials and conductors at all angles except near-forward scatter.

3.5 Signal-to-noise Ratio

3.5.1 Definition

The radar equation allows us to describe the effectiveness of a radar system in terms of the scattered signal power reaching the receiver. For this, we need only a few numbers characterizing the system components. But the key to characterizing system capabilities is not the absolute level of signals received by the radar receiver. Rather, it is the relationship of the received echo power to that of the omnipresent noise present in the receiving system. For this purpose, the important system performance parameter is the ratio of signal power to noise power, which we refer to as the *signal-to-noise ratio* (SNR).

In most cases, it does not matter how a particular value of SNR is achieved. The primary concern is that the received signals exceed some minimum value with respect to the noise, above which we can consider observations of echo signals to be reliable. For a system in which the SNR is insufficient for some purpose, we might meet the needed minimum value by, for example, increasing transmitter power, improving receiver sensitivity, increasing antenna size, restricting operation of the radar to certain low-noise sites, or a combination of these. Radar design allows many different implementations to reach a needed performance level, and each radar system represents a balancing of components suited to that particular application. When we speak of "designing a radar," we often refer to the relative choices of subsystem and component values that will result in a minimum SNR subject to technological constraints imposed by the radar operating platform and one or more key subsystem components.

We formally define the value of SNR as,

$$\text{SNR} = \frac{P_\text{r}}{N} \tag{3.18}$$

where P_r is the echo signal power received by the radar and N is the noise power at the same point within the system. Often the reference point is chosen to lie at the connector between the receiver and antenna, but any convenient accessible location will serve.

While the definition above is applied in the analysis of all systems, be aware that a complete analysis considers the effect of any signal manipulation, beyond simple amplification, before calculating the final SNR. Ultimately, in signal detection and measurement, we are concerned with measurement uncertainties in our observations of processed echo signal. As a first example of this point of view, we can define an SNR based on average signal power with respect to a measure of the noise power fluctuations; the standard deviation of the noise, (σ_N), is a frequent choice,

$$\mathrm{SNR}_{\sigma_N} = \frac{\langle P_r \rangle}{\sigma_N} \tag{3.19}$$

An unfortunate standard, but confusing, set of notation uses variants of the Greek σ for both RCS and for the standard deviation of a quantity. It is important to keep in mind the context of each quantity as both of these can appear in a single equation. It will be even more so in later chapters when we introduce normalized RCS σ^0, yet another quantity with similar symbolic description.

Combining the radar equation as presented previously, Eq. (3.1), and the definition in Eq. (3.18), we express SNR in the following form, useful for general systems,

$$\begin{aligned} \mathrm{SNR} &= \frac{P_r}{N} \\ &= \frac{P_t G_t A_r}{N(4\pi)^2 R_1^2 R_2^2} \langle \sigma \rangle \end{aligned} \tag{3.20}$$

where, for now, we have made use of $\sigma_N \sim \sqrt{N}$, which is applicable to noise estimates from individual measurements. Intuitively, SNR is the critical parameter in system performance, as it suggests how easily a radar echo can be recognized in the sea of surrounding noise. As we see from the system parameters used in this formulation, SNR is closely linked to antenna structure, individual subsystem characteristics, and assumed or known properties of the scattering object.

3.5.2 A Model for Receiver Noise

There are many potential sources for noise in a radar system. Some noise results from thermal excitation of electrons within the receiver circuitry of the radar, some arises from natural sources such as the cosmic background radiation, some is produced by man-made interference, and yet still more results from nonlinear system effects such as quantization for digital processing. Typically, we measure or estimate the noise contribution inherent in the equipment and from as many of the other sources as possible. This information is then combined into a single equivalent noise source. The degree that noise is contributed by the radar equipment can be controlled as part of the system design. In operation, the noise level can be *estimated* and used to adjust critical parameters in the receiving system.

Figure 3.5 presents a linear model for system noise in which the receiver electronics are treated as perfect devices, while all noise at the output of the amplifying and filtering system of the receiver

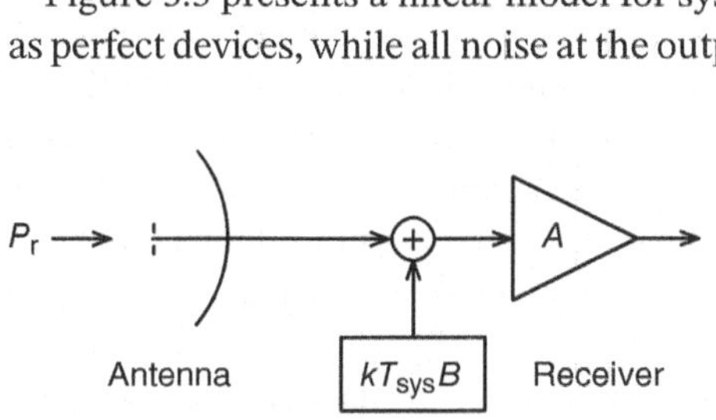

Figure 3.5 Noise model for receiving system. Noise in a receiver can be modeled as the contribution from a matched load at a system temperature T_{sys}, limited in bandwidth to the most restrictive bandwidth at any point of interest in the system. T_{sys} is made up of contributions from the external environment, which are received by the antenna, T_{ant}, from losses within the antenna and the transmission system, and those from the internal electronics of the receiver itself, T_{rec}. The latter is scaled to the value required to match the total electronic noise were it to originate at the input to the first amplifier. $T_{sys} = T_{ant} + T_{rec}$.

is referenced to a fictitious equivalent noise source located at the input of the receiver. The noise power at the system output is restricted to that which falls within the system bandwidth, so we only need to consider the noise within the bandwidth of the receiver output, which is N in our calculations. By an adaptation of Planck's Law describing natural radiation due to thermal effects, the strength of this noise can be expressed in terms of an equivalent temperature, T_{sys} (K). The equivalent noise power is given by,

$$N = kT_{sys}B \quad \text{(W)} \tag{3.21}$$

where $k = 1.38 \times 10^{-23}$(J/K) is Boltzmann's constant, and B is the most restrictive bandwidth within the receiving system.[8]

The model noise source is located in front of the receiver in the signal flow, i.e., at the so-called "receiver front end," we can think of the amplifier input as subjected to a uniform noise spectrum, independent of the frequency. From this point of view, a receiver with bandwidth B admits only a fraction of the possible noise into the system, resulting in the noise power N given in Eq. (3.21). The wider the bandwidth, the greater the total noise power at the output *and* vice versa. Extending this idea, we define the noise spectral density of the noise as:

$$N_0 = kT_{sys} \quad \text{(W/Hz)} \tag{3.22}$$

In Eq. (3.22) the noise is present at the input of the amplifier as a *temperature*, separating the noise power spectral density from the bandwidth. For this reason, it is common to speak of the equivalent temperature of an amplifier or a system as a basic statement characterizing noise performance. Obviously, $N = N_0B$ is the total equivalent noise at the input of the ideal amplifier of the model.

It is also useful to separate the contribution of the noise originating in the environment external to the receiver from that within the receiver itself. The first of these two components is made up of external noise associated with the antenna and transmission line system, while the second is inherent to the receiving system per se. For convenience, here, we combine the latter two sources and characterize the noise from these by the equivalent temperature, T_{ant}. To this we add the noise from the electronics of the receiver itself, T_{rec}. We treat the sum of these as originating at the input to the first amplifier. Thus, we can express system noise described above as:

$$T_{sys} = T_{ant} + T_{rec} \tag{3.23}$$

This subject is developed more completely in Sections 3.6 and 6.4.

The antenna temperature T_{ant} is the result of natural radiation impinging on the antenna from astronomical sources above the local horizon and from blackbody radiation of the natural environment, including the atmosphere. The strength of this noise source depends on wavelength and the antenna pattern weighting of the various sources. Antenna temperatures can vary between a minimum of a few kelvins to many thousands of kelvins. This parameter can be determined by measuring, or in some cases estimating by calculation, the power $kT_{ant}B$ at the antenna terminals in the absence of other signals. This must be done with the antenna otherwise in its normal operating environment.

In radars, intended for remote sensing and related applications, the primary source of noise is from the natural background, from emissions by electrical machinery and electronic systems, and sometimes from interference arising from other users of the radio spectrum, whether intentional or not. Usually, such noise is treated on an *ad hoc* basis.

Typically, $T_{sys} \simeq 10$–1000 K for practical radar systems designed for environmental applications. The background noise from natural sources contributing to T_{ant} ranges from $\approx$3 K for the open

8 With k in joules/kelvin, B must be in hertz, and T_{sys} in kelvins. N, then, in watts.

sky, to ≈300 K for the surface of Earth, to higher values for some celestial objects such as Venus, $T_{\text{Venus}} \approx 640$ K and the sun, $T_{\text{sun}} \approx 10^5$ K.

Designers of military radars are also very concerned with deliberate interference intended to degrade a radar's performance—a third source of noise. Such interference is referred to as *jamming*. A closely related problem of interest in military systems is the effect of signals intended to produce false data output from the radar. This last signal type is referred to as *spoofing*. Obviously, these latter forms of "noise" are very different in structure from that originating from stochastic processes in the receiver and in nature. We do not consider these further.

3.5.3 Measuring Noise Levels

In the noise model, the receiving electronics are assumed to be completely noiseless, as the system noise is modeled as a source injected at a point just ahead of the first amplifier. The electronic contribution to the receiver temperature is incorporated in T_{rec}, above, an effective temperature that expresses the effect of all random sources of signal within the receiver electronics in terms of a single source at the receiver front end, as discussed above. This quantity is conveniently measured in the laboratory by several methods, including the use of instrumentation manufactured for the purpose. A simple method that demonstrates the meaning of the definition is to inject a signal of calibrated strength from a matched source at the receiver's front end, and then observe the signal level required to double the power output of the receiver relative to that in the absence of a signal; $kT_{\text{rec}}B$ is equal to the injected power.[9]

Some systems incorporate facilities for noise injection as a standard feature of the design. For example, a receiver might be calibrated by reference to a pure "tone," i.e., a low-level, calibrated, highly stable, single frequency signal injected into the receiver front end while the radar is in operation. If the frequency of the injected signal is chosen to appear in an unused portion of the received spectrum, e.g., either above or below the predominant frequency band occupied by the echo, then it will cause no significant interference to the echoes. Such a calibration tone can be used as a solid basis of comparison for determining absolute echo strength as the two signals pass through the same system. Alternatively, a calibrated source of broadband noise can be used for the same purpose. In this method, known generally as a "Dicke Switching,"[10] the noise source is modulated, usually between "On" and "Off." The resulting systematic change in the system noise level is readily detected and also can be used to calibrate the received power level.

3.6 Noise in Circuits

3.6.1 Nyquist's Law

Referring to the left-hand portion of Figure 3.6, what is the noise voltage associated with a resistance R at a temperature T. From *Nyquist's Law*, the mean-square voltage due to random thermal processes is,

$$\langle V^2 \rangle = 4kTRB \tag{3.24}$$

9 Some care is required here, as the signal source can also be a source of unwanted noise. The degree to which this occurs depends on the internal configuration and the methods by which the connections are made. It is also necessary, for accurate results, to consider the effects of a variation in gain across the amplifier/receiver passband.
10 After the radio astronomer who first implemented this type of system.

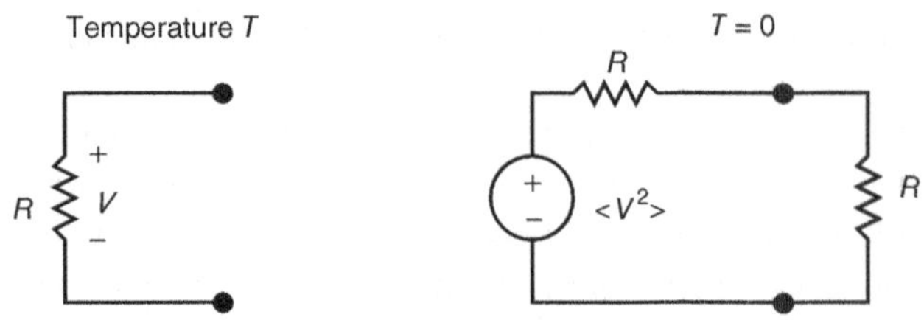

Figure 3.6 Circuit models for thermal noise. Open-circuited resistor at temperature of T(K) on left has a fluctuating terminal voltage due to random motion of electrons within the resistive material. Thevenin equivalent of resistor is a "cold" resistor with an external series voltage source with mean-square output of V^2 as shown in circuit with matched load at right. Value of open-circuit voltage for a warm resistor is from Nyquist's law.

Table 3.4 Example resistive noise calculations. Note the sole dependence of the thermal noise on temperature. Although the output voltage changes by a factor of 10^4, in proportion the resistance, the power available to a matched load depends only on the physical temperature (see Figure 3.6).

T (K)	B (kHz)	kTB (mW)	R (Ω)	$\langle V^2 \rangle^{1/2}$ (μV)
300	5	2×10^{-19}	5×10^5	6.4
300	5	"	5×10^1	0.064
30	5	2×10^{-20}	5×10^5	2.0
30	5	"	5×10^1	0.02

where B is the bandwidth. We find the power transferred to a matched load from the equivalent circuit at the right of Figure 3.6,

$$\frac{\langle V^2 \rangle}{4R} = kTB \tag{3.25}$$

which is *independent* of R.

Table 3.4 gives a few example values for $\langle V^2 \rangle^{1/2}$ and kTB for various values of R, T, B.

3.6.2 Noise in Amplifiers

Now consider the noise in an amplifier, as suggested by the black box of Figure 3.7. The noise power fed into the amplifier is $N_{in} = kT_sB$; we denote the output noise power N_{out}.[11]

Since an amplifier at a temperature greater than 0 K must impose some additional noise on the amplified signal, $N_{out} > AN_{in}$, where A is the amplifier gain. We can write N_{out} as,

$$N_{out} = (kT_{eq}B) \cdot A = k(T_s + T_{amp})B \cdot A \tag{3.26}$$

From the above expression, the amplifier temperature, T_{amp}, is treated as though the excess noise was added at the input to the amplifier, as shown in Figure 3.8.

The *noise figure*, F, of an amplifier is the SNR at the amplifier input divided by the SNR at the output, i.e., the ratio of the SNR ratios. Referring to Eq. (3.26) and Figure 3.8, we express the noise figure as,

11 Note that if $R_s = R_{rad}$, $T_s = T_{ant}$, where R_{rad} is the radiation resistance of the antenna and T_{ant} is the antenna "temperature."

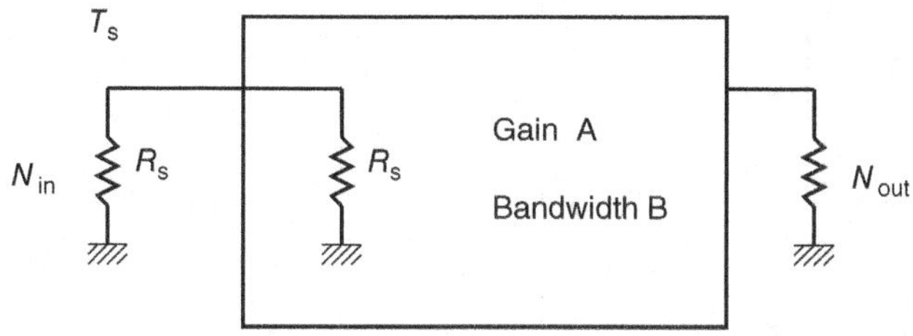

Figure 3.7 Input and output noise in amplifiers. Input power N_{in} is supplied from a matched source with resistance R_s at temperature T_s as shown; N_{out} is a matched load.

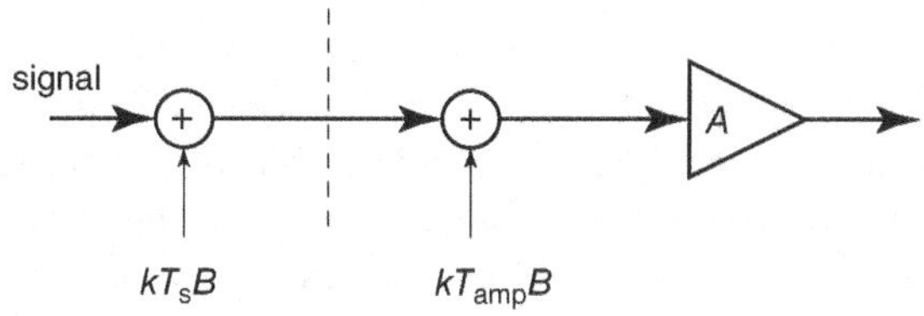

Figure 3.8 Model for amplifier noise.

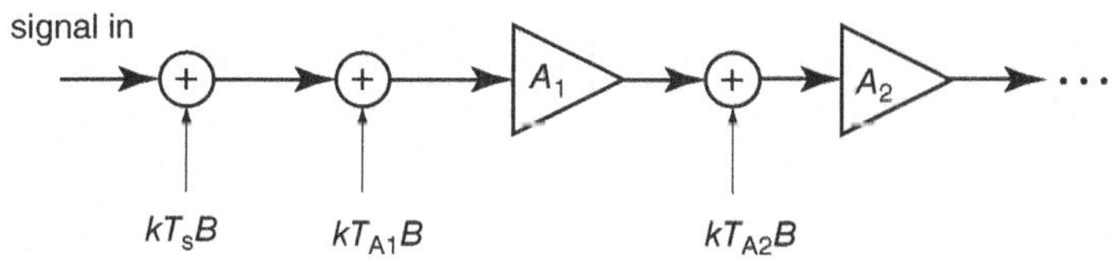

Figure 3.9 Amplifier chain.

$$F = \frac{(S/N)_{\text{in}}}{(S/N)_{\text{out}}} = \frac{S/(kT_s B)}{S/(kT_{\text{eq}} B)} = 1 + \frac{T_{\text{amp}}}{T_s} > 1 \tag{3.27}$$

Note that the definition of noise figure depends on T_s, which is conventionally taken as 293 K, corresponding approximately to room temperature.[12]

For low-noise amplifiers (LNAs), it is much easier to work with temperature alone, rather than noise figure. For example, for the amplifier chain shown in Figure 3.9, we can write the SNR as,

$$\begin{aligned} \text{SNR} &= \frac{S \cdot A_1 \cdot A_2 \cdots}{kB((((T_s + T_{A1})A_1 + T_{A2})A_2 + \cdots)\cdots)} \\ &= \frac{S}{kB(T_s + T_{A1} + T_{A2}/A_1 + \cdots)}. \end{aligned} \tag{3.28}$$

Notice that the noise contribution from each amplifier stage is effectively reduced by the gain of the preceding amplifiers, and the effective noise temperature referred to the input to the amplifier system is given by the factor in parentheses in the denominator.

12 We note that the rounded value of 290 K is also used.

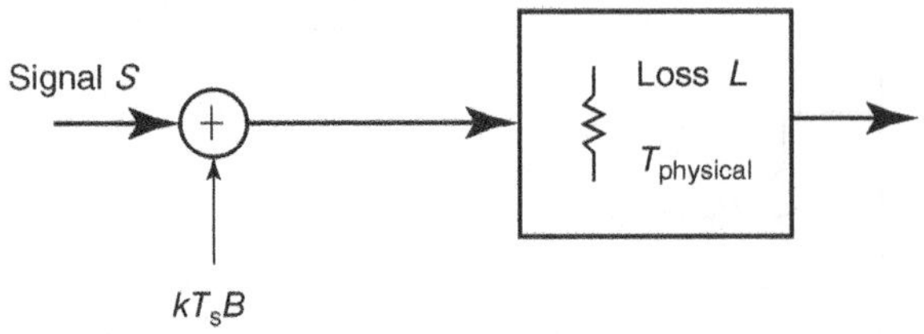

Figure 3.10 Noise in attenuators. Attenuators have the dual effect of reducing signal strength while often increasing noise by increasing coupling to the local environment.

3.6.3 Dissipative Elements

How do we account for the effects of "lossy" elements, such as attenuators, transmission lines, etc.? In Figure 3.10, L represents the fraction of power *lost* to attenuation. It is straightforward to show that the signal power remaining at the output of the attenuator is $S(1-L)$. The effect on the noise power is more slightly more complicated, though, since the attenuator both absorbs some of the input noise while simultaneously emitting noise source owing to the resistive losses incorporated in its construction. That is, the losses built into the attenuator for the purpose of reducing the signal level also provide a path for some internal thermal energy to leak into the signal path. Consequently, an attenuator emits electrical noise by the same mechanism as our warm resistor in Figure 3.6 and Eq. (3.24).

Letting T_s be the equivalent noise temperature of the source—this is the same as for the amplifier—and $T_{physical}$ be the actual physical temperature of the attenuator, the noise power at the attenuator output is,

$$N_{out} = \underbrace{kT_s(1-L)B}_{\text{noise passed by attenuator "}L\text{"}} + \underbrace{k\mathrm{T}_{phys}LB}_{\text{noise added by "}L\text{"}} \tag{3.29}$$

Note that the second term in Eq. (3.29), accounting for the noise added by the attenuator, is proportional to the loss. A totally absorbing attenuator, i.e., $L = 1$, has a noise output to a matched load of $k\mathrm{T}_{phys}B$, which is the same as that of an *arbitrary* resistor. At the same time, $L = 1$ implies that no signal or noise power from the source passes through the attenuator, so the output for this condition comprises only "attenuator" noise.

Applying the definition of noise figure from p. 41 and the result above, the noise figure for the attenuator is,

$$\begin{aligned} F = \frac{(S/N)_{in}}{(S/N)_{out}} &= \frac{S/(kT_sB)}{S(1-L)/(kB(T_s(1-L)+\mathrm{T}_{phys}L))} \\ &= \frac{T_s(1-L)+\mathrm{T}_{phys}L}{T_s(1-L)} \\ &= 1+\frac{\mathrm{T}_{phys}L}{T_s(1-L)}. \end{aligned} \tag{3.30}$$

If we let α denote the fraction of power passed by the attenuator (i.e., $\alpha = 1-L$), then we have a second commonly used expression for the attenuator noise figure, namely,

$$\begin{aligned} F\Big|_{\mathrm{T}_{phys}=T_s} &= 1+\frac{1-\alpha}{\alpha} \\ &= \alpha^{-1} \end{aligned} \tag{3.31}$$

which applies when the physical temperature of the attenuator and the noise temperature of the source are the same, as is often the case for laboratory measurements. Applying Eq. (3.31) to a 10 dB attenuator ($\alpha = 0.1$), the noise figure is also 10 dB.

We can find the equivalent temperature of the attenuator by rewriting the output SNR as,

$$(S/N)_{\text{out}} = \frac{S(1-L)}{k(T_s(1-L) + \text{T}_{\text{phys}}B} = \frac{S}{k\left(T_s + \text{T}_{\text{phys}}\left(\frac{1}{1-L}\right)\right)B}. \tag{3.32}$$

Hence, the equivalent front-end temperature of the attenuator is

$$\text{T}_{\text{attn}} = \text{T}_{\text{phys}}\frac{L}{1-L} \to \infty \qquad \text{as}\, L \to 1. \tag{3.33}$$

3.7 The dB Table—A Design Tool

Equation (3.20) gives the signal-to-noise of a radar system as a function of subsystem performance parameters and noise arising from the radar electronics and the natural environment. While it is straightforward to change the values of each term and arrive at SNR estimates for a wide variety of system configurations, conceptualization of a final value dependent on a string of products and quotients can be awkward owing to the multiplication and division required. Radar engineers use an alternate structure to organize and visualize the calculation of SNR in connection with the radar equation. Mathematically equivalent to the form given in the SNR equation above, a "dB Table," in which each quantity is expressed logarithmically in decibels rather than in natural units, greatly simplifies the arithmetic. Although, historically, the use of dB tables goes back to communication link design, it is equally applicable to radar system design, and allows for a convenient, straightforward visualization for evaluation of the radar equation.

This approach has three advantages: (i) the multiplications and divisions become additions and subtractions, so that it is a simple matter to estimate SNR rapidly in one's head without relying on computational aids; (ii) the use of logarithms makes it particularly easy—trivial, actually—to determine the effects of changes in the subcomponents on overall system performance; and (iii) many components used in radar system design by convention are specified directly in dB, so that a further calculation of the dB specification of a part from natural units is not required.

As an example, the gain of a microwave amplifier might be described as "25 dB," rather than as "×316.2." If one were to consider changing the amplifier to one with an order of magnitude less gain, e.g., 15 dB, it is easier to subtract 10 dB in the logarithmic calculation than it is either to multiply by (216.2/316.2) or to redo the entire calculation using 216.2 in place of 316.2—check the numbers! This example begs the question as to how one returns to real units...a problem which is obviated, i.e., in engineering terms, "short-circuited," by expressing the needed SNR is dB as well. The latter is accomplished by appending the appropriate unit to "dB," as we illustrate below.

Construction a dB table for radar problems follows from our expression for the SNR of a received radar echo. Writing Eq. (3.20) in logarithmic form,

$$\text{SNR (dB)} = 10\log\left[\frac{P_r}{N}\right] \tag{3.34}$$

$$= 10\log[P_r] - 10\log[N] \tag{3.35}$$

where we keep in mind that the decibel *always* represents a dimensionless ratio of powers. When expressed in decibels, though, the ratio decomposes into the difference between the terms, as above.

Expanding the terms for the numerator and denominator again leads to sums, with differences as well. This property of the logarithm allows us to create a table of these terms and treat them separately.

Considering first our result for the received echo power, Eq. (3.1), and expressing in logarithmic form, we again obtain,

$$10\log[P_\mathrm{r}] = 10\log\left[\frac{P_\mathrm{t}G_\mathrm{t}}{4\pi R_\mathrm{t}^2}\cdot\sigma\cdot\frac{A_\mathrm{r}}{4\pi R_\mathrm{r}^2}\right] \tag{3.36}$$

$$= 10\left[\log[P_\mathrm{t}G_\mathrm{t}] - \log[4\pi R_\mathrm{t}^2] + \log[\sigma] - \log[4\pi R_\mathrm{r}^2] + \log[A_\mathrm{r}]\right] \tag{3.37}$$

In this instance, the quantity on the left-hand side is of the form "log[power]" rather than a ratio, which certainly is not in decibels. It is, however, the first term in Eq. (3.35). Now, considering our statement of the system noise, Eq. (3.21), and expressing the noise in logarithmic form,

$$10\log[N] = 10\log[kT_\mathrm{sys}B] \tag{3.38}$$

$$= 10\left[\log[k] + \log[T_\mathrm{sys}] + \log[B]\right] \tag{3.39}$$

where again, the result is not in decibels. The difference of Eqs. (3.37) and (3.39), expressed in natural units, remains the ratio of the echo power to noise power, as this was our starting point, Eq. (3.34), which is the logarithm of a ratio.

Engineers give meaning to the individual terms in expressions such as those in Eq. (3.37) by modifying the dB notation slightly for instances in which the quantity represented not a pure power ratio. To see how this works, consider the first term in Eq. (3.37), which can be expanded further as,

$$10\log[P_\mathrm{t}G_\mathrm{t}] = 10\log[P_\mathrm{t}] + 10\log[G_\mathrm{t}] \tag{3.40}$$

Here the first term on the right is of the form "log[power]" while the second is "log[gain]." Recalling that the definition of antenna gain is the ratio of the power per unit area at distance radiated by a particular antenna to that radiated per unit area at the same distance by an ideal isotropic antenna, we see that "gain" is a ratio of powers, and hence can be expressed properly in decibels. So, we can write,

$$G_\mathrm{t} = 10\log[G_\mathrm{t}]\ (\mathrm{dB}) \tag{3.41}$$

The log[power] term is problematic, though, as power per se is not a ratio of powers. This is handled by an engineering convention that treats power expressed in decibels as normalized by its unit. For example, if power is given in watts, then, for the purpose of expressing this in logarithmic form, we consider the quantity of 10 W to be the ratio of 10 W to 1 W, or 10 : 1. Noting that the reference unit is 1 W, we express this in dB as:

$$10\log\left[\frac{10W}{1W}\right] = 10\ \mathrm{dBW} \tag{3.42}$$

where the "W" appended to dB signifies the use of 1 W as the reference power. That is, the notation "0 dBW" corresponds to a power of 1 W. One can make use of other reference values. For instance, the designation "dBm" signifies 1 mW.

In radar parlance, it is also common to designate units of area in a manner similar to that of absolute power, as this occurs in describing both RCS and the effective area of antennas. Table 3.5 provides a few examples.

Table 3.5 Unit normalized logarithmic designations. Unit appended to "dB" specifies the unit physical value corresponding to the zero reference for the decibel scale.

Physical quantity	Physical unit	Log designation
Power	watts	dBW
	milliwatts	dBmW
	microwatts	dBμW
Area	sq. meters	dBm^2
	sq. centimeters	$dBcm^2$

Table 3.6 Example dB table for SNR. Result column indicates subtotal for echo power, P_r and noise, N, and resulting SNR. Column "Parameter" lists factors from the radar equation and the Planck representation of the noise; terms are from Eqs. (3.35), (3.37), and (3.39). Column "Value" provides examples for a typical, commercially available marine radar. Columns "10 Log10" and "Units" give evaluated parameters expressed in decibels. Value for the ratio on last row is obtained by converting +27 dB to a decimal number.

Result	Parameter	Value	Unit	10 Log10	Unit
P_r	P_T	1	kW	30	dBW
	G_T	10^4		40	
	$4\pi^{-1}$	0.08		−11	
	R_t^{-2}	$R_t=10^4$	m	−80	
	σ	1	m^2	0	dBm^2
	A_R	0.1	m^2	−10	
	$4\pi^{-1}$	0.08		−11	
	R_r^{-2}	$R_t = 10^4$	m	−80	
			Subtotal	−122	dBW
N	k	1.38×10^{-23}	J/K	−229	
	T_{sys}	1000	K	30	
	B	100,000	Hz	50	
			Subtotal	−149	dBW
SNR	$P_r - N$			+27	dB
	P_r/N		SNR = 500 : 1		

Table 3.6 illustrates the construction and use of a dB table by showing an example calculation of received echo power, P_r, noise power, N, and SNR using a dB table. It is conventional in such tables to omit any reference to the units for dB values except when the unit is power or area as above; otherwise, the units only need be consistent with each other.

The example of Table 3.6 summarizes the performance of the radar system by its SNR expressed in dB. Transmitted and received power are specified relative to a reference power of 1 W. The contribution of each factor in the radar equation to the total is apparent, and the effect on the final SNR of an adjustment to the value for each parameter is easy to calculate in dBW. For example, a potential increase in the transmitted power by a factor of two increases the entry in the dB table to 33 from 30 dBW. This would allow the antenna area to be reduced to 0.05 from 0.1 m^2, or to −13 dBm^2, while maintaining the original +27 dB SNR. Or perhaps one is interested in the potential increase in operating range at the same SNR. By displaying these sensitivities directly, use of the tabular form often helps the designer trade-off one parameter for another, while at the same time linking the physical properties of the system components with the geometry to predict the overall performance.

The trade between peak power and antenna area is quite common as various platforms have different physical or other limitations, in some instances making an increase in transmitter power less expensive overall than an increase in antenna aperture, with other situations favoring other trade-offs. The impact of alternative implementation strategies is immediately apparent through the radar equation in either the multiplicative form or in the dB table form, whether the table exists in a computer or on the back of an envelope.

4

Radar Cross Section

Radar design for a particular application requires knowledge of the scattering properties of the anticipated targets. We now expand on this subject, introduced in Chapter 3, and ask more precisely, "How is radar cross section related to the physical characteristics of a scattering object?" First, we review the definition of radar cross section (RCS). Next, in order to develop a better feel for the scattering process, we use geometric optics to calculate the scattering from three canonical targets. Finally, we discuss a few general points regarding scattering from complex objects and actively assisted radar echoes.

4.1 RCS Parameters

4.1.1 RCS from Wave Intensity

Recall that the bistatic radar equation relates received power to parameters of the transmitting and receiving systems. The target properties are expressed in terms of an effective area of wave interception, normalized as though the target reradiates, or *scatters*, isotropically. With this definition of the target's RCS, it is a simple matter to write out the expression for received power from first principles, as in Eq. (3.1), which is repeated here for convenience,

$$P_\mathrm{r} = \frac{P_\mathrm{t}G_\mathrm{t}}{4\pi R_\mathrm{t}^2} \cdot \sigma \cdot \frac{1}{4\pi R_\mathrm{r}^2} \cdot A_\mathrm{r} \tag{3.1}$$

In order to bring out the physical meaning of the factors in Eq. (3.1) apropos the RCS, we identify the wave intensities at the target and at the receiver,

$$\mathcal{S}_\mathrm{inc} = \frac{P_\mathrm{t}G_\mathrm{t}}{4\pi R_\mathrm{t}^2} \qquad \{\mathrm{W/m^2} \text{ at the target}\} \tag{4.1}$$

$$\mathcal{S}_\mathrm{scat} = \frac{P_\mathrm{t}G_\mathrm{t}}{4\pi R_\mathrm{t}^2} \cdot \sigma \cdot \frac{1}{4\pi R_\mathrm{r}^2} \qquad \{\mathrm{W/m^2} \text{ at the receiver}\} \tag{4.2}$$

where $\mathcal{S}_\mathrm{scat}$ and $\mathcal{S}_\mathrm{inc}$ are power densities at the target and receiver, respectively. When we are concerned primarily with the strength of the incident or scattered waves, we refer to these as *wave intensities.*

The power densities are related by the RCS and the range to the receiver,

$$\mathcal{S}_\mathrm{scat} = \mathcal{S}_\mathrm{inc} \cdot \sigma \cdot \frac{1}{4\pi R_\mathrm{r}^2} \tag{4.3}$$

so, we can express the RCS as a ratio of scattered to incident wave intensity or, equivalently, the ratio of scattered to incident field strength squared,

$$\sigma = 4\pi R_r^2 \frac{\mathcal{S}_{\text{scat}}}{\mathcal{S}_{\text{inc}}} \tag{4.4}$$

which is a slightly different form of Eq. (3.4). This last expression can be taken as a far-field definition of RCS expressed in terms of the incident and scattered power densities at the target and receiver.

4.1.2 Target Referenced Scattering Coordinates

In general, RCS is a function of several variables representing the geometry, wavelength, and polarization of the incident field and the received field and, in some instances, incident power. Symbolically,

$$\sigma = \sigma(\theta_{\text{scat}}, \phi_{\text{scat}}, \mathbf{p}_{\text{scat}}; \theta_{\text{inc}}, \phi_{\text{inc}}, \mathbf{p}_{\text{inc}}; \lambda; \mathcal{S}_{\text{inc}}) \tag{4.5}$$

where the vectors $\mathbf{p}_{\text{scat}}$ and $\mathbf{p}_{\text{inc}}$ represent the polarizations of the incident and scattered waves. The angles are in standard polar form, as defined in Figure 4.1 relative to a target-fixed coordinate system.[1] For a given radar, the incident wavelength, polarization, and power density are fixed by the transmitting system, while the received polarization is selected by the receiving antenna. For the purpose of organizing the multidimensional data required to describe a typical scattering object, we think most often of σ as being a function of the variables of observation, θ_{scat}, ϕ_{scat}, and $\mathbf{p}_{\text{scat}}$, with the corresponding variables of incidence treated as fixed parameters, as indicated in Eq. (4.5). In the great majority of cases, the potential variation with $\mathcal{S}_{\text{inc}}$ can be neglected.

For monostatic geometries, the description of σ is simplified since for backscatter, $\theta_{\text{scat}} = \theta_{\text{inc}}$ and $\phi_{\text{scat}} = -\phi_{\text{inc}}$.

4.1.3 Bistatic, Phase, and Scattering Angles

Some terminology is needed to describe the scattering geometry. By convention, the angle measured *from the direction of incidence* is called the *scattering angle*, α, as shown in Figure 4.2. The

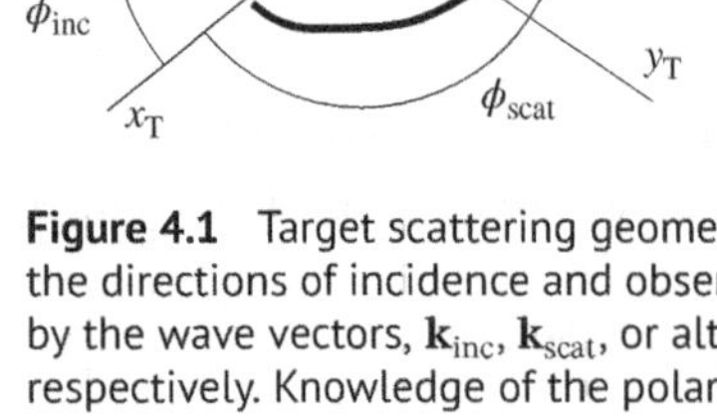

Figure 4.1 Target scattering geometry. Geometrically, the scattering properties of an object, T, depend on the directions of incidence and observation with respect to the object, and the polarization. These are given by the wave vectors, $\mathbf{k}_{\text{inc}}$, $\mathbf{k}_{\text{scat}}$, or alternatively, by the system of polar angles θ_{inc}, ϕ_{inc}, and θ_{scat}, ϕ_{scat}, respectively. Knowledge of the polarization is implicit here.

1 This is basically the same system as that pictured in Figure 6.1, where in both instances the coordinate system is tied to the physical structure of the target here, and to the structure of the antenna in Chapter 6.

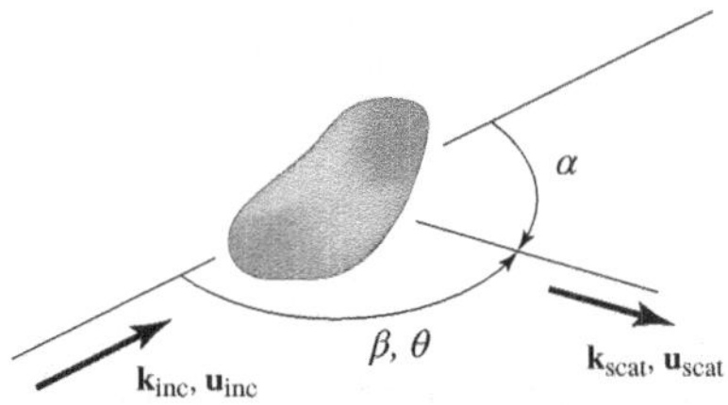

Figure 4.2 Definition of scattering angles. Scattering angle α is the deviation of a scattered wave direction from the direction of the incident wave, i.e., the angle between the incident and scattered waves, $\mathbf{k}$, or unit, $\mathbf{u}$, vectors; the bistatic angle, β (alternatively θ), is measured from the backscatter direction. In classical problems the latter is referred to as the *phase angle*, the same terminology that refers to the variation in appearance and the strength of scattered sunlight for various "phases" of the moon.

complement of this angle, e.g., the same angle measured *from the backscattering direction* is called the *bistatic scattering angle*, or the *phase angle*. This latter angle is usually denoted β in the scientific literature, but often appears as unsubscripted θ in the radar literature. Monostatic radars observe targets in backscatter only, i.e., $\beta = \theta = 0$, or $\alpha = \pi$.

4.2 Geometric Optics Calculation of RCS

Geometric optics, or "GO," refers to the treatment of wave propagation, reflection, refraction, and scattering in the "high-frequency," or "short-wavelength," limit in which the transport of electromagnetic energy can be described by a system of rays, or continuous lines, defining the paths followed by radiated energy. "Ray theory" and "ray optics" are synonymous with geometric optics. Here, we need to be somewhat precise with the definition, as we wish to accurately calculate the variation of power density along ray paths as a step in obtaining the RCS when the radar wavelength is very short, as compared with the size of the scattering object or feature.

The method of geometric optics can be summarized by a set of six rules governing its application, listed below:

1. Geometric optics applies strictly in the limit as wavelength $\lambda \to 0$. In practical application, the wavelength needs to be only much less than that of the structure scale of the medium through which it propagates, or the scale of structure on objects with which the wave interacts.
2. Electromagnetic energy flows along lines, called rays, for which the changes in direction are governed by the laws of refraction.
3. Energy flowing within a tube of rays is confined to the tube, although tubes can cross.
4. In lossless media, energy flow is conserved, otherwise the rate of energy loss along a ray is that of a uniform, homogeneous, block of the material in which the ray is located.
5. Rays are reflected specularly from locally smooth surfaces, i.e., at all points along the reflecting surface the angle of reflection equals the angle of incidence; the reflected ray remains in the plane determined by the incident ray and the surface normal at the reflection point (see Figure 4.3). The power in the reflected ray matches the reduction in power transmitted across the reflecting interface.
6. Rays are transmitted across interfaces between dissimilar materials in accordance with Snell's law, $n_{inc} \sin[\theta_{inc}] = n_{trans} \sin[\theta_{trans}]$, where n_{inc}, n_{trans} are the refractive indices on "incident" and "transmission" sides of the interface, respectively, and the θs are the corresponding angles of incidence and transmission.

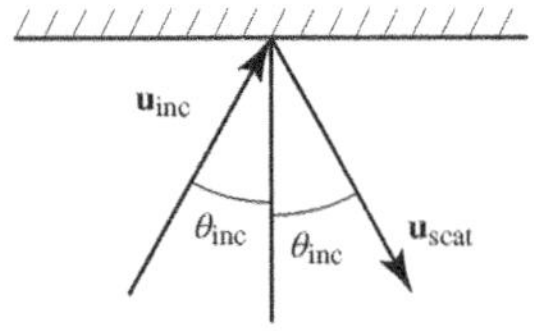

Figure 4.3 Specular reflection. This is the condition that angle of reflection equals the local angle of incidence, θ_{inc}, at every point on the interface; additionally, the reflected ray, $\mathbf{u}_{scat}$, remains in the plane determined by the incident ray, $\mathbf{u}_{inc}$, and the normal to the reflecting surface. A similar result is obtained for the wave-optics solution, but it applies only for reflection from an infinite plane.

The procedure following from the six rules above has only two parts: (i) the use of individual rays to trace the flow of energy; and (ii) the use of the shape of tubes, or bundles, of rays traced from the source to some point of interest to compute the power density along ray paths. In the remainder of these sections, we apply this two-pronged procedure to determine example, monostatic RCS of three simple geometric objects. The RCS follows from the definition in Eq. (4.4) once the incident and scattered power densities, $\mathcal{S}_{inc}$, $\mathcal{S}_{scat}$ are determined.

4.2.1 Three Examples

4.2.1.1 RCS of a Plane

We first illustrate the application of GO to scattering by calculating the RCS of an infinite, perfectly reflecting plane—i.e., a very large mirror, if you like. Consider the monostatic radar geometry in Figure 4.4, in which a transmitter and receiver are colocated beneath the plane at the location "T–R." An inverted right circular cone with apex at the radar and axis of length R perpendicular to the plane defines a bounding surface of rays emanating from the transmitter. At the plane, energy within the inverted cone is spread over the diameter of its base, $2w$, which lies in the plane. Specular reflection of the illuminating rays at the plane reverses their upward direction; at the same time, the reflected rays are deflected outward at the angle from the surface normal at which they arrived. The reflected rays form the frustum of an upright cone, cf. Figures 4.3 and 4.4, where we take the frustum's base to be located at distance at R below the plane, so it contains location T–R. It follows that the axis of the frustum is perpendicular to the plane and that the included angles of the two cones are the same, so that point T–R is at the center of the frustum base which is twice as wide as the illuminated area of the plane, or $4w$. Because the plane is perfectly reflecting and, by the laws of specular reflection for all rays internal to the inverted cone, all the energy flowing upwards in the inverted cone also flows downward within the frustum.

With regard to the definition of RCS, the incident power density is at the location of the reflecting plane, which in this instance is the radar "target"; the distance to the plane is $R_t = R$. Similarly, the scattered power density is that at the location of the radar receiver at located at distance $R_r = R_t = R$ from the plane. The ratio of the scattered to incident power density can be written, on average, in terms of the powers in well-defined areas. That is,

$$\frac{\mathcal{S}_{scat}}{\mathcal{S}_{inc}} = \left.\frac{P_{scat}/A_{scat}}{P_{inc}/A_{inc}}\right|_{P_{inc}=P_{scat}} = \frac{A_{inc}}{A_{scat}} \tag{4.6}$$

where P_{scat}, P_{inc} are the total scattered and incident powers, respectively, and the notation for the areas is similar. Eq. (4.6) tells us that when energy is conserved, we need to work only with the areas. In this instance, the ratio of the areas is,

$$\frac{A_{inc}}{A_{sc}} = \frac{\pi w^2}{4\pi w^2} = \frac{1}{4} \tag{4.7}$$

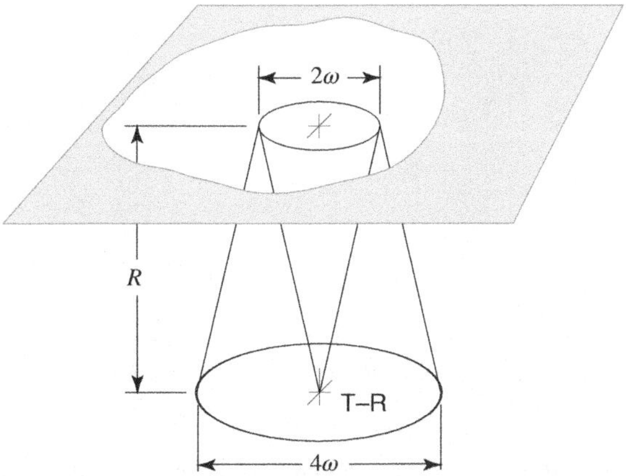

Figure 4.4 Apropos RCS of an infinite plane. The geometric optics RCS of a plane can be calculated from the ratio of the illuminated, or incident, area to the area where the corresponding reflection is observed. In this instance, the ratio is $(2w)^2/(4w)^2$; see Eq. (4.7).

That is, the effect of our arbitrary choice for the incident cone angle is unimportant insofar as the ratio of average power densities is concerned.

Our actual interest is in the power density in the immediate vicinity of the radar, rather than the average over some larger area. In Figure 4.4 the radar is represented by a single point at location T–R. In practical problems, the area of a receiving antenna has only limited linear extent, which is much less than R, so that the cone angle is small, since $\arcsin(2w/R) << 1$. With this in mind, we can finish the calculation by letting w shrink until it just covers the receiver. Mathematically, we use the approximation afforded by the use of the formal limit as $w \to 0$. Applying Eq. (4.4) to calculate the RCS, we find that,

$$\sigma_{\text{plane}} = \lim_{w\to 0} 4\pi R^2 \cdot \frac{\pi w^2}{4\pi w^2} = \pi R^2 \tag{4.8}$$

That is, the RCS of a plane is proportional to the square of its distance from the radar. It is, in fact, the area of a circle with radius equal to the range R.

Notice that the intermediate radius of the cone evaluated on the target plane, w, canceled in the final calculation, so that the result is independent of the assumed value of w. In particular, our result holds as $w \to 0$, at which point the area around the radar shrinks to a point. This is the condition under which we should interpret σ_{plane}. Otherwise, we are computing an average value over some large area of no special interest to us. In many instances, the intermediate steps do not result in such simple mathematical forms as this illustration, so the use of the limit is not always so simple as in the example here.

Although the scattering object may be large—an infinite plane in this example—our GO calculation applies strictly to scattering from a very small region immediately above T–R. From this, it should be apparent that size of the scattering object is not equal to the RCS.[2] While the computation is based on a tube or ray bundle of vanishing size, the result for the RCS actually arises from the differential behavior of the rays in the region surrounding the central ray as it is based on the

2 Two extreme scales are important in this example: (i) the plane is infinite, while (ii) the active scattering area is very small. Neither of these matches the area of the result in Eq. (4.8), which depends only on the separation between the radar and the reflecting plane.

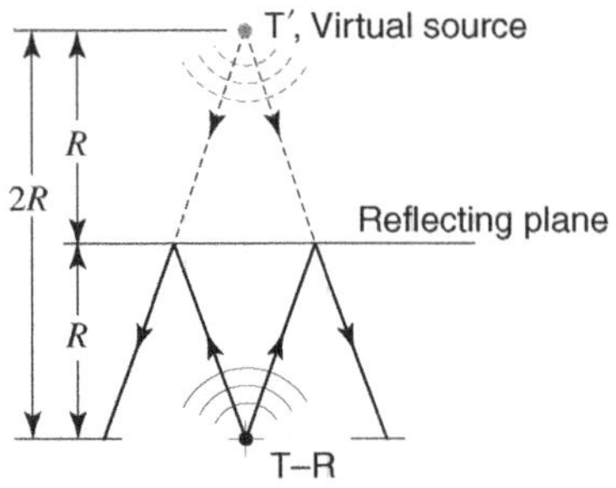

Figure 4.5 Reflection of the transmitter in a plane. Seen as a mirror image in the plane, the transmitter appears as a source, T′, located at a distance R above the plane.

ratio of areas within the ray bundle incident at the target and at the receiver, and it is this ratio that is physically important in determining the RCS.

Viewed another way, our example implies that, for geometric optics, we do not require the entire plane in order to produce the calculated cross section. Our final calculation would be the same if we were to assume only a small circular disk centered on the reflecting point, rather than a surface extending to infinity. Apparently, only a small portion of the plane around the perpendicular connecting the plane with the radar is needed. Geometric optics cannot answer the question as to how large an area is actually required. We address this question later, in Section 4.2.4.

Scattering from a plane is closely related to propagation from the mirror image of the transmitter in the surface of the plane, as illustrated in Figure 4.5. From the figure, unfolding the reflection of the radar in the plane places the transmitter at T′. Calculating the propagation loss from T′ to the radar at R, we see that we obtain the same result as before for the "scattered" field resulting from the radar transmitter at T–R,

$$P_\mathrm{r} = \underbrace{\frac{P_\mathrm{t} G_\mathrm{t} A_\mathrm{r}}{4\pi (2R)^2}}_{\text{Propagation from T}'} = \underbrace{\frac{P_\mathrm{t} G_\mathrm{t}}{(4\pi)^2 R^4} \cdot \pi R^2 \cdot A_\mathrm{r}}_{\text{Scattering from a plane}} \tag{4.9}$$

Thus, we have a cross-check on the calculation and a reference point of view for visualizing the RCS of more complex objects. In this instance, the mirror-like plane leads to an RCS that grows rapidly with distance from the radar site at T–R. This is a consequence of the reflection laws applied to a flat surface, for which the spreading of the reflected rays is the same as that of the incident rays, as shown in Figure 4.5. Note, though, that the growth of $\sigma(R)$ is overcome by the increasing distance between T and R, as illuminated in Eq. (4.9). On the other hand, scattering by curved surfaces changes the angles of the scattered rays as compared with those incidents, thereby modifying the strength of the received echo, increasing or decreasing the result depending on whether or not the scattered rays spread more or less rapidly than those incidents. Our next two examples illustrate this for the usual instance in which the echo strength is decreased.

4.2.1.2 RCS of a Cylinder

Analysis of backscattering from a cylinder is similar to that for the plane. In this instance, the curve of the cylinder about its axis adds slightly to the complexity of the analysis while strongly modifying the result. Figure 4.6 gives two views of the geometry, one of the plane containing the axis of the cylinder, the other looking along that axis. As before, the radar is located directly beneath the axis of the cylinder at the position T–R. Parallel to the axis, the geometry is the same as for the reflection from a plane, panel (a). Looking along the axis, however, as in panel (b), the curvature

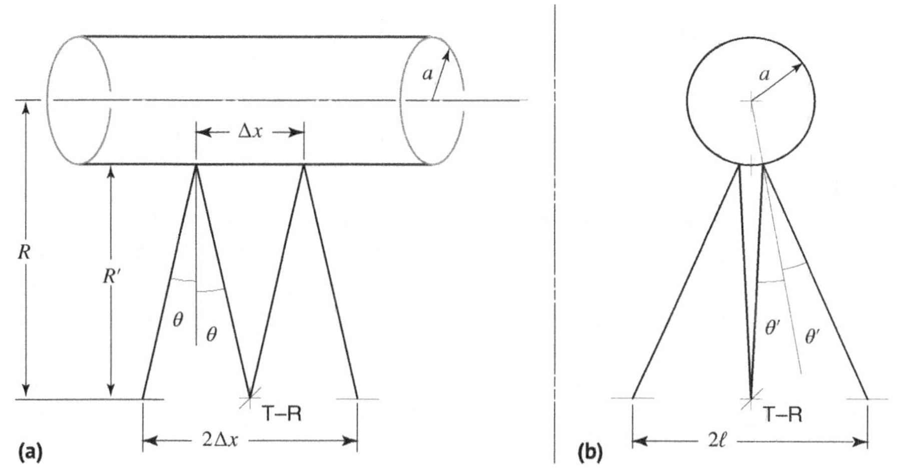

Figure 4.6 Apropos RCS of an infinite cylinder. The geometric optics RCS of a cylinder can be calculated from the ratio of the illuminated, incident area to that of the corresponding area of reflection. (a) View normal to cylinder axis. Angle θ is the specular angle of incidence and reflection in the plane of the paper parallel to the cylinder axis. Note two possible measures of the distance, R, R', to the lower surface of the cylinder from T to R. (b) View along axis of cylinder. Angle θ' is the angle of incidence and reflection from the curved surface of the cylinder in the plane perpendicular to the axis of the cylinder. Spreading of the rays transverse to the cylinder axis is increased significantly with respect to that of a plane (cf. Figure 4.4).

of the reflecting surface exaggerates the transverse spreading of the rays relative to that of a plane. Details are given in Figure 4.7, where we have called the one-sided width of the transverse spreading ℓ. In this instance, we need to distinguish between two measures of distance between the radar and the target, as shown in panel (a). The first, R, is taken as the distance from the radar to the center of the cylinder; the second, R', is the distance from the radar to the closest point on the cylinder so $R' + a = R$.

We are now ready to find the RCS. From Figure 4.7, assuming small angles, we can relate the values of δ and ϕ to the radius a and the range R',

$$R'\delta \doteq a\phi, \qquad \delta \doteq \frac{a\phi}{R'} \tag{4.10}$$

Note that this approximation is valid to second order in ϕ. With Eq. (4.10) and Figure 4.6b we can write out the half-width of the transverse spreading, ℓ,

$$\ell \doteq a\phi + R'(2\phi + \delta) \tag{4.11}$$

$$= 2R'(\phi + \delta) \tag{4.12}$$

$$= 2R'\left(1 + \frac{a}{R'}\right)\phi \tag{4.13}$$

Treating both A_{inc} and A_{scat} as rectangular with dimensions $2x \times 2\ell$, the ratio of the areas is, to second order,

$$\frac{A_{\text{inc}}}{A_{\text{scat}}} = \frac{\Delta x \cdot 2a\phi}{2\Delta x \cdot 2 \cdot 2R'(1 + a/R')\phi} \tag{4.14}$$

In the limit that $\phi \to 0$, we obtain,

$$\sigma_{\text{cyl}}\bigg|_{\phi,\Delta x\to 0} = 4\pi R'^2 \frac{a}{4(R')(1 + a/R')} \tag{4.15}$$

Taking a second limit, as $R' \to \infty$,

$$\sigma_{\text{cyl}}\Big|_{R' \gg a} = \pi a R' = \pi a R \tag{4.16}$$

4.2.1.3 RCS of a Sphere

The geometry of the rays is the same for the sphere as for the transverse section of the cylinder (Figure 4.7), except that, for the sphere, the figure is one of revolution about the line between the radar and the center of the sphere, as suggested by Figure 4.8. The radius of the area of scattering is the same as the spreading of the rays scattered by the cylinder transverse to the axis. Denoting this ℓ, as before,

$$\ell = 2R'(1 + a/R')\phi \tag{4.17}$$

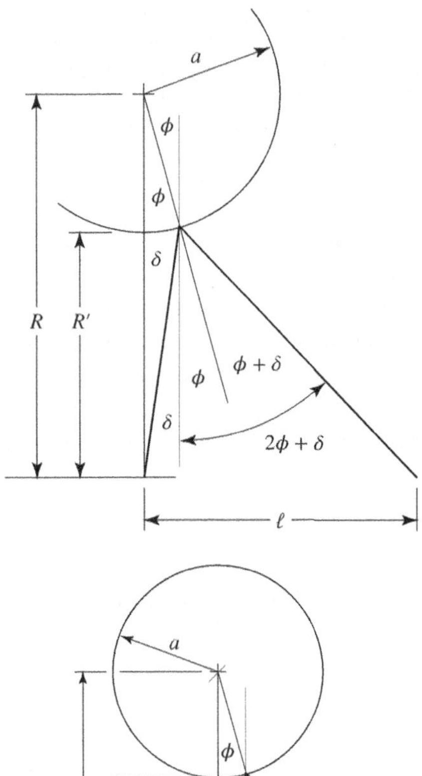

Figure 4.7 Detail in plane normal to cylinder axis. View is along the axis of the cylinder, which is located at cross at height R. Offset line through ray intersection with arc of the cylinder is vertical, paralleling line dropped from center of arc. Horizontal distance of reflected ray transverse to the axis of the cylinder, ℓ, is calculated from angles δ and ϕ (cf. Figure 4.6, see Eq. (4.7)).

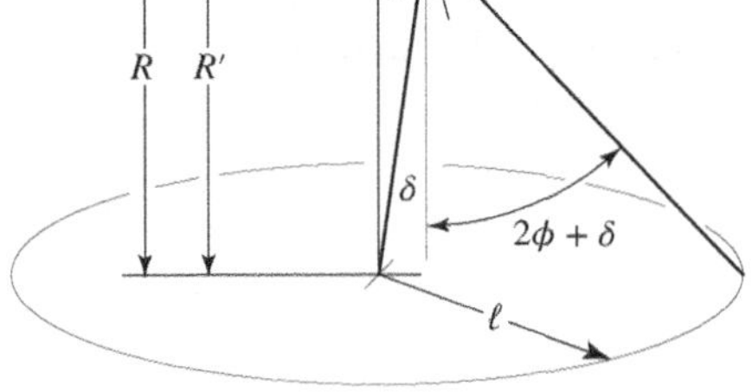

Figure 4.8 Apropos RCS of a sphere. Geometric optics RCS of a sphere is calculated following examples of plane and cylinder. The angular relations are the same as that of the cylinder normal to the cylindrical axis, except that the geometry here is a figure of revolution about the vertical axis through the center of sphere (see Figure 4.7 and Eq. (4.7)).

We write down the ratio of the areas in terms of the areas of two circles, a small one of radius $a\phi = \delta R'$ on the lower surface of the sphere, and a larger one of radius ℓ in a plane containing the radar.

$$\frac{A_{\text{inc}}}{A_{\text{scat}}} = \frac{\pi a^2 \phi^2}{\pi 4 R'^2 (1 + a/R')^2 \phi^2} \tag{4.18}$$

Taking the limit as $\phi \to 0$,

$$\sigma_{\text{sph}}\bigg|_{\phi \to 0} = 4\pi R'^2 \cdot \frac{\pi a^2}{4\pi R'^2 (1 + a/R')^2} \tag{4.19}$$

and again for a sphere at a very great distance,

$$\sigma_{\text{sph}}\bigg|_{R \to \infty} = \pi a^2 \tag{4.20}$$

4.2.2 Summary of RCS for Plane, Cylinder, and Sphere

The geometric optics results for our three objects, then evaluated as the distance between the point T–R and the object of interest becomes very large, are,

$$\sigma_{\text{plane}} = \pi R^2 \tag{4.21}$$

$$\sigma_{\text{cyl}} = \pi a R \tag{4.22}$$

$$\sigma_{\text{sph}} = \pi a^2 \tag{4.23}$$

where R is the range and a is the radius of object. Certainly, this is an easy form for remembering the result, since it can be written in terms of the dimensions of the scattering object (one, two, or three). The result above can also be interpreted more generally, since in every case, the RCS is given by,

$$\pi \times (\text{product of two limiting radii of wave curvature}) \tag{4.24}$$

where the curvature is set either by that of the object, a, or that of the wavefront, R.

What do we mean by the "product of two limiting radii of wave curvature" when one of the curvatures is a wavefront? What is "radius of curvature of a wavefront," anyway? Obviously, from Eqs. (4.21) to (4.23), it is the distance from the source, since we are thinking in terms of spherical expansion of waves from a point. This is consistent with ray theory, since a curve drawn perpendicular to a uniform collection of rays emerging from a point source is also a sphere, as suggested by Figure 4.9.

Within the regime of geometric optics, it is helpful to be able to visualize the interaction of the wavefront with a scattering object. For this, consider a wavefront incident on a plane, as in Figure 4.10. This figure is generated by inverting the shape of that portion of the wavefront which has passed the plane. This is the wave-optics equivalent to Figure 4.5 for ray optics.

4.2.2.1 Interpretation of the Limiting Radii of Wave Curvature

Another way of visualizing the results for our three objects is to consider all three as limiting cases in which aspects of the wave surface interactions are dominated either by the curvature of the incident wave or by the surface curvature at the sub-radar point. The RCS is determined by the smaller of the radii of curvature—that of the incident wave or that of the reflecting object. Since the transmitter always generates a spherical wave, we compare the radius of the wave at a

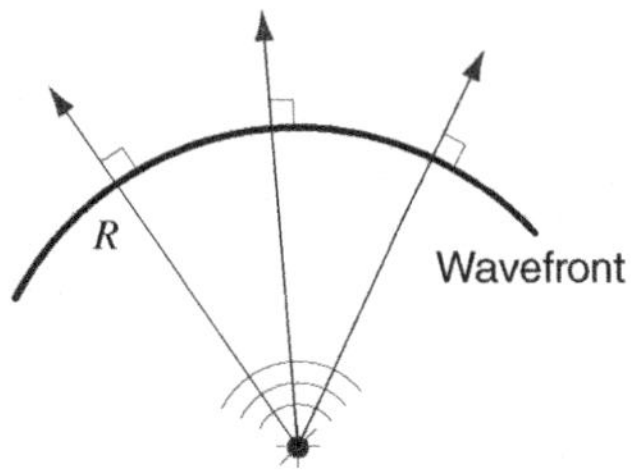

Figure 4.9 Illustrating the connection between rays and a wavefront. In homogeneous media, a wavefront is a sphere of radius R, i.e., the distance or "range" when R is measured in standard units or the accumulated "phase" along the ray when R is measured in units of accumulated phase or wavelength. In homogeneous media, all points on a wavefront have the phase. Geometric optics rays are normal to the wavefronts everywhere.

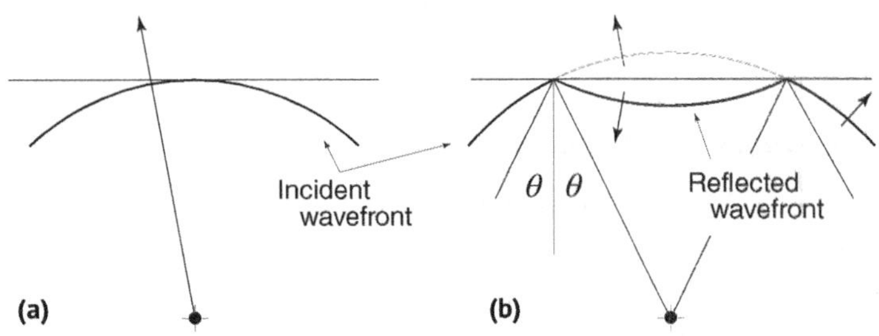

Figure 4.10 Reflection of a wavefront. The reflection of a wavefront can be visualized by applying the law of specular reflection to each contact point of the wavefront with a surface. (a) An advancing, incident wavefront that has reached a horizontal reflecting surface located above the source. (b) The same incident wavefront a short time later. In the absence of the reflecting surface the central portion of the wavefront would have progressed to a position indicated by the dashed line. At this instant the "reflected wavefront" is the mirror image of the segment of the wavefront that has impinged on the reflecting surface. Segments of the wavefront on either side of the reflected portion have not yet reached the reflecting surface. Under GO, similar drawings may be generated for reflections from curved surfaces, as well.

specific distance R with the radius of the scattering object at the sub-radar point for a monostatic radar, or at the specular point for a bistatic radar implementation.

To visualize how this works, think about first our example of scattering from a plane, as though the scattering were from a sphere as a curved surface, e.g., a sphere with a very large radius. Rays from a source at finite distance with $R \ll a$, will be reflected as though from a very large plane, similar to the illustration in Figure 4.10. As the radius of the incident wave is much less than that of our sphere in this instance. The result is that the interaction is very similar to that with a flat plane. As a result, the RCS is close to $\pi R \cdot R$.

Similarly, for the cylinder, the radius of the incident wave is the limiting curvature in the direction parallel to the axis, but when $a \ll R$ the curvature of the cylinder is limiting in the direction, it transverse to the axis. The effect of the curve of the cylinder is readily understood by comparing Figure 4.6a with the detail of Figure 4.7. with the result that $\sigma_{\text{cyl}} = \pi a \cdot R$. Finally, for the sphere (assuming $a \ll R$) the radius of the sphere, a, is limiting, with the result that $\sigma_{\text{sph}} = \pi\, a \cdot a$.

The above interpretation also follows from geometric optics. Consider the sphere, for which the general result, Eq. (4.25), before letting $R' \rightarrow \infty$, is,

$$\sigma_{\text{sph}} = 4\pi R'^2 \cdot \frac{\pi a^2}{4\pi R'^2 (1 + a/R')^2} \tag{4.25}$$

For a radar well-removed from the sphere, $R' \gg a$, Eq. (4.25) gives the result in Eq. (4.20). For $a \gg R'$, however, the result is different as it reduces to $\sigma_{sph} = \pi R'^2$, which is the result for the plane. The issue in question is one of relative scales. When the radar is remote from the sphere, the incident wavefront measured over the width of the sphere is essentially flat. On the other hand, as the source approaches the sphere very closely, at some value of $R' \ll a$, the curvature of the sphere diminishes in importance relative to that of the incident wave. Consider, for example, an instance in which, say, $R' \approx 0.01a$. An observer at a radar directly beneath a sphere with radius of 100 times the distance to the sphere's surface will "see" a sky largely filled by the sphere, which would appear very much as a plane. In such a circumstance, any rays encountering a sphere immediately above a radar will be reflected and received essentially as though from a plane, as is demonstrated by analysis. Examination of Eq. (4.15) shows that a similar result is obtained for a cylinder.

The physical phenomenon underlying the results summarized in Eqs. (4.21)–(4.23) is the dependence on the radius of curvature of the size of the interaction region between the incident wave and the object. Comparing two objects at great distance from a radar, a larger object with smaller relative curvature (smaller a^{-1}) results in a larger distance of interaction measured in the plane under consideration, while a smaller object with larger relative curvature (larger a^{-1}) results in a smaller distance of interaction. The greater curvature implies a lesser RCS.

4.2.2.2 RCS of Compound Surfaces

We also need to consider the more general topic of GO results from scattering the RCS of points on a surface with compound curvature. For this, we assume a monostatic geometry with a plane wave incident on the point of interest. We wish to examine the effects of the shape only. Figure 4.11 provides two views of a single feature illustrating the geometry. As indicated by the opposing wave vectors, $\mathbf{k}_{inc} = -\mathbf{k}_{scat}$, the example applies to backscatter. Every point on a well-behaved surface has local compound, or *Gaussian curvature* (GC), described by maximum and minimum radii of curvature, a_1 and a_2, that are always found in orthogonal planes. For a radar at a great distance from such a surface and operating at a sufficiently short wavelength to accommodate GO, i.e., $R \gg a_1$, $a_2 \gg \lambda$, the backscatter RCS for a wave incident along the normal to the surface is,

$$\sigma_{GC} = \pi \, a_1 \cdot a_2 \tag{4.26}$$

which is a general geometric optics result. The RCS for a sphere is a special case of the more general result with the radius $a = a_1 = a_2$ at every point on the surface. It can be shown that Gaussian

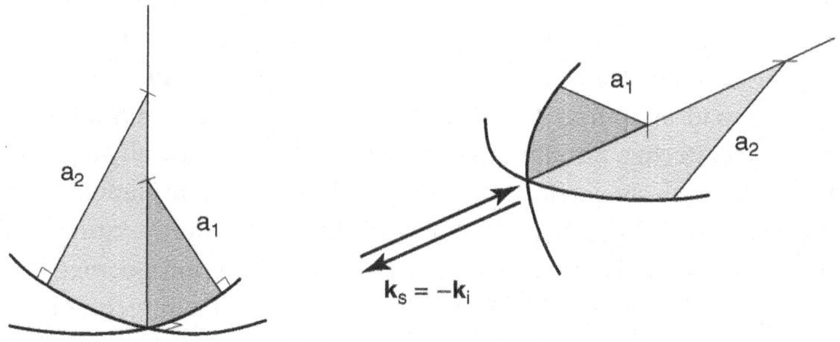

Figure 4.11 Apropos the RCS of doubly-curved surfaces. Figure shows two views illustrating the same surface feature. Two gray areas represent orthogonal planes with distinct radii of curvature a_1, a_2 anchored to points on a common axis, which is in turn orthogonal to the surface. Surface shape at every point on a well-behaved surface can be described in terms of two such radii in mutually orthogonal planes. The orientation of the orthogonal planes is not fixed with respect to the particular coordinate system in use. Rather, the orientation of the planes and the values of the radii of curvature are properties of the surface.

curvature exists at every point on a well-behaved surface. A more general, albeit still restricted example, would be points on the surface of a triaxial ellipsoid.

Such well-behaved surfaces are relatively rare in nature. For example, the concept of compound curvature and GO is not easily applied to vegetated surfaces at short wavelengths. Still, there are many examples for which the above is useful. Scattering from surfaces with roughness scales much smaller than the radar wavelength can be approximated by scattering from a smoothed model of the actual object. Small pebbles on the surface of a desert have little effect on scattering at meter wavelengths, for example.

4.2.3 Extensions of Geometric Optics RCS

4.2.3.1 Results for Dielectric Surfaces

We can apply geometric optics to well-behaved natural surfaces—which in general are not perfect conductors—by accounting for the loss of reflected energy that results from transmission of the incident wave into the material. In order to obtain the RCS of a partially reflecting object, we need only reduce the GO cross section by the appropriate *power reflection* coefficient. The needed physical approximation is derived from the Fresnel power reflection coefficient for planar surfaces of uniform material. For backscatter such surfaces, the result does not depend strongly on the polarization. The power reflection coefficient is,

$$\rho_{\mathrm{pwr}} = \left|\frac{\eta_1 - \eta_0}{\eta_1 + \eta_0}\right|^2 \tag{4.27}$$

with the wave impedance, $\eta_1 = \sqrt{\mu_1/\epsilon_1}$, calculated from the permeability, μ_1, and dielectric constant, ϵ_1, of the surface material, η_0 is calculated from the corresponding values for free space, and ρ_{pwr} is always a real number. As an example, if $\rho_{\mathrm{pwr}} = 0.5$, the RCS of an infinite plane would be $0.5\pi R^2$. In our three examples of simple GO scattering from simple shapes, we assumed total reflection of the incident rays at the surfaces of the plane, cylinder, and sphere, i.e., $\rho_{\mathrm{pwr}} = 1$. From the above, for actual objects, the power reflection coefficient falls between 0 and 1.

The procedure of applying a simple scaling factor to adjust the strength of scattering relative to a perfect reflector contains a hidden limitation. While the power reflection coefficient, ρ_{pwr}, correctly describes the reflected power; it does not account for the power that is not reflected, which can be either converted to heat or transmitted into the material. In some situations, power transmitted into the material can subsequently escape the object and be reradiated in the direction of the source. Retro-reflecting road signs redirect "signals" radiated by automobile headlamps in the general direction from which they arrive, where they are sensed as relatively very bright objects to the driver's eyes. In this example, the road marker is designed to make use of total internal optical reflection to effect the desired backscattering. Similar phenomena also occur naturally at radio wavelengths, particularly in icy objects such as glaciers, and the surfaces of some planets. The recognition and inclusion of such effects is an integral part of RCS analysis and interpretation.

Application of Eq. (4.27) is valid when there is sufficient internal loss or inherent geometrical conditions to prevent power entering a transparent object from returning a significant echo to the radar receiver. Under such conditions, penetrating rays can be neglected insofar as the calculation of RCS is concerned, and only the externally reflected rays need be considered. These are the conditions that lead to Eqs. (3.15)–(3.17).

4.2.3.2 Geometric Optics of Bistatic Scatter

The approach to analysis of backscatter echoes can be extended to an oblique, or "bistatic" scattering geometry. We use the general rules of geometric optics for the analysis and calculations, but the reference areas must be redefined for oblique geometries. The case of oblique scattering from a plane is illustrated in Figure 4.12. The incident area, A_{inc}, is the actively scattering area to the line of sight from the transmitter to the reflecting surface. Similarly, the secondary spreading of the power after interacting with the surface is referenced to the distance between the reflecting plane and the observer, while the reference area for scattering, A_{scat}, is perpendicular to the line of sight from the target to the observer. Note that for this geometry, the obliquity of the incident rays must be considered when setting up the calculation (cf., Figures 3.1 and 4.1; Eqs. 3.1 and 4.5).

4.2.4 Relationship Between RCS and Physical Target Size

We have not yet addressed the question, "Is there a rule connecting the physical size of a scattering object with the size of its accompanying RCS?" Or, from the point of view of interpreting remote sensing observations, "What can we infer regarding the physical size of a scattering objects from measured values of their RCS?" In the specific examples of the plane, cylinder, and sphere, we have seen that there is a systematic progression in the strength of the RCS with the size of the object. But both the plane, which is infinite in area, and the cylinder, which is infinite along its axis, produce only finite radar echoes, as does the finite-size sphere. So the answer to the first question is a "No," and to the second, that the relationships are not immediately obvious. In fact, the relationship between the size of a physical object and its RCS can be complex. This topic is an ongoing, central issue in radar remote sensing research.

For objects that are suitable for analysis by GO, though, some broad generalizations apply:

Large RCS ⇒ Large object size
Small RCS ⇒ Small object size, or
Small ρ_{power}, or
Unfavorable orientation

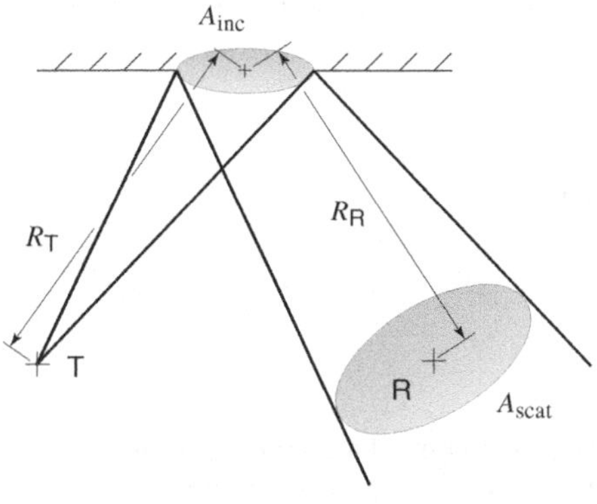

Figure 4.12 Apropos RCS for oblique geometry. The geometric optics approach for bistatic geometry employs the ratio of the scattering to incident areas, $A_{\text{scat}}/A_{\text{inc}}$. Propagation distances, or "bistatic ranges," R_T and R_R, typically are not the same; the obliquity of the incident rays on the scattering surface must also be considered.

The first of these generalizations is related to the size of the active scattering area of large objects; the second is exemplified by objects and systems that enhance RCS—either intentionally or unintentionally.

In *nature,* a large RCS is generally associated with large size, while low values of RCS are associated with smaller features. A primary exception is that large values of RCS can also be produced by relatively small targets, such as fragments of large objects, and also by concave corners that selectively return energy in the direction of the source, and ice.

4.2.4.1 Physical Optics Description of Backscattering from a Sphere

As suggested earlier in our discussion of the GO RCS of a plane, cylinder, and sphere, only a small portion of the scattering object is required to produce the scattering effect. The success of GO in practical problems suggests that the actual area of active scattering is not large, as is apparent from our use of the mathematical limit, for example, in Eq. (4.8), which excludes the entire extent of the plane exterior to the limiting area—which is vanishingly small! Although geometric optics tells us that the active scattering region is "small," GO cannot quantitatively address the actual size of the region when the wavelength is finite.

An approach to finding the physical size of the active scattering area is suggested by Figure 4.13, which shows a small "cap" centered on the sub-radar point located on the left side of a sphere. Still, assuming that the wavelength is very small as compared with the radius, "What is the size of the spherical cap needed to produce the same RCS as the entire sphere?" For this we use the *physical optics (PO)*, or *high frequency*, approximation for short wavelengths to expand our GO results. This is accomplished by including the concept of *phase* and *interference phenomena* into the calculations, which allows us to incorporate the integrated effects of the incident wave over an area or volume of interest into the solution.

We adopt PO to scattering by visualizing a slightly more complex process than ray theory. Applying PO, we recognize that the incident signal from T generates a distribution of local currents covering a scattering object, and that the reflected signal received at R is the net effect of a large number of elementary waves independently radiated from each point on the scattering surface. Thus, rather than treating each point of the scattering surface as a mirror—as in GO—we treat each point on the sphere's surface as an induced source of radiation driven by the incident wave.

For example, in Figure 4.13, consider that a wave incident on the surface of a highly conducting metal sphere generates currents in the conducting surface, and that each filament of current on

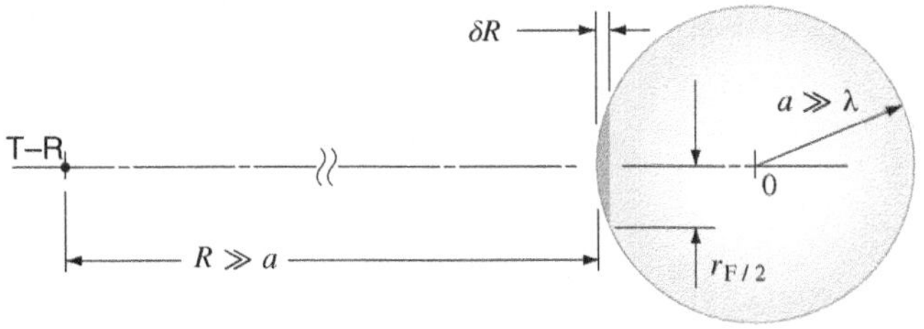

Figure 4.13 Apropos the size of the active scattering region. Only a fraction of a sphere's surface, a spherically shaped "cap," here indicated by the darkly shaded area at left, is required to produce the calculated RCS of the entire sphere. The size of the cap is specified by its radius r_F, measured from the axis of rotation between T–R and the center of the sphere at **0**. The gray area to the right of cap does not, in the whole, contribute to the RCS. The area of the cap becomes vanishingly small as $\lambda \to 0$, i.e., in the GO limit (see Eq. (4.7)). A small, strong, active scattering area is often referred to as a radar "glint," following optical terminology.

the surface reradiates the incident energy. The echo signal received at R results from the reradiation. In such a process, wavelets only add constructively when they are essentially in time phase at the receiving location. Reasoning from this model, for a monostatic radar, all such reflected or reradiated wavefronts must originate from points that are about the same distance from the radar, as suggested by the gray cap in Figure 4.13. While points to the right of the cap also reradiate, it can be shown that at sufficiently short wavelengths the mutual interference from points within the light region the right of the cap results in a near perfect net cancellation of these fields a receiver.

PO analysis reveals the radius of the cap to be,

$$r_{F/2} = \frac{1}{2}\sqrt{a\lambda} \tag{4.28}$$

where $r_{F/2}$ is one-half the diameter of the first *Fresnel zone* on the sphere.[3] The corresponding value of the extra path length relative to the on-axis path is,

$$\delta R = \frac{\lambda}{8} \tag{4.29}$$

We find the active scattering area from Eq. (4.28),

$$A_{F/2,\,\text{sphere}} = \pi\frac{a\lambda}{4} \tag{4.30}$$

Equations 4.28–4.30 are valid when $R \gg a \gg \lambda$, and apply to the backscatter surface reflection only.

Although presaged by the geometric optics analysis, the wave-optics calculation of the active scattering areas is a truly remarkable result. Consider the actively scattering area on a smooth portion of Earth's surface such as a desert dry lake, or playa, measured by a microwave satellite altimeter located 800 km above Earth's surface. Such dry lakes are often remarkably smooth on vertical scales of 10 cm, and thus, to first order are a reasonable approximation to a smooth plane when probed, say, with a radar operating at a wavelength of 1 m. In this instance, the radius of the active scattering area would be approximately 600 m, limited solely by wave-optics effects, as can be shown by PO analysis of backscatter from a plane along the lines illustrated above for the sphere. In contrast to the example of Earth's surface viewed from nearby, if we consider as an example a hypothetical, "large," smooth sphere the size of the Moon, $a \simeq 1738$ km, as observed from Earth, also at 1 m–λ, the radius of the actively scattering cap on such a sphere would be a little less than 1 km. In either instance, a relatively small circular area cut from the respective spheres would have the same RCS as the entire body. In the example of the moon, a small circular area with diameter of about 0.1 percent of the radius, if properly oriented with respect to the radar, would have the same RCS as does the entire sphere! The similarity in active areas—600 versus about 1000 m—reflects the similarity in the controlling radius of curvature for the two examples—800 versus about 1700 m. It matters not that the controlling radius is the distance from the radar in the example of the playa and the radius of the moon in the example of a distant Earth-based radar observer.

4.2.4.2 Scattering from Complex Objects

Geometric optics scattering from complex objects can often be understood in terms of the superposition of returns from many small scattering areas similar to the backscattering cap on the sphere. Such areas, or "glints," are not restricted to backscatter though, as they occur around points on the

3 This nomenclature refers to the area on a surface encompassed by the line of the first phase reversal with respect to that of the line-of-sight ray. This is the standard measure of size in wave diffraction and propagation problems. The values are calculated only to the second order.

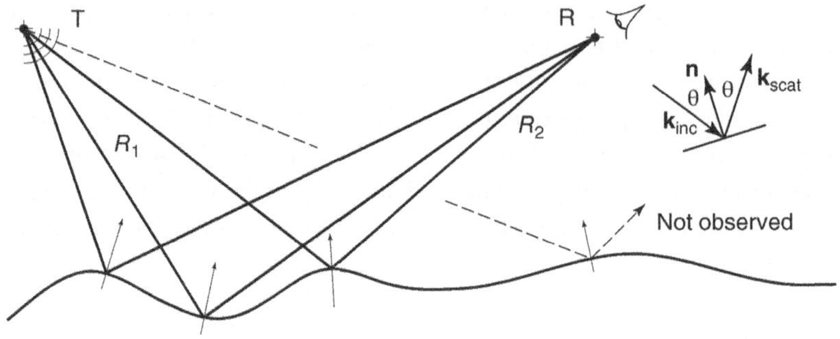

Figure 4.14 Apropos RCS of complex surface. Scattering from well-behaved complex surfaces of homogeneous material can be calculated by applying geometrics optics theory locally. The dominant reflections occur at those points for which the scattering is locally mirror-like, or "specular," with equal angles of incidence and reflection, as shown in the insert; note four lines drawn normal to the surface for reference. Signals reflected from around "specular points" on the surface are referred to as "glints." The region of the surface labeled "not observed" is devoid of points with sufficient slope to turn a ray from T to R.

object from which there are specular reflections connecting the transmitter and receiver. This is illustrated in Figure 4.14 where the location of each glint is the point on the surface where a mirror tangent to the surface would reflect the transmitted signal toward the receiver. The local RCS for each point can be characterized by extension of physical-optics analysis to oblique, bistatic geometries. Equations (4.28) and (4.30) are a rough approximation of the glint size, while an adaptation of Eq. (4.26) provides the oblique scattering cross section. More accurate expressions can be quite complex owing to the geometry, although the physics is a straightforward adaptation of that for the backscattering sphere to the separated paths of incidence and reflection. Glints similar to those suggested in Figure 4.14 are readily observed by the eye when viewing the surface of a lake or the ocean in bright sunlight. Each sparkle on the surface is a location where a specular reflection path connects the position of the sun with the eye of the observer.

The concept of glints is useful in situations where the wavelength is small compared to the size of the object of interest. For example, radio-wave glints on the skin of an aircraft illustrate a common radar backscatter geometry and closely mimic the behavior of their optical counterparts. In this instance, the glints occur at locations where the aircraft skin is normal to the radar line of sight. For the geometry in Figure 4.15 these points are located on the starboard tips of the propeller, cowl, wing, and horizontal stabilizer, as indicated by the darkened, shaded areas. The individual RCS of each glint can be estimated from Eq. (4.26). given $a_1, a_2 \gg \lambda$. The "glint" over the cabin (g') is somewhat different, though in that typically the curve at the top of the windscreen is a very thin metal section, so that the radius of curvature in the vertical direction (i.e., perpendicular to the page) is extremely small and PO fails because $a \not\gg \lambda$; this is nevertheless a location of strong scattering for a horizontally polarized electric field incident. As a result, we cannot use the smooth curve formula developed here to estimate the scattering from this feature.

In general, the concept of glints is a useful tool for developing an intuitive understanding of a very common scattering mechanism. An awareness of the phenomenon can be developed by observing the scattering from a objects as different as blades of grass and sunlight on almost any cultural object.

Not discussed above is how to combine our knowledge of scattering from individual objects to find or interpret scattering from an ensemble of objects. For example, in Figure 4.14, the scattering cross section of the surface in which many glints from a given area cannot be separated from each other. This subject is treated in Chapter 5.

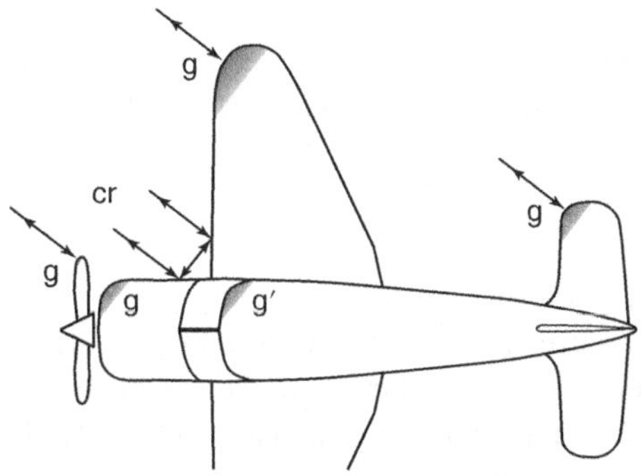

Figure 4.15 Glints and corner reflectors on aircraft. Figure is drawn for a distant radar located at upper left. Total scattering from complex objects often comprises numerous glints (*g*). The largest glint backscattering cross sections are from right wingtip and right side of the cowling, as these points have the largest compound radii of curvature. Weaker returns arise from the tail and upper surface of cabin. RCS of the propeller is weak and is temporally modulated by rotation of the blades. The corner reflector effect (cr) contributes to echo in some areas, illustrated here for the wing root (see Section 4.3.1). Total echo power is coherent sum of its components.

4.3 RCS Enhancement

4.3.1 Corner Reflectors

Passive means of providing a standard RCS target are employed when the complexity associated with the use of active electronic devices is an impediment or otherwise impractical, but some types of ground truth are needed. Calibration of the scattering cross section and as a means of providing a geographic datum are two examples where passive radar targets are useful.

Calibration of RCS is an important task requiring strong echoes for testing and for interpretation of remote sensing observations. Likewise, geometric registration of mapped scenes by placement of identifiable, discrete scatterers at known locations in the area of interest is another. Such benchmarks also can be used to calibrate RCS measurements within a mapped scene. These example applications require the use of compact devices designed to produce strong echoes from relatively small structures, typically only several to tens of wavelength in size. In the construction of such devices we consider the problem of "scattering by design."

Corner reflectors are concave rectilinear structures of either two or three, typically orthogonal, sides with a backscatter RCS similar to that of a perfectly conducting flat plate of the same projected area seen face on (i.e., the $\mathbf{k}_{inc} \perp$ surface of plate and $\mathbf{k}_{inc} = -\mathbf{k}_{scat}$). Unlike a flat plate, within certain limits, echoes from corner reflectors are not sensitive to the direction of incidence; however, in that they provide predictable moderate- to high-values of RCS that vary slowly over a wide range of monostatic viewing angles. The same is true for the dihedral device, but only in the plane perpendicular to the single corner. The optical equivalent of the radar corner reflector is the "corner cube" often incorporated into surveying equipment and optical benches.[4]

Enhanced backscattering by concave corners also occurs commonly in remote sensing observations of built-up areas, such as cities and their suburbs. In these instances, the corners occur in constructed features such as window frames, doorways, stacked compartmented industrial storage areas, and many other ways. In addition to enhanced backscattering, corner reflectors also modify

4 Extensive discussions of the characteristics of corner reflectors are available. For example, Ruck (see Bibliography).

echo polarization in predictable ways, thus providing helpful clues in distinguishing among possible built artifacts and natural structures. In nature, corner-like reflections can occur in terrain as different as forests and in areas of exposed rock such as badlands, although the scattering behavior in natural occurrences is not deterministic (cf. Figure 5.3).

4.3.1.1 Dihedral Corner Reflectors

A rectangular *dihedral corner reflector* (DCR) is the simplest form. The DCR consists of two highly reflective flat plates joined at right angles, as shown in Figure 4.16. This is the form that would be obtained from a section of adjoining walls in the corner of a rectangular room. Geometric optics provides good approximations for short wavelengths with respect to the dimensions of the DCR, i.e., for $\lambda \ll a, b$. Rays perpendicular to the dihedral axis, arriving from any direction between the normals of the two plates, are redirected toward their source after two reflections. Following the rays through the two reflections and the paths internal to the DCR, it is straightforward to show by following the rays in Figure 4.16 that the round-trip phase is constant for all backscatter rays within a given range of angles of arrival. Thus, a planar wavefront entering the aperture is also planar after the turnaround.

Given the uniform phase of the reflected wavefront, it is straightforward to determine the backscatter RCS. Under geometric optics, the effective capture of the DCR is the same as the physical retro-reflecting area; the power extracted from the scattered wave and redirected toward the source is proportional to this area, A_{dcr}. As the losses in a structure are small by design, essentially all the energy intercepted by the DCR is reradiated through the same aperture by which it arrived. This can be seen by expanding the rays' paths in Figure 4.16 to explore the limits of intercepted rays and their path reversals.

In this instance, it is straightforward to obtain the RCS of the DCR, since we can show that its radiation is direct backscatter from an aperture with area A_{dcr}. From the definition of RCS,[5] if the power intercepted by A_{dcr} were reradiated isotropically, the RCS would be A_{dcr}. Because the DCR radiates preferentially in backscatter, however, rather than isotropically, we need to include the radiation gain of the DCR. From the definition of RCS, we see that the correction factor for the backscatter enhancement is, simply, the radiation gain of the corner reflector. That is,

$$\sigma_{dcr} = A_{dcr} \cdot G_{dcr} \tag{4.31}$$

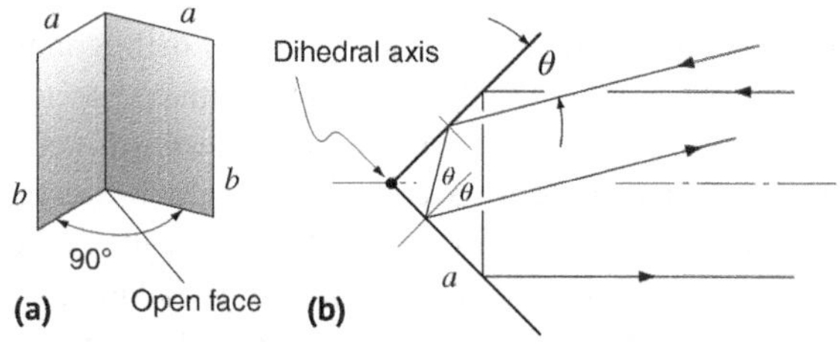

Figure 4.16 DCR. (a) Concave dihedral reflector comprising two highly reflecting plates of size $a \times b$; plates meet at 90°. (b) View along join of dihedral plates. Rays impinging on dihedral plates are reflected according to Snell's law, with equal angles of incidence and reflection. Rays arriving from any direction within the dihedral opening and striking both plates exit the structure in a direction parallel to and opposite that of their arrival. Exception is for $\theta = 0$, when only a single plate is illuminated.

---

5 See Chapter 3, p. 47, also compare Eq. (3.4) with Eq. (4.4).

where σ_{dcr} is the RCS of the corner reflector, and A_{dcr} is the projected area of the DCR as seen by an incoming ray, and G_{dcr} is the corresponding gain. Applying Eq. (4.31), we have,

$$\sigma_{\text{dcr}} = 4\pi \frac{A_{\text{dcr}}^2}{\lambda^2} \tag{4.32}$$

This result is the same as the RCS of a flat conducting plate with area A_{dcr} and lateral dimensions much greater than a wavelength, when viewed perpendicular to its surface. The difference between the flat plate and the DCR being that within the plane perpendicular to the line connecting its two parts, the DCR backscatters radiation arriving from any angle, whereas the flat plate backscatters only in the direction perpendicular to its surface. Note that $\sigma_{\text{dcr}} \propto (A_{\text{dcr}})^2$ because the effect of the DCR area enters this result twice—once as a collector of incident radiation and a second time as an effective radiator in the backscatter direction.

All that is missing now in order to complete our analysis of the DCR is an expression for A_{dcr}. Figure 4.17 shows the geometry, which is abstracted from Figure 4.16 with emphasis on ray paths entering and exiting at a finite angle ϕ to the normal to the face of the DCR. Tracing rays as a function of the angle with respect to the plate bisector provides a means of determining the physical area over which the dihedral reflections operate to backscatter the incident radiation. In the figure, the entry–exit ray path illustrates the widest possible separation for an entry ray just grazing the topmost point on the upper plate; from the geometry, this condition produces the widest possible separation, w, between the entry and exit segments of the path. Expressing this in terms of the grazing angle of the entry ray, θ, is easy to show that for $|\theta| < \pi/4$,

$$w = a\frac{\sin 2\theta}{\cos \theta} \tag{4.33}$$

where the restriction on θ is needed to ensure that the ray reflected at the corner remains with the DCR. Then using $\theta + \phi = \pi/4$, substituting in the above, and simplifying,

$$w = \sqrt{2}a(\cos \phi - \sin \phi) \tag{4.34}$$

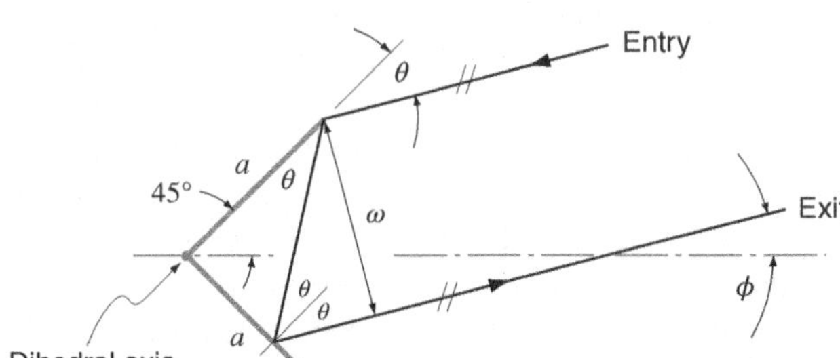

Figure 4.17 DCR ray-path geometry. View is in direction of dihedral axis; dihedral plates shown in gray are joined at 90°. Entry ray at upper right touches upper plate just below the top-most point on the DCR; incident and reflected rays at that point are at angle θ with respect to the plane of upper plate. Reflected entry ray crosses the axis of reflector to intersect the lower plate at angle θ with respect the normal. Reflection from lower plate produces the exit ray antiparallel to the entry ray. Only parallel rays entering between entrance and exit rays shown can contribute to backscatter at angle illustrated; rays above the entrance ray pass over the corner reflector while rays below exit ray pass to the right of the upper corner reflector surface after reflection from the lower plate. Note that $\theta + \phi = 90°$. Cf. left-hand side of Figure 4.16 to locate a.

Considering the height of the DCR, b, we can calculate the area of the aperture from,

$$A_{\text{dcr}}\Big|_{|\phi|\le\frac{\pi}{4}} = \sqrt{2}ab(\cos\phi - \sin\phi) \tag{4.35}$$

$$= ab\sqrt{2}\sqrt{(1-\sin 2\phi)} \tag{4.36}$$

which is valid for all twice-reflected rays within the DCR. Note that the maximum value is $A_{\text{dcr}} = \sqrt{2}ab$, which is consistent with the aperture as seen looking into the DCR along the line of symmetry.

What occurs when rays enter the DCR parallel to either of the dihedral plates? This is a special case in which the entry ray is not reflected from the upper plate, Figure 4.17, but is perpendicular to the lower plate, where it is reflected directly back to the source. So, by inspection, the area is just that of a single dihedral plate,

$$A_{\text{dcr}}\Big|_{|\phi|=\frac{\pi}{4}} = a\times b \tag{4.37}$$

Finite but generally weak scattering occurs for values of $\pi \ge |\phi| > \pi/4$, that is, outside the design range above in which the DCR functions as a retro-reflector, with the obvious exception of rays arriving perpendicular to the backs of the dihedral plates.

We use the results above to calculate the geometric optics RCS of the DCR. Applying Eq. (4.32) and making use of Eq. (4.36) for the doubly-reflected backscatter rays, and of Eq. (4.37) for the edge-on condition, gives us,

$$\sigma_{\text{dcr}} = \begin{cases} 8\pi\dfrac{(ab)^2(1-\sin 2\phi)}{\lambda^2}, & 0\le|\phi|<\dfrac{\pi}{4} \\[2ex] 4\pi\dfrac{(ab)^2}{\lambda^2}, & \phi=\pm\dfrac{\pi}{4} \end{cases} \tag{4.38}$$

In actuality, diffraction broadens the $\phi = \pi/4$ glint. Although Eq. (4.38) correctly predicts the maximum strength of the glint, this expression provides only the indication that the RCS spike at $\phi = \pm\pi/4$ is very sharp. In the plane perpendicular to the axis of the dihedral join, i.e., the plane of the paper, the angular spread of the backscatter spike is similar to that of a flat plate with the same aperture, also as a result of aperture diffraction. For these reasons, the variation of RCS near $\phi = \pi/4$ will be less sharp than is indicated by the expression above. Diffraction also broadens the backscatter beam in the plane containing the dihedral join, i.e., the plane perpendicular to the paper; in this instance the effect, is less noticeable owing to the angular spreading of the backscatter response by geometric design.

The effect of the DCR on polarization is another salient feature of backscattering by corner reflectors. In the main beam of the DCR, i.e., for $0 \le |\phi| < \frac{\pi}{4}$, the backscattered polarization is the *same* as that incident where, for example, a right-hand circularly polarized (RCP) wave incident results in an RCP echo, which is opposite the effect of backscattering from a smooth conducting surface. The specular glint at $\phi = \pi/4$ has the same polarization characteristics as a smooth surface, though, so RCP incident results in a left-hand circular polarization (LCP) echo. In the first instance, within the main beam, the double bounce accounts for the unchanged result. That is, the polarization is changed twice, with the second reflection undoing the effect of the first. The $\phi = \pi/4$ glint, however, behaves like the single reflection because that is precisely what it is.

As mentioned above, DCR effects appear in natural settings, as well as in connection with engineered or other constructed artifacts, and are important as a result of their large RCS enhancement. In some engineering applications, they suffer from the restriction that strong backscattering occurs only when the radar line of sight is near the perpendicular to the dihedral axis.

4.3.1.2 Trihedral Corner Reflectors

A right *trihedral corner reflector* (TCR) comprises three mutually orthogonal plates forming a three-dimensional corner. Geometrically, the corner is identical to the intersection of two walls with the ceiling or floor in a rectilinear room. Three simple examples with square, triangular, and circular-arc plates are shown in Figure 4.18, which differ only in the manner in which the edges are shaped; the length of the plate intersections, a, is the same for all three. The addition of the third plate increases the size of the solid angle over which backscatter rays can exist, as compared with the dihedral corner, which is limited to a plane. At the same time, the total cross-sectional area is increased, leading to an increase in the backscatter RCS.

Rays arriving from directions within angles of roughly 25–30° of the axis of symmetry, depending on the type of TCR, are efficiently backscattered. Within this substantial solid angle, incident rays are redirected toward their source after two reflections.

The backscattering cross section is calculated by the same method as for the DCR, except now three reflections are involved within the solid angle of effective operation. The expression for calculating the RCS of a TCR has the same form as above for the dihedral case,

$$A_{\text{tcr}} = A_{\text{tcr}} \cdot G_{\text{tcr}} \tag{4.39}$$

where the quantities have the same meaning of reflector area and gain as before.

Table 4.1 gives values for the backscattering RCS of the three examples in Figure 4.18.

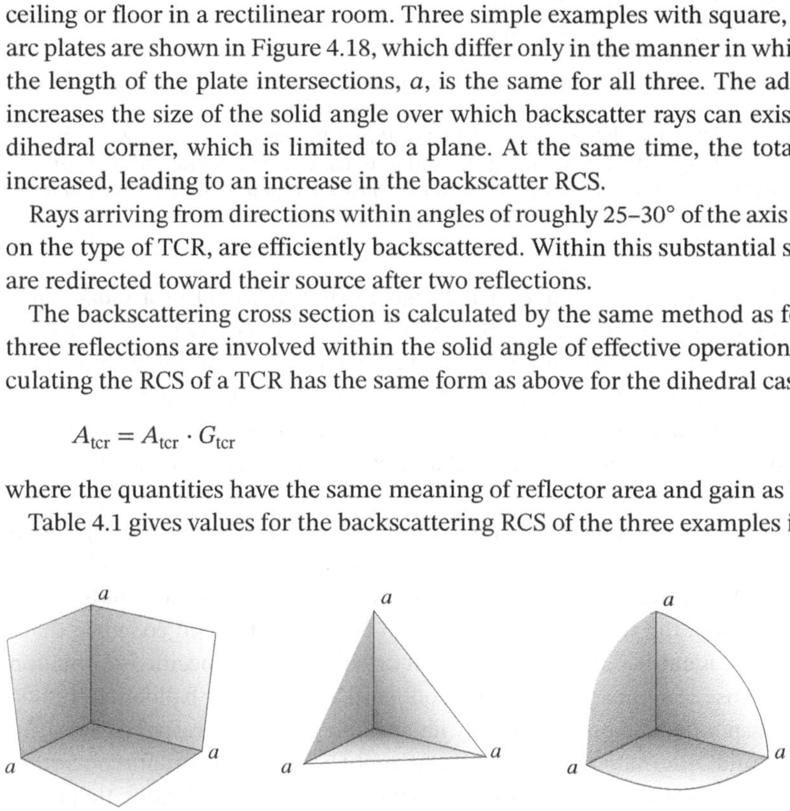

Figure 4.18 Three TCRs. From left to right, square, triangular, and circular TCRs with right-angle intersections as seen from near the axis of symmetry. Sizes are described by the length of the plate intersections, which are the same for the three examples.

Table 4.1 Backscatter RCS for three trihedral corner reflectors. Terms "square," "triangular," and "circular," refer to the shape of the plates forming the TCR. Values shown are maxima that are obtained on the axis of symmetry with respect to the plates.

	Square	Triangular	Circular
RCS	$\frac{12\pi a^4}{\lambda^2}$	$\frac{4}{3}\frac{\pi a^4}{\lambda^2}$	$\approx \frac{5\pi a^4}{\lambda^2}$

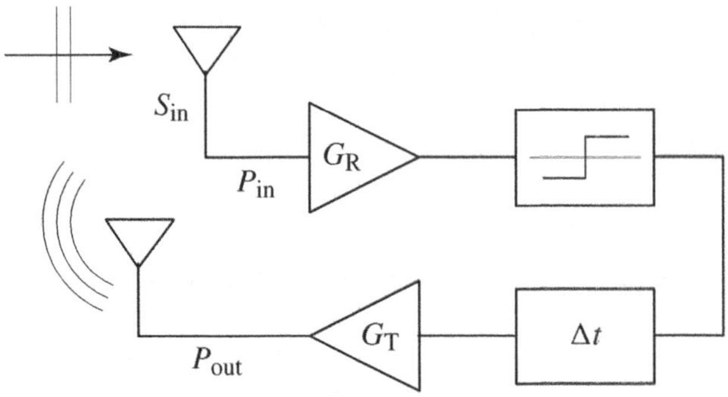

Figure 4.19 Simplified radar transponder block diagram. A radar transponder emits a signal in response to received signal, S_{in}, often transmitting a particular waveform or modulation for purposes of easy identification. The emitted signal, P_{out}, may be on the same or a different frequency, may be essentially simultaneous or delayed, and may be modulated for identification or to convey information to a cooperating radar. In common applications of these devices, the output power is constant, albeit of limited time duration.

4.3.2 Radar Transponders

Radar transponders are transmitters whose output is triggered by the arrival of a radar pulse. Transponders are used to enhance the strength of the received radar echoes by use of active circuitry. Figure 4.19 shows the idea. The transmitter is activated by reception of a signal, and means must be provided to isolate the receiver from the transmitter output or otherwise suppress the trigger function during the transponder "reply" transmission. For this reason, it is common for the reply signal to be radiated and received at a frequency that is offset from that of the radar.

In the block diagram of Figure 4.19, we show only a simple retransmission of the received signal occasioned by reception of a signal that exceeds a predetermined signal strength, although more complex systems are often used. For example, the receiver may be designed to respond only to coded radar waveforms so that a trigger signal is generated solely in response to inputs with a particular modulation. Similarly, the transponder output can be modulated and coded so that a particular transponder can be identified, for example, or in order to convey specific information, for example, altitude in the case of a transponder carried onboard an aircraft or balloon. In remote sensing applications, a transponder can provide readily identifiable radar calibration signals from known locations on the ground. Alternatively, a transponder might be used to relay data concerning local environmental conditions in the area of interest, e.g., wind vectors or rainfall rates, that might be combined with radar signals in interpreting the observation field.

What is the RCS corresponding to a transponder signal? The answer varies with the particulars of each device type. In almost all cases, the effective RCS of the transponder is much greater than that of the passive target alone, since a modest transmitter can emit more power than is typically backscattered by even a very powerful radar located some distance away.

In calculating the transponder RCS, we begin with the Friis equation for propagation from the transponder to the radar receiver,

$$\frac{P_r}{A_r} = \frac{P_{out}\, G_{xpndr}}{4\pi R_r^2} \tag{4.40}$$

where P_{out} is the power radiated by the transponder and G_{xpndr} is the gain of the transponder antenna. From this, the strength of the received "echo" power density depends on the transponder output and the range and does not depend on the strength of the illuminating radar signal, so long as it is sufficient to trigger a response.

For the device in our simplified block diagram, Figure 4.19, we can calculate the power density received for triggering signals received above the triggering threshold. From the definition of RCS, Eq. (4.4), then,

$$\sigma_{\mathrm{xpndr}} = \frac{\frac{P_{\mathrm{out}}\, G_{\mathrm{xpndr}}}{4\pi R_{\mathrm{r}}^2}}{\frac{P_{\mathrm{t}}\, G_{\mathrm{t}}}{4\pi R_{\mathrm{t}}^2}} 4\pi R_{\mathrm{r}}^2 = \frac{P_{\mathrm{out}}\, G_{\mathrm{xpndr}}}{P_{\mathrm{t}} G_{\mathrm{t}}}\, 4\pi R_{\mathrm{t}}^2 \tag{4.41}$$

We have the peculiar result that the RCS of a transponder depends on the transmitted power of both the radar and the transponder, and varies with range in a manner similar to the RCS of a plane surface (cf. Eq. 4.21). This result is just a quirk of applying a definition intended for passive scatterers to active devices.

5

Practical Effects in Scattering

Our look at geometric-optics solutions to some canonical scattering problems provides us with a feel for the meaning of radar cross section (RCS) in terms of the relative spreading of the scattered energy by simple objects such as plates, spheres, corners, and undulations in "well-behaved" natural terrains. In remote sensing applications, we encounter a larger variety of scattering phenomena, most of which require techniques to determine RCS that are more advanced than geometric optics. As the study of electromagnetic scattering is far more complex than we can address here, we provide an overview of scattering effects commonly encountered in remote sensing applications. Following this, we illustrate the importance of mutual interference among scattered signals from unresolved diverse elements within a single radar echo. With this as a starting point, the reader can further investigate this subject or special topics using the literature of electromagnetic scattering found in specialized volumes and journals.

5.1 Phenomena

In working with remote sensing data, it is important that we are aware of the variety of scattering phenomena that control the scattering characteristics, ranging from simple ordinary objects to complex, extended terrains. In the simplest instances, these are direct extensions of the geometric-optics model for surface reflections from smooth surfaces. In many important cases, though, calculation of scattering from familiar objects requires application of Maxwell's complete equations governing the interaction of electromagnetic fields with matter. This is commonly the case when the wavelength and object of interest are of comparable sizes. Such interactions, e.g., the scattering by a freshly fractured igneous rock of a wave within a factor of three of the rock's average dimension, are a difficult computational problem, well beyond application of geometric optics. Between these extremes are several important processes and mechanisms that commonly play significant roles in scattering electromagnetic energy and are governed by the basic physics of light and electromagnetism. We group these as listed below:

1. Geometric-optics internal rays
2. Multiple reflections
3. "Creeping" waves
4. Polarization conversion
5. Diffraction and "wave optics"
6. Resonant scattering effects
7. Effects on nonlinear and anisotropic materials

Additionally, mutual interference of wave fronts scattered by large objects, while more an effect of propagation rather than scattering, plays an important role in determining the observed characteristics of scattering from extended objects.

5.1.1 Internal Rays

Electromagnetic waves readily penetrate dielectric materials, which, to varying degrees, we may think of as the stuff of transparent or partially absorbing objects—crystal balls, raindrops, dry terrains, glaciers, trees and vegetation, birds, walls and dry rock, items constructed of composite materials, to name a few. In the case of dielectric objects with characteristic dimension $D \gg \lambda$, we are able to apply ray concepts to penetrating or "internal" rays that exist *inside* the object, in addition to those that remain *exterior* to the object, as in our earlier examples. We observe, of course, only those penetrating rays that also emerge from within the object, which we refer to as *emergent rays*.

Some important types of scattering are controlled by the internal structure of an object, as this determines the behavior of the internal rays. In objects with sufficiently low internal ray adsorption, or *loss*, the emergent rays can be as, or more important than, those reflected externally. The combination of the emergent and external rays observed by the receiver can make for complex scattering behavior. Perhaps the most common example of this—arguably the most wonderful—is the rainbow, which is made up of bundles of emergent rays that form intense beams of light at certain angles from the sun. More intriguing in our time, because the phenomenon is not well understood, is the enhanced radar backscattering from the icy surfaces of some satellites in the outer solar system, the polar caps of Mars, and, apparently, ice-filled craters in the polar regions of Mercury. Similar enhanced backscattering phenomena are exhibited by the glaciers of, e.g., Greenland, and elsewhere. These occurrences in ice are believed to result from rays that undergo numerous interactions with discrete scattering centers within the ice.[1]

Figure 5.1 illustrates the physical situation involving internal and external rays. *Internal rays* result from partial transmission into the object of an incident external ray, where it is (i) partially absorbed, (ii) either partially or totally internally reflected and/or scattered within the object, and (iii) finally emerges from the interior. Typically, the space within the object is filled with internal rays, each with a distinct point and direction of emergence. Refraction at the surface of the object plays an important role in determining the path of the internal ray. There is always the possibility that an internal ray emerges in the direction of the source, as suggested by Figure 5.1. At the same time, external rays continue to be reflected from the exterior of the object, as discussed previously (see Section 3.4 and Chapter 4). In general, there can be one or many internal and external rays connecting a radar transmitter and the receiver, depending on the actual shape, internal structure, and, practically, the absorptive properties of the object.

At the receiver, the total field comprising the coherent sum[2] of all arriving internal and emergent rays must be considered in calculating the complete geometrical optics RCS. For objects that partially absorb electromagnetic waves, the importance of the internal rays depends on the rate of absorption and the total length of the internal ray path. Internal rays can often be neglected when

1 This can occur because the internal loss in water ice allows radio signals to travel hundreds to thousands of wavelengths with little attenuation, where the actual distances depend primarily on the conditions of temperature and composition.

2 That is, the sum of the fields associated with each ray, taking into account the relative phase differences of the signals arriving along each ray. This is necessary since radio receivers measure the sum total of all impinging waves, inherently making use of the wave coherence in the detection process.

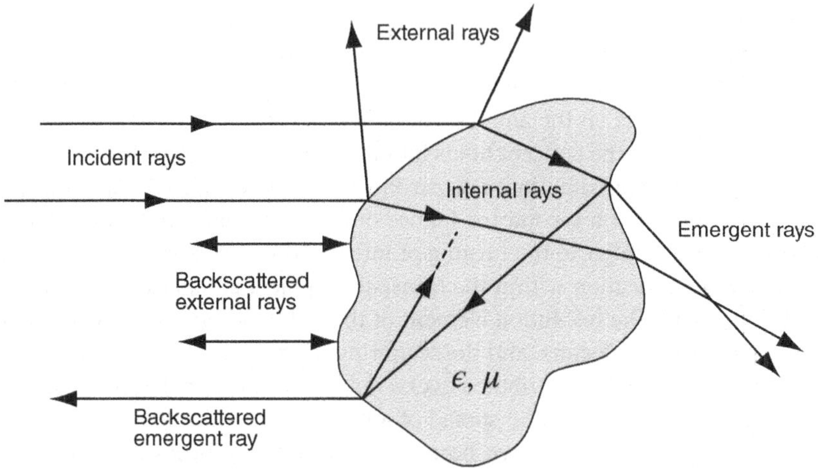

Figure 5.1 Internal and external rays: scattering from a large, $D \gg \lambda$, dielectric object can appear as a complex plethora of electromagnetic rays, while individual rays follow orderly rules. Incident rays enter dielectric materials to emerge after some degree of adsorptive and transmission losses at the points of internal reflection; the strength of each successive emergent ray is reduced by both effects. But it should be noted that if the internal angle of incidence is greater than the critical angle for the material, total internal reflection can occur, rendering the transmission loss zero in these instances. If the absorptive losses are large, potential emergent rays are lost, i.e., unobservable for practical purposes. Internal reflections within low-loss materials can contribute significantly to the total RCS for both backscatter and bistatic geometries. The total RCS depends on the coherent combination of rays in the direction of the observer–receiver. For backscatter, the bistatic angle $\theta = 0$, but in general $\theta \neq 0$. Note the special case of a backscattered emergent ray and the example of obliquely scattered external rays. Straight-line internal ray paths occur within homogeneous materials.

the integrated losses along the internal path exceed a few factors of e^{-1}, in which case the penetrating energy, effectively, is converted to heat inside the object. For instances in which the internal losses are low, just the opposite can occur, with the received energy in emergent rays dominating that of external reflected rays. Each case must be examined individually in order to determine whether the strength of the residual internal rays is important in comparison with that of the various external rays.

Thus, the importance and prevalence of internal rays are illustrated by several examples. The spectacular display of the rainbow is formed by a combination of internal reflection and color dispersion within nature's nearly spherical raindrops. A similar phenomenon occurs in cut jewels with facets designed to form families of rays from light arriving from sources located over a wide range of solid angles from object.

At the wavelengths of remote sensing radars, water ice in terrestrial glaciers and other locations in the solar system produces exceedingly strong backscatter, for reasons that are not yet well understood but almost certainly involve internal reflection and scattering from fragments and irregularities in the structure of the ice. Similarly, manufactured "corner cubes," based on total internal reflection from three internal planes, are optical counterparts of the corner reflectors of Chapter 4.

5.1.2 Multiple Reflections

The phenomenon of multiple discrete reflections from various unresolved portions of a scattering object is referred to generally as *target multipath*. Figure 5.2 indicates how this can occur for an example, comprising a collection of three buildings plus four other smaller objects, all of which are sufficiently close together that echoes from the individual elements cannot be separated in a hypothetical observation. Similarly, Figure 5.3 provides an example suggestive of a natural setting in which radar observations of a group of small boulders or the limbs in a section of tree are seen only as composites owing to limits on the angular resolution of the system. The example of Figure 5.2 is drawn for backscatter, but the concept holds for bistatic scattering as well.

In terms of the geometric-optics approximation, rays striking one proportion of a large target can be scattered in directions such that a second (or higher order) scattering event turns them toward the receiver. This process is similar to the turning of internal rays of Figure 5.1. But here, the systematic process of internal reflection within the transparent object is replaced by random scattering from nearby features with the resolution element of the radar system. In the examples shown, there are few pure 90° corner reflectors, and the scattering from the curved building front and the large circular object in Figure 5.2 is distributed over a range of bistatic angles, and not just the backscatter rays shown. Some rays within the spread of angles will produce backscatter rays after a second reflection, although none are shown here. Typically, but not necessarily, double-bounce rays are weaker than are those with only a single reflection or scattering. It is important to keep in mind that not all rays in this general situation are returned to the radar, as shown by the ray striking the structure at upper right, and then passing out of the picture at the top.

For complex cultural features such as large buildings, there are many more possible paths than those suggested by Figure 5.2. For example, window sills, doorways, pilasters, and other structural and decorative features all contribute to the scattering process. In addition, in modern buildings, the facade is often made of glass, which generally is as transparent to radio waves as to light, or more so. In these situations, the complex structure of the building interior, the mechanized equipment, and the interior furnishings become effective elements of multiscattering.

Figure 5.3 illustrates two common natural features that support multipath scattering. Trees provide backscattering paths that are similar to those in corner reflectors. In this instance, the scattering/reflecting elements are primarily combinations of the surfaces of the trunks—which are

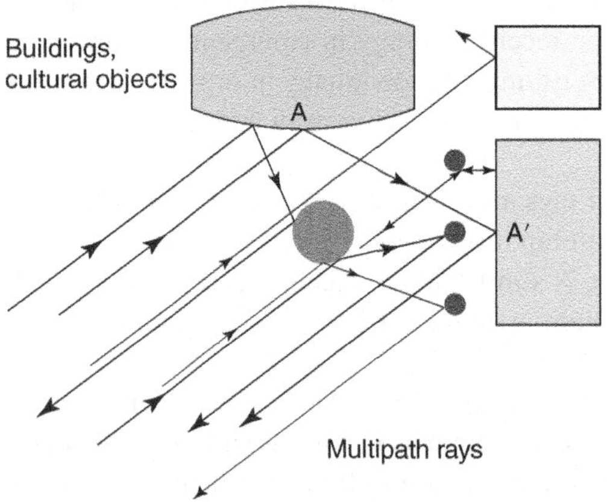

Figure 5.2 "Target" multipath. Multiple reflections within a scattering area increase the complexity of RCS, here illustrated by multiple backscatter rays caroming between buildings and cultural objects. Ray path AA′ is a typical two-point link between incident and backscatter rays, but more complex paths are possible. Note that every path is also reciprocal, as indicated by two-way path in the middle right of the figure. Multipath phenomena tend to broaden the angular range of overall scattering and to fill in nulls in the primary scattering mechanisms. Under certain conditions, e.g., reflections from close corners, multipath effects can greatly enhance the RCS. Scattering in complex cases is often treated probabilistically.

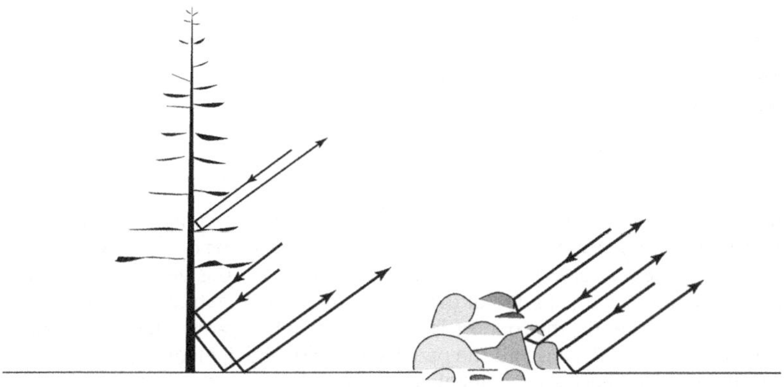

Figure 5.3 Multipath scattering in nature. The great variety of surface and object orientations in nature provides numerous multiple scattering paths, including backscattering paths such as those illustrated. Such paths include links between terrain and tree trunks as well as tree branches and trunks. Important lateral paths exist between branches. Similarly, rough, irregular, usually rocky terrain also provides scattering paths over a wide range of bistatic angles, as well as backscatter paths.

cylindrical rather than plane—with the surrounding terrain and the lateral branches. Lateral and oblique paths, not shown, also exist between combinations of branches. Rough terrain and rock exposures can also support multipath scattering. Again, while the sketches in Figure 5.3 depict only the supplementary angles associated with backscatter, many additional bistatic paths typically exist in such structures.

Although it is convenient to think of the paths depicted in Figure 5.3 as "like corner reflectors," keep in mind that the scattering elements in this instance—branches, rocks, etc.—have curved or even irregular shapes. Thus, relative to a manufactured device, the efficiency of the scattering is greatly reduced while the range of angles over which multiple reflections obtain is significantly increased. Both effects are the result of greater angular dispersion of the scattering/reflection from the individual elements with respect to that of the plane surfaces of typical corner reflectors.

5.1.3 Bragg Scattering

Bragg scattering is the rubric attached to a wave-wave interaction resulting redirection or other modification of an incident wave.[3] It is result of a phase match(es) between the periodic variations of an incident wave and those of a two-dimensional (2D) or three-dimensional (3D) structure. This type of interaction is often referred to as "resonant" scattering, as the strength of the scattered wave is increased by coherent additional energy from successive cycles of the incident wave. But no energy is stored within the system, unlike traditional second-order resonant processes.

Figure 5.4 illustrates how Bragg scattering works. Consider an incident signal with $\lambda/2 \leq \Lambda$, where λ is the length of the incident wave and Λ is the period of the surface undulations. For the Bragg scattering condition,

$$\lambda = 2\Lambda \cos \phi \tag{5.1}$$

3 Named after W. H. and W. L. Bragg, this type of scattering was first recognized and applied in connection with X-ray diffraction studies of crystals.

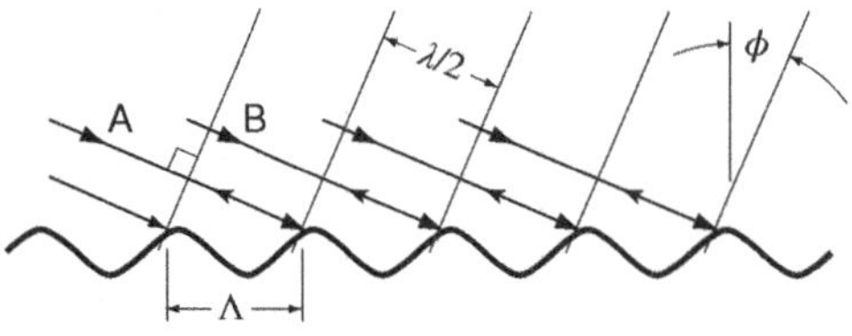

Figure 5.4 Bragg scattering. A train of waves incident on a periodic surface, illustrated here for backscatter by a sinusoidal surface, produce pronounced scattering effects at certain combinations of wavelength, λ, and grazing angle, ϕ, for which the differential phase paths, modulo 2π, are equal. Consider the two paths A and B, which terminate at points separated by one period of the surface oscillations, while the difference in the two path lengths differ by $\lambda/2$. Consequently, the round-trip distance from the transmitter-to-surface-to-transmitter differs by a single wavelength. Because all the paths shown are equal, modulo (λ), the contributions of these points to the received backscatter signal add in phase.

all round-trip path lengths are equal, modulo (λ). For example, round-trip paths of A and B differ by λ, while the path to the right of B differs from that of A by two λ, etc. As can be seen from the geometry and Eq. (5.1), the range of wavelengths that can satisfy the Bragg condition is $\lambda \leq 2\Lambda$.

While we have illustrated the geometry for paths spaced by Λ along the horizontal, the complete problem must consider the effect of variations within horizontal length Λ. Letting $s(x;\lambda)$ be the incremental scattering along the horizontal for a periodic surface, with the received signal proportional to the sum of the incremental effects, we have,

$$S_{\text{rcvd}} \propto \int_{x_1}^{x_1+N\Lambda} s[x(\text{mod}\,\Lambda);\lambda]dx \tag{5.2}$$

$$\propto N \int_{x_1}^{x_1+\Lambda} s[x(\text{mod}\,\Lambda);\lambda]dx \tag{5.3}$$

where S_{rcvd} is the strength of the received signal, the constant of proportionality is that of the radar system and the range, and we assumed the total length of the undulating surface to be $N\Lambda$ for convenience. The linear increase in the received signal strength with the length of the scattering surface is the result of the periodic geometry observed with a wavelength satisfying Eq. (5.1). The coherence of the scattering results in the received power increasing even more rapidly, since the received power goes as the squared magnitude of the field,

$$P_{\text{rcvd}} = |S_{\text{rcvd}}|^2 \tag{5.4}$$

In laboratory applications, observations of Bragg scattering are used, for example, to find the orientation and spacing of atoms in crystalline structures. In studies of the environment, Bragg scattering is used to study the spectral distribution of a geophysical field, e.g., the wave spectrum of the sea, or the properties of natural rough surfaces. The coherent nature of the coupling between the probing radar signal and the corresponding surface periodicity results in a high sensitivity to weak undulations and high spatial frequency resolution.

5.1.4 "Creeping" Waves

Creeping waves occur when a portion of the energy in a propagating wave field attaches to a conducting surface. In the practical case of finite surfaces, especially surfaces with finite curvature, the attached fields radiate a portion of their energy into space. This radiation is incremental with distance along the surface, with the result that there is a steady diminution of the strength of the surface wave in the direction of surface wave propagation. The surface waves in such instances propagate at speeds that are slightly less than the speed of light in the medium above the surfaces; this is the reason for the rubric, "creeping waves."

The illustration in Figure 5.5 shows how this occurs for a wave attached to the surface of a right circular cylinder. The strength of radiation from the creeping wave is proportional to the strength of the wave, so that the strength of the radiated field falls exponentially with distance along the creeping wave path. Even so, these waves can be remarkably strong in backscatter after rotating through 180°, 540°, or even larger angles, around an object. Such waves can also be important in forward scatter after rotations that are multiples of 360°. As with the other phenomena in this section, creeping waves do not usually exist alone, but exist simultaneously with the other scattering mechanisms and are only one component of the total RCS. Radiation from creeping waves is most important in scattering from smoothly curved surfaces with large radii of curvature, $r_c \geq \lambda$. Creeping waves also are important in scattering by spheres, but these are difficult to visualize in detail.

5.1.5 Diffraction and "Wave Optics"

Diffraction, or *wave optics*, effects are always present in nature. The full development of wave optics is much more complex than simple calculations of curvature.[4]

These effects are created by the fundamental wave nature of propagation, first described by Huygens, in which a given wave can be described as the composite result of an infinite ensemble of spherical wavelets growing from each point on a previously existing wavefront closer to the original

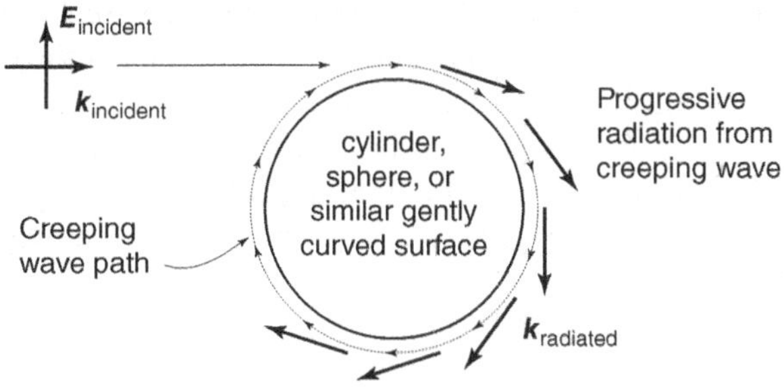

Figure 5.5 Creeping waves. Wave field $E_{incident}$ propagating in direction $k_{incident}$ is incident on a cylinder equator of a sphere. The vertically polarized component of the incident wave satisfies the EM boundary conditions at the top of the cylinder, which results in attachment of some wave energy to the cylindrical/spherical surface. The attached creeping wave, subsequently, is guided by the surface and flows clockwise around the cylindrical axis perpendicular to the page. The guiding process is a weak one, so that energy continuously leaks from the creeping waves to radiated waves in directions $k_{radiated}$, approximately tangential to the local surface. Radiation from creeping waves fills the bistatic pattern in geometric shadow zones. Creeping waves that have traveled around the surface by odd multiples of 180° contribute significantly to the backscatter radiation. Not shown: the incident wave also excites creeping waves in the counterclockwise direction. The total radiation from such waves can be found from the superposition of two wave fields.

4 The subject of diffraction is concerned with the radiative and interference processes in bending waves around corners. It also addresses the more subtle questions dealing with the weighting of the individual components of elemental wavelets in creating the resultant total field. Section 3.1.3, touches on the importance of wave curvature, while Section 6.2.2 describes how wave interference effects control the shapes of antenna beams.

source. Fresnel completed the scalar theory by adding the concept of *phase* to Huygen's idea. The Huygens–Fresnel theory can be derived from J. C. Maxwell's comprehensive equations describing the exact behavior of electromagnetic fields.

In complete formulations of EM fields, individual "rays" lose their identities as discrete objects, as they only approximate the flow of energy. The differences are particularly striking in the vicinity of material edges, where the presence of an edge results in the diversion of energy into the shadow region, i.e., into the volume geometrically hidden from the source. We apply this theory in a limited form when discussing the dependence of antenna patterns on the distribution of aperture fields, Section 6.2.2. Here, we provide a simple example illustrating the difference between the character of ray- and wave-optics results for scattering from a simple object. Keep in mind that wave optics applies in situations where the wavelength of EM wave is of a scale comparable to the structure or feature size of a structure of interest, whether that object is a wave guide, an antenna, or an object probed by radar.

Figure 5.6 shows an example polar diagram representing the intensity of energy flow about the backscatter direction from a small disk a few wavelengths across. While most of the energy is concentrated in a strong beam centered on the specular direction, there is significant energy in an accompanying set of lesser lobes, called *sidelobes*. This beam structure—the finite-width beam centered on the specular direction in combination with the presence of the minor lobes—is a consequence of interference among wavelets reradiated from each part of the disk in response to the transmitter illumination.

There is also a diffracted beam of comparable strength in the forward-scatter direction induced by the wave blocking effects of the disk. The forward-scatter beam is always present, independent of the shape of the blocking object (disk or otherwise); for very large, complex objects, scattering in the forward beam is typically much stronger than in other directions. On the other hand, geometric optics applied to normal incidence backscatter from a flat plate predict only a uniform, confined beam of zero angular width, and only a shadow in the forward-scatter direction. The full-wave solution reduces to the geometric-optics result as the wavelength vanishes. But for any finite wavelength, the shadow is always filled beyond a calculable distance behind the disk. Two hybrid theories, the *geometric theory of diffraction* (GTD) and the *uniform theory of diffraction* (UTD), employ

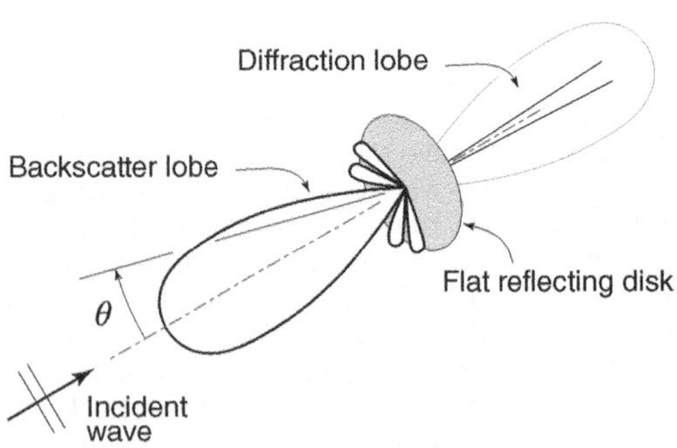

Figure 5.6 Wave optics effects. Diffraction effects are always present in nature. These account for lobe structure not predicted by G.O. Such effects are important and must be considered when the size of the scattering object is of order about one tenth to a few wavelength in size. Diffraction also produces a forward-scattering lobe, suggested in figure, which is always present. For disks, forward and backscatter lobes are same width and shape, as indicated by the grey forward-scatter lobe; for large complex objects, the forward diffraction lobe can be much stronger than backscattering. Note: Babinet's principle tells us the relationship between the diffraction from blocked and unblocked portions of a diffracting screen.

canonical results for edges to construct adjuncts to strict ray theory near the ray optic shadow boundaries. These are in common use for analysis of radio wave scattering problems.

5.1.6 Resonant Scattering Phenomena

Resonance effects in scattering structures can result in effective scattering cross sections that are much larger than the physical size. Such effects occur primarily in objects comprising "low-loss" material with very slight dissipation of electromagnetic energy. Metallic structures are the common examples, but structures of predominantly dielectric materials are also possible.

Common examples of manufactured objects are metal trusses, other open lattice frames, and building features, such as the leaded window lattice. Open cavities, such as vehicles or other container-like enclosures are another possibility, but uncommon. The requisite element is the presence of interrupted conducting paths on which resonant currents can be excited by incident waves. Scattering from such structures will be highly dependent on frequency.

Similar possibilities exist for layered dielectrics. These can be visualized by extension of familiar exercises in EM theory treating the reflection of EM waves passing though various slabs of dielectric materials.

5.1.6.1 Scattering by Antennas

General result: The RCS of an antenna is calculated by a method similar to that based on Eq. (4.31) for corner reflectors. In this instance, assume that we have knowledge of the antenna gain, rather than the area of the corner reflector. Again using Eq. (3.7), but this time expressing the RCS in terms of the gain, we have for any antenna operating at wavelength λ,

$$\begin{aligned}\sigma_{\text{ant}} &= A_{\text{ant}} \cdot G_{\text{ant}} \\ &= \frac{\lambda^2 G_{\text{ant}}^2}{4\pi}\end{aligned} \tag{5.5}$$

This expression is convenient since antenna performance is most often expressed in terms of G_{ant}. Note the explicit dependence of the result on λ, which is explicit, while the dependence of G on λ is implicit.

Half-wavelength dipole: The most common example of a resonant scattering is a half-wavelength dipole antenna. Applying Eq. (5.5) to an ideal, lossless device, and letting λ_0 be the wavelength at the center frequency of the dipole, we have,

$$\sigma_{\text{ant}} = \frac{1}{\pi}\left(\frac{3}{4}\lambda_0\right)^2 = 0.18\lambda_0^2 \approx \frac{\lambda_0^2}{5} \tag{5.6}$$

This is a remarkable result in that the area of physical dipole is only $2\lambda_0/2 \cdot d = \lambda_0 d$, where d is the diameter of the dipole element, which typically is the range $0.001 < d/\lambda_0 < 0.1$. for d in this range, the effective area of the dipole exceeds the physical area by factors of 4 : 1 to 400:1. Such is the effectiveness of the resonance phenomenon in intercepting and radiating the incident energy, and it demonstrates another exception of the rule-of-thumb that RCS is comparable to physical size.

5.1.7 Polarization Conversion

Polarization effects may be the most familiar to us of these various phenomena. In the context of scattering, though, we are concerned primarily with *conversion of polarization* from one state to another; e.g., the change in polarization state that results from the interactions of the illuminating radar wave with the scattering object. Some scattering objects convert vertically polarized incident waves into horizontally polarized scattered energy, for example. Similarly, backscattering of a circularly polarized incident wave usually results in significant scattered power with the opposite circular polarization. For example, right-hand circular polarization (RCP) can be converted into left-hand circular polarization (LCP), in addition to other polarization states. In general, RCS is a function both of the transmitted, illuminating polarization and of the received polarization. Knowledge of a scattering object's effect on polarization is important both for predicting RCS for purposes of system design and in the interpretation and analysis of radar observations (see Section 3.1, p. 28).

Perhaps the best example of polarization conversion is that of scattering by a thin wire or linear dipole. Because thin conductors carry currents most readily along their axis, radiation from wires—defined conductors whose as radius is much less than a wavelength—is strongly polarized with the electric field and the axis of the conductor in a common plane. Thus, for a wave incident on a wire, only the electric field component parallel to the wire is scattered efficiently. Consequently, regardless of the polarization of the incident wave, the predominant component of the electric field of the scattered wave is found in a plane containing the wire. The exception being instances in which the incident wave is perpendicular to the wire, and the scattering is very weak, indeed. In scattering from wires, incident circular polarization is converted to linear, while incident linear polarization is selectively converted to a linear orientation in the plane of the scattering wire.

Figure 5.7 illustrates the case of a linearly polarized incident wave obliquely incident on a screen of parallel wires, where the radius of the wire $a \ll \lambda$. The resulting scattered wave is preferentially polarized parallel to the wires independently of the incident polarization. This occurs because only the parallel polarized component of the incident wave interacts strongly with the screen. This can be visualized by considering the boundary conditions for EM fields, which require the field to terminate on a conductor at right angles; stated another way, the electric field parallel to a perfect conductor must be zero at all times. The result is that a field component perpendicular to the wires can pass through the screen with very little disturbance. The field component parallel to the wires in the screen, on the other hand, must be zero on the wire—a very large disturbance indeed! By superposition[5] the incident parallel field must be canceled by an equal and opposite induced parallel field component. This canceling field is accompanied by strong induced currents within the wires of the scattering grid, and an associated radiation field polarized parallel to the wires within

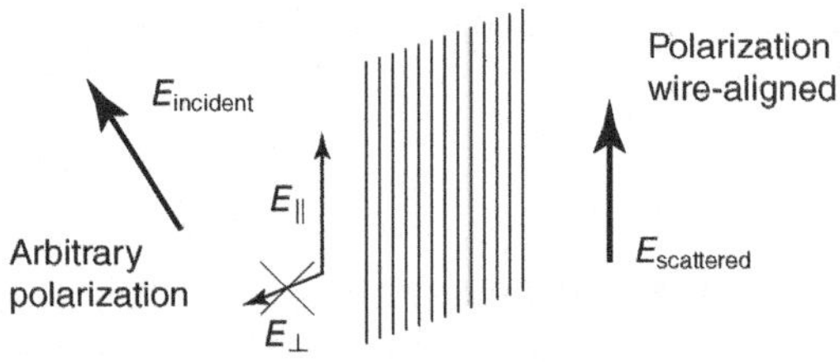

Figure 5.7 Polarization effects. Scattering can modify the polarization of the incident wave so that the scattered energy is partially or totally cross-polarized, i.e., orthogonally polarized, with respect to the incident wave. Thus, major contributions to the total RCS can appear in polarizations other than those present in the illumination. In the example shown, a wire grid strongly scatters electromagnetic waves with the electric field selectively oriented parallel to wires in the grid, so $E_{scattered}$ is strongly aligned with $E_{\parallel}$, thus enhancing this component of the RCS while any component of the scattered wave perpendicular to the wires, i.e., aligned with $E_{\perp}$, is suppressed.

5 This reasoning is permitted by the linearity of Maxwell's equations.

the grid. If you wish, think of the wire grid as an antenna system comprising parallel wires driven by the induced currents required to satisfy Maxwell's equations (see Figure 5.7).

Several scattering mechanisms act to alter the polarization—the example above is just one of many. These can be exploited to enhance observations in a number of ways, based on known or inferred properties of the interest. In addition, the presence of polarizing effects in exploratory observations can be used to constrain geophysical and habitat models of natural environments.

5.1.8 Natural Shapes and Artifacts

Nature is replete with scattering phenomena. As radar signals employ wavelengths in the range 0.001–1000 m, the structures probed are on scales of interest in the geosciences and field biologists. Beyond this direct sensitivity to features over a wide range of sizes, *radar images* constructed from observations with a high degree of spatial resolution reveal patterns within the natural environment. Perhaps the most familiar are radar weather maps constructed from the observed intensity in 2D and 3D. Less familiar are maps of ocean topography constructed from the time of flight of nidar-directed radar signals from Earth orbiting satellites. Variations in sea level, patterns of ocean currents, and the topography of the sea floor are all accessible from such results.

Bragg scatter is often used to understand radar reflections from slightly rough surfaces, as the strength of the radar echo depends on the amplitude of the Fourier component of the surface that resonates with the illuminating wave. Combining such measurements at several wavelengths constrains the spectrum of the surface roughness, and can be diagnostic of various geophysical processes associated with surface type, formation, and weathering.

Studies of vegetated areas can make use of several observables. Vegetation canopies can act as nearly isotropic reradiators if they are dense, and the physical structure of a trunk roughly perpendicular to the ground can form a very bright dihedral corner reflector. Each of these mechanisms has distinct signatures in polarization of the reflected waves, so that many studies of vegetated surfaces employ measurement of the full polarization properties of the canopy.

5.2 Interference Effects

5.2.1 Interference in Scattering from Large Objects

The first three topics under "Phenomena" (Sections 5.1.1–5.1.3), as mentioned earlier, deal with physical situations that include backscattering along two or more paths. Because the scattered signals of interest are the result of the same illuminating signal, scattered signals traveling in the same directions are inherently phase coherent, i.e., in step with one another, and can coherently interfere. That is, echoes from various points on a distant target, received by the same antenna, share a common frequency and differ only by fixed amplitude ratios and phase differences. In the instance of Bragg scattering, the echo strength from a given region of a periodic surface is built up by coherent addition of signals from distinct sections of the target surface. Multipath scattering from complex scattering areas, such as those in Figures 5.2 and 5.3, on the other hand, are made up of a plethora of paths that are either unknown for practical reasons or unknowable.

Thus, interference effects are an important feature of scattering from large complex structures or from areas containing multiple *scattering centers*, by which we mean the localized points or compact features on or within a complex object that contribute significantly to the RCS. Such points or features can be of either natural origin or artifacts. Examples are the localized reflections from an aircraft wingtip and tail, Figure 4.15, or a rock pile, Figure 5.3. When viewed with radar systems that are unable to resolve such individual elements, the composite signal from all elements taken together is treated as a coherent, localized source of scattered energy, e.g., as from the aircraft in the first instance and the tree and rock pile in the second. Radar measurements that lack sufficient resolution to discriminate among such individual features observe the linear sum of the signals from the ensemble of localized objects. For example, when observing an aircraft, a radar with resolution coarser than the largest dimension of the plane will always observe the sum of the signals from the various glints. When working in such a situation, in order to determine the net RCS, we must consider the relative phase differences of the signals, i.e., the interference effects among signals from the various glints. The interference pattern depends on the details of size and shape of the overall object, and also on the orientation of the object with respect to the radar. Analysis of this problem requires that we consider two independent phenomena: (i) the field strength and phase of signals from the individual scattering elements making up the larger object, and (ii) propagation and orientation effects that control the relative phases with which signals from various glints are combined.

5.2.2 Two-element Interference Geometry

We now illustrate the basic effects of orientation and spacing on the composite RCS of several objects by examining the scattering pattern of a pair of hypothetical isotropic scattering centers. In order to simplify the analysis, we assume scattering cross section of our isotropic "targets" to be $\sigma = 1$, and also neglect all interactions between the two. Our immediate interest is in the effects of the arrangement with respect to the locations of the transmitter and receiver of the two objects on the scattering pattern. We extend the simple two-object results to more general cases in a later discussion.

The geometry is shown in Figure 5.8. The scattering centers are separated by distance ℓ, as shown. All distances are referenced to the midpoint of the line connecting the scattering centers, which we designate S1 and S2. Incident illumination in the form of a plane wave arrives from a transmitter located at a very great distance in direction θ_{inc}. Since S1 and S2 are identical, and we neglect their interactions, the corresponding scattered waves differ only in their point of origin (S1 or S2) and initial phase, as determined by their respective distances from transmitter. Similarly, the direction from S1 and S2 to the receiver is specified by angle θ_{scat}. Any difference in the distances from S1 and S2 along direction θ_{scat} also contributes to the final phase path differences at the receiver.

To be more specific, the total scattered field at some distant point is the sum of the fields from the two spheres, with due allowance for the differences in the phase paths. Then, relative to a path from source to observer via the reference point,

1. Excitation of S1 [S2] is advanced [retarded] by $(k\ell/2)\sin\theta_{\text{inc}}$.
2. Scattering by S1 [S2] is retarded [advanced] by $(k\ell/2)\sin\theta_{\text{scat}}$.

Assuming an $e^{+j\omega t}$ time variation and that the distances from the transmitter and receiver are $\ggg \ell$, so any differences in path lengths are negligible,

$$E_{\text{tot}} = \underbrace{E_0 e^{+j\frac{k\ell}{2}(\sin\theta_{\text{inc}} - \sin\theta_{\text{scat}})}}_{\text{for S1}} + \underbrace{E_0 e^{-j\frac{k\ell}{2}(\sin\theta_{\text{inc}} - \sin\theta_{\text{scat}})}}_{\text{for S2}} \tag{5.7}$$

where $E_0 = |E_{0_1}| = |E_{0_2}|$ incorporates the excitation by the incident field, scattering by S1 and S2, the variation of the field strength with distance, and the mean path length. Rewriting the above in trigonometric form, the interference patterns are given by,

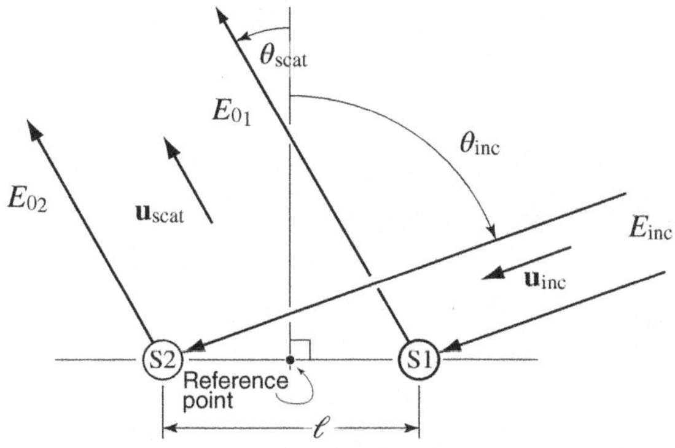

Figure 5.8 Geometry of dual scattering paths. Scattering by laterally offset twin objects, S1 and S2, exhibits strong wave interference effects. Incident field with wave vector $\mathbf{u}_{inc}$ and strength E_{inc} originate at a source, located at upper right; receiver at upper left captures fields with wave vector $\mathbf{u}_{inc}$ and strength $E_{0_1} + E_{0_2}$, which have the same magnitude but, in general, differ in their phase. Both source and receiver are located at great distances with respect to the separation of S1 and S2, $\ggg \ell$ (see Eq. (5.7) and its text description). Subject to the condition of ideal isotropic scattering, the resulting interference patterns are a function of only the wavelength λ, the separation distance ℓ, and the angle of incidence θ_{inc}.

$$E_{tot}(\theta_{scat}; \lambda, \ell, \theta_{inc}) = 2E_0 \cos\left[\frac{k\ell}{2}(\sin\theta_{inc} - \sin\theta_{scat})\right] \qquad (\theta_{scat} \neq -\theta_{inc}) \tag{5.8}$$

$$= 2E_0 \cos\left[k\ell \sin\theta_{inc}\right] \qquad (\theta_{scat} = -\theta_{inc}) \tag{5.9}$$

Equations (5.8) and (5.9) represent, respectively, the bistatic or general case and the special case of backscatter.

In order to appreciate the behavior of the interference pattern, consider first the locations of the scattering lobes, defined as the direction of the local maximum of the scattered signal amplitude for E_0 constant. For fixed θ_{inc}, these are found at angles θ_{scat} for which the net phase difference along paths containing S1 and S2 is an integer multiple of π, or,

$$\frac{k\ell}{2}(\sin\theta_{inc} - \sin\theta_{scat}) = n \cdot \pi \tag{5.10}$$

where λ is the wavelength, n is zero or one or more positive or negative integers for which the value of θ_{scat} is real, and θ_{inc} is treated as a parameter of the problem. Solving for $\sin\theta_{inc}$ directly gives,

$$\sin\theta_{scat} = \sin\theta_{inc} + \frac{n\lambda}{\ell} \tag{5.11}$$

Values of θ_{scat} that satisfy Eqs. (5.10) and (5.11) are the directions for which the interference pattern has maximum intensity. Equation (5.11) is known as the *grating equation* because it originally appeared in the analysis of diffraction gratings. Examining Eq. (5.8) again, zeroes of the bistatic pattern occur at values of θ_{scat} for which the argument is an odd multiple of $\pi/2$. These are given by solutions of,

$$\sin\theta_{scat} = \sin\theta_{inc} + \frac{(2n-1)}{2}\frac{\lambda}{\ell} \tag{5.12}$$

For backscatter, $\theta_{scat} = -\theta_{inc}$, and the grating equation particularizes to,

$$\sin\theta_{inc} = \frac{n\lambda}{2\ell} \tag{5.13}$$

Solutions for θ_{inc} from the above provide the angles of incidence where the backscatter is a maximum. Similarly, for the backscatter zeroes,

$$\sin\theta_{inc} = \frac{(2n-1)}{4}\frac{\lambda}{\ell} \tag{5.14}$$

which is readily found from Eq. (5.12), again by setting $\theta_{scat} = -\theta_{inc}$.

5.3 Solution of the Grating Equation

5.3.1 Bistatic Scattering

Solutions to the grating equation can be understood by use of graphical analysis. Figure 5.9 provides a plot of the two sides of Eq. (5.11) on the same set of axes so that solutions are located at intersections of the two curves, with θ_{scat} read from the abscissa. Here, the left-hand side of the grating equation is a simple sinusoid, while the right-hand side represents a set of level lines with differing offsets in the ordinate. Values of the right-hand side depend on the angle of incidence, θ_{inc}, which is fixed for a given illumination, the distance between the spheres ℓ, the wavelength, λ, and the index, n. For λ fixed and specified θ_{inc}, ℓ, the level lines of the right-hand side have a vertical spacing λ/ℓ with an initial vertical offset of $\sin\theta_{inc}$, measured from the origin to the line with index $n = 0$. Solutions for θ_{scat} are read on the abscissa corresponding to the intersections of the curve of $\sin\theta_{scat}$ with the level lines. These solutions represent the angular locations $\theta_{scat,n}$ of maxima in the interference pattern of scattering derived here for two identical spheres. Note that the solution will have the same shape for scattering by any two identical scatterers, and for any linear array of identical scatterers with the same values of λ and ℓ.

A few features of this solution are apparent immediately from Figure 5.9. First, the effect of varying the angle of incidence is to shift the level lines upward or downward. The rate of shift also depends on θ_{inc}, with the greatest rate of change occurring for θ_{inc} near 0 and π, and the least rate of change for a given shift occurring for θ_{inc} near $\pm\pi/2$ and $\pm 3\pi/2$. Next, the angular spacing between solutions increases as the baseline between the spheres decreases, and decreases as the baseline increases. Finally, there is only a finite number of solutions for real values of θ_{scat}, since for real-valued solutions we require that $|\sin\theta_{inc} + n\,\lambda/\ell| < 1$. Decreasing the wavelength or increasing the baseline, or both, reduces the vertical spacing between the horizontal lines, thereby increasing the number of solutions and decreasing the spacing between maxima.

Normalizing Eq. (5.8), we can plot the relative pattern $|E_{tot}(\theta_{scat})/E_0|^2$, where the normalization results in convenient values. With this, we see only the effect of the arrangement of the spheres with respect to one another. An example is shown in Figure 5.10. The pattern is that which would be mapped out for fixed θ_{inc} as the direction $\mathbf{u}_{scat}$ rotates around the line between the spheres. Note that Eq. (5.11) with $n = 0$ shows that there is a maximum in the direction of an observer located at $\theta_{scat} = \theta_{inc}$. Also, there is always a maximum in the direction of incidence, i.e., in the direction of an observer at angle $\theta_{scat} = \pi - \theta_{inc}$. Recognizing that the interference pattern is a figure of revolution about the baseline, we see that these two particular directions ($\theta_{scat} = \theta_{inc}, \pi - \theta_{inc}$) correspond to *specular reflection from the axis connecting the two spheres.* Although in a 2D plot these two directions seem quite different, both lie on a cone for which the geometrical phase difference introduced by the locations of the spheres relative to the transmitter is canceled exactly as a consequence of their location with respect to the observer. Thus, the special direction, $\theta_{scat} = \theta_{inc}$, measured from the plane perpendicular to the axis between the spheres, is always a direction of a maximum, regardless of the values of the angle of incidence.

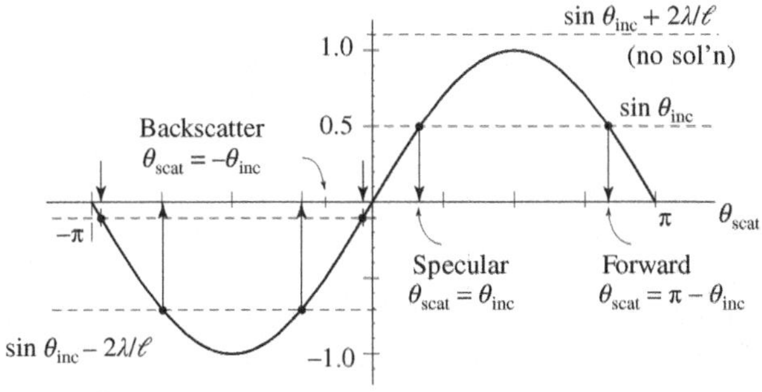

Figure 5.9 Solutions of the grating equation. Bistatic solutions for the six angular locations of the maxima of the interference lobes are located at the intersections of the dashed level lines and the sinusoidal curve, indicated by points, "•." Vertical arrows indicate corresponding values of θ_{scat}. Solutions exist only for $n = -2, -1, 0$ (level line for $n = 0$ not labeled). Curve is drawn for $\theta_{inc} = 30°$ and separation $\ell = 1.65\lambda$. For parameters here, backscatter direction is not located at a maximum of an interference lobe, while forward scatter is always so located. Coordinate system is that of Figure 5.8 (cf. Figure 5.10.)

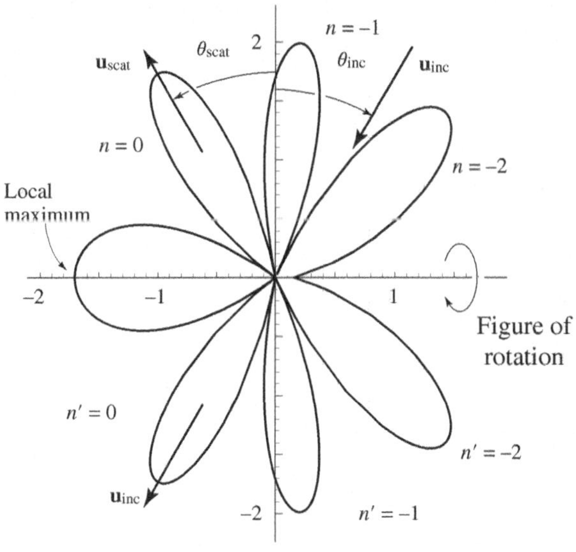

Figure 5.10 Composite pattern of a pair of two isotropic scatterers. Plot shows the bistatic RCS of a two-element array of isotropic scatterers positioned as in Figure 5.8. Linear scattering pattern is a figure of revolution about the axis of the scatterers; there is always a maximum in the specular direction with respect to the axis, as well. Locations of maxima are those illustrated by graphical solution in Figure 5.9. Values of $n = 0, -1, -2$ refer to order of solution in grating equation; primed values represent solutions with $|\theta_{scat}| > \pi/2$. Note: (i) symmetry about axis of rotation; (ii) lobe and minimum at $\theta_{scat} = \pm\pi/2$ are local values of the pattern only (cf. Figure 5.9; see Eq. (5.7) and its text description).

5.3.2 Backscattering

Referring to Eq. (5.13), a graphical solution for backscattering can be constructed along the same lines as before. But now the reference lines from the right-hand side are symmetrically distributed with respect to the abscissa, while the line spacing is $\lambda/2\ell$, or one-half that for bistatic scattering, Figure 5.11. As a result of these differences, there are about twice as many solutions in the backscatter case as there are for the general case. The backscattering pattern is sketched in Figure 5.12.

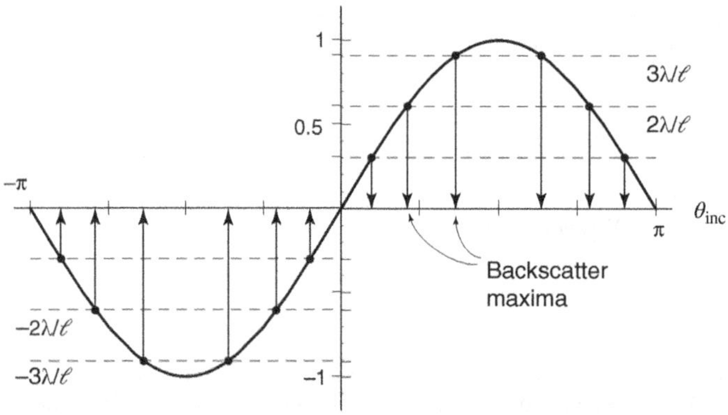

Figure 5.11 Backscatter solution of the grating equation. When particularized to backscatter, the scattering lobes become more narrow and usually twice as numerous as for a general bistatic pattern. Because the steepest slope of $\sin\theta_{inc}$ occurs for $\theta_{inc} = 0$, the narrowest backscattering lobes are located approximately perpendicular to line connecting spheres, Figure 5.12. Lobes are symmetric about both the line $\theta_{inc} = 0$ and baseline connecting the two scatterers. Compare with Figure 5.9, also Eq. (5.11) with Eq. (5.13) as parameters are the same; $\theta_i = 30°$ and separation $\ell = 1.65\lambda$.

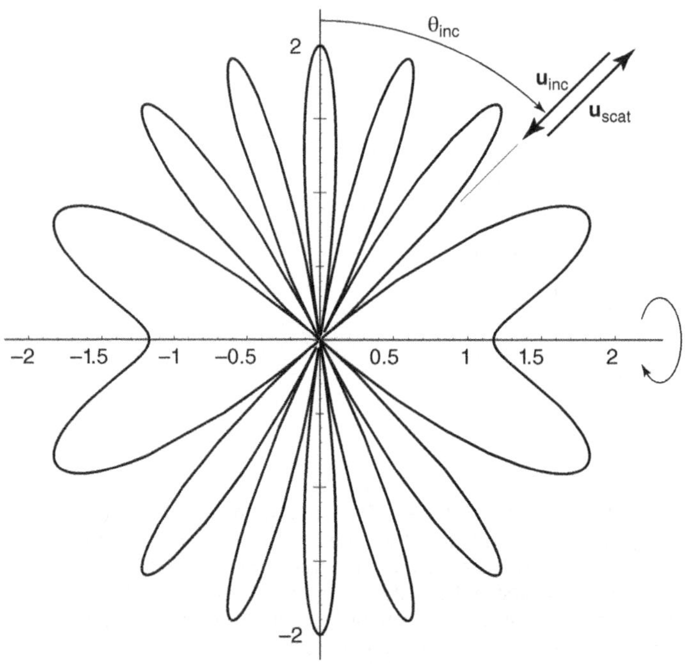

Figure 5.12 Interference pattern for backscatter from points. $\theta_{scat} = -\theta_{inc}$. Plot gives the backscatter RCS of a two element array of scatterers, normalized by the RCS of a single element. Curve is backscatter counter part of Figure 5.10, with separation $\ell = 1.65\lambda$. Compare number and width of lobes with bistatic pattern, Figure 5.10. Pattern has two symmetries, one about the horizontal axis connecting the two scatterers and a second perpendicular to the first. All maxima have equal strength except along the axis of the spheres; pattern is a figure of rotation about the horizontal, but not the vertical axes.

An important difference between Figures 5.10 and 5.12 is the interpretation of the coordinates with respect to the direction of scattering. In Figure 5.10, for bistatic scattering, the direction of incidence is fixed, and the pattern depicts the relative value of the scattering cross section as the observer changes angular position with respect to the axis between the two spheres. In the second instance, depicted in Figure 5.12, the illuminating source and the observer are co-located, e.g., at a monostatic radar site, so that as the receiver and transmitter move together. Thus, for backscatter, the pattern represents the result of moving a monostatic radar around the fixed reference point of the two spheres. Alternatively, the radar can be considered fixed and the axes of the two spheres rotated in the plane of the paper with respect to a fixed position of the radar in the same plane. Such differences in the point of view and interpretation of scattering diagrams are typical in discussions of differences between bistatic and monostatic observations.

5.3.3 RCS for the Two-sphere Case

We have evaluated the normalized interference pattern for the two noninteracting spheres. What is the absolute scale? To find this, we need only recognize that $\sigma \propto |E_{\mathrm{scat}}/E_{\mathrm{inc}}|^2$. With this, we can plot $|E|^2$ and relabel it for the appropriate scale. At the maxima,

$$E_{\mathrm{tot}} = 2 \times E_0 \tag{5.15}$$

Because squaring of the scattered field strength quadruples the RCS, in this instance, it follows that,

$$\sigma_{\mathrm{max}} = 4 \cdot \sigma_{\mathrm{sphere}} \tag{5.16}$$

The calculated minima are still given by,

$$\sigma_{\mathrm{min}} = 0 \tag{5.17}$$

so even in practical cases, we can expect the actual minima to be very small.

As mentioned above, mutual sphere-sphere interactions are usually negligible for $\ell \gtrsim 10a,\ 10\lambda$, whichever is less.

5.4 Scattering from Multiple Independent Objects

The scattering behavior of a larger, more complex, but still unresolved collection of objects can be understood by extension of the behavior of the two spheres, as mentioned earlier, insofar as the objects scatter independently. Suppose that N objects are involved. Then, as is the case for a pair of objects, the RCS will depend on the manner in which waves scattered from the various objects interfere at the location of the observer. As before, we approach this problem by examining the manner in which the complex phasors representing the fields from the N objects combine at the location of the observer. At the same time, we briefly introduce an alternative notation that is widely used in the broader scattering community.

We can write down an expression for the sum of signals with differing amplitude and phase in a straightforward way. Letting ϕ_n and $\theta_{\mathrm{path}-n}$ represent, respectively, (i) the phase change associated with the interaction of the electromagnetic wave—our radar signal— with the nth scattering object and (ii) the phase along the total path to the nth object. For simplicity, we make the strength of the wave-object interactions the same for all objects. With this, the RCS is given by,

$$\sigma_N = \left| \sum_{n=1}^{N} c \qquad A e^{j\phi_n} \qquad e^{j\theta_{\mathrm{path}-n}} \right|^2 \tag{5.18}$$

where $cA\exp[j\phi_n]$ is the van de Hulst *complex scattering amplitude* or "CSA." This last quantity is related to σ_n by,

$$\sigma_n = N \cdot c^2|A|^2 \tag{5.19}$$

The scattering literature is divided as to notation. The CSA is the natural quantity for calculation of scattering, and it is the form in which formal analytical and numerical results for scattering are usually given. On the other hand, in the engineering literature, use of σ became the standard notation for RCS at a time when measurements of phase were difficult, at best. When phase data became available, the heuristic form with $\sqrt{\sigma}$ and the possibility of a finite, i.e., deterministic, nonzero, phase shift, ϕ_n, appeared.[6]

The expression for total RCS as the square of a sum, Eq. (5.18), is always correct. Often, though, some information for completing the calculation is lacking. Most frequently, the missing piece is knowledge of phase, which, to be useful, must be precise to a fraction of a radar signal cycle. When phase is random, e.g., if the $\{\sigma_n\}$ were to represent a collection of falling raindrops for which we are very unlikely ever to know the location of the individual drops—our scatterers— a statistical approach is required. The result of a collection of scatterers when the unknown phases are uniformly distributed over 2π is,

$$\langle \sigma_N \rangle = \sum_{n=1}^{N} \sigma_n \tag{5.20}$$

which generally applies when the positions of the raindrops are uncertain on the scale of about one-half wavelength or more, i.e., $\Delta p_n/(2\lambda) \gtrsim 1$, where Δp_n is the uncertainty in the path length from the transmitter to the nth scatterer and thence to the observer. Such a calculation is accurate in situations where the phase is known to be random. We refer to a calculation in which the phases of the individual scattering elements are added as an *incoherent* sum of the scattering cross sections σ_n. As we shall see, in such cases, we adopt the total power received from the collection of raindrops as our measure of the volume and density of the rain. Similar summations occur frequently in estimates and calculations of the total RCS of individual complex scattering objects, as well as of collections of objects. Equation (5.20) is a surprisingly useful even for relatively small values of N.

5.4.1 Example: Scattering from Five Points in a Plane

We illustrate the convergence of the coherent to the incoherent sum by reduction of Eqs. (5.18)–(5.20) with the following example, in which we take a small number of independent scatterers and examine the statistical behavior of their sum. Although this can be calculated as an ensemble average, it is more instructive in developing a physical understanding to do this for a fixed configuration and then to introduce the path variations by changing the aspect angle of the radar relative to the configuration.

---

6 Equation (5.18) is a form commonly employed in the scientific literature of scattering, and you will certainly see it while working in this field. The alternative use of $\sqrt{\sigma_n}$ for the amplitude is extremely awkward because the amplitude scaling is at odds with the natural normalizations when analyzing the scattering properties of various objects, and because the phase is simply tacked on without relationship to the physics of scattering. While awkward, use of $\sqrt{\sigma_n}$ is an expeditious way to proceed when starting with measurement-based knowledge of the RCS together with concomitant measurements of phase. When engaged in analytical work or using results from the literature of scattering, however, it is much more straightforward to use the CSA. In any event, the expression for the combination of signals from a collection of scatterers has the features of Eq. (5.18).

Assume that we are observing a group of five scatterers with the configuration shown in Figure 5.13. The points are chosen in sequential pairs from a uniform probability distribution on the interval [−5, 5]. The three points denoted by hexagons are an illustrative subset with roughly equal spacings. The differences in scattering from the subset of three and the complete set of five points illustrate the transition from the pattern of the simple pairs of points in our examples above, to more complex situations shown in Figures 5.14 and 5.15.

Figure 5.14 shows the monostatic scattering pattern in the x–y-plane for the subset of three hexagons in Figure 5.13. Clearly, increasing the set of scatterers by addition of a single extra point, to three from two, results in a much more complex pattern structure than that in Figure 5.12. The symmetries that existed with respect to the axis of only two spheres have disappeared, although a reflection symmetry about the center of the system remains. Also, minima in the three-point pattern typically reach a level about 10 dB below the maxima of the pattern, unlike the two-point pattern with its systematic occurrence of perfect nulls. The only deep minimum of about −25 dB lying on the line between about 2 and 8 o'clock has been truncated at the −10 dB level in the center of the plot. This change in the behavior of the minima arises from the small likelihood that three equal-length vectors with arbitrary phases will add to zero, as this can happen only when the phasors form an equilateral triangle.[7]

What might we expect for characteristic lobe spacing, width, and magnitude? Consider, from Eq. (5.13) and Figures 5.11 and 5.12, that for a given baseline between two scattering objects the narrowest backscattering lobes occur perpendicular to the baseline. We can use these particular lobes to estimate the range of lobe structure that we might expect to see in a backscattering pattern from all five points in Figure 5.13. Since the most closely spaced points must contribute to the widest backscatter lobes, while the most widely separated points must contribute to the narrowest

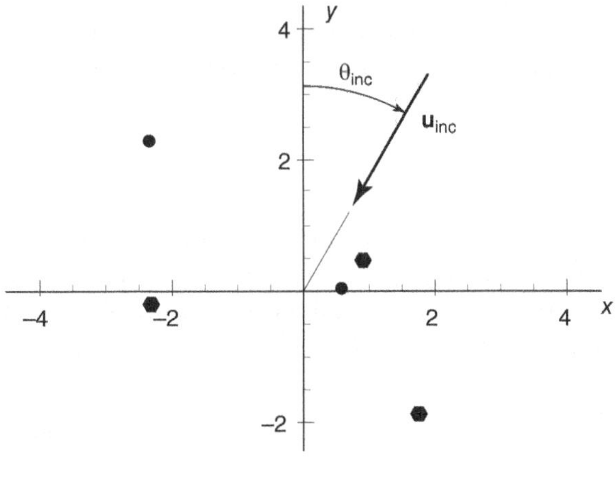

Figure 5.13 Locations of five scattering points. Points at randomly chosen locations (x, y) = (0.91, 0.47), (−2.34, 2.29), (0.58, 0.04), (1.76, −1.87), (−2.32, −0.20) for example calculation of scattering. Three points denoted by hexagons are used in calculating Figure 5.14; all five points are used to generate Figure 5.15 (see Eq. (5.7) and its text description).

7 Obviously, this simple statement applies only to our particular example in Figure 5.14. In a more complex situation in which the scattering amplitudes are unequal and the number of scatterer exceeds three, the configuration of phasors can be much more complex.

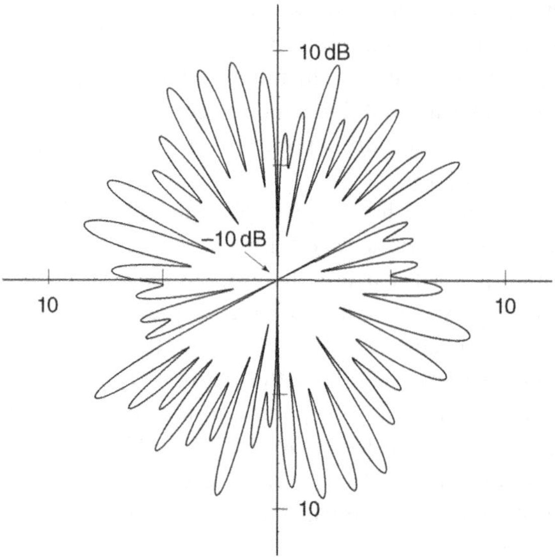

Figure 5.14 Monostatic scattering from array of three points. Plot gives the backscatter RCS of a three-element array of scatterers, normalized by the RCS of a single element. Identical scattering objects are located at the positions of the three hexagons in Figure 5.13, $(x, y) = (0.91, 0.47), (1.76, -1.87), (-2.32, -0.20)$; wavelength $\lambda = 1$ on the scale of Figure 5.13. RCS is in dB relative to the scattering from a single point; i.e., the 0 dB level corresponds to an RCS for a single scatterer. Lobe widths and peak values are consistent with expectation for the subset of three points denoted by hexagons in Figure 5.13. Two very deep minima extending to ≈ -25 dB, located in the trough between approximately 2 and 8 o'clock, have been truncated at the −10 dB level (see Eq. (5.7) and its text description).

lobes, we can estimate the least and greatest lobe widths from the extremes of the baselines present in the scattering array. For a given baseline, as an approximation to the characteristic width of the most narrow lobe, we use one-half the angular distance between the two adjacent nulls, which we calculate by using values $n = 0, 1$ in Eq. (5.14).[8]

$$\theta_{\frac{1}{2}} \approx \frac{\lambda}{4\ell} \tag{5.21}$$

What are the maximum and minimum baselines, $\ell_{\text{max}}, \ell_{\text{min}}$, for the array of three scatterers? From the plan view in Figure 5.13, the maximum spacing between any two points in the triad occurs for the pair $((1.76, -1.87), (-2.32, -0.20))$, while the minimum spacing occurs between $((0.91, 0.47), (1.76, -1.87))$. The distances are, respectively,

$$\begin{aligned} \ell_{\text{max}} &= (4.08^2 + 1.67^2)^{1/2} = \sqrt{4.41} \simeq 4.4 \\ \ell_{\text{min}} &= (0.85^2 + 2.34^2)^{1/2} = \sqrt{2.34} \simeq 2.5 \end{aligned} \tag{5.22}$$

For example, the values of $\theta_{\frac{1}{2}}$ are,

$$\frac{1}{4 \cdot 4.4} \quad \text{and} \quad \frac{1}{4 \cdot 2.5} \tag{5.23}$$

8 We note that $\theta_{\frac{1}{2}}$ approximates a quantity referred to as "full-width, half-power," meaning the full angular width between the points on the scattering lobe at which the RCS is one-half its peak value.

In degrees,

$$3.3^\circ < \theta_{\frac{1}{2}} < 5.7^\circ \tag{5.24}$$

This appears reasonable as a measure of the typical lobe width in Figure 5.14. In this instance, the broadest lobes occur roughly in the second and fourth quadrants corresponding to the direction along the axis of the three points, while the thinnest lobes are generally in the first and third quadrants, which are roughly perpendicular to the longest axis among the three points.

Figure 5.15 shows the calculated interference pattern associated with the five points of Figure 5.13. Notice that the broadest lobes have expanded while the thinnest lobes have contracted relative to those of the three-element array pattern in Figure 5.14. These changes arise from the closer spacing between the two points in Figure 5.13 located at $x = 0.58, 0.91$, and the wider spacing between the points at the extremes, $x = -2.34, 1.76$. For complex patterns, the width of the thinnest lobes is always controlled by the greatest dimension within the pattern, while the structure of broader lobes depends on the detailed locations of the remaining points.

5.4.2 Strength of RCS for Multiple Scatterers

In our example, there are either three or five scatterers of equal cross sections. If the phases of the scattered waves were to add in phase for some particular direction, then $E_{\text{tot}} = NE_{\text{scat}}$, where $N = 3, 5$. This situation is referred to as *coherent* scattering, in order to emphasize the orderly

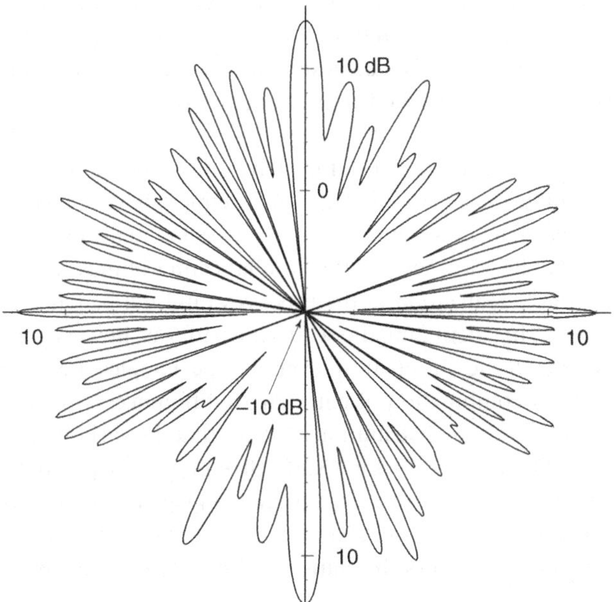

Figure 5.15 Monostatic scattering from array of five points. Plot gives the backscatter RCS of a five-element array of scatterers, normalized by the RCS of a single element. Lobe widths and peak values are consistent with expectation for the array of points in Figure 5.13 (see Eq. (5.7) and its text description).

arrangement of the phases. For aligned phasors, since $\sigma \propto |E|^2$, the resulting aggregate cross section for identical scatterers increases as the square of the number of scattering objects. So we have,

$$\left.\frac{\sigma_{\text{max}}}{\sigma_{\text{single}}}\right|_{3\text{ objects}} = 3^2 = 9 \tag{5.25}$$

$$\left.\frac{\sigma_{\text{max}}}{\sigma_{\text{single}}}\right|_{5\text{ objects}} = 5^2 = 25 \tag{5.26}$$

Note that the coherent sum produces the maximum possible aggregate cross section. Then, since ratios of 9 : 1 and 25 : 1 correspond to logarithmic changes of 9.5 and 14 dB, respectively, and comparing the logarithmic increases with the peak values seen in Figures 5.14 and 5.15, we see that the logarithmic increase of the scattering cross section of three and five identical objects summed coherently is in fair agreement with the largest values found on the corresponding plots.

Referring again to Eq. (5.20), the average, or "expected value," of the RCS of a random aggregation of a number of equal-strength scatterers is proportional to their sum,

$$\frac{\langle\sigma\rangle}{\sigma_{\text{single}}} \simeq N \tag{5.27}$$

Applying this rule to our examples,

$$\left.\frac{\langle\sigma\rangle}{\sigma_{\text{single}}}\right|_{N=3} \simeq 3 \tag{5.28}$$

$$\left.\frac{\langle\sigma\rangle}{\sigma_{\text{single}}}\right|_{N=5} \simeq 5 \tag{5.29}$$

These values correspond to increases that are close to 5 and 7 dB, respectively. Again, examining the typical values of the lowest scattering peaks in Figures 5.14 and 5.15, i.e., the values of the lowest local maxima in the scattering diagrams, we see that these are consistent in a general way with our numerical examples of scattering from identical objects located at three and five randomly selected points.

Why might we expect the statistical calculations above to work for such a small number of scattering objects? The answer lies in the nature of the observations. The two examples above are based on a random set of points. This randomly selected set is but one of many possible realizations of the random process. Calculation of standard quantities, such as the familiar center of mass or the moment of inertia of these points, from a single realization of the process would not be particularly reliable indicators of the average behavior. In calculating the scattering patterns, however, we examine the difference in path lengths among all the objects from over all aspect angles. Because each aspect angle results in a slightly different combination of path length differences, this process effectively generates a collection of pseudo-random distributions. We see that the variations in the resulting scattering patterns are predicted by Eqs. (5.18) and (5.20). Given typical separations between points that are greater than a wavelength, the use of a large number of viewing directions is sufficient to effectively sample the net result. Such phenomena provide an important basis for understanding material in later chapters where we discuss the statistics of remote sensing observations of natural surfaces.

5.5 Results for a Large Number of Scatterers

5.5.1 Distribution of the Field Amplitude and Phase

Remote sensing often applies to situations where the number of scatterers involved in a given problem is very large. We discussed the examples of multiple glints from natural terrain above. There are many others. Surveys of forests, crop fields, and glaciers come to mind, as does the monitoring of flights of birds, the use of weather radar to estimate cloud cover and precipitation rates, and radar astronomy studies of the surface of Titan and the Rings of Saturn. In each of these instances, practicality and need to combine so that the resulting radar system design results in the inclusion of multiple scattering centers within a single resolution element. For example, if the goal is to estimate the number of pears ripening on trees within an orchard, it is impractical to design a system that could isolate and identify each individual pear, but an accurate statistical measure of the expected yield in a statistical sense could be quite useful. At the same time, it can also be impractical to rely on simulation as a basis for understanding a general phenomenon. Under these circumstances, it becomes important to have statistical formulas available both to support radar design studies and for analysis of observational results.

As it happens, at the same time that the increasing size of the RCS calculations makes the problem more analytically and computationally difficult, the statistics of large numbers come into play, thereby allowing us to obtain rather accurate numbers for the expected scattering even in very complex instances. Consider the physical situation involving a large number of discrete objects, as sketched in Figure 5.16. For convenience, these might be considered as "hydrometeors," also known as raindrops(!). As above, there are many other possibilities, though. The phasor resultant, E_{tot}, and its subcomponents for the echo from an ensemble of such scatterers might look something similar to those depicted in Figure 5.17, but where the number of phasor components is very large. The total field of Figure 5.17 is given by,

$$E_{\text{tot}} = \sum_{n=1}^{N} E_n \tag{5.30}$$

in which E_n represents fields from individual scatterers.

If, in the above, λ is small as compared with the typical spacings among the randomly located scatterers, then, effectively, the phase of each scatterer is distributed uniformly over 2π. That is, when the scatterer positions are taken as randomly located on the scale of a wavelength—although

Figure 5.16 Geometry for backscatter from a large number of discrete objects. Radar transmitter and receiver are located in direction $\mathbf{u}_{\text{scat}} = -\mathbf{u}_{\text{inc}}$. Figure defines coordinates and path differences for backscattering from a large number of scattering objects. Round-trip phase difference for scattering from objects at positions r_i and r_j is calculated from geometry.

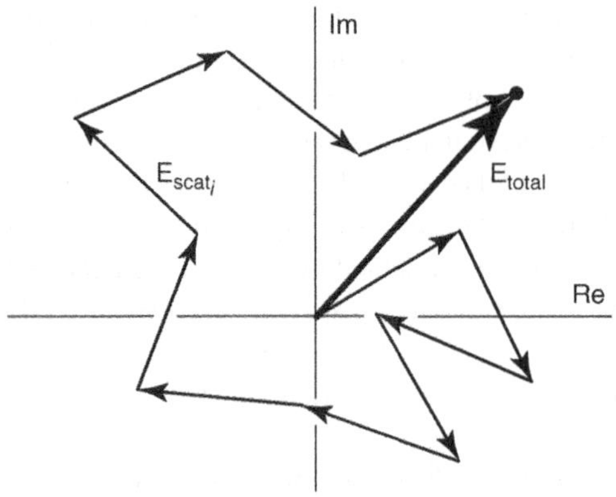

Figure 5.17 Phasor components of scattering from a large number of discrete points. Scattering of harmonic waves from a large number of points can be represented in an Argand diagram as the phasor sum of the contributions from each of the individual scatterers. Here, the eleven phasors shown are of equal length, representing equal-strength scattering from each object; in general, the phasor lengths are unequal, varying with the strength of scattering from each object in the ensemble.

individual objects are not necessarily separated by a wavelength—then for backscatter at least, all possible phases of the echoes from individual objects will be equally likely.[9]

That is,

$$p(\phi) = \frac{1}{2\pi}, \quad 0 < \phi < 2\pi. \tag{5.31}$$

where here, for example, ϕ is the net phase of the field from each scattering object. With only these assumptions E_{tot} can be shown to obey a Rayleigh probability distribution function (PDF) when the number of scatterers is large.

In order to derive the distribution function for the total field, we first express E_{tot} as a complex sum,

$$E_{\text{tot}} = \sum_{n=1}^{N} E_n e^{j\phi_n} \tag{5.32}$$

where the net phase for the field from the nth scatterer is ϕ_n and, for now, take $E_n = E_0$, a constant, so that all the scattering amplitudes are the same. Expanding the complex field by use of Euler's theorem, the total field becomes,

$$E_{\text{tot}} = E_0 \sum_{n=1}^{N} (\cos\phi_n + j\sin\phi_n) \tag{5.33}$$

In the above, the assumption that the ϕ_n are uniformly distributed over 2π, as in Eq. (5.31), implies that $\cos\phi_n$, $\sin\phi_n$ are random variables (RV) with zero mean. By the central limit theorem (CLT) the sum of the real and imaginary terms in Eq. (5.33) individually becomes a Gaussian RV for sufficiently large N. Thus, for $N \to \infty$,

$$E_{\text{tot}} = X + jY \tag{5.34}$$

9 This is because, for backscatter, the rate of change of path length with respect to change in the range to a particular scattering object is maximum; for forward scatter the total path is invariant with object position. For general geometry, the rate of change is a function of the bistatic angle.

where quantities X, Y are uncorrelated Gaussian RV with zero mean. In order to represent the results in terms of the magnitude and phase of the polar coordinates corresponding to the variables X, Y, we change variables to

$$r = \sqrt{X^2 + Y^2}, \quad \theta = \tan^{-1} Y/X \tag{5.35}$$

and transform the joint PDF for X, Y to a joint distribution of r, θ,

$$p(X, Y) = p(X)p(Y) \rightarrow p(r, \theta) \tag{5.36}$$

Calculation of the echo power requires the variance of X, Y. Applying the definition of variance and using the zero mean of $\cos \theta_n$, for θ_n still distributed uniformly over 2π,

$$X = E_0 \sum_{n=1}^{N} \cos \theta_n; \quad \mathrm{Var}(X) = E_0^2 \cdot N \cdot \mathrm{Var}(\cos \theta_n) \tag{5.37}$$

Applying the definition,

$$\mathrm{Var}(X) = \mathsf{E}[(X - \bar{X})^2] \tag{5.38}$$

where E is the expected value, and $\mathsf{E}[X] = \bar{X}$,

$$\mathrm{Var}(X) = N\,E_0^2 \cdot \frac{1}{2\pi} \int_0^{2\pi} \cos^2\phi \, d\phi = \frac{N\,E_0^2}{2} \tag{5.39}$$

Similarly,

$$\mathrm{Var}(Y) = \frac{N\,E_0^2}{2} \tag{5.40}$$

Making use of the CLT and the absence of correlation between X and Y,

$$p(X)p(Y) = \frac{1}{\pi N E_0^2} e^{\frac{-X^2+Y^2}{N\,E_0^2}} \tag{5.41}$$

Changing variables now, per Eqs. (5.35) and (5.36),

$$p(r)p(\theta)drd\theta = \frac{1}{\pi N E_0^2} e^{\frac{-r^2}{NE_0^2}} r\, dr\, d\theta \tag{5.42}$$

This expression provides the PDF for r, θ in a separable form, from which we have,

$$p(r) = \frac{2}{NE_0^2} r\, e^{\frac{-r^2}{NE_0^2}}, \quad 0 \leq r \leq \infty \tag{5.43}$$

and

$$p(\theta) = \frac{1}{2\pi}, \quad \theta_0 \leq \theta < \theta_0 + 2\pi \tag{5.44}$$

Obviously, perhaps, a similar argument in which the E_n are taken as RVs, instead of a constant, leads to a similar result, namely, that $|E_{\text{tot}}|$ is a Rayleigh-distributed RV.

Usually, in real systems, the field quantity NE_0^2 is not accessible to us directly. Instead, we work with voltages at the terminals of antennas that are proportional to the field variables. Recognizing this, we rearrange the variables slightly in order to make the terminology correspond more closely to actual observational situations. Take,

$$NE_0^2 = \langle |E_{tot}|^2 \rangle = V_0^2 \tag{5.45}$$

and rescale the amplitude,[10]

$$V = \sqrt{2}r \tag{5.46}$$

We now have a Rayleigh PDF for the resultant voltage,[11]

$$p(V) = \frac{V}{V_0^2} e^{\frac{-V^2}{2V_0^2}}, \quad V \geq 0 \tag{5.47}$$

as the voltage distribution for an echo signal scattered by a randomly located array of objects, such as that discussed above, provided that the span of spacings is large compared with a wavelength.

Some useful ancillary quantities are

$$\overline{V} = \left(\frac{\pi}{2}\right)^{\frac{1}{2}} V_0 \qquad \text{Mean value} \tag{5.48}$$

$$\overline{V}^2 = 2V_0^2 \qquad \text{Mean square} \tag{5.49}$$

$$V_{\text{med}} = (\ln 4)^{\frac{1}{2}} V_0 \qquad \text{Median value} \tag{5.50}$$

Given the above, the Var(V) is found readily,

$$\text{Var}(V) = \overline{V}^2 - (\overline{V})^2 = 2V_0^2 - \frac{\pi}{2}V_0^2 \tag{5.51}$$

5.5.2 Distribution of the RCS

Having determined the distribution of the field for scattering by a large collection of scatterers, what is the corresponding distribution of the RCS resulting from the ensemble?

$$\sigma \propto \frac{E^2}{2} \propto \frac{V^2}{2} \tag{5.52}$$

By the usual rules for the distribution of a function of an RV.

$$p(\sigma) = p(V)\left|\frac{dV}{d\sigma}\right| = p(V)\frac{1}{|V|} \tag{5.53}$$

Substituting in the results above,

$$p(\sigma) = \frac{1}{\overline{\sigma}} e^{D\frac{-\sigma}{\overline{\sigma}}}, \quad \sigma \geq 0 \tag{5.54}$$

which is an exponential distribution with the expected value of the cross section,

$$\sigma = N\overline{\sigma} \tag{5.55}$$

---

10 $r\,dr \rightarrow \frac{V\,dV}{2}$, canceling the 2.

11 The choice of V ~ voltage is for convenience in later discussions.

where again $\overline{\sigma}$ is the mean of the RCS distribution. That is, the expected value of the total RCS is the sum of the expected values of the RCS of the individual scatterers.

5.5.3 Applicability of the Rayleigh Distribution

Rayleigh statistics for signal amplitudes apply to any large distribution of randomly located scattering objects, provided that each complex bivariate scattered wave amplitude is independent of all others. While this is a useful approximation in many cases, very detailed measurements of scattered wave statistics suggest that other distributions better describe the observations. This remains an area of investigation today.

The Rayleigh distribution applies not only to all of the remote sensing examples above, but in general to any situation in which there occurs a collection of several or more glints within a resolution element from complex objects or natural scenes. Stated more precisely, Rayleigh amplitude and exponential RCS type scattering distributions are characteristic of composite echoes from a large number of *statistically independent scatterers*, taken as a group, lacking a dominant member of the population of scattering objects. Inherent in this requirement is that the members of the group scatter independently, and that any scattering interaction among the various objects in the ensemble can be neglected. This applies to widely diverse systems of radar targets.

Later in these notes, the same distributions occur again in statistical descriptions of receiver noise, when we discuss the radar detection problem. These distributions are central to understanding a number of important issues in radar systems, so it is well to have them clearly in mind.

Probability theory makes a distinction between theoretical or probabilistic *expectations* obtained from theory, such as we calculated for echo strength and RCS above, and experimental averages of observations, such as arise in practical situations. Whether or not these two are the same, or even similar, depends on the particular physical processes and systems that are involved. For example, the average of a constant voltage measured with a quantized meter will never be correct unless the value of the voltage happens to coincide with a quantization level of the device. On the other hand, the average of a fluctuating voltage measured with the same meter can be very accurate provided that the fluctuations range over several quantization levels and that a sufficiently large number of observations is included in the average. In the first instance, the expectation from probability theory and measured statistical averages are typically different, whereas in the second instance, they are very close to the same. Processes for which the theoretical expectation and measured result from a single realization of the random process are the same are said to be *ergodic*. While the formal conditions for ergodicity are complex, as a practical matter, ergodicity holds in the analysis of most radar problems. Our example calculation of the peak and average values of the scattering patterns in Figures 5.14 and 5.15 illustrates how this works. In the remaining chapters, formal calculations of probability are often compared with statistical results from simulation or observations. It is a good idea to remain aware of the distinctions between calculation and observation, and to consider for yourself the validity of such calculations each time they are encountered.

6

Antenna Parameters

Antennas perform a critical function in the operation of any radar system—the radiation and reception of electromagnetic energy. We have already seen that this results in the characteristics of the transmitting and receiving antennas appearing as two factors in the bistatic radar equation and as the square in the more common, monostatic equation in which only a single antenna is employed by both the transmitting and receiving systems. Thus, a factor of 2 : 1, say, improvement in antenna gain results in a factor of 4 : 1 increase in received signal power. Similarly, a reduction of antenna performance by one half has the effect of reducing radar system performance in terms of received echo power to just one fourth that of the original.

In addition to its role affecting the strength of a radar's signal, an antenna also collects noise from the environment in the forms of natural emissions and interference. Both natural radiation and interference can be received through the antenna "sidelobes," often from directions well away from the areas and directions of direct interest. Consequently, it is not uncommon in remote sensing applications for interference from transmitters unrelated to the radar system to be of concern in the operation of the radar and, therefore, of concern in the design of the radar with the goal of mitigating interference problems. Similarly, depending on the application and the siting of the radar, aspects of the antenna pattern structure may be of concern with respect to thermal emission from the environment. Overall optimization of the antenna design, including the gain and shape of the main beam as well as the structure of the pattern outside the primary beam, is important—often critical—to successful system operation.

Finally, the antenna system determines the polarization of the radiated signals and the degree of coupling to the polarization to received signals. For many remote sensing applications, the degree to which the polarization is controlled and understood is essential to understanding and accurate interpretation of the remote sensing observations. Strict orthogonality of transmitted and received polarizations is particularly important when observational objectives include determination of the complete polarization signature of a scattering area.

For the reasons above, a sound understanding of a remote sensing radar system requires that we understand the characteristics of the associated antenna system(s). In this chapter, we discuss the basic characteristics of antennas, their definitions, and the basic relationships between these characteristics and the antenna structure. We also develop the connections between an antenna's pattern and the degree to which noise from the external environment is presented to the receiving system. We defer a discussion of the more involved question of polarization until Chapter 16, where we develop this subject in some detail.

6.1 Antenna Parameters

Antenna effects enter the radar equation for received power, Eq. (3.1) in the form of the product,

$$G_t \cdot A_r \tag{6.1}$$

thereby comprising an important physical element of any radar system.

Consider a radiating aperture described in the coordinate system given in Figure 6.1. Assume that this antenna is the transmitting element of a radar system. The *near fields*, closer to the aperture than about $2D^2/\lambda$ have a complex structure, while the *far fields* fall off as spherical waves at the rate R^{-1}, so the power in the far field decreases as R^{-2}. In the far field, we describe the flow of energy by a simple formula for the *power density*, $P(\theta, \phi)/R^2$, in units of W/m^2, where $P(\theta, \phi)$ represents the variation of power density with direction θ, ϕ at a fixed range R. On the other hand, if dA represents the effective aperture of a receiving antenna, then the output power available at the terminals of this antenna is,

$$P_r = P(\theta, \phi) \cdot \frac{dA}{R^2} = P(\theta, \phi) \cdot d\Omega \tag{6.2}$$

where $d\Omega$ is the solid angle[1] subtended by the effective aperture of a receiving antenna at a distance R. From the above, it is clear that the units of $P(\theta, \phi)$ are watts/steradian. Rewriting Eq. (6.2), we can display and define the important engineering parameters of the transmitting antenna,

$$P_r = \underbrace{\frac{P(\theta, \phi)}{P_{rad}/4\pi}}_{D(\theta,\phi)} \cdot \underbrace{\frac{P_{rad}}{P_t}}_{\rho_{eff}} \cdot \frac{P_t}{4\pi R^2} \cdot dA \tag{6.3}$$

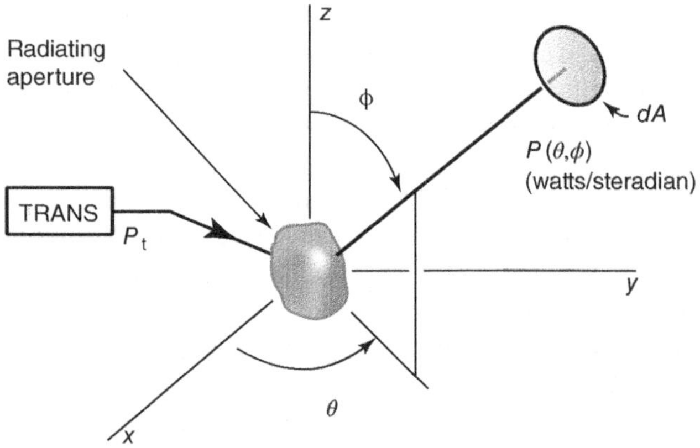

Figure 6.1 Antenna coordinates. Antenna radiation/reception patterns are typically expressed in polar coordinates fixed to the antenna structure. Here, we let θ vary from 0 to π and ϕ from 0 to 2π. Beam antennas often have the z-axis aligned with maximum response, but in practice, this choice is subject to wide variation.

1 Solid angle is defined as an area pendicular to a line of sight normalized by the square of the distance to that area. Hence, $d\Omega = dA/R^2 = \sin\theta d\theta d\phi/R^2$ is the solid angle subtended by dA at distance R. The unit of solid angle is the *steradian*. The solid angle of sphere seen from its center is 4π steradians, which is the area of the surface of any sphere divided by its radius squared.

The two new factors identified above, $D(\theta, \phi)$, and ρ_{eff}, are physically meaningful parameters of antenna performance, described below. These are obtained by multiplying Eq. (6.2) by unity in the form P_t/P_t and P_{rad}/P_{rad} and then rearranging the result, where P_t is the power supplied to the antenna by the transmitter and the total radiated power, P_{rad}, is found from the integral of the radiated power density over the solid angle subtended by an enclosing sphere,

$$P_{rad} = \int_{4\pi} P(\theta, \phi) d\Omega \tag{6.4}$$

The *directivity* of the antenna is the radiated power per unit steradian relative to the mean radiated power density on a sphere. As indicated in Eq. (6.3), this is given by,

$$\begin{aligned} D(\theta, \phi) &= \frac{P(\theta, \phi)}{P_{rad}/4\pi} = \frac{4\pi P(\theta, \phi)}{P_{rad}} \\ &= \frac{\text{power density in } \theta, \phi \text{ direction}}{\text{isotropic density of radiated power}} \end{aligned} \tag{6.5}$$

Observe that the directivity of an antenna depends only on the *radiated* power and is independent of the *input* power. Because it is insensitive to the actual power radiated, directivity is used in theoretical analyses and in performance estimates based on radiated power alone. It is useful in evaluating the relative importance of sidelobes with respect to the strength of the main beam, for example.

Similarly, the *radiation efficiency* of an antenna connects the radiated power density to the antenna input power,

$$\rho_{eff} = \frac{P_{rad}}{P_t} \tag{6.6}$$

This ratio is always less than unity since there is always some power dissipated by heating of the conducting and lossy dielectric components of the antenna structure.

The *gain* of an antenna links the input power directly to the radiated power density. It is defined as the radiated power per steradian relative to the antenna input power radiated isotropically. That is, from the above,

$$\begin{aligned} G(\theta, \phi) &= \rho_{eff} \cdot D(\theta, \phi) = \frac{P(\theta, \phi)}{P_t/4\pi} \\ &= \frac{\text{power density in } \theta, \phi \text{ direction}}{\text{isotropic density of input power}} \end{aligned} \tag{6.7}$$

Note that the difference between Eqs. (6.6) and (6.8) lies in the use of radiated power, P_{rad}, and input power, P_t, respectively, as the reference quantity for normalizing the pattern. "Gain" is used in the radar equation, and commonly in radio signal link calculations, because the input power is more readily accessible by measurement or specification than is radiated power.[2]

Sometimes you will see the radar equation written in a more explicit, detailed form such as,

$$P_r = \frac{P_{trans} L_1 L_2 \cdots L_n \cdot G_t \cdot A_R}{(4\pi)^2 R^4} \cdot \sigma \tag{6.8}$$

2 Note the difference in the use of the term "gain" here from the sense in which it applies to the active gain of amplifiers. Amplifier gain is the effective increase in the power of a signal by control of power sources. Antenna gain describes a passive concentration of available radiated power into a particular angular region of space.

In this expression, P_{trans} represents power at the output of a transmitter, while the L_i represent losses in various elements between the transmitter and its antenna, and also between a receiving antenna and the attached receiver. For calculation of radiated power, the detail shown in Eq. (6.8) is essentially a bookkeeping issue. In the development, aforementioned, all transmission losses are lumped into

$$P_t = P_{trans} \cdot L_1 \cdot L_2 \cdots \cdots L_n \tag{6.9}$$

where we have taken the measure of transmitter power, P_t, as the power delivered to the antenna terminals, or in the case of waveguide connections, to the antenna connection flange, either of which serves as a well-defined reference location for determination of the gain and antenna efficiency. Consequently, here the transmitter power, P_t, is that reaching the antenna reference connector. Use of this convention avoids issues not directly related to the performance of the antenna structure itself.

Problems with keeping track of losses in an overall system arise because real systems are assembled from real components. A transmitter engineer specifies the transmitter power as that at the terminals or waveguide flange coming out of his box, and responsibly so, since he may have no idea what will be connected at this point. Then the system designer adds transmission lines and matching networks to transfer the output power from the transmitter "box" to the antenna, which could be designed and constructed by yet a third party. That third party, the antenna engineer, guarantees the performance of the antenna from the reference terminals available to him or her—the antenna terminals or waveguide flange. It is the system designer who must account for the added elements about which neither the transmitter nor the antenna engineer necessarily has any knowledge. Hence, the multitude of loss factors, the L_i, each accounting for some additional effect—none of which improves performance—in the radar equation. Similar considerations apply to the receiving side. In practical systems, these are extremely important issues that can easily account for factors of two or more in performance.

6.2 Aperture Antennas

Aperture antennas are structures possessing a well-defined "window," or *aperture*, through which emitted and received radiation passes. Usually, the aperture is determined by the physical characteristics of the antenna, and is apparent on inspection. Typically, the apertures are many wavelengths in size—a minimum of 5–10 λ—in at least one dimension. This is not always the case since, for example, the radiating open end of a λ-scale waveguide also falls into this category of device. Perhaps the most common types of aperture antennas are waveguide "horns," named for their similarity in appearance and function to acoustic horns, and "dishes," usually parabolic reflecting structures with displaced primary radiation sources known as "feeds." In modern radar applications, planar and conformal arrays must also be considered. These are collections of small (λ-scale) radiating elements arranged over some well-defined area or volume, and excited or sampled coherently to form transmitting and receiving beams. Such arrays can be very compact and lightweight for aircraft weather radars, say, or they can be massive structures for complex applications requiring rapid beam changes or the formation of multiple simultaneous beams from ship- or shore-based systems.

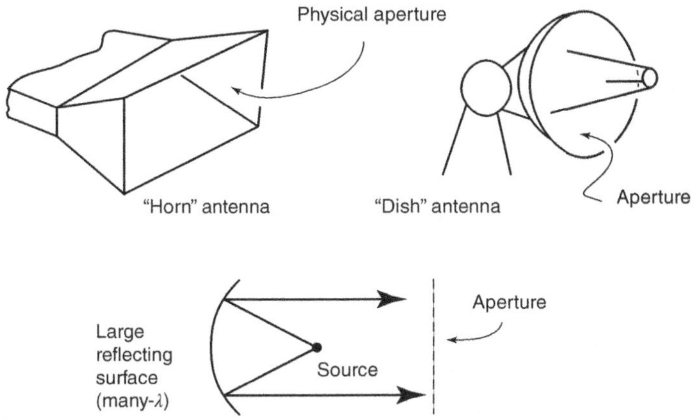

Figure 6.2 Aperture antennas. Example aperture antennas take various forms. The horn antenna was developed at Stanford; these structures provide direct transition from waveguide to space. Dish antennas make use of a secondary reflector to convert spherical wave from a source "feed" antenna to plane wave across the aperture. There are many other variations, however.

Figure 6.2 shows examples from the three types just mentioned. For such antennas, we make use of the relationship, Eq. (3.7), connecting gain and effective aperture to calculate the gain based on the size of the physical structure, which we extend here,[3]

$$G = 4\pi\frac{A_{\mathrm{eff}}}{\lambda^2} = 4\pi\eta\frac{A_{\mathrm{phys}}}{\lambda^2} \tag{6.10}$$

where the new parameter η is the *aperture efficiency* of the antenna and A_{phys} is the actual area of the physical aperture. The aperture efficiency represents the effectiveness, for its physical size, of a given antenna in collecting energy from an incident wave, and, alternatively, the ability of the antenna to concentrate radiated energy in a particular direction when transmitting. When expressed in this manner, the aperture efficiency implicitly incorporates the losses represented by ρ_{eff} in the relationship between gain and directivity.

For typical aperture antennas $\eta \approx 0.5$–0.6, but can be in the range of 0.8–0.85 for advanced designs. The aperture efficiency obtained depends largely on the distribution of the electromagnetic field within the aperture, and approaches unity when the field strength is uniform across the radiating area. As the characteristics of the antenna radiation pattern, such as beamwidth, gain, and other performance parameters, also depend on the aperture field distribution, aperture efficiency must be traded off against overall system performance. The presence of conductive and other losses in the antenna contributes to a loss of aperture efficiency. Usually, the dissipative losses are relatively low for aperture antennas. In any event, dissipative losses rarely have a significant effect on the pattern.

The proportionality between effective aperture, A_{eff} for receiving and the gain, G, Eq. (6.10), implies that the shapes of the transmitting and receiving patterns of an antenna are the same. Thus, in a direction for which the gain is maximum, the effective receiving aperture is also maximum. Similarly, for directions in which the transmitted signal is relatively weak, the effectiveness of an antenna in receiving signals arriving in the opposite direction is also relatively weak, in the same proportion as for transmitting. In typical usage, the terms "gain," "area," "aperture," and "efficiency" are intended to refer to maximum values of these quantities. Still, one can speak of the

3 Proofs of this remarkable relationship can be found in standard introductory works on antennas, e.g., *Stuzmann & Thiele* or *Balanis* (see Bibliography).

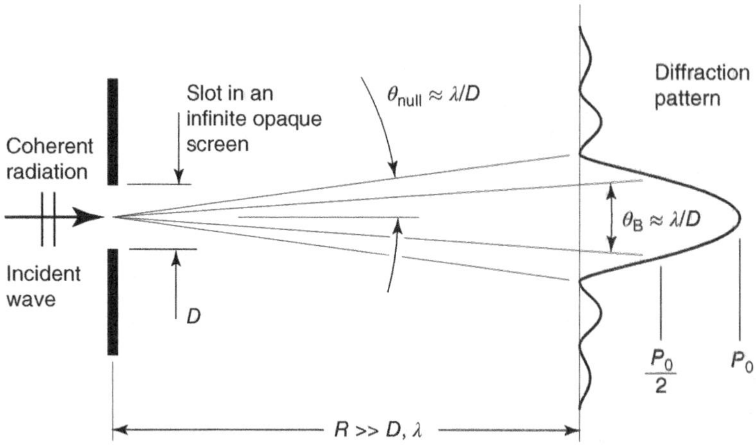

Figure 6.3 Diffraction from aperture antennas. Radiation patterns of aperture antennas can be calculated by application of diffraction theory, such as that used to determine diffraction pattern of a slot. For a rectangular aperture, such a slot in an opaque screen, as mentioned earlier, the angular distance from the beam maximum to the first zero of the pattern, θ_{null}, is the same as the angular distance between opposite points on either side of the beam at which the power pattern is one-half its maximum, θ_B.

"gain or the effective aperture of the second sidelobe," for example, in referring to the effectiveness of a minor feature in the antenna pattern.

6.2.1 Beamwidth of Aperture Antennas

By design, the electromagnetic fields in the aperture of large antennas approximate those of the opening of an otherwise opaque diffracting screen. From this observation, it is convenient to think of the patterns of such antennas as the diffraction pattern of the aperture fields. To first order, this provides accurate approximations of the far-field response of aperture antennas. Figure 6.3 provides a sketch of such a pattern in relation to the source aperture. A large aperture, diameter $D >> \lambda$, results in a sharply limited beam accompanied by sidelobes. For uniformly illuminated rectangular apertures, the minimum in radiated power located at the smallest angular distance from the maximum, referred to as the first null, occurs at angle λ/D from the peak of the main beam. Coincidentally, the full-width at half-power (FWHP) has the same angular dimension, as illustrated. Stated another way, for a rectangular aperture, the angular distance from the beam peak to a point at which the radiated power is one-half the maximum power happens to be one-half the distance from the peak to the first null. If the pattern is expressed in decibels relative to the peak, the one-half power level corresponds to a decrease in the antenna's response by 3 dB on the log scale. On the other hand, the null corresponds to an infinite logarithmic decrease! So the angular distance from the direction of peak gain to the point of a 3 dB reduction is the same as the angular distance from the 3 dB loss point to the point of infinite logarithmic loss! This comparison indicates just how sharp the beams of aperture antennas can be, and the necessity for accurate pointing of the antenna beam with respect to the angular scale of λ/D.

6.2.2 Pattern Dependence on Aperture Fields

More general relationships between quantities such as the beamwidth and the locations of the nulls depend on the aperture shape. We can calculate the antenna illumination pattern for a given aperture using *Huygens' principle*, which states that each point in the radiating aperture may be considered the source of a spherical wavelet, and that the total field at any position in space is the sum of all such wavelets, integrated with due respect to phase over the radiating aperture. The idea is shown in Figure 6.4.

As an example, the nulls of an aperture antenna are the result of destructive interference among radiated wavelets from each incremental element of area within the aperture. The first null of a uniformly illuminated aperture occurs, for example, at the angle for which the phase paths from opposite sides of the antenna differ by one wavelength, Figure 6.5. Real antennas do not possess infinitely deep nulls, of course, because the aperture field distributions are not formed perfectly and also because the aperture fields are perturbed by structural elements not considered in simple models, but the angular scale illustrated by the example is meaningful nevertheless.

We now generalize Huygens' principle to find the shape of the pattern in arbitrary directions from the aperture. As suggested by Figure 6.4, if the field at distance ρ from each infinitesimal element of the aperture is proportional to an amplitude and phase propagation factor,

$$\frac{e^{\left(-j\frac{2\pi}{\lambda}\rho\right)}}{\frac{2\pi}{\lambda}\rho} \tag{6.11}$$

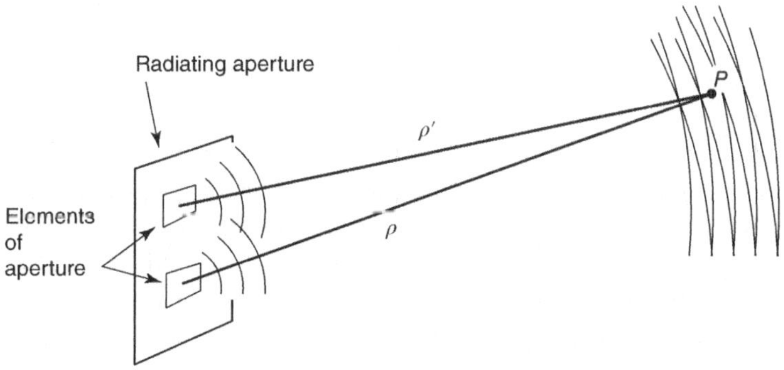

Figure 6.4 Huygens' principle. The total field present at an observation point P is the phasor sum of the fields from spherical waves centered at each infinitesimal element of a radiating aperture. Radiation from two such aperture elements results at a distance, e.g., at point P, in the interference of two spherical waves. The wave intensity at P can be either enhanced or reduced, depending on the initial phases of the radiating elements and the difference in the path lengths, ρ and ρ'.

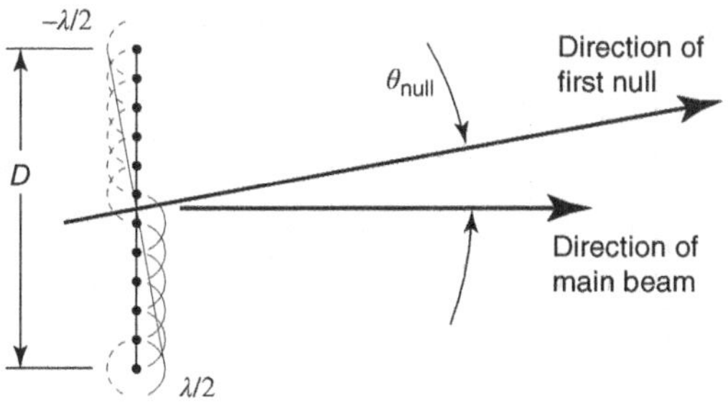

Figure 6.5 Phase delay at first null. Small arcs with free space radii of $\lambda/2$ represent radiation from each point in the aperture. For uniformly illuminated rectangular antennas, shown here in cross section, the formation of the first null occurs at the angle for which there is one cycle of phase variation across the aperture. This results in complete cancellation of the radiation by mutual interference among waves radiated from different parts of the aperture, as can be seen by comparing pairings corresponding to positive and negative phase. For the geometry illustrated, for each point in the aperture with a given positive value of phase there, is a corresponding point with the same value but negative sign. The result is complete pairwise cancellation of the fields, resulting in the radiation pattern null.

adding up the fields from each elemental area in the aperture, the total field at an observation point P can be represented by,

$$\mathrm{E}(P) = \int_{\text{aperture}} f(x,y)\frac{e^{-j\frac{2\pi}{\lambda}\rho}}{\frac{2\pi}{\lambda}\rho}\,dxdy \tag{6.12}$$

where $f(x, y)$ is the complex representation of the excitation at the aperture, x, y are the coordinates in the plane containing the aperture, and ρ is a function of the location P and the radiating point x, y in the aperture. Although simple in form, Eq. (6.12) can be difficult to evaluate analytically. We are able to compute patterns more easily if we use the *Fraunhofer approximation*, which applies in most practical instances in which the observation point P is at a large distance from the aperture. Under this condition, there is a Fourier transform relationship between the aperture shape and the antenna pattern.

To see how the Fourier relationship develops, consider the geometry of a one-dimensional aperture as shown in Figure 6.6, where each aperture element dx radiates a spherical wave described by Eq. (6.11) to an observation point P. We can express ρ as an exact function of the source position in the aperture, x, by applying the Pythagorean theorem to calculate ρ and ρ_0, and then taking the difference of these. The result is,

$$\begin{aligned}\rho^2 &= (x - x_P)^2 + \rho_0^2 - x_P^2 \\ &= \rho_0^2 - 2x \cdot x_P + x^2\end{aligned} \tag{6.13}$$

where the distance R along the z axis was eliminated in the subtraction. For P located at great distances from the aperture, we can consider the rays arriving from the aperture to be essentially parallel. In order to take advantage of this condition, first assume small $x \ll x_P$, which as a practical matter is always the case[4] even for small values of x/ρ_0 when $R \ggg x_{\max}$, where $x_{\max}$ is the maximum extent of the aperture from the z-axis. So, ignoring x^2 in comparison with $2x \cdot x_P$,

$$\rho^2 \doteq \rho_0^2 - 2x \cdot x_P \tag{6.14}$$

$$\rho \doteq \rho_0\left(1 - \frac{2x \cdot x_P}{\rho_0^2}\right)^{1/2} \tag{6.15}$$

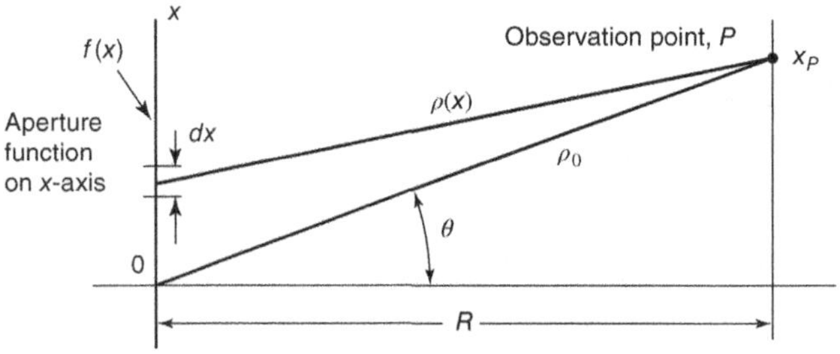

Figure 6.6 Fraunhofer geometry. Assuming that both the aperture size and the displacement of the point P from the origin are small compared to the observation distance, the total field may be calculated as the Fourier transform of the aperture illumination function $f(x)$.

---

4 This is a fine point. We are most interested in the variation of the pattern with angle, rather than with x_P per se. Our approximation, Eq. (6.14) fails when $|x| > |x_P|$. But the expression in Eq. (6.17) remains valid as a result of our assumption that $R \ggg x_{\max}$. This is because ρ is not sensitive to $|x - x_P|$ when $\theta \approx 0$.

and, since $x \cdot x_P << \rho_0^2$,

$$\rho \doteq \rho_0 - \frac{x \cdot x_P}{\rho_0} \tag{6.16}$$

Now, since for $x_P \lll R$, $\sin\theta \doteq \frac{x_P}{\rho_0}$, we can express ρ in terms of the angle from the z-axis as,

$$\rho \doteq \rho_0 - x\sin\theta \tag{6.17}$$

Thus, applying Huygens principle, our integral for the field at P becomes,

$$\mathrm{E}(P) = \int_{-\infty}^{\infty} f(x)\,\frac{e^{-j\frac{2\pi}{\lambda}(\rho_0 - x\sin\theta)}}{\frac{2\pi}{\lambda}\rho}\,dx \tag{6.18}$$

Again, when $x_P \ll R$ we can ignore the small amplitude variation caused by variations in the denominator across the aperture. We can also remove the constant phase represented by the term ρ_0 in the exponential since this does not affect the shape of the pattern. With these changes,

$$\mathrm{E}(\theta) = \frac{\lambda}{2\pi\rho_0} e^{-j\frac{2\pi}{\lambda}\rho_0} \int_{-\infty}^{\infty} f(x) e^{-j2\pi x\frac{\sin\theta}{\lambda}}\,dx \tag{6.19}$$

Thus, the field pattern E(θ)—expressed as an angular function—is seen to be the Fourier transform of the illumination function $f(x)$, with $\frac{\sin\theta}{\lambda}$ as the argument. The corresponding development in 2D with the external constants suppressed results in,

$$\mathrm{E}(\theta_x, \theta_y) \propto \iint_{-\infty}^{\infty} f(x, y) e^{-j2\pi\left(x\frac{\sin\theta_x}{\lambda} + y\frac{\sin\theta_y}{\lambda}\right)}\,dxdy \tag{6.20}$$

where the new quantities, θ_x and θ_y are the from the z-axis measured in the xz and yz planes, respectively.

So we see that for any illumination function, the pattern may be derived by considering its transform. Since E(θ_x, θ_y) is a field quantity, its magnitude must be squared to give the power pattern, $|\mathrm{E}(\theta_x, \theta_y)|^2$. For example, in one dimension, if the illumination function is a "pulse" function of length ℓ, i.e., rect(x/ℓ), the corresponding field pattern is sinc(ℓ sin θ/λ), while the power pattern is $|\mathrm{sinc}(\ell / \sin\theta/\lambda)|^2$. In this example, we could just as well have squared the "sinc" as this is a real function. This is a special instance of the general rule, and is not usually the case except in simple examples.

We can compare the radiation patterns of the rectangular and circular apertures, by analyzing their respective Fourier transforms. For the rectangular aperture in two dimensions, the results above apply independently in each dimension, as can be seen from Eq. (6.20), and as illustrated in Figure 6.7. The angular beamwidth of a uniformly illuminated rectangular aperture is $\lambda/W \times \lambda/H$, where W and H are the aperture width and height, respectively; note that the beam is more narrow/wide in the plane of the greater/lesser dimension.

The FWHP beam from a uniform circular aperture is slightly larger than 5/4 greater than that of a rectangular aperture of the same nominal size. Referring to Figure 6.8,

$$\theta_B \simeq 1.22\frac{\lambda}{D} \quad \text{FWHP} \tag{6.21}$$

$$\theta_{\text{null}} \simeq 1.22\frac{\lambda}{D} \quad \text{peak-to-null} \tag{6.22}$$

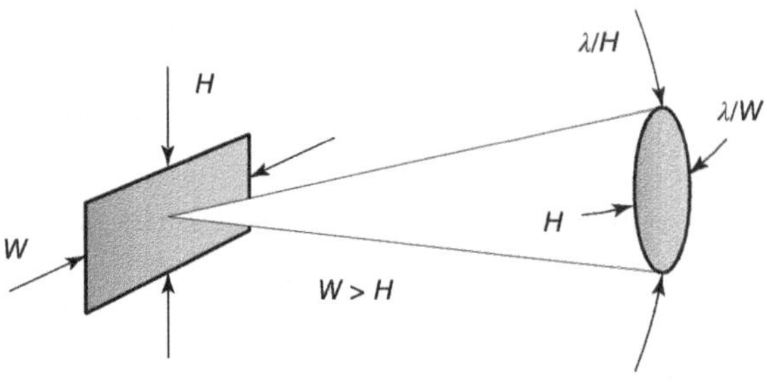

Figure 6.7 Beamwidths of a rectangular aperture.

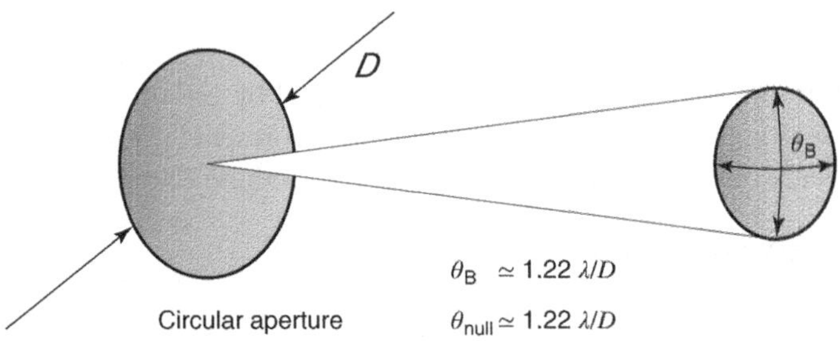

Figure 6.8 Beamwidth of a circular aperture. Eliminating the corners of a uniformly excited rectangular aperture to form a circle one of the same maximum dimension has the effect of broadening the resulting antenna beam by 22%. At the same time, the beam becomes rotationally symmetric about the perpendicular to the aperture. The equivalence between the FWHP width of the beam and the distance to the first null also applies to circular beams from uniformly illuminated apertures.

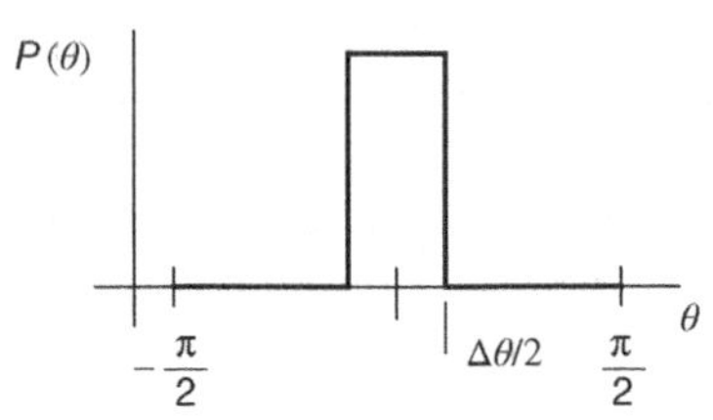

Figure 6.9 "Top Hat," or pulse antenna pattern. This type of pattern represents an idealized antenna for which the gain is constant within an angular distance of $\Delta\theta/2$ of $\theta = 0$. Viewed in three dimensions, the pattern is a cone with a cap of constant radius from the origin. There are no sidelobes.

Other shapes typically lead to more complex results. Many useful results for beam shape can be expressed in terms of well-known Fourier transform relationships between the aperture field distribution and the far-field pattern.

6.2.3 Gain and Directivity of a Narrow Beam

It is helpful to have some simple approximations for directivity of an antenna. If we assume that a narrow beam can be modeled as a pulse function of angle θ, then all the power is confined in space to a cone. Figure 6.9 gives the idea where the solid angle of the beam is taken to be $\Delta\Omega$.
The power flux in the beam is,

$$P_{\max}(\theta, \phi) \simeq \frac{P_{\text{rad}}}{\Delta\Omega} = \frac{P_{\text{t}}\,\eta}{\Delta\Omega} \tag{6.23}$$

from which we find the directivity,

$$D = \frac{4\pi}{\Delta\Omega} \tag{6.24}$$

Assuming that we have an estimate of the efficiency, η, we can also estimate the gain,

$$G \simeq \frac{P_t \eta}{\Delta\Omega} \cdot \frac{1}{P_t/4\pi} = \frac{4\pi \cdot \eta}{\Delta\Omega} \tag{6.25}$$

Following this model, a rectangular beam of angular dimensions $\theta_1 \times \theta_2$ has directivity,

$$D = \frac{4\pi}{\theta_1\theta_2} \simeq 4\pi\frac{W}{\lambda} \cdot \frac{H}{\lambda} = \frac{4\pi A}{\lambda^2} \tag{6.26}$$

which corresponds exactly to Eq. (3.7) for an ideal antenna with a uniformly illuminated rectangular aperture, and to Eq. (6.10) for an antenna either with loss or for which the aperture is not employed with maximum efficiency.

For a more realistic Gaussian beam, the model is less definitive,

$$D \simeq \frac{\pi^2}{\theta_\beta^2} \simeq \frac{4\pi}{\theta_\beta^2} \simeq \frac{16}{\theta_\beta^2} \tag{6.27}$$

where θ_β is the measure of the gaussian width. All of these approximations are found in the literature.

6.3 Examples

6.3.1 Gain of Large Aperture Antennas

A few examples can help give a feel for the accuracy of the approximations just discussed, as well as for the performance of aperture antennas. First, consider the Big Dish on the hill behind Stanford campus.[5] The diameter of the Big Dish is 46 m. Using this value and an efficiency of 50%, the gain at 300 MHz is,

$$G_{46\,\mathrm{m}} = \left.\frac{4\pi\eta A_{\mathrm{phys}}}{\lambda^2}\right|_{\eta=0.5,\,\lambda=1\mathrm{m}} \simeq 0.5 \cdot 20,884 \simeq 10^4 \quad (40\ \mathrm{dB}) \tag{6.28}$$

Alternatively, we can estimate the directivity from the estimated beamwidth. Using one half-power beam diameter from Eq. (6.21) and estimating the beam area as $\Delta\Omega \approx (\pi) \times \left(\frac{1}{2} \cdot \mathrm{FWHP}\right)^2$,

$$D_{46\,\mathrm{m}} \simeq \frac{4\pi}{\pi(0.61\lambda/D)^2} = 2.3 \times 10^4 \quad (43.6\ \mathrm{dB}) \tag{6.29}$$

which is a little more than a factor of two larger than the result in Eq. (6.28), primarily because we have not accounted for the aperture efficiency. If we assume a 50% antenna efficiency, as in Eq. (6.28), and apply this to our estimate of directivity, then we find,

$$G_{46\,\mathrm{m}} = 2.3 \times 10^4 \cdot \frac{1}{2} \quad (40.6\ \mathrm{dB}) \tag{6.30}$$

5 This antenna was constructed by Stanford Research Institute, now SRI-International, and was the largest antenna of its type at the time it was erected. Errors in the shape of the antenna surface and the use of a wire mesh surface limit its operation to frequencies below about 2 GHz.

This second estimate is close to the first, since the primary assumptions are now the same. Note, however, that we did not account carefully for aperture distribution estimating D from the ratio of the beamwidth, and this accounts for the remaining difference. As illustrated, one can expect to get within about 1 dB with a good idea of the efficiency and these approximations.

Gains can be considerably larger than in the Big Dish example. Consider the NASA Deep Space Network (DSN) 70 m antennas. There are three of these located in California, Australia, and Spain, so that they separated around the globe by approximately 120° of longitude. This arrangement guarantees almost continuous viewing of any point in the sky with low-to-moderate declination angles by at least one antenna.[6] This facility is equipped to operate as a radar at wavelengths of 0.036 m. The California site is referred to as "Mars" because the antenna and associated systems were built originally to support the NASA Viking Missions to the Red Planet. For the Mars system,

$$G_{70\,\mathrm{m}} = 0.6\frac{\pi^2 D^2}{\lambda^2}\bigg|_{\lambda=0.036} \simeq 22.3\times 10^6, \quad (73.5\ \mathrm{dB!}) \tag{6.31}$$

The more than seven orders-of-magnitude gain of this antenna is truly impressive; it is more than four orders of magnitude greater than the "Big Dish" at Stanford. The DSN 70 m antennas can operate at wavelengths of 1 cm or shorter, although the effective diameter is somewhat less than the full 70 m at these higher frequencies.

Antennas need not be the monsters just discussed—most, obviously, are not. For comparison, consider a small dish operating at the same wavelength at the DSN Mars Station.

$$G_{0.5\mathrm{m}} \approx 0.5\frac{\pi^2 D^2}{\lambda^2}\bigg|_{\lambda=0.036} \simeq 951 \quad (29.8\ \mathrm{dB}) \tag{6.32}$$

We can see that the benefit of the additional aperture is considerable.

6.3.2 A Radar Example: The Big Dish "Looking" at the Moon

With a better understanding of the antenna parameters and having just calculated a few representative gains for aperture antennas, we can do an interesting radar example by asking what is the strength of a moon echo using the Stanford Big Dish as a monostatic radar? Since the Big Dish antenna has significant gain/directivity, even at 1 m wavelength, the moon is an easy target for a radar using this antenna, even with transmitter powers of only 1 kW or so.

For this problem, we use the radar equation in the form Eq. (3.11), but we separate the radar cross section (RCS) from the other factors:

$$P_\mathrm{r} = \left\{\frac{P_\mathrm{t} G_\mathrm{t}^2 \lambda^2}{(4\pi)^3 R^4}\right\}\sigma \tag{6.33}$$

For purposes of the example, assume that $\lambda = 1$ m, $R = 400,000$ km, and the transmitter power into the antenna is 100 kW. We show the factors in the radar equation are shown below, organizing the calculation in decibels. First, we evaluate the quantity in braces,

6 The DSN 70 m antennas are not quite uniformly spaced. The greatest spacing is between Australia and Spain, which results in a small coverage gap there.

$$
\begin{aligned}
P_{\rm t} &= 10^5 && 50\ \text{dbW} && (80\ \text{dBm}) \\
G_{\rm t}^2 &\simeq 10^8 && 80\ \text{dB} && \\
\lambda^2 &= 1 && 0\ \text{dB} && \\
(4\pi)^3 &= 1984 && -33\ \text{dB} && \\
R^4 &= 2.56 \times 10^{34} && -344\ \text{dB} && \\
P_{\rm r}|_{\sigma=1} &= 2.0 \times 10^{-25} && -247\ \text{dbW} && (-217\ \text{dbm})
\end{aligned}
\tag{6.34}
$$

With this, we can now write an expression for the received power both in terms of decibels and as a simple product, depending on whether or not we prefer addition or multiplication. Here, we give the result in the four common units of radio frequency power.

$$
\begin{aligned}
P_{\rm r} &= \sigma \times (1.95 \times 10^{-25}) && \text{W} \\
&= \sigma \times (1.95 \times 10^{-22}) && \text{mW} \\
&= 10\log(\sigma) - 247 && \text{dBW} \\
&= 10\log(\sigma) - 217 && \text{dBm}
\end{aligned}
\tag{6.35}
$$

Our only question with regard to the radar "looking at the moon" now is, "What value do we use for σ?" We estimated this earlier, but repeat it here. The formula is Eq. (3.15),

$$\sigma_{\rm moon} = \simeq 0.07\pi(1.736 \times 10^6)^2 = 6.63 \times 10^{11}\ \text{m}^2 \qquad (118\ \text{dB}) \tag{6.36}$$

From this, either adding the dB table or multiplying out the numbers,

$$P_{\rm r} \simeq 1.3 \times 10^{-13}\ \text{W} \qquad (-129\ \text{dbW}) \tag{6.37}$$

How "much" signal is this? One way to think about it is to ask what voltage is present at the input terminals of the receiver. RF systems typically employ 50 Ω as a standard impedance. We might consider the voltage produced across a matched load, as suggested by Figure 6.10. Working this out for $P_{\rm r}$, as mentioned earlier, $v_{\rm rms} = 2.5\mu$V. This is the voltage produced by our lunar echo at the terminals of the receiving antenna. Similarly, the transmitter voltage at 50 Ω terminals of the transmitting antenna is $\simeq 2,200$ V. The ratio of the receiving to transmitting voltages is about 1.1×10^{-9}. The ratio of the receiving to transmitting power is much less, only about 1.3×10^{-18}. Keep in mind that, for our example, the received power is not particularly weak nor is the transmitter power particularly strong. Both of these can be orders of magnitude more extreme than the values chosen here. Design of systems that can accommodate such an enormous dynamic range is a primary challenge to radar engineers.

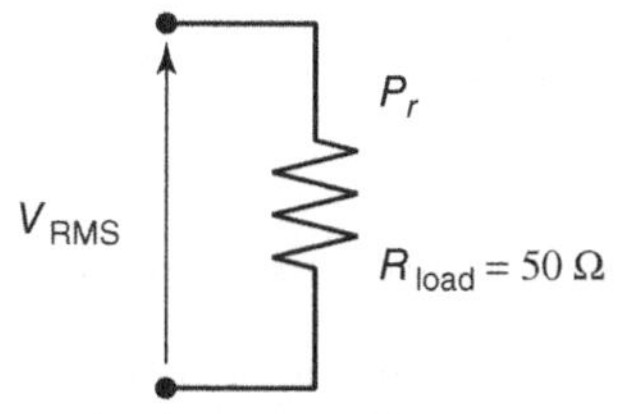

Figure 6.10 Noise across 50 Ω resistor.

6.4 Noise in Antennas

We are now in the position of complete the last link in our understanding of the origins of noise in radar equation. Thus far, we have treated random noise receiver noise at the system level by including all noise effects in the system noise temperature, T_{sys}, which we refer to the input of the receiving system, Section 3.5.2. We have also characterized electronic noise arising from active and passive circuits elements, Section 3.6. We must also consider, as part of T_{sys}, noise that originates externally to the radar receiver and enters the system through the antenna. Sources of such noise include thermal radiation from the earth, atmosphere, and sky, as well as the cosmic background radiation. Noise from these sources is in addition to the random thermal processes in the receiving system considered thus far.

6.5 Natural Sources Characterized

6.5.1 Brightness

Consider the source/receiver configuration shown in Figure 6.11. The radio *brightness* of the source is defined to be the electromagnetic flux density (power per unit area) at the receiver per unit solid angle of source. By dimensional analysis,

$$\left(\frac{\text{Power}}{\text{m}^2}\right)_{\text{at collector}} \times \left(\frac{1}{\text{m}^2/R^2}\right)_{\text{at source}} = \frac{\text{Power} \cdot R^2}{\text{m}^2 \cdot \text{m}^2} = \text{Brightness}.$$

From this we can also see that an equivalent definition of brightness is power per unit area of source per unit solid angle of receiver. A variant of this concept is *monochromatic brightness* defined as brightness per unit frequency. Often, the term "brightness" is only used, without the monochromatic qualifier. Usually, the intention is clear from the context.

Compare the definition of brightness with that of *flux density* or, equivalently, *wave intensity*, which is simply power per unit area. Brightness is suitable for extended sources, while flux is suitable for point sources, as shown in Figure 6.12. In the figure, each cross-sectional area (S, S', S'')

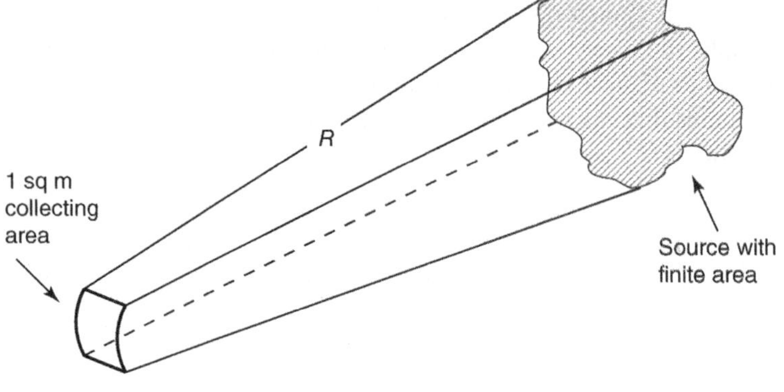

Figure 6.11 Radiation from a natural source. Radiation from an extended natural source is made up of a superposition of independent random wavefronts radiated from each point on the source, as seen from the observer. In this situation, the average power density seen by a distant observer is the sum of the power densities from each element of the source. This is conveniently described in terms of the total power density at the observer, or power per unit area, e.g., in units of W/m^2. N.B. This definition is independent of R.

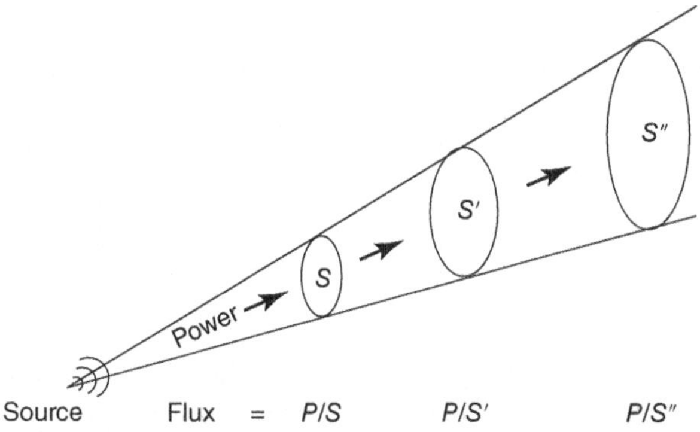

Figure 6.12 Power flux from a point source. Radiation from a point source results in a spherical wavefront in which the power density falls off as R^{-2}. This is equivalent to constant total flux or power within a cone whose vertex is at the source. For a series of cross-sectional areas within a cone, such as S, S', and S'' illustrated. The power intercepted by all three areas is the same.

intercepts the same total power, but the flux density decreases as energy propagates away from the source. The power intercepted by a receiver at any point is the flux density times the effective aperture of the receiver, a relation which we have already used many times when calculating the power received by an antenna.

6.5.2 Blackbody Radiation

A *blackbody* is an idealized object that absorbs all electromagnetic energy at all wavelengths impinging upon it. Since an object in thermal equilibrium emits energy at the same rate that it absorbs energy, a blackbody also radiates electromagnetic energy at all wavelengths. Although no blackbodies exist in nature, many objects (e.g., the sun) behave approximately like blackbodies over a range of frequencies. From Planck's Law, the total power per unit frequency radiated isotropically from a unit area of blackbody at temperature T is

$$P_\nu = \frac{2\pi h\nu^3}{c^2}\frac{1}{e^{h\nu/kT}-1}\ \mathrm{W\,s\,m^{-2}} \tag{6.38}$$

where ν is the frequency, h is Planck's constant (6.63×10^{-34} J s), and k is Boltzmann's constant (1.38×10^{-23} J K^{-1}). The MKS unit used for P_ν is the jansky,[7] where 1 jansky = 1 Jy = 10^{-26} W m^{-2} Hz^{-1}. At long wavelengths, i.e., those for which $h\nu \ll kT$ or, equivalently, $\nu \ll 2 \times 10^{10}\ T$, the Rayleigh–Jeans approximation gives,

$$P_\nu \approx \frac{2\pi h\nu^3}{c^2}\frac{kT}{h\nu} = \frac{2\pi kT}{\lambda^2} \tag{6.39}$$

What is the monochromatic brightness of a blackbody? To find this, consider the radiation from a spherical blackbody of radius a. Conservation of energy requires that the power emitted from the blackbody be equal to the power intercepted by a concentric sphere of radius R. Since energy is radiated from a blackbody isotropically, we can write

$$4\pi a^2 P_\nu = 4\pi R^2 P'_\nu \tag{6.40}$$

7 Named after Karl G. Jansky, who discovered early in 1932 that the galaxy is a source of radio emissions and hence became the "Father of Radio Astronomy."

where P_ν is the power per unit area per unit frequency leaving the surface of the blackbody and P'_ν is the power at any point a radius R. Rearranging, we have,

$$\frac{P'_\nu}{P_\nu} = \left(\frac{a}{R}\right)^2 \tag{6.41}$$

and substituting for P_ν from Eq. (6.38),

$$P'_\nu = \frac{2\pi h\nu^3}{c^2} \frac{1}{e^{h\nu/kT} - 1} \left(\frac{a}{R}\right)^2 \text{ W m}^{-2}\text{ Hz}^{-1}. \tag{6.42}$$

Equation (6.42) gives the received flux density per unit frequency at radius R. Then, by definition, the brightness is the received flux density per unit frequency per unit solid angle of source,

$$\begin{aligned} B_\nu &= \frac{2\pi h\nu^3}{c^2} \frac{1}{e^{h\nu/kT} - 1} \left(\frac{a}{R}\right)^2 \frac{1}{\pi a^2/R^2} \\ &= \frac{2\pi h\nu^3}{c^2} \frac{1}{e^{h\nu/kT} - 1} \left(\frac{a}{R}\right)^2 \frac{R^2}{\pi a^2} \\ &= \frac{2h\nu^3}{c^2} \frac{1}{e^{h\nu/kT} - 1} \text{ W m}^{-2}\text{ Hz}^{-1}\text{ sr}^{-1}, \end{aligned} \tag{6.43}$$

which differs from Eq. (6.42) dimensionally and by a factor of π in magnitude. The corresponding Rayleigh–Jeans approximation is,

$$B_\nu = \frac{2kT}{\lambda^2}, \qquad \frac{h\nu}{kT} \ll 1. \tag{6.44}$$

6.5.3 Antenna Temperature

How do we calculate the power captured from an extended source? In Figure 6.13, we show a narrow beam antenna whose pattern is confined to a solid angle $d\Omega$. The beam intercepts an extended

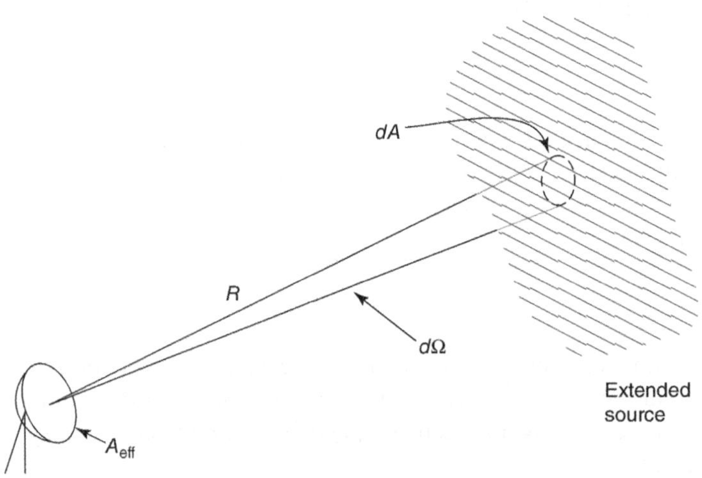

Figure 6.13 Power received from an extended source. The portion of an extended source, i.e., one which subtends a solid angle much larger than the antenna beam, depends on the combination of the antenna pattern and the variation of the source with angular direction from the antenna site. The power received by a narrow beam, high-gain antenna depends primarily on the strength of the source in the direction of the beam. Not depicted are the antenna sidelobes, which also contribute significantly.

brightness source, which could be either (i) noise background, or (ii) the signal of interest to radio astronomers! If the antenna is pointed toward the (θ, ϕ) direction, how much power is received in bandwidth ΔB?

$$\begin{aligned}
P_r(\theta, \phi) &= \frac{1}{2} B_\nu(\theta, \phi) d\Omega A_{\text{eff}}(\theta, \phi)\Delta B \\
&= \frac{1}{2}\frac{2kT(\theta, \phi)}{\lambda^2} d\Omega A_{\text{eff}}(\theta, \phi)\Delta B \\
&= \frac{kT(\theta, \phi)G(\theta, \phi)d\Omega\Delta B}{4\pi},
\end{aligned}$$

where

$$G(\theta, \phi) = \frac{4\pi A_{\text{eff}}(\theta, \phi)}{\lambda^2}$$

is the gain of the antenna. Note that an extra factor of 1/2 appears in the equations above because an antenna can extract power in only a single polarization, while noise power is randomly generated and is equally partitioned between two—any two—orthogonal polarizations.

The total noise power received by the antenna is,

$$\begin{aligned}
P_{r_{\text{noise}}} &= \frac{k\Delta B}{4\pi} \int_{4\pi} G(\theta, \phi)T(\theta, \phi)d\Omega \\
&= k\Delta B T_{\text{ant}}
\end{aligned} \tag{6.45}$$

where

$$T_{\text{ant}} = \frac{1}{4\pi} \int_{4\pi} G(\theta, \phi)T(\theta, \phi)d\Omega. \tag{6.46}$$

Considering the discussion in Sections 3.5.2 and 3.6, it should be clear that T_{ant} corresponds to the physical temperature of a matched resistor, connected across the input terminals of the receiver, that would produce the same input noise as an actual antenna. This temperature varies with the orientation of antenna with respect to the earth and sky, as changes in the antenna orientation correspond to changing in the weighting of the noise contributions with direction in Eq. (6.46).

Significant contributions to the effective antenna temperature, T_{ant}, can come from a "large" product GT over small "areas of sky" or a "small" GT over large "areas of sky." In actuality, the integral expression, Eq. (6.46), tells us precisely the result of this trade-off between the importance of the main beam and the sidelobe contributions to the total noise. In practice, for typical high-gain antenna structures, the net contribution to T_{ant} from the solid angle contained within the main beam and that from the remainder of all directions outside the main beam are very roughly the same. For this reason, in critical applications, certainly, one cannot ignore the structure of the antenna response outside the main beam.

6.5.4 Example Calculation of SNR

We are now in a position to calculate the signal-to-noise ratio (SNR) for an antenna connected to an amplifier. This is given by writing out the expression for SNR,

$$\text{SNR} = \frac{P_r}{kT_{\text{sys}}B} = \frac{P_t G_t G_r \lambda^2}{(4\pi R)^2} \frac{1}{kT_{\text{sys}}B}, \tag{6.47}$$

where we now include T_{ant} as part of T_{sys},

$$T_{\text{sys}} = \underbrace{T_{\text{ant}}}_{\frac{1}{4\pi}\int_{4\pi} G_r T_{\text{sky}} d\Omega} + \underbrace{T_{\text{feed}}}_{T_{\text{phys}}\frac{L}{1-L}} + \underbrace{T_{A1} + \frac{T_{A2}}{A_1} + \frac{T_{A3}}{A_1 A_2} + \cdots}_{\text{amplifier dependent}}. \tag{6.48}$$

Due to cosmic background microwave radiation left over from the "Big Bang," $T_{\text{ant}} \geq 2.8$ K. For fairly efficient systems, values for T_{sys} might be 20, 50, and 100 K for amplifiers based on the use of masers, HEMTs, and FETs, respectively.[8] It might also need to be necessary, however, to consider other sources of loss not directly associated with the antenna/receiver system. For example, clouds, raindrops, atmospheric gases, etc., all can act as attenuators for satellite signals. Some of the signal power goes into heating the clouds every so slightly, while some of the inherent heat from the clouds is radiated into the receiving antenna. This loss can be incorporated into our model as another loss, L, as before.

Suppose you are running a high-performance, high-efficiency ground station with $T_{\text{sys}} = 30$ K, operating at a wavelength of a few centimeters, and that about one half of the contribution to T_{sys} comes from the sky noise background in the main beam of the antenna. At centimeter wavelengths and shorter, scattering and absorption of radio wave signals by clouds can be a significant factor in system performance. A thin stratus cloud might give 1 dB absorptive loss (i.e., $L = 0.2$). Supposing that this is so. Then,

$$P_{\text{sig}} \to 0.8 P_{\text{sig}} \tag{6.49}$$

Keeping in mind that an attenuator in an amplifier system also contributes to the system noise in proportion of the loss, Section 3.6.3, a cloud that absorbs energy also radiates noise in proportion of the amount of signal power absorbed. How much does the cloud radiate, and how does this affect our system? As an illustration, assume that the cloud covers the entire sky and is located at an altitude near the freezing level in the atmosphere, so temperature of the cloud is ≈ 273 K. As far as our system is concerned, the cloud is simply a 273 K attenuator in our signal path.[9] Thus, for a 1 dB loss, the cloud contributes about 0.2×273 K to our system temperature. This corresponds to a system temperature with the cloud in the path, that is,

$$\begin{aligned} T_{\text{sys}} &= 15 + 0.8 \times 15 + 0.2 \times (\approx 273)\ \text{K} \\ &\approx 82\ \text{K} \end{aligned} \tag{6.50}$$

The ratio of SNRs with and without the cloud is,

$$\frac{(S/N)_{\text{cloud}}}{(S/N)_{\text{no cloud}}} = \frac{0.8/84}{1/30} \approx -6\ \text{dB} \tag{6.51}$$

which has increased as the result of thermal radiation concomitant with the loss of signal energy in the cloud. As a result, even though we have only about a 1 dB loss of signal power, the SNR decreases by 6 dB, most of which is the effect of the cloud on system temperature.

---

8 *M*icrowave *a*mplification by *s*timulated *e*mission of *r*adiation, high *e*lectron *m*obility *t*ransistor, and *f*ield *e*ffect *t*ransistor.

9 Consider Figure 3.10 and Eq. (3.29). The figure is drawn in the context of signal traveling on wires. The essence of the problem, though, is the degree of loss encountered along the path. For the antenna problem, the situation is slightly different since antenna noise is collected from a range of directions of arrival. The extended nature of an overcast renders the effect of the cloud essentially equal over the entire beam.

On the other hand, what if T_{ant} is very large? So that,

$$T_{\text{ant}} \gg T_{\text{phys}} \frac{L}{1-L} + T_{\text{amp}} \tag{6.52}$$

where T_{amp} is the noise temperature of the receiver amplifier chain. Under this condition, $T_{\text{sys}} \approx T_{\text{ant}}$ is dominated by noise that has arrived by propagation from remote locations, rather than noise that is generated within your receiver. This is the case for AM radio ($f_c \approx 1$ MHz). In this situation, as long as $P_r/(kT_{\text{ant}}B) \gg 1$ you can use almost any receiving system. The dominance of antenna noise is the reason why a very short antenna ($\ll \lambda \approx 300$ m) works well enough, although the efficiency of the short antennas is very low, and these antennas typically have a large capacitive reactance that makes it very difficult to match the antenna to the receiver. In this situation, the poor performance of the antenna is of little consequence because the SNR is set basically by the ratio of the signal power to noise power *arriving* at the antenna. Because both signal and noise power impinging on the antenna are equally "ill-coupled" to the receiver amplifier, the receiver noise makes little difference.

A similar situation arises in the operation of remote sensing radars when scanning the Earth. When a radar "looks down" from an elevated platform such as an aircraft or satellite, a narrow antenna beam can be filled by the terrestrial "scene." That is, the entire angular range subtended by the antenna beam and, often, the principal sidelobes, intercepts the ground. Under this condition, the antenna temperature typically approximates that of the ground, or very roughly ≈ 300 K in temperate regions. Thus, the minimum system temperature, T_{sys}, is set at a fairly high level by the antenna temperature T_{ant}. Not even an ideal, noiseless receiver can reduce T_{sys} below the level set by radiation from the ground.

7

The Analytic Signal and Matched Filter Detection

Now that we have an overview of the physical elements of radar and scattering from various objects of interest, it's time to return to the analysis of radar and investigate features of radar signals more carefully. In particular, we need to formalize and understand more fully the trade-offs between use of differing signal waveforms in terms of radar performance. That is, we might ask, "What are the rules for general waveforms governing the degree to which we can simultaneously determine range and range rate, and how is this consideration related to signal-to-noise ratio (SNR) and measurement ambiguities?" Previously we introduced pulse trains and CW signals that are commonly used in radar systems, and saw that the different waveforms allowed for the measurement of different properties of the radar echoes. Now we ask, "What about all the other possible waveforms?"

7.1 The Analytic Signal

We are now at a point where we can generalize some important relationships between simultaneous measurement of range, i.e., time delay, and range-rate, i.e., Doppler shift, or frequency. We hinted earlier that it is possible to make rather powerful statements in this regard. In fact, the results can be stated in a few theorems and corollaries related to filter theory. Knowledge of fundamental limitations on signals applicable to measurements of signal characteristics is extremely useful to the radar designer and analyst.

Before we can proceed efficiently, we need to develop a few basic tools to make the job more straightforward. We begin with the *analytic signal*, which is an extension of the commonly employed phasor notation of basic circuit theory to a represent a broad class of time-varying signals. Sinusoidal steady state analysis in the time domain makes use of Euler's theorem to write,

$$e^{j\Omega t} = \cos \Omega t + j \sin \Omega t \tag{7.1}$$

which makes it convenient to identify,

$$\cos \Omega t = \Re[e^{j\Omega t}] \tag{7.2}$$

$$\sin \Omega t = \Im[e^{j\Omega t}] \tag{7.3}$$

where $\Re$ and $\Im$ denote the real and imaginary parts of the associated quantity. Complex Fourier analysis makes use of the closely related expressions, to connect representations of actual harmonic forms directly to the complex decompositions of Eq. (7.1).

$$\cos \Omega t = \frac{e^{j\Omega t} + e^{-j\Omega t}}{2} \tag{7.4}$$

$$\sin \Omega t = \frac{e^{j\Omega t} - e^{-j\Omega t}}{2j} \tag{7.5}$$

In Eqs. (7.4) and (7.5), cosine and sine functions in the time domain are represented by the sum and difference of two phasors oppositely rotating in the complex plane at rate Ω radians s^{-1}. This is a well-known, direct link between the "real" sine and cosine representations and the "complex" phasor point of view.[1]

The analytic signal formally combines the complex phasor and time domain viewpoints for time-varying signals. This is achieved by making use of the Fourier decomposition of a general signal into a system of phasors to construct a representation that is similar to but more powerful than the simple $\exp[j\Omega t]$ adaptation of Euler's representation. To see how this is accomplished, consider the Fourier transform pair,

$$S(f) = \int_{-\infty}^{+\infty} s(t)e^{-j2\pi ft}\, dt \tag{7.6}$$

such that,

$$s(t) = \int_{-\infty}^{+\infty} S(f)e^{+j2\pi ft}\, df \tag{7.7}$$

subject to the usual integrability constraints of Fourier theory. From these relations it readily follows that the transforms of real-valued signals are Hermitian, i.e., a real signal $s(t)$ has a Fourier transform for which $S(f) = S^*(-f)$. Consequently, knowledge of either the positive or negative frequency side of the transform provides the information needed to write down the other, so either side is sufficient to characterize a real signal. The analytic signal incorporates the Hermitian property in constructing a compact representation of a general signal, as described below. The particular value at $f = 0$, i.e., the average or "DC" signal component, is often unimportant; in any event $S(0)$ must be treated as special case when reconstructing the signal from only one side of the spectrum.

7.1.1 Definition of the Analytic Signal

We define the *analytic signal*, $z(t)$, as:

$$z(t) = 2\int_{0}^{\infty} S(f)e^{+j2\pi ft}\, df = x(t) + jy(t) \tag{7.8}$$

where $S(f)$ is the Fourier transform of the real signal $s(t)$. We recover the real, observable signal from the analytic signal representation by taking the real part,

$$s(t) = \Re[z(t)] = x(t) \tag{7.9}$$

1 Much of the engineering analysis is built on the abstract characterization of the sinusoidal steady state condition. This is convenient for linear systems because the real and imaginary parts of the complex representation are not mixed by linear operations.

The result in Eq. (7.9) can be demonstrated easily by evaluation of Eq. (7.7), making use of the Hermitian property of $S(f)$ to eliminate the integration for $f < 0$.

In working with the analytic signal, keep in mind that $z(t)$, is a time-varying, complex waveform whose projection on the real plane describes $s(t)$ exactly. The imaginary part of $z(t)$ is not arbitrary, however. In several ways, the analytic signal is very similar to the treatment of cos Ωt and sin Ωt as parts of the complex exponential, Eqs. (7.2), (7.3). In fact, when applied to cosinusoidal waveforms, the analytic signal reduces to the familiar form of Eqs. (7.4) and (7.5).

The analytic signal is most widely used to represent narrowband signals. In the frequency domain, for $s(t)$ real, the amplitude spectrum of a narrowband signal might appear something like that of Figure 7.1 where the significant portions of the spectrum are concentrated on either side of frequencies $\pm f_c$. Because $s(t)$ is real $S(f)$ is Hermitian, and the form of the spectrum can be expressed as the sum of a positive and negative frequency image,

$$S(f) = \tilde{S}(f - f_c) + \tilde{S}^*(-f - f_c) \tag{7.10}$$

where $\tilde{S}(f)$ is some narrowband function. More precisely, $\tilde{S}(\Delta f) = 0$ for $\Delta f \geq B/2$, where $\Delta f = f - f_c$ and $f_c \gg B/2$ stated another way, $\tilde{S}(f)$ has a finite valued spectrum only near $f = 0$, with the bandwidth B much less than the frequency offset from $f = 0$.

Following through for the analytic signal representation for narrowband signals,

$$z(t) = 2\int_0^{\infty} S(f)e^{+j2\pi ft}df \tag{7.11}$$

$$= 2\int_{-f_c}^{\infty} \tilde{S}(f)e^{j2\pi(f+f_c)t}\,df$$

$$= e^{j2\pi f_c t} \cdot 2\int_{-f_c}^{\infty} \tilde{S}(f)e^{j2\pi ft}\,df \tag{7.12}$$

The result above shows that the analytic signal for a narrowband signal can be represented in terms of the significant portion of the inverse Fourier transform surrounding the reference frequency, f_c. We will see that in many instances f_c has physical meaning as the frequency of a *carrier* signal "modulated" by the information represented by the integral in Eq. (7.12).

As the integral in this expression represents a general, non-sinusoidal component of the time-varying part while the factor of two comes from folding of the Fourier spectrum in the definition of the analytic signal, we include the factor of two with the integral in Eq. (7.12) and give this product a distinct symbol, $Z(t)$, so that,

$$z(t) = Z(t)e^{j2\pi f_c t} \tag{7.13}$$

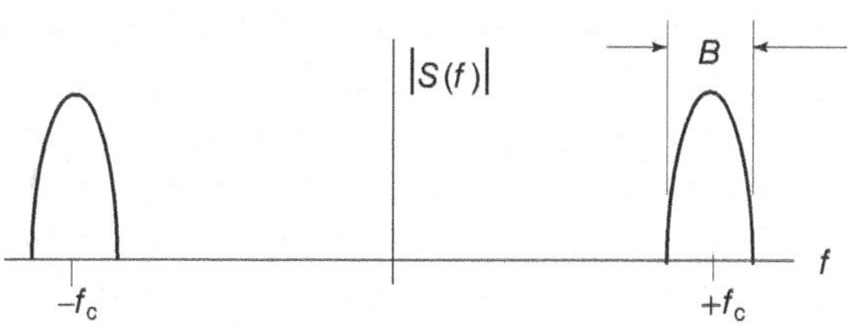

Figure 7.1 Spectrum of a narrowband signal: signal is concentrated in the vicinity of frequency f_c.

Under our assumption that $\tilde{S}(f)$ vanishes except near the origin, it is common practice to express the integral as,

$$Z(t) = X(t) + jY(t) = 2\int_{-\infty}^{+\infty} \tilde{S}(f)e^{j2\pi ft}\, df \tag{7.14}$$

where we extend the lower limit to $f = -\infty$ for convenience of notation. The only true constraint on our taking this step is that $f_c \geq B/2$.

Given $Z(t)$, we recover $s(t)$ by combining the real and imaginary parts of $Z(t)$ with the Eulerian components of $e^{2\pi f_c t}$ to obtain the form in Eq. (7.13). Then, taking the real part,

$$s(t) = \Re[Z(t)e^{j2\pi f_c t}] = X(t)\cos 2\pi f_c t - Y(t)\sin 2\pi f_c t \tag{7.15}$$

where the reference frequency f_c is the "carrier," previously identified, while the functions $X(t),\ Y(t)$ are referred to as the *in-phase* (or "I") and *quadrature* (or "Q") components of $s(t)$.[2]

Function $Z(t)$ can also be described in terms of its magnitude $M(t)$ and phase $\Phi(t)$,

$$Z(t) = M(t)e^{j\Phi(t)} \tag{7.16}$$

where,

$$M(t) = \left\{X(t)^2 + Y(t)^2\right\}^{\frac{1}{2}}, \qquad \Phi(t) = \tan^{-1}\left\{\frac{Y}{X}\right\} \tag{7.17}$$

Because variations in $M(t)$ effectively expand and contract a plot of $x(t)$ (or $z(t)$), the magnitude is thought of as an *envelope* surrounding the carrier signal. At the same time, variations in the phase, $\Phi(t)$, describe the advance and retardation of the complex signal modulation with respect to the uniformly rotating carrier phasor, $\exp[j2\pi f_c t]$. These changes in magnitude and phase, taken together as the time-varying function $Z(t)$, are referred to as the *complex envelope* of $s(t)$. From the point of view of information transmission, the variations represented by $Z(t)$ are a generalized form of information signal modulation applied to the carrier signal $\cos 2\pi f_c t$.

In the above, the carrier frequency f_c is often described and thought of as the "center frequency" of a narrowband signal, but is otherwise undefined. That is, the question "How is f_c defined, specifically?" is not addressed. While the term "center" is a useful characterization when $B \ll f_c$, in fact, the choice of f_c is *essentially arbitrary* in that the analysis leading to the analytic signal does not depend on any particular choice of this frequency, provided that $f_c \geq B/2$. It is better, perhaps, to think of $e^{j2\pi f_c t}$ simply as a *reference phasor* to which I and Q are compared. Usually f_c is chosen to lie within the modulation bandwidth B about f_c, but the definition of the analytic signal does not require this. On the other hand, all other information regarding $s(t)$ is contained in $Z(t)$, the complex envelope. Once defined, $Z(t)$ can be applied to any "carrier" without loss of information. As we shall see, most radio frequency receiving systems operate by manipulating f_c while leaving $Z(t)$ effectively unchanged. "Coherent" laser receiving systems work in essentially the same way.

Although the definition of the analytic signal is ambiguous with respect to the meaning of the carrier frequency, in any particular system f_c is well-defined by the designer's choice of system architecture. To illustrate this point, consider how the component functions $X(t),\ Y(t)$ might be obtained in practice. Figure 7.2 shows one method for recovery of the in-phase component, $X(t)$, by use of a cosinusoidal reference signal to transform the nominal signal spectrum, $S(f)$, to

2 The function $X(t)$ is the real part of $Z(t)$ rather than that of $z(t)$; similarly, $Y(t)$ is not the imaginary part of $z(t)$. The real form $s(t)$ is recovered from $X(t),\ Y(t)$ by use of Eq. (7.15).

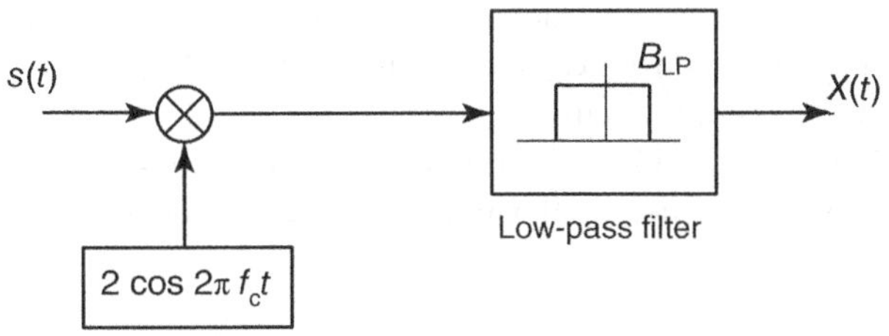

Figure 7.2 Method for extracting $X(t)$. Mixing with cosinusoid at the reference frequency, f_c, followed by low-pass filter provides the real part, $X(t)$, of the complex envelope, $Z(t)$. Imaginary part of the envelope, $Y(t)$, is obtained in a similar way by replacing mixing signal $\cos 2\pi f_c t$ with $\sin 2\pi f_c t$.

"baseband," i.e., to a frequency range that includes or is close to $f = 0$, and then applying a low-pass filter to the result. Analytically, using Eq. (7.15) to represent $s(t)$, the multiplication, or "mixing," of the $s(t)$ by the function $2\cos 2\pi f_c t$ in Figure 7.2 corresponds to,

$$2\,s(t)\cos 2\pi f_c t = 2\left(\frac{1}{2}X(t) + \frac{1}{2}X(t)\cos(2\cdot 2\pi f_c t)\right) - \frac{1}{2}Y(t)\sin(2\cdot 2\pi f_c t) \tag{7.18}$$

after which the filter removes the terms at frequency $2f_c$, assuming that $B_{LP} < 2f_c$. Thus, mixing of $s(t)$ with $2\cos 2\pi f_c t$ followed by low-pass filtering is sufficient to recover the I component. A similar operation in which the reference signal is $2\sin 2\pi f_c t$ yields the Q component $Y(t)$. Combining $X(t)$ and $Y(t)$ according to Eq. (7.14) yields the complex envelope $Z(t)$. In this example, assuming that f_c lies within or near the center of the narrowband spectrum, it should be clear that $X(t),\ Y(t)$ are defined by the combination of the characteristics of the signal $s(t)$ and the *choice* of the center frequency, f_c, in any particular system. There is nothing in the structure of the analytic signal that dictates the use of a particular value of f_c, although for practical reason of minimizing B_{LP} a choice of f_c near the center of the spectrum of $s(t)$ is usually convenient.

The function of the low-pass filter merits further discussion. In the event that $2B_{LP} \geq B$ *and* the choice of the frequency f_c places the spectrum $\tilde{S}(f)$ within the $2B_{LP}$ pass band, then the signal characteristics are fully preserved. It can happen, however, that $2B_{LP} < B$, in which case, the low-pass filter becomes the limiting filter in the system and thereby determines B for the output signal. That is:

i. $2B_{LP} \geq B$ leaves $s(t)$ completely unchanged;
ii. $B > 2B_{LP}$ inserts B_{LP} as a critical filter defining a new $s(t)$, represented by $X(t),\ Y(t)$ with bandwidth $2B_{LP}$.

The critical condition between these two cases is $B/2 = B_{LP}$. In the first instance, the low-pass filter acts something like an anti-aliasing filter in removing the high-frequency mixing products; in the second instance, however, the low-pass filter additionally establishes the observable bandwidth of $Z(t)$.

In either instance, $X(t)$ and $Y(t)$ obtained at the output of the low-pass filter have maximum frequencies limited to the lesser of $B/2$ or $B_{LP}/2$. As both the I and Q components of the complex envelope are themselves real signals with Hermitian transforms, their individual bandwidths are never larger than B_{LP}. The maximum bandwidth of the original signal, $s(t)$ is twice this, or $2B_{LP}$. When the I and Q components are combined in phase quadrature in the complex envelope, $Z(t)$, however, the result is no longer Hermitian and the full bandwidth is B, with the information distributed between $\pm B_{LP}$. Thus, we see that the information content of $s(t)$ is equally divided between the I and Q components, both of which are required to construct $Z(t)$ with bandwidth B. The latter situation often occurs in practice as it provides a convenient method to filter a high-frequency signal while simultaneously obtaining the I and Q components.

It's worthwhile to think through these issues. The arbitrary nature of f_c can lead to confusion before this analysis is fully understood. At the same time, the flexibility of the definition gives the analytic signal wide applicability.

7.1.2 The Hilbert Transform and the Analytic Signal

The development above proceeds by recognizing that for real signals only one half of the Fourier transform is required to represent a generalized, time-varying waveform. For reasons of convenience and convention, the positive frequency half of the spectrum is chosen as the basis of the compact analytic signal form.[3] An alternative point of view with deeper mathematical roots is found in the Hilbert transform of signals, which provides a basis for the analytic signal set entirely in the time domain.

The Hilbert transform $\mathcal{H}(s)$ of a signal $s(t)$ can be defined in terms of the convolution of a given $s(t)$ with the function $1/(\pi t)$, which we refer to as the *Hilbert kernel*,

$$\hat{s}(t) = \mathcal{H}(s) = s(t) \star \frac{1}{\pi t} \tag{7.19}$$

or, in integral notation,

$$\hat{s}(t) = \mathcal{H}(s) = \frac{1}{\pi} \int_{-\infty}^{\infty} \frac{s(\tau)}{t - \tau} d\tau \tag{7.20}$$

We can show that $z(t) = s(t) + j\hat{s}(t)$. That is $\hat{s}(t)$ is the imaginary component of the analytic signal.

Since the Hilbert transform is defined by convolution, we can find the connection with the frequency domain easily. Letting $\mathcal{F}()$ represent the Fourier transform of the Hilbert kernel,

$$\mathcal{F}\left(\frac{1}{\pi t}\right) = -j \ \mathrm{sgn}(f) \tag{7.21}$$

where sgn() represents,

$$\mathrm{signum}(x) = \begin{cases} 1, & x > 0 \\ 0, & x = 0 \\ -1, & x < 0 \end{cases} \tag{7.22}$$

Hence, the Fourier transform of $\hat{s}(t)$ is,

$$\mathcal{F}(\hat{s}(t)) = -j \ \mathrm{sgn}(f) \cdot \mathcal{F}(s(t)) \tag{7.23}$$

$$= -j \ \mathrm{sgn}(f) \cdot S(f) \tag{7.24}$$

How is this transform related to the analytic signal? In order to see the connection, consider the Fourier transform of the sum, $s(t) + j\hat{s}(t)$,

$$\mathcal{F}(s(t) + j\hat{s}(t) = S(f) + j(-j \ \mathrm{sgn}(f)) \, S(f) \tag{7.25}$$

$$= S(f) + \mathrm{sgn}(f)) \, S(f) \tag{7.26}$$

$$= S(f)(1 + \mathrm{sgn}(f)) \tag{7.27}$$

3 Possibly because the idea of "negative" frequencies is so bothersome to many of us.

But the function sgn(f) merely reverses the sign of $S(f)$ for $f < 0$ as compared with $f > 0$. With this, the final result from Eq. (7.27) is,

$$\mathcal{F}(s(t) + j\hat{s}(t) = \begin{cases} 2S(f), & f > 0 \\ S(f), & f = 0 \\ 0, & f < 0 \end{cases} \tag{7.28}$$

The frequency domain representation of the operations in Eq. (7.27) is illustrated in Figure 7.3.

Our previous definition was,

$$z(t) = 2\int_0^{\infty} S(f)e^{j2\pi ft}df \tag{7.29}$$

which is equivalent to that obtained using the Hilbert transform,

$$z(t) = s(t) + j\hat{s}(t) \tag{7.30}$$

In forming $z(t)$ with the Hilbert transform, we see that the imaginary component of $z(t)$ is always given by $\mathcal{H}(s) = \hat{s}(t)$, thus forming a natural quadrature component to the original real signal.

Either way, the analytic signal is the complex signal formed in the frequency domain from the positive frequencies or in the Hilbert domain by convolution with $(\pi t)^{-1}$. Three properties of the analytic signal can be seen from the structure of the Hilbert kernel used in convolution with a real signal. First, the average value of $s(t)$ cannot contribute to the imaginary part of $z(t)$, owing to the antisymmetry arising from the change in sign of the Hilbert kernel at the origin. The second, that the reference phase for the analytic signal, i.e., $\phi = 0$, is tied to the very long term average of $s(t)$, follows directly from the first. Because the integrated differences of the remote tails of $1/t$ vanish only for the average value of $s(t)$, contributions to the convolution integral for regions that are

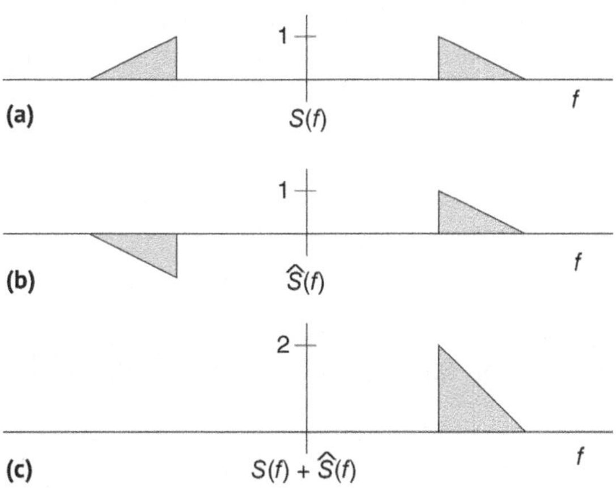

Figure 7.3 Graphical construction of $z(t)$ from Hilbert transform. (a) Fourier transform of $s(t)$, $S(f)$, as mentioned earlier. (b) $\hat{S}(f)$ with sign of positive and negative frequency branches reversed. (c) Combination of first and second lines yields Fourier transform of $z(t)$. Graphical sequence leads to result in Eq. (7.28), demonstrating that the analytic signal is a linear combination of $s(t)$ with its Hilbert transform. Typically, as mentioned earlier, $S(f) = 0$, or is negligible.

remote from the origin of the kernel control the phase reference. Finally, recall that convolution with a closely spaced pair of impulse functions of opposite signs closely approximates the first derivative of a function.[4] The local behavior of the Hilbert transform in the vicinity of $t - \tau = 0$ approximates first derivative of $s(\tau)$, albeit the values are smoothed by the long hyperbolic decay tail of $(\pi t)^{-1}$. From this we see that the local amplitude of quadrature component of the analytic signal, $y(t) = \hat{s}(t)$, is associated closely with rapid changes in the amplitude of the real part, $x(t) = s(t)$. At the same time, the local amplitude on broader scales is coupled to the local variations in the real part over a range of temporal scales owing to the anti-symmetrical, smooth decay of the kernel on either side of $t = 0$.

The usefulness of the analytic signal is that it allows us to separate the time variation of the signal amplitude and phase from the underlying, steady increase of phase of an intrinsic carrier, which in some sense is the "average" signal. At the same time, frequency domain analysis shows that by defining the complex envelope of a signal of finite bandwidth, the role of the carrier can be separated from that of the modulation. Thus, a signal with a modulation bandwidth of 10 MHz, for example, provides the same time response and requires the same ultimate, 10 MHz bandwidth for transmission and reception, regardless of whether the carrier is 25 or 2500 MHz.

In practice, the carrier frequency is often well-known from our knowledge of how particular signals are generated. At the same time, our earlier analysis, Section 7.1.1, has shown that the choice of f_c as a reference for determining the complex envelope, $Z(t)$, is somewhat arbitrary, and primarily requires that we be consistent in our usage for a given problem or system. In considering this, keep in mind that actual systems employ two, three, and sometimes more distinct stages of amplification at different frequencies. These systems all employ physical means to shift frequency that are well-modeled by our operations on the analytic signal, with no intrinsic loss of information due to this process at any point.

7.1.3 Analytic Signal Representation of Filters

Application of the analytic signal to representation of filters is straightforward and easy, albeit a bit messy to work out. We give only the results here. The complex envelope of a filter with real impulse response $h(\tau)$ is given by,

$$W(\tau) = 2\int_{-\infty}^{\infty} \tilde{H}(f)e^{+j2\pi f\tau}\, df \tag{7.31}$$

corresponding to the analytic signal representation of $h(\tau)$,

$$w(\tau) = W(\tau)e^{+j2\pi f_c \cdot \tau} \tag{7.32}$$

With this, we can calculate the complex envelope of the filter response by convolution of the envelopes of the input and impulse response,

$$Z_{\text{out}}(t) = \frac{1}{2}\int_{-\infty}^{+\infty} W(\tau)Z_{\text{in}}(t-\tau)\, d\tau \tag{7.33}$$

Thus, the complex envelope of the filter output is the convolution of the complex envelopes of the input signal with that of the filter impulse response.

4 Convolution with an impulse doublet, created by letting the separation between the impulse functions vanish, gives the first derivative exactly. Bracewell (see Bibliography).

The real output conventionally obtained by convolution of $s_{\text{in}}(t)$ with $h(\tau)$ now is obtained from $Z_{\text{out}}(t)$ by restoring the carrier and taking the real part,

$$s_{\text{out}}(t) = \int_{-\infty}^{+\infty} h(\tau)s_{\text{in}}(t-\tau)\,d\tau = \Re[Z_{\text{out}}(t)e^{j2\pi f_c t}] \tag{7.34}$$

7.2 Matched Filter Detection

For reasons both practical and theoretical, we would like to know the form of the filter that, for a specified input waveform, results in the maximum output SNR.[5] For convenience we express the result in terms of the analytic signal form for the filter output, Eq. (7.33), as mentioned earlier.

Suppose that a particular radar transmits a signal $s(t)$. The echo from a so-called *point* target is then proportional to $s(t - T_0)$. The receiver shifts the frequency and adds more delay—presumably while introducing only negligible distortion—and, through the process of amplification, rescales the result. Through all this, the shape of the complex signal *envelope* remains unchanged within a constant factor. The receiving system adds noise to the signal, as we have seen before. It almost always happens that the maximum range and other performance characteristics of a radar are limited by the weakest signal that can be usefully recovered, which in turn is limited by amount of noise present. Consequently, in many applications it is very essential to maximize the obtainable SNR at every step in the reception process.

Pictorially, suppose that we were to transmit a waveform with a very simple complex envelope, for example the simple truncated exponential decay shown in Figure 7.4, where we assume that the signal is finite only for time $0 \geq t \geq t_d$. Then, within a scale factor, the received waveform from a point target might appear as that in Figure 7.5, where we have lumped all the signal delays into the parameter T_0 and added noise. Our question is, "What is the filter that maximizes $|s_{\text{out}}|^2 / < \sigma_{n_{\text{out}}} >$?"

The solution is referred to as the *matched filter* because of the relationship of its form to that of the original signal waveform. The impulse response of this filter is a time-delayed, reversed form of the original signal. That is, the matched filter impulse response is given by,

$$h(\tau) = s(t_d - \tau) \tag{7.35}$$

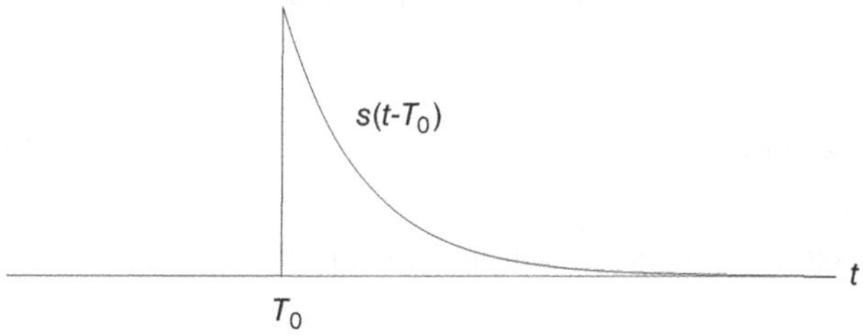

Figure 7.4 Example signal, $s(t)$. Decaying curve represents envelope of a simple analytic signal with magnitude decreasing exponentially with time, truncated at $t = t_d$.

5 We restrict the answer to linear, time-invariant (LTI), systems, however.

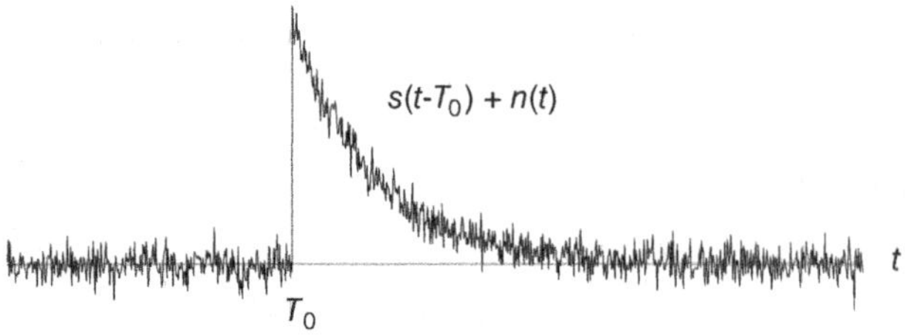

Figure 7.5 Example echo signal in presence of noise. Plot shows signal of Figure 7.4 delayed by time T_0 with noise added to the original waveform, Figure 7.4. Shift in time, T_0, accounts for total delay from the instant transmission of $s(t)$ began, comprising propagation delay to scattering object plus delays in receiving system prior to echo signal's arrival at terminals of filter.

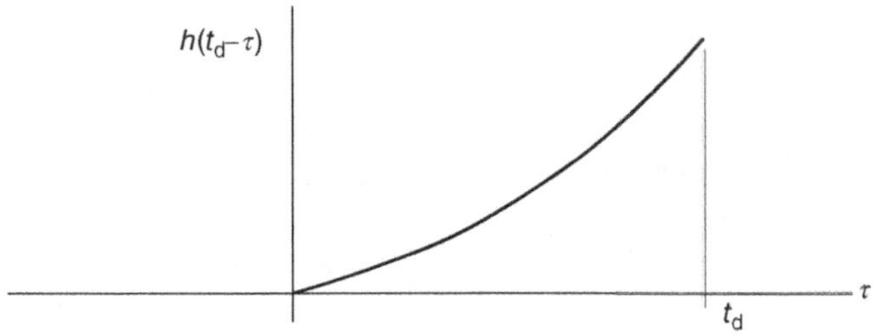

Figure 7.6 Impulse response of filter matched to $s(t)$. Impulse response shown corresponds to signal of Figure 7.4. Note that the impulse response of the matched filter is obtained by delay of the time-reversed transmitted signal waveform. Referring to Figure 7.4, the impulse response above can be generated by flipping $s(t)$ left-to-right, and then shifting the result to the right by t_d.

where τ is the time from application of the impulse and the time delay t_d is required for a causal filter implementation.[6] This filter is shown in Figure 7.6, for our example, input waveform in Figure 7.4.

Neglecting the scale factor, in terms of the real signal representation, the output of the matched filter is given by,

$$s_{out}(t) = \int_{-\infty}^{+\infty} s(\tau - T_0)h(t - \tau)\, d\tau = \int_{-\infty}^{+\infty} s(\tau - T_0)s(t_d + \tau - t)\, d\tau \tag{7.36}$$

where T_0 is the overall radar propagation and receiver delay, as mentioned earlier. This output is maximum when $t = t_d + T_0$,

$$s_{out}(t)\Big|_{max} = s_{out}(t = t_d + T_0) = \int_{-\infty}^{+\infty} s^2(\tau)\, d\tau \tag{7.37}$$

Analysis of this filter shows that SNR is maximized when the output signal peaks, with a value,

$$\text{SNR}\Big|_{max} = \frac{2E}{N_0} = \frac{2E}{kT_{sys}} \tag{7.38}$$

In the expression for maximum SNR the new quantities are the total received echo signal energy, E, and the noise power spectral density, N_0. Boltzmann's constant, $k = 1.38 \times 10^{-23}$ J/K, and the

6 The output of an actual filter cannot precede in time the input stimulus that "causes" it. Thus, a necessary requirement for a *causal filter*, i.e., a filter that can be implemented in practice, is that the output must lag the input.

system's equivalent noise temperature, T_{sys}, are as before. The derivation and more complete discussions can be found elsewhere.[7] That a matched filter maximizes output SNR is well-known in communication theory, where it provides a primary basis for system design and performance analysis.

This remarkable result—that the output *SNR is independent of signal waveform and bandwidth*—provides the radar designer with considerable flexibility in structuring the overall system. As example, limitations on maximum transmitter output power can be traded against increases in signal duration. At the same time, transmitted signal waveforms with inherently high temporal resolution can be selected without concern for possible reductions in SNR often associated with the use of large signal bandwidths. In instances where pulse bursts are employed in conjunction with a filter coherently matched to the burst, rather than to individual pulses, *the energy in the entire burst determines the SNR.*

It is fairly straightforward to visualize the operation of a matched filter. Dropping the propagation time delay, for convenience, and considering a general waveform,

$$\begin{aligned} \mathcal{F}[s(t)] &= S(f) \\ \mathcal{F}[h(\tau)] &= H(f) = S^*(f)e^{-j2\pi f t_d} \\ \mathcal{F}[s_{out}(t)] &= S_{out}(f) \end{aligned} \tag{7.39}$$

Applying the convolution theorem, we make use of an expression for the matched filter output spectrum as the product of the Fourier transform of the transmitted signal and the filter impulse response,

$$\begin{aligned} S_{out}(f) &= S(f) \cdot H(f) \\ &= S(f) \cdot S^*(f)e^{-j2\pi f t_d} \\ &= |S(f)|^2 e^{-j2\pi f t_d} \end{aligned} \tag{7.40}$$

As the last step, we use the inverse Fourier transform of the output signal to return to the time domain.

$$s_{out}(t) = \int_{-\infty}^{+\infty} |S(f)|^2 e^{j2\pi f(t-t_d)}\, df \tag{7.41}$$

From Eq. (7.41), it is clear that the output is maximum at $t = t_d$, since the value of the exponential is unity for $2\pi f(t - t_d) = 0$. Any value of $t \neq t_d$ changes the integrand from a positive-definite function to an oscillatory one, thus reducing the value of the integral. From this we see that the maximum value of $s(t)$ occurs at time t_d. Additionally, *the filter output is maximum with respect to all other input signal waveforms of the same energy* because the filter cancels all variation in phase within the integral. The underlying mechanism for this result can be seen from Figure 7.7, which shows the effect of conjugating the transfer function in construction of the filter; any choice of phase $\phi(t)$ other than the conjugate reduces the real part of the Fourier integral over frequency, thus reducing $s_{out}(t = t_d)$. Stated another way, the conjugation is perfect only when the filter impulse response matches the transmitted signal exactly.

Finally, observe that the filter weight at each frequency is proportional to "local" SNR in either the time or frequency domain. In the frequency domain, the greatest weight is given to those portions

7 Derivation of the matched filter is found in many standard texts on communication and information theory: an early reference is Turin [1960].

of the spectrum with the highest SNR, while spectral regions devoid of signal content are weighted by zero. Similar behavior is seen in the time domain, where phase variations are again removed by the conjugating property of the filter at the instant of complete alignment of the signal with the filter impulse response.[8] A typical variation of the filter power transfer function if given in Figure 7.8.

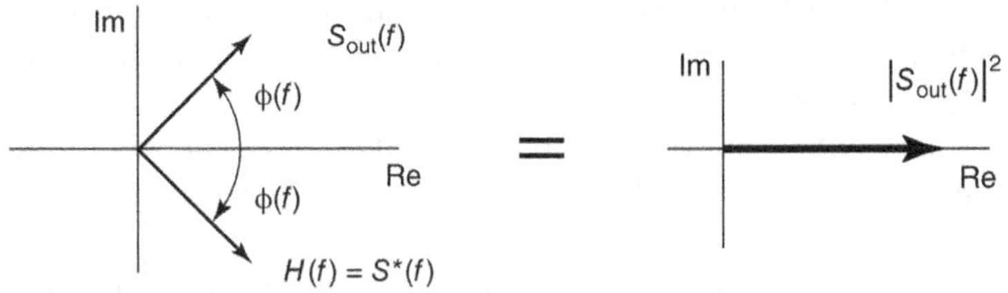

Figure 7.7 Conjugate phasors. Product of two conjugate phasors is the square magnitude of the amplitude; the magnitude of the product for any other phasing of two equal length phasors is complex and less than this. Conjugate form appears in the matched filter transform integral, showing that the performance of the filter is achieved by providing maximal cancellation of any structural phase variations in the echo signal at every frequency in the filter output.

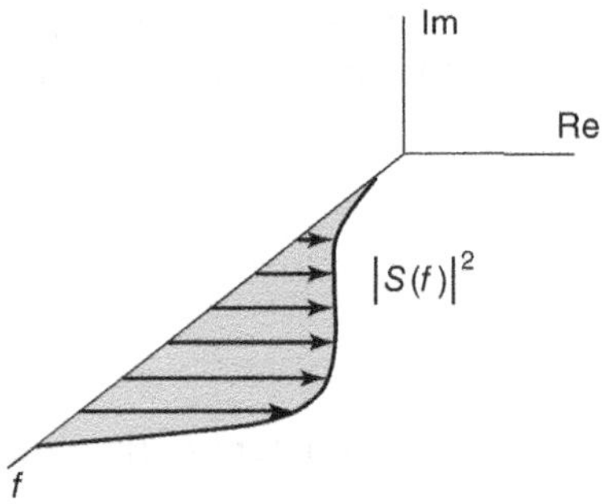

Figure 7.8 Power transfer function for matched filters. Since the matched filter is assumed to operate in white noise, the power transfer function weights the filter output in proportion to the input SNR, at each frequency in the signal spectrum.

8 Readers familiar with filter theory, recalling that the noise here is uniform with frequency, will see this characteristic as reminiscent of Wiener filters, which employ a similar weighting based on the local SNR within the frequency spectrum.

8

Radar Detection

Our work with signals to this point has been concerned primarily with explaining the physics of radar problems. That is, how do we describe, quantitatively, the launching, propagation, scattering, and reception of an electromagnetic wave. Or, given such a wave, what are the effects of the medium through which it passes. We also understand the manner in which the geometry of a given problem underlies the characteristics of the received signal, specifically the time delay and Doppler shift of the received signals. Finally, we understand the contribution of noise arising from the environment and from electronic systems on radar signals, and how the relative strength of radar signals with respect to such noise can be determined.

Given our present kitbag of tools, we are in a position to understand and design simple radar systems capable of a broad range of measurements. Our task in this chapter is to extend the basic tools in our bag, so that we are equipped to explore more fully the limits of performance of radar systems. The most challenging radar applications and experiments always occur at the limits of performance—under conditions in which the echoes are weak, usually at maximum range. For example, in order to be useful, systems built to warn of Earth approaching asteroids need not only to determine the presence of such objects but also to provide reliable information on their physical characteristics at the earliest possible moment, in other words, at the greatest possible range. But what determines the greatest range? Surely it is the range at which the echo signals can just be distinguished from the system noise. But at what level of reliability, and after how many, and how extended a set of observations? Given a detection of an approaching object, the next question might be, "Is this object following a path that is potentially dangerous or of no consequence to us?" And after that, "What is the size of the detected object?" These are important, basic questions for which answers are required.

8.1 Detection and Estimation

Radar systems "see" remote objects by illuminating them with electromagnetic radiation, and then subsequently recording and analyzing the echoes. In nearly all applications, we wish to infer various properties of the objects being studied on the basis of their radar echo characteristics. Typically, we ask the following two questions about the object in sequence:

i. Is there a reflecting object present?
ii. What are the properties of the object?

The first of these two questions, together with its solution, is known as the "detection problem" of radar, while the second question and its solution, addressing physical parameters such as the RCS of the scattering object, is known as the "estimation problem."

Although distinct, the questions of detection and estimation are closely related even when applied to the same object, as in the example of the asteroid. They arise, however, from substantially different types of applications and require different analytic approaches. In the detection problem, we are confronted with a signal from the receiver and ask whether or not this signal contains an echo, perhaps from a particular scattering object, or is it entirely the result of background noise or interfering signals—perhaps reflections from other, uninteresting scatterers. In the estimation problem, we are aware *a priori* that the received signal contains reflected energy from an object, and we wish to make use of this echo to learn more about the object. This might include quantitative descriptions of the object's properties, such as its size, composition, orientation, position, and velocity, or some other characteristic. These characterizations take the form of *statistical estimates* of the various characteristics of interest.

8.1.1 Examples of Detection Problems

The simplest detection problem is that in which we transmit a single radar pulse, and then examine the receiver output for an indication that an echo is *or* is not present. Ornithologists might be interested in detecting birds migrating at night. Automotive engineers might find radar detection of preceding vehicles practical as part of a collision avoidance system in an automobile, where the observation of importance is whether or not there is a car directly in front of you.[1] Often, the nature of the detection problem is such that complex systems employing the use of multiple pulses—diverse in time and/or direction—are required in order to sense the presence of weakly scattering objects in a region of interest.

In detection problems, we usually ask whether or not objects are present within the region observed by the radar illumination beam. For many uses, the answer may be a simple binary "yes" or "no" obtained from a single radar pulse. But for other applications, the needed information may be the relationship of the yes's and no's to echo time delay and arrival angle. Examples of these more complex problems include weather radars and air traffic control (ATC) radars. In one type of weather radar, the sky is scanned for water droplets large enough to fall as precipitation. Individual droplets are not of particular interest, but the locations of collections of large number of droplets and their spatial distribution are important for identifying areas of intense rainfall. By knowing the direction to and distance from the antenna of each detected patch of rain, a map of the distribution of rainfall can be produced. Such maps are commonly seen in commercial weather reporting. Here, the detection question we ask is "Is it raining, yes or no, at each point in the area of interest?" By way of contrast, the ancillary estimation question is, "What is the volume of liquid water and its rate of fall?" (Figure 8.1).

Another detection problem is that of ATC. In this application, a radar system located near an airport along an airway probes the airspace and "looks for," i.e., *seeks to detect*, airborne objects large enough to be aircraft, Figure 8.2. The location of each plane is determined, and the result is presented to air traffic controllers as an aid in managing the flow of air traffic. Here, the detection question is "Are aircraft present and, if so, where are they located?" In the ATC problem, there is

---

1 This can be generalized, of course, to the question of whether or not there is another vehicle on a collision course with your own vehicle. But, this more general question brings in the estimation problem for distance and velocity.

Figure 8.1 Weather radar. These systems typically display the areal distribution of precipitable water droplets falling as rain. Such images are a widely used tool for determining the present state of the weather and for forecasting future weather, and are seen regularly as part of weather forecasts both professionally and for the public. Airborne versions of these instruments are used by pilots to avoid severe weather conditions; an operating aircraft weather radar is required equipment for airliners when thunderstorms are forecast along the route of flight.

no need to estimate the scattering properties of each plane, as the aircraft type and other related facts are communicated by other methods (see Figure 8.2).

Yet another detection example is found in medical ultrasound imaging systems, Figure 8.3. Although in this instance, acoustic rather than electromagnetic waves are used, the detection problem is quite similar to that of an electromagnetic system. Here, the purpose is to establish the presence, say, of obstructions to flow of blood within an artery. If the reflections from obstructions are sufficiently different in their mechanical scattering properties from the background tissue, we can determine their presence without requiring invasive medical procedures. The detection question is, "Are there obstructions present in the veins or arteries?" A related estimation problem addresses the mass density of the material in any obstruction detected.

8.1.2 Examples of Estimation Problems

Estimation, as indicated above, is concerned not so much with whether or not an object is present but with particular physical properties of that object.[2] One example of an estimation problem might be the planetary astronomy question, "How many particles are found in the Rings of Saturn?" Here we know already that the ring system is made up of discrete particles, but we wish to know the sizes of the particles and how many particles are present. One approach to this measurement is

2 A word about terminology: Strictly speaking, inferences made under noisy conditions should always be recognized as "estimates," as used here; this is standard usage in radar applications. When such estimates become sufficiently accurate that the errors are negligible for a particular purpose, we tend to speak of them as "determinations."

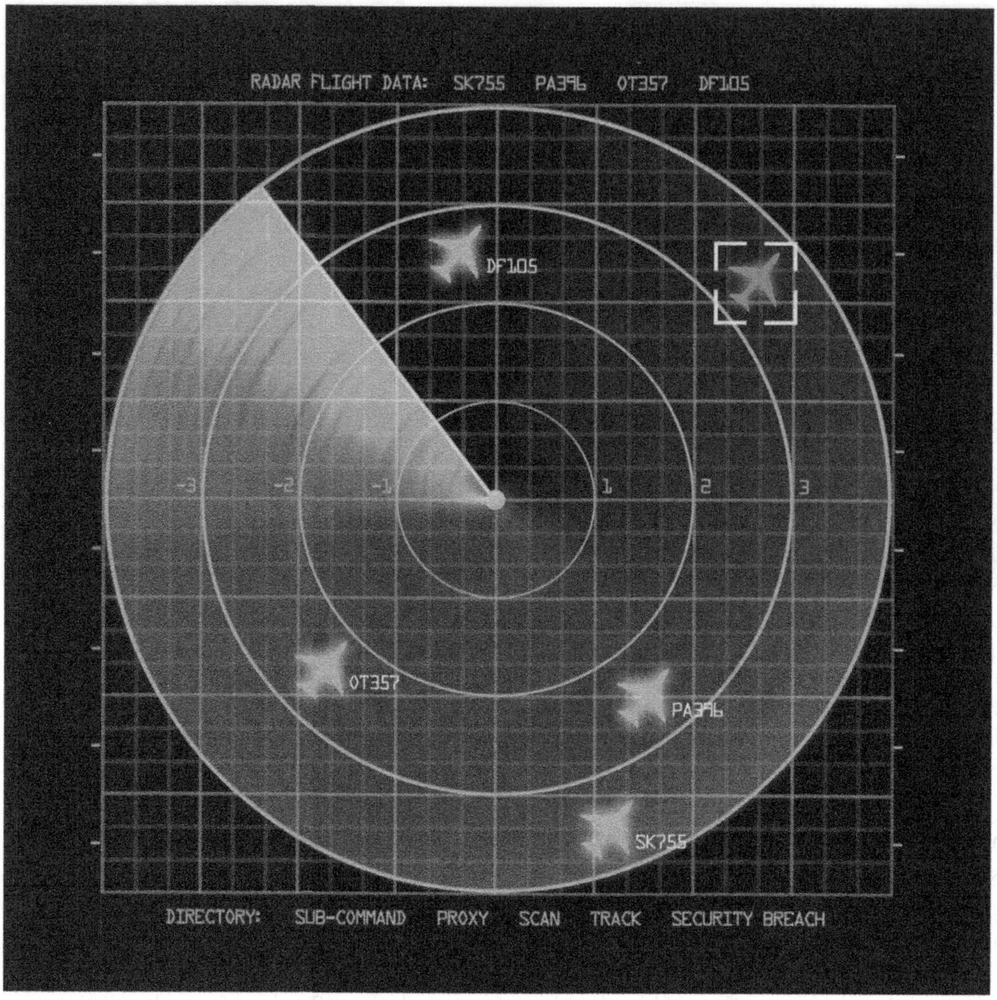

Photomelon/Adobe Stock Photos

Figure 8.2 Air traffic control. The control of commercial air traffic and all flights in bad weather is based on radar tracking of individual aircraft. In this application, the ATC radar system is aided by transponders carried by each airplane. Use of the transponder assures that the aircraft will be detected any time it is above the radar horizon of the ground system. The transponder also provides increased detection margins needed when the aircraft is located within or behind heavy rain. In addition, the transponder signal is coded with an ATC identifier and the aircraft altitude.

to use a radar beam to illuminate the ring system, and then measure the strength of the echo, Figure 8.4, since we know in general that the larger and more numerous the reflecting particles present, the larger the echo intensity, we know that echo strength is related to our objective. By using our understanding of the theoretical scattering properties of similar, model particles, we are able to establish a quantitative estimate of the number of particles required to produce a particular observed echo. In this situation, the question is not whether the rings exist, a "yes" or "no" detection question, but rather, "What is the distribution of material and the total amount of material in the ring system?"

Closer to home, we might be interested in estimating the quantity of biomass in various forests of the world. Specialized radars can simultaneously estimate the height of forest canopies and the

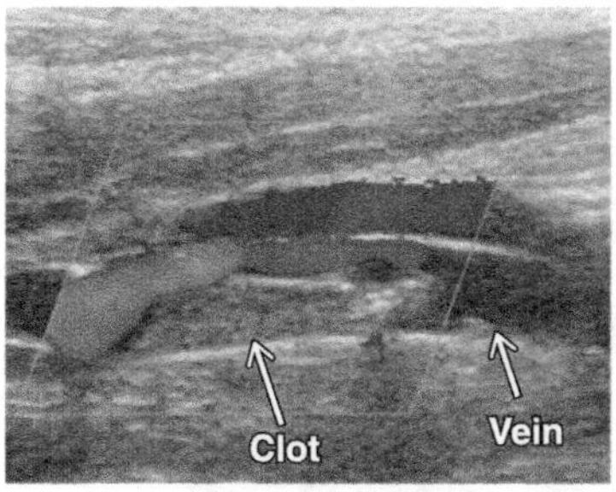

Figure 8.3 Medical ultrasound. Use of ultrasound, for example, to detect an obstruction to blood flow with an artery or vein, is based on the adaptation of radar principles to medical electronics. Such images are obtained with sensors threaded through the blood vessel to the site of the clot. *Source*: SAE International. https://www.medicaldesignbriefs.com/component/content/article/13138-ultrasound-technique-breaks-down-blood-clots (last accessed Feb. 2, 2026).

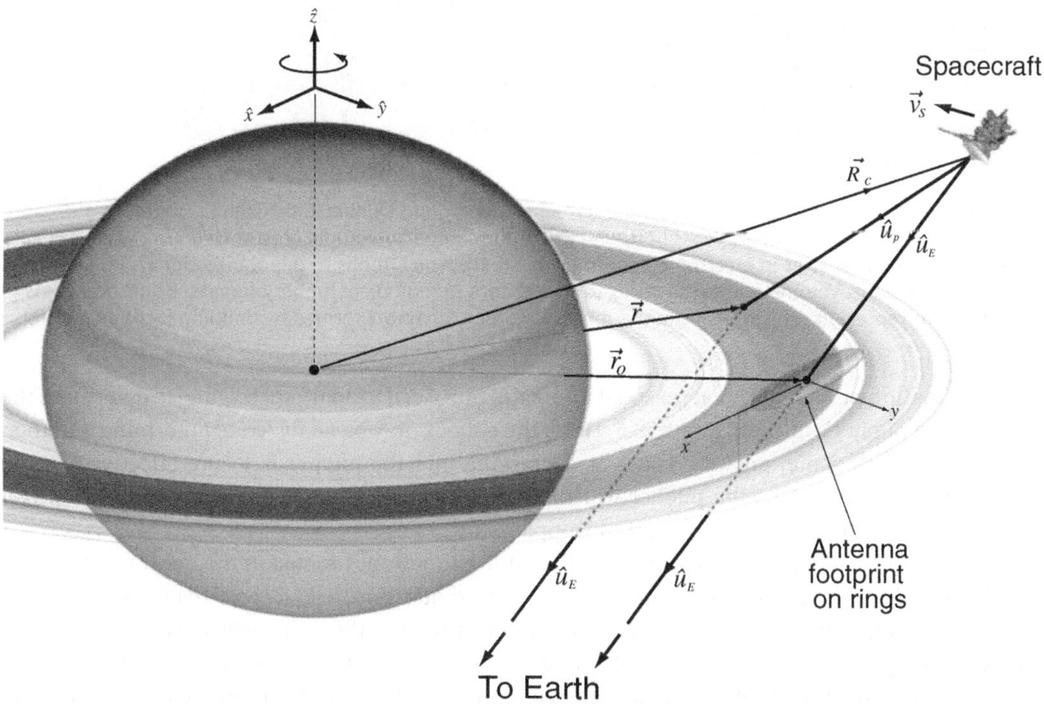

Figure 8.4 Planetary astronomy. Earth- and spacecraft-based radar studies of other planets provide unique views of their surfaces and ring systems. A great deal of our detailed knowledge of Saturn's Rings was obtained by using radio transmissions from Voyager 1 spacecraft, and receiving the echoes on Earth. Saturn's Rings have also been studied by use of specialized Earth-based radar systems. NASA's Cassini mission to Saturn employed radar studies of the rings and the surface of Saturn's satellite Titan.

size of and number density of trees, Figure 8.5. Although strikingly different in scale and details of implementation, this problem is a close cousin of Saturn Ring mass estimation problem. In tropical regions where cloud cover is a common, near-year-round feature of the climate, radar is often the only practical instrument for this purpose.

Figure 8.5 Boreal forest. Radar image of the boreal forest around Quebec City, Canada, contains information on its total biomass. Most of the image is dominated by volume scatter (see green color as an estimate of volume scatter in electronic edition to this text), which tends to depolarize signals due to multiple reflections. Volume scattering tends to occur more in regions of higher biomass. Brighter returns in the city areas represent reflections from the dihedral corner reflectors formed by building faces and streets.

An estimation problem of great seasonal interest is the use of a radar "speed gun" to determine the velocity of a pitcher's fastball, Figure 8.6. Here, the echoes may be analyzed to determine whether, during a baseball game, pitching speed is decreasing and the pitcher is losing effectiveness, or, alternatively, increasing as the pitcher warms to his task! The desired quantity is the actual ball velocity, obtained from a measurement of Doppler shift.

In summary, the difference between detection and estimation is rooted in the nature of the question being asked. For detection, we address a yes or no question, asked repetitively, perhaps, and possibly in a specific context, such as location. For estimation, the various observable parameters of the echo signal described in Chapter 2 and the information that we can extract from these are an important feature of the problem; such information is used to learn about physical characteristics of objects sensed by the radars.

8.2 Radar Detection

8.2.1 Approach

We now turn to the quantitative analysis of the detection problem. The immediate problem is that of analyzing under what conditions we can determine the presence or absence of an echo from a scattering object, given that the radar system output includes an intrinsic noise background signal

Figure 8.6 Speed guns. In sports, "Is the pitcher's fastball still holding up in the 5th, 6th, or 7th inning?" (see Figure 2.5 for block diagram.)

independent of any echo? The most direct approach to this problem is to base the decision on a comparison of possible echo output signals with a predetermined value, referred to as a threshold. Those signals exceeding the threshold are counted as "yes," i.e., an echo is deemed to present; those signals which fail to exceed the threshold are counted as "no," i.e., no echo is deemed to be present. In either instance, however, that is whether the threshold is exceeded or not exceeded, the deemed "yes" or "no" may or may not be coincident with the presence or absence of an echo.

Consider the way in which radars operate. If a target is present, some small portion of the initially transmitted radar energy is received from a scattering object of interest. If we somehow know that a target is present, we can calculate the received signal strength using the radar equation, based on an assumed or known value for the target RCS. The echo signal at the output of the receiver, however, is unavoidably corrupted by noise arising from physical phenomena of various sorts. Because "noise" is inherently a process involving irregular, unpredictable variations in the signal, a signal comprising noise alone takes on a wide range of levels from near zero to those exceeding the strength of weak radar echoes.

Our problem is to determine from an observation at *any single instant in time*, if a particular value of the output signal strength could be due primarily either to the echo from a scattering object or, alternatively, due to the chance occurrence of that output value resulting from noise alone. Conversely, observation of a weak output could result either from the absence of an echo signal or, alternatively, to an unfavorable noise condition that tends to cancel the echo signal. Obviously, when the echo signals are very large with respect to the noise fluctuations, we do not expect to have any difficulties in deciding whether or not an echo is present, i.e., between "yes" and "no." Operations with signals that are either relatively weak or comparable in strength to the noise are another matter altogether. In such instances, we can never be correct all the time, but we can decide how to choose among well-defined trade-offs based upon the known operating parameters of the radar. For this, we need to apply the theory of probability and our knowledge of the statistical properties of noise. This allows us to develop objective criteria on which to base our judgment.

8.2.2 Noise Models

Noise is unavoidable in radar systems. Although many individual sources may contribute to the noise present at the output of the radar receiver, the primary sources are those discussed previously. These are electronic noise arising from the discrete nature of charge conductors and imperfections in the electronic devices (Sections 3.5 and 3.6), thermal noise coupled directly to the receiver through lossy elements in transmission lines and signal conductors (Section 3.6.3), and noise from external natural sources introduced through the antenna system (Section 6.4). In most cases, noise can be modeled as a single composite source added to the signal at the antenna port, as in Section 3.5.2. For convenience, we must add the effect of weakly scattered echo signal background for the continuum of scatterers present in some radar environments, although we do not analyze this case.

8.2.3 Signal Detectors

Consider the following signal path in a radar receiver. A signal is captured by the antenna, equivalent noise is added, the composite signal is amplified and filtered. The amplified signal is then processed electronically, or "detected" to form the receiver output, as shown in Figure 8.7. The first part of this system is the same as that discussed in connection with Figure 3.9. The only difference here is that the detector function has been added at the right-hand side. The purpose of the detector is to convert the rapidly oscillating form of the echo signal to a more tractable variable that provides a measure of the signal strength. This conversion can be accomplished in a number of ways. To see the need for this and how it is accomplished here, consider that in this block diagram the output of the amplifier is at the same frequency as the input signal. The detector illustrated operates by inverting the negative portions of the signal to obtain its absolute value, thereby forcing the signal to be always positive, although the signal still oscillates rapidly with two peaks for every

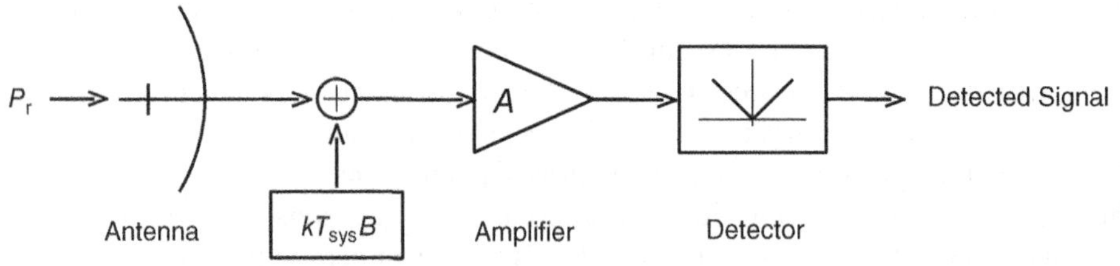

Figure 8.7 Block diagram of a simple radar detection system. System comprises an antenna for signal reception, an additive noise source representing the composite of all noise within the system, the receiver consisting of radio frequency filters and amplifiers, and a signal processor that conditions the received signal prior to a decision as to whether or not an echo signal is present. The processor shown is a piecewise-linear function that takes the absolute value of the input signal. In practice a low-pass filter follows the nonlinear operation, not shown (see Figure 8.8).

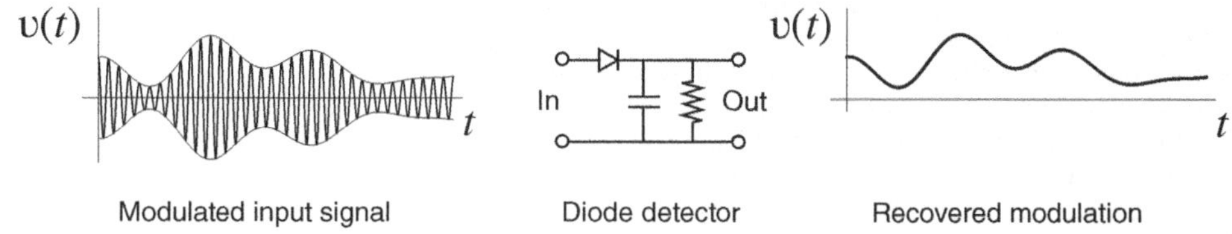

Figure 8.8 Recovery of a signal envelope by use of an analog AM detector. Modulated input signal (left) consists of high-frequency "carrier" signal with superposed, relatively slow variations in signal strength with time. Diode detector charges the capacitor to peak input voltage with each positive cycle of the carrier; pulse-charging effect of carrier is removed by filtering effect of resistive discharge of capacitor, which is designed to have much a time constant much longer than that carrier period. Recovered modulation (right) is an accurate representation of the input signal envelope. This process is insensitive to phase modulation of carrier. In terms of the analytic signal, the output here is the same as the magnitude of the complex envelope (see Section 7.1, Figure 7.16).

cycle of the carrier frequency. Averaging of these pulses over several or more cycles of the input frequency provides a smoothed output that varies on the time scale of the averaging period. That is, the detector converts the radio frequency signal to a series of pulses with heights proportional to the instantaneous signal amplitude, smoothed by a low-pass filter.

The detector of Figure 8.7 is very similar to that found in an ordinary amplitude-modulated (AM) radio, comprising only a diode which rectifies the receiver signal, followed by a low-pass filter which removes the residual carrier frequency oscillations in order to obtain the signal envelope, or amplitude, as a function of time. The amplitude contains the information of interest, the audio program, that was encoded onto the carrier in the radio transmitter, Figure 8.8. Similarly, it is most often the amplitude of the radar echo that contains primary information pertinent to the presence or absence of an object within the radar's antenna beam.

In Figure 8.7, note that the amplification and filtering steps are linear, while detection requires a nonlinear operation in order to remove the rapid residual oscillations in signal output. Historically, the detection function was accomplished by the use of thermionic and crystal rectifiers constructed of naturally occurring materials. These include simple envelope detectors—functionally the equivalent of diodes with low-pass filtered outputs—which calculate the magnitude of the signal envelope, Figure 8.8 again, and square-law detectors, which determine the average squared amplitude of the signal. With the availability of digital computation, algorithmic approaches now provide a wide range of refinements in detector design. Many types of specialized detector algorithms are in use in radar systems today. Table 8.1 compares the transfer functions of the "linear" and "square-law" detectors, and a more general digital form.

8.2.4 Received Signal Envelope as a Function of Time

Suppose we plot the envelope of the radar receiver signal output as a function of time, assuming that echoes from one or more scattering objects appear briefly at various times, as in Figure 8.9. The signal at a particular time $t_{echo} = 385$ (the units are arbitrary here), corresponding to the range delay to a scattering object, increases in amplitude above the random fluctuations caused by receiver noise. But other fluctuations in the noise may mimic this behavior. For example, is there a second reflecting object at a later time, $t_? = 80$, in the interval shown? While not as large as the echo from the object at t_{echo}, it is nonetheless larger than the typical background and must be considered as a possible second echo (see Figure 8.9). If we had a rule for judging the variations in the envelope signal in comparison to the expected noise level in the system, then we could evaluate this situation

Table 8.1 Four example detector types and their transfer functions. Simple diodes pass signals of only a single polarity. Linear detectors extract the envelope, or magnitude, of a signal, while square-law detectors extract power. Numerical processing of sampled signals can provide arbitrary functions of the input waveform, including those that are not symmetric; piecewise-continuous form is example.

Example detectors	Signal transfer functions
Diode, or rectifier	$v_{out} \propto v_{in}, \quad v_{in} \geq 0$
Linear, or envelope	$v_{out} \propto \lvert v_{in} \rvert$
Square-law	$v_{out} \propto v_{in}^2$
Piecewise-linear	$v_{out} \propto \begin{cases} a_1 \lvert v_{in} \rvert, & 0 \leq v_{in} < v_1 \\ a_2 \lvert v_{in} \rvert, & v_1 \leq v_{in} \\ \dots \end{cases}$

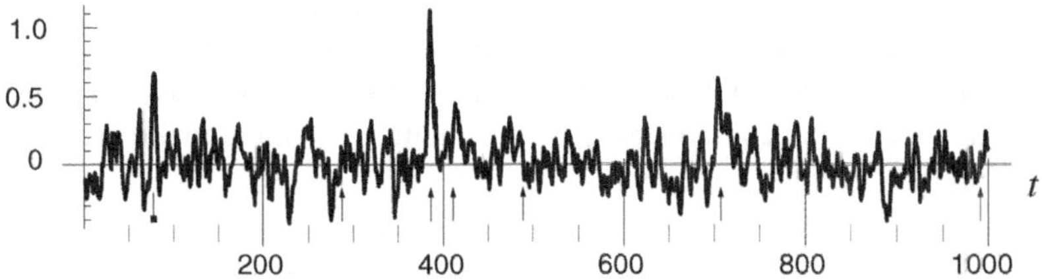

Figure 8.9 Six pulse signals in noise. Trace shown represents band-limited, time-varying signal comprising noise plus six pulses. Width of pulses is ≈ 10 units on the scale shown; independent noise variations occur at about the same rate. Pulses are located as shown by arrows, with amplitudes of, from left to right, 0.15, 1.2, 0.30, 0.03, 0.6, and 0.07; rms noise fluctuations are about 0.13. Obviously, only the two strongest pulses located at positions 385 and 705 are readily apparent, although one could argue that the signal at 413 is also visibly affecting the waveform. Note, however, that large signal at location ≈ 70, marked with a small square; no pulse is present at or close to this location. The small square indicates only a larger than usual fluctuation of the noise, comparable in strength to the signal pulse at location 705. Detection theory offers a systematic approach to deal with this uncertainty as to which of the receiver output increases are treated as though they represent echoes, and which are treated as noise.

objectively. In other words, for a particular signal-to-noise ratio (SNR) at the receiver input, the quality of our determination as to the presence or absence of an echo depends on the setting of the decision threshold. For given, well-defined circumstances, the manner in which we employ probability theory, the statistical characterizations of signals and noise, and the decision rules based on them form the basis of detection theory.

The "radar threshold detector" is the objective, automated version of the decision-making exercise we have just examined in Figure 8.7. It is implemented in a simple and straightforward manner by feeding the output of the "signal detector" to a device for making the signal threshold decision,

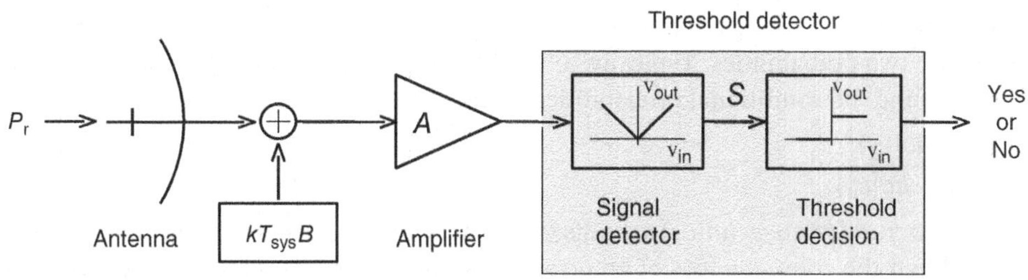

Figure 8.10 Receiver with threshold detector. In practice, a low-pass filter follows the signal detector, not shown.

where the level of the threshold is to be determined by the overall operating conditions for the radar. This is illustrated in Figure 8.10.[3]

8.3 Detection Criteria

8.3.1 Approach

We now address the question of how to set the detection threshold for a simple receiver, such as that in Figure 8.10. In order to accomplish this, we examine some implications of all possible choices. As an input to the decision, we use the output of the signal detector, v_d, in Figure 8.10, but without specifying the particular form of this detector. That is, we consider the problem in a general way that can be applied to any choice of the signal detector transfer function without restricting ourselves to the linear detector of the figure.

Suppose that average behavior of output v_d can be described by its conditional probability density functions as follows. The distribution of v_d depends on whether or not a target is present in the receiver output. If there is no echo present, so that only noise is observed, then the probability that v_d lies within $\pm dv_d/2$ of v_d is given by,

$$p(v_d|no\_echo)dv_d \tag{8.1}$$

where variable *no_echo* indicates the absence of an echo, and the notation "$p(|)$" indicates the probability density of v_d given *no_echo*. On the other hand, if an echo is present so that the output signal is a combination of the echo signal plus noise, then the probability density function (pdf), is,

$$p(v_d|echo) \tag{8.2}$$

where variable *echo* indicates that an echo is present. By the nature of the problem, there is no way to know ahead of time whether or not an echo is present—in fact, the purpose of the observation in the detection problem is to determine when an echo appears as best we can.

Given our ignorance of the echo, we cannot know which of the two distributions, that of Eq. (8.1) or Eq. (8.2), applies. Although limited in this way, we can still do something quite useful. We ask the question, given a single measurement v_d, to what degree is it more "likely" that the particular value of v_d we observe is drawn from the first or from the second of the two conditional pdfs above? In other words, *for a given, observed output signal*, v_d, what is the relationship between the

3 The terminology can be confusing, as the combination of devices comprises the "radar threshold detector," while in a circuit's context, the device for sensing the threshold crossings is also referred to as a threshold detector. So the term "detector" is sometimes ambiguous.

probabilities that an echo is present and that it is absent? We can base our choice of alternatives on a comparison of the two probabilities. Typically, a "likelihood ratio," LR, is used for this purpose. That is, we turn around the conditional probabilities to calculate,

$$\mathrm{LR} = \frac{p(echo|v_\mathrm{d})}{p(no\_echo|v_\mathrm{d})} \tag{8.3}$$

where now $p(echo|v_\mathrm{d})$ and $p(no\_echo|v_\mathrm{d})$ are pdfs for the presence and absence of an echo, respectively, given v_d. When the consequences of an error in our choice are the same for the *echo* and *no_echo* outcomes, we choose the hypothesis (i) that an echo is present when the ratio exceeds unity, and (ii) that the signal comprises only noise when the ratio is less than unity. If the consequence of missing an echo were greater than that of mistaking noise for an echo, say, then we might choose a decision point less than unity. This sort of hypothesis testing with likelihood ratios is a well-established and highly effective testing procedure. Note, however, that the likelihood ratio depends on the probabilities of the presence of a true echo conditioned on the observed value of v_d, while we know only the statistics of v_d conditioned on the existence of the echo.

We overcome the limitation above by application of Bayes' Theorem, which relates conditional and joint probability densities,

$$p(y|x) = \frac{p(x|y)p(y)}{p(x)} \tag{8.4}$$

This allows us to recast the likelihood ratios as follows. For similar outcomes, when an error occurs, we adopt the hypothesis that an echo is present when,

$$\frac{p(v_\mathrm{d}|echo)p(echo)}{p(v_\mathrm{d})} > \frac{p(v_\mathrm{d}\,|no\_echo)p(no\_echo)}{p(v_\mathrm{d})} \tag{8.5}$$

Since the distribution function $p(v_\mathrm{d})$, and in fact all probabilities, are always positive, this reduces to selecting the target-present hypothesis when,

$$\frac{p(v_\mathrm{d}|echo)}{p(v_\mathrm{d}|no\_echo)} > \frac{p(no\_echo)}{p(echo)} = S_0 \tag{8.6}$$

where S_0 is called the threshold value. The ideal procedure, then, is as follows. First, we compute the probability of the measurement v_d conditioned on each hypothesis, after which we select the target-present hypothesis if the ratio of the probabilities exceeds the inverse of the corresponding ratio of hypothesis occurrence.

The reasoning above is often applied in communication theory to implement optimal detection receivers. If it is equally probable that one or a zero is sent over a noisy channel, the prior probability of each is 0.5, and the threshold value is unity; and we simply choose the hypothesis that has greater probability given the measured value of v_d. If the decision criterion is determined in this way, it is called a *maximum likelihood criterion.* If we know that the two symbols are sent with unequal but known probabilities, the threshold is set by their ratio, and the criterion is known as *Bayes' criterion.* The threshold may also incorporate the "cost" of each outcome, that is, it may be better to be right about some outcomes than others. For example, if we were designing a system to detect an asteroid or a nuclear missile that is about to impact somewhere on Earth, we might argue that neglecting to warn people to take shelter would be worse than occasionally having them do so just in case. If these costs are well-known, then we can properly set the threshold value.

For radar detection, very often we either do not know the probability that a target is present, or we know it only very roughly. Consequently, we are unable and cannot deterministically calculate the optimum threshold stated in Eq. (8.3). Rather, we use what is called the Neyman–Pearson approach to set the conditional probability of detection based on a specific model for the probabilities.

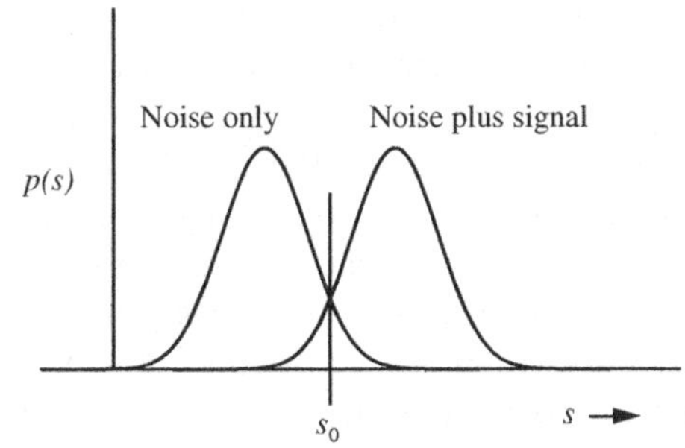

Figure 8.11 Probability density functions for echo plus noise and noise-only hypotheses. Two notional curves illustrate pdfs of radar receiver output signals when only noise is present (leftmost) and when noise and an echo signal are both present (rightmost). Curves give the probability that the output signal will be found within interval $v_d \pm dv_d/2$. Salient feature of these two signals is the shift of the pdf to higher values of v_d when an echo is present. The extended tails of the two pdfs always overlap as a result of the broad distribution of the thermal noise, which affects both cases. Vertical line at S_0 is representative of a decision threshold. Pdfs here are not specific to a particular signal detector type.

8.3.2 Decision Probabilities and the Neyman–Pearson Criterion

The Neyman–Pearson criterion follows from examining the possible outcomes of a radar detection trial, given the radar system parameters and a potential decision threshold. Plotting the pdf of the radar system output signal for the instance of noise alone and for an echo plus noise on a single graph allows us to visualize readily how well we are able to distinguish between the two cases based on the receiver output. The distribution under each hypothesis is a function of the receiver output signal v_d, which typically represents either amplitude or power depending on the type of detector used. Figure 8.11 illustrates the general nature and relationship of the curves. The probability of each outcome is obtained by integrating the pdfs to obtain the area under each distribution on either side of the threshold, S_0. Applying this procedure for various values of S_0 provides us with a method for optimizing the choice of the detection threshold.

As the signal for noise alone is always weaker than in the presence of an echo, the mode[4] of the noise-only distribution occurs at lower values of v_d than does the mode when an echo is present. Hence, the two distributions are displaced with respect to each other and, in general, also have different shapes. A central feature of the plot is the overlap between the two distributions for v_d. Assuming that the echo signal is somewhat stronger than the noise, as illustrated, then, when the signal is very weak, i.e., when the observed output distribution is close to the left-hand curve, it most likely arises from the noise-only hypothesis. On the other hand, as the output increases, the echo-present hypothesis becomes more likely. In practice, however, a decision must be made at each instant of observation, rather than over the extended period of time required to establish the form of the pdf. For this reason, the fundamental problem requires a sequence of independent judgments each based on a single measurement as to whether or not an echo is present. That is, at each instant of time, we observe a value of v_d and must reach as decision, as in Figure 8.9. We base our judgment on the output signal strength relative to the threshold S_0. The value of v_d is more likely to exceed the threshold when an echo is present and to fall below when an echo is absent. The particular value chosen must consider the "costs" and "benefits" of the outcome. Our choice for S_0—in combination with the overall system design—determines the detection characteristics of the system.

4 The "mode" of a distribution is most likely value or occurrence, i.e., the peak of the pdf.

In order to understand how this works and the implications of the overlap in the pdfs, we examine the behavior of the radar system as we vary the location of the decision threshold S_0. We have only a single threshold we can use, because we do not know ahead of time to which of the two distributions we will be subject to. In order to understand this better, consider the meaning of the threshold S_0 in the context of the pdfs plotted in Figure 8.11. For any choice of S_0, there are four possible outcomes to each observation, corresponding to whether or not an echo *is actually* present, and, in each of these instances, whether we *deem*—i.e., judge that is it most likely based on our threshold criterion—that an echo is present or absent. These possibilities are listed in Table 8.2. The probabilities of the four outcomes are not independent, however. Rather, they are coupled through the system performance parameters and the choice of S_0. We evaluate the probabilities of each outcome by integrating the density functions as a function of our decision threshold.

Assume first that there is a signal present, so that $p(v_d|echo)$ might be represented by the example density function on the right in Figure 8.12. The pdf gives the probability of the measurement v_d when an echo signal is present, while the threshold at S_0 marks the value of the detector input separating decisions of *echo* ($S > S_0$) and *no_echo* ($v_d \leq S_0$). The area under the curve to the right of the threshold S_0 in Figure 8.12 represents the probability that the "detection" is actually correct, that is, the probability P_D that we deem an echo present when this is actually so. On the other hand, the area to the left of S_0 gives the probability, P_{MD}, that we miss the detection because a combination of the specific noise realization during the measurement interval and the echo waveform fails to exceed the threshold S_0; that is, an echo is indeed present, but we incorrectly deem that the output is the result of noise alone.

Table 8.2 Possible decision outcomes. Two actual conditions and two possible decisions lead to four possible outcomes. Upper left: Echo deemed present when actually present is a "Correct detection" (D). Upper right: Echo deemed present when not present is a "False alarm" (FA). Lower left: Echo deemed absent when actually present is a "Missed detection" (MD). Lower right: Echo deemed absent when actually absent occurs when "Noise [is] Correct" (N).

	Actual condition	
Decision	**Echo present**	**Echo absent**
Echo deemed present	Correct detection (D)	False alarm (FA)
Echo deemed absent	Missed detection (MD)	Correct noise (N)

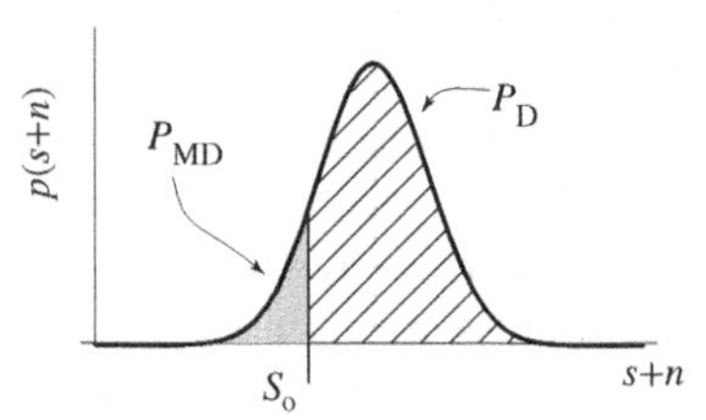

Figure 8.12 PDF of signal-plus-noise with detection threshold. $p(s+n)$ is pdf for composite signal v_d made up of echo signal in the presence of noise. As pdf is conditioned on the presence of the signal, unity area under entire curve corresponds to actual presence of an echo signal. Detection threshold S_0 divides curve into region, P_{MD}, defined by condition $s+n \leq S_0$ in which signal is *deemed* absent, and region P_D, defined by condition $s+n > S_0$ in which signal is *deemed* present. $P_{MD} + P_D = 1$.

If an echo is not present, then the situation is as shown in Figure 8.13. Although the shape of the pdf is different here, the value S_0 must be the same as that in Figure 8.12, since there can only be one threshold in a given system. Integrating the parts of this pdf to the right and left of the detection threshold gives the probabilities that we judge an echo present or absent when, in actuality, it is absent. The area under this curve to the right of the threshold represents the probability of a *false alarm*, P_{FA}. That is, while no echo is present, an unusually large excursion of noise signal leads us to choose incorrectly the alternative that echo signal is present. Finally, the area to the left of S_0 gives the probability, P_{FA}, that we correctly deem that no echo is present. Reexamination of the two pdfs and use of a common threshold shows that the probabilities of the four outcomes in Table 8.2 are interrelated and cannot be specified independently, Figure 8.14.

We choose the threshold value based on some desired balance of the four possible outcomes. In the Neyman–Pearson approach, one typically first selects a level for the probability of false alarm, and then solves for the corresponding threshold value, S_0. This procedure determines the detection probability for the system, the chance that we will make the correct choice that an echo is present when an echo is indeed present.

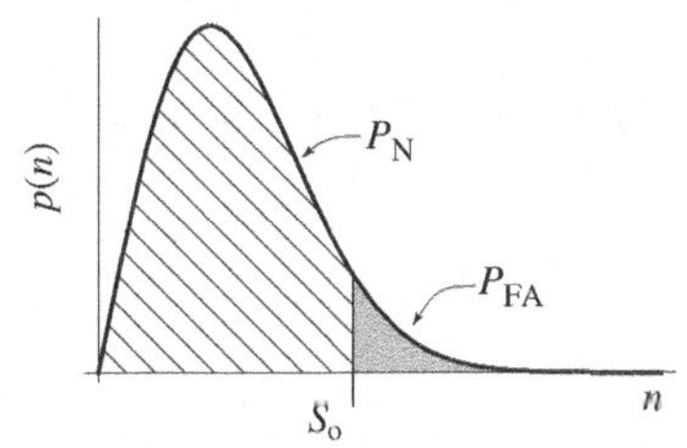

Figure 8.13 Noise envelope distribution and detector threshold shown on same scale. The area under the curve to the right of the threshold represents the probability that we claim a detection if there is no signal present, the area to left is the probability that we have correctly decided that there is no signal present.

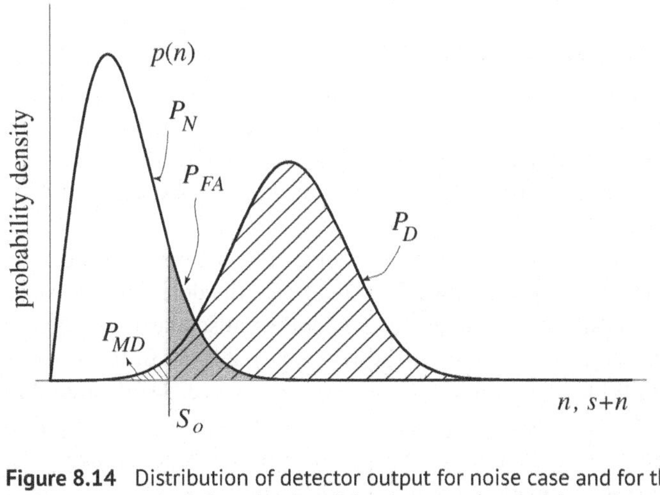

Figure 8.14 Distribution of detector output for noise case and for the case with signal present. Areas under both curves to the left and right of the detector threshold are the probabilities for all four possible decision outcomes. The right-most outline illustrates the effect on $p(s + n)$ of increasing echo strength.

Formally, the probability of a false alarm is related to the threshold value S_0 through,

$$P_{\mathrm{FA}}(S_0) = \int_{S_0}^{\infty} p(v_{\mathrm{d}}|no\_echo)dv_{\mathrm{d}} \tag{8.7}$$

In instances where the integral cannot be evaluated analytically, it is necessary to iterate Eq. (8.7) in order to find the value of S_0 corresponding to a particular choice of P_{FA}.

In practice, for most setting, "action" on a radar observation only follows a detection or when an error is made—for example, in the counting of some event. Since actions typically involve the expenditure of resources, there is a high level of interest in the rate at which false alarms occur in a given system.[5]

$$f_{\mathrm{FAR}}(S_0) = \frac{P_{\mathrm{FA}}(S_0)}{t_{\mathrm{pulse}}} \tag{8.8}$$

where $t_{\mathrm{pulse}} \approx B^{-1}$ is the effective pulse duration, as before, and B is the pulse bandwidth. Setting the false alarm rate thus refers to the process of choosing S_0 so that the probability of a false alarm gives the desired rate, f_{FAR}. This is average rate per unit time at which the detection threshold will be exceeded by fluctuations of noise alone.

Having selected a threshold value on the basis of PFA, we calculate the probability of a correct detection in a similar manner,

$$P_{\mathrm{D}}(S_0) = \int_{S_0}^{\infty} p(v_{\mathrm{d}}|echo)dv_{\mathrm{d}} \tag{8.9}$$

Recalling that the pdf, $p(v_{\mathrm{d}}\,|echo)$, is conditioned on the strength of the echo signal, Eq. (8.9) tells the probability that we expect to correctly identify an echo of a specified signal strength relative to the system noise level in the event that an echo is actually present. Here, the concept of a detection "rate" is not very meaningful, as calculation of this rate would depend on the probability that an echo is present—usually something that we do not know. On the other hand, noise is always present, so the FAR makes sense in a practical way.

In the engineering design of a radar system, what if the value for P_{D} just calculated is not acceptable? Then we must either adjust the design or arrive at some compromise between P_{D} and P_{PFA} in the design requirements. Since it is impossible to set each probability independent of the others, an engineering compromise is needed. Either some flexibility in the desired specifications must be allowed, or else the system parameters themselves (and hence the shape of the pdfs) must be reexamined.

In other words, because both probabilities depend on a single choice of S_0, we have a probability of detection that is a function of the false alarm rate. If we do not care for the detection probability resulting from our choice of false alarm rate, we may choose another false alarm probability and thereby obtain a new detection probability. We compare these two probabilities until we arrive at a workable compromise that is helpful for the problem we are trying to solve. As illustrated in Figure 8.14, a choice of a threshold S_0' to the right of S_0 greatly reduces P_{FA} while reducing P_{MD} somewhat. Increasing S_0 to S_0', in the absence of an increase in signal strength, reduces the probability of false alarms while significantly lowering the probability of detection.

The selection of an operating threshold, which follows from the choice of probabilities for false alarm rate and detection, can be different for each application of the radar system. Substantially

5 In a military context, an "action" can be very expensive indeed. In the days of the "Cold War," roughly 1946–1989, a false detection of a ballistic missile launch could have been disastrous!

different operational goals and measurement objectives typically require very different values for the two probabilities. As an example drawn from national security, suppose that we are implementing a radar system for the detection of Earth-crossing asteroids.[6] In this instance, it is important to set the probability of detection very high, for a failure to detect even a single asteroid on a collision course with Earth could lead to cataclysmic consequences. Under such circumstances, we might tolerate a higher false alarm rate from the radar in order to increase the probability of detection, and depend on additional information from other sources to determine whether or not each deemed detection is valid. On the other hand, too high a false alarm rate inevitably overloads auxiliary systems, possibly resulting in an inappropriate response, such as that in the story of Chicken Little!

This type of approach can be applied in many ways. Suppose we are using a medical imaging system to locate dangerous tumors within a human body. In this instance, we might want to set the probability of false alarms low so as to avoid unnecessary therapy because many curative procedures carry significant risk to the patient. On the other hand, we do not want to set the FAR so low as to miss cases of disease—this is always a trade-off of a desired outcome versus an undesirable outcome. Another case might be that of a weather radar, in which we might want to balance both types of errors. Accurate daily forecasts carry economic and personal benefits by allowing us to make better use of our time. But an inaccurate forecast of dangerous weather is not a disaster, although it might be inconvenient. Accurate real-time tracking and path prediction of mesoscale and smaller severe storms—which is accomplished on the basis of Doppler weather radar—is critical to safety in many parts of the world, and can tolerate little error. In yet another area of everyday life, many of us might agree that automobile intrusion alarms have often P_{FA} set much too high! Each type of detection problem exists in its own context, leading to different criteria as to what is the proper balance between the probabilities of false alarm and detection. As we can see from these examples, the general detection criteria described here can be applied to a wide range of problems, and are quite general as to the observed or theoretical signal distributions applicable to a given problem.

8.4 Statistical Representations of Signals and Noise

We now apply the general theory to some specific pdfs often encountered in radar systems, and also examine the practical effects of choosing particular decision criteria and thresholds. In order to develop accurate detection algorithms, we must have accurate knowledge of the behavior the receiver output both when echo signals when targets are present in the antenna beam and when only noise is present, in the absence of echoes. This requires consideration of the signal detector output, v_d, in both situations.

Receiver outputs are actual voltages, so for this purpose, it is instructive to work with a "real" representation of the time-varying waveform. We already know from our work with the analytic signal that real waveforms can be described in terms of independent in-phase and quadrature (I and Q) components (see Section 7.1.1, esp. p. 122). Drawing from this, but using different notation, let the echo signal alone as a function of time be represented as,

$$s_s(t) = a(t) \cos 2\pi f_c t + b(t) \sin 2\pi f_c t \tag{8.10}$$

6 These are leftover remnants from the formation of the solar systems whose orbits about the sun cross that of Earth.

where $a(t)$ and $b(t)$ vary with time over intervals of B^{-1}, the reciprocal of the signal bandwidth. In Figure 8.9, B^{-1} is an estimate for the characteristic rate of fluctuation in the signal waveform. Alternatively, describing the signal in polar form,[7]

$$s_s(t) = A_s(t)\sin(2\pi f_c t + \phi_s(t)) \tag{8.11}$$

where,

$$A_s(t) = \sqrt{a^2(t) + b^2(t)} \tag{8.12}$$

$$\phi_s(t) = \arctan\left[\frac{b(t)}{a(t)}\right] \tag{8.13}$$

and for both representations, the subscript "s" indicates the echo signal in the absence of receiver noise and interference. Over short time intervals, $\Delta t < B^{-1}$, we can treat the variables a, b, or A_s, and ϕ_s, as constants, drawn from random distributions. That is, we treat the waveform, Eq. (8.10), as a sequence of smoothly connected random, i.e., unpredictable, signal voltages. We need to understand the distribution of these voltages for the times when an echo is present and for times when no echo is received and the output comprises only noise. The various distributions of the receiver form a basis for evaluating the various detection probabilities, as indicated above.

8.5 Representation of the Noise Waveform

The noise portion of the signal is represented as a narrowband process using the same form as the echo signal, as mentioned earlier, but using different variable names in order to distinguish between the random character of the noise and the usually deterministic, predictable character of the echo signals. For noise,

$$n(t) = x(t)\cos 2\pi f_c t + y(t)\sin \pi f_c t \tag{8.14}$$

where $x(t)$ and $y(t)$ most are often well-approximated by Gaussian random processes of zero-mean and variance N. That is,

$$\langle n(t)\rangle = \langle x(t)\rangle = \langle y(t)\rangle = 0 \tag{8.15}$$

and

$$\mathrm{Var}(x(t)) = \mathrm{Var}(y(t)) = \frac{1}{2}\,\mathrm{Var}(n(t)) = N \tag{8.16}$$

Interpreting this quantity physically the noise variance N and signal power are in same units, so N represents the average noise power across a 1 Ω resistor. Exploiting this gives us a convenient way of describing N as power, as long as we realize that the load convention, and that in an actual circuit the load impedance usually has a value different from unity.

The envelope or amplitude of the noise-only signal, again assuming that x and y vary slowly on a time scale of B^{-1}, is

$$A_n(t) = \sqrt{x^2(t) + y(t)^2} \tag{8.17}$$

7 Drawing on our work with the analytic signal, radar waveforms can be represented in the complex plane as a rotating phase with time-varying amplitude A_s and phase ϕ_s.

which has a Rayleigh distribution,

$$p(A_n) = \frac{A_n}{N^2} \exp \frac{-A_n^2}{2N^2} \tag{8.18}$$

This is the same pdf as that for the scattered field magnitude from a large collection of scatterers, Eq. (5.47), with the exception that the variables now represent ensembles of random signals from a variety of sources, rather than a random superposition of scattered electric fields. An example of the Rayleigh distribution is given in Figure 8.13, and is also the noise pdf in Figure 8.14.

The phase of the noise-only signal

$$\phi_n(t) = \text{arc tan}\left(\frac{y(t)}{x(t)}\right) \tag{8.19}$$

is uniformly distributed over the interval $(0, 2\pi]$ (see Eq. 5.31). Because the noise process is accurately represented as a zero-mean Gaussian process, simply knowing the average noise power level suffices as a complete statistical description of the signal in the case of noise alone.

In many applications, it is also useful to know the spectrum of the noise. The analytic signal representation of the noise and its spectrum are related by the transform pair,

$$\tilde{\phi}(\tau) = 2\int_{\infty}^{\infty} \tilde{\Phi}(\omega)d\omega/2\pi \tag{8.20}$$

$$\tilde{\Phi}(\omega) = \frac{1}{2}\int_{\infty}^{\infty} \tilde{\phi}(\tau)d\tau \tag{8.21}$$

In many circumstances, it is useful to think of the noise signals in actual system as the result of smoothing ideal "white noise" by passing it through a filter. While noise is fractal in nature and, as such, is much too complex to describe simply. Nevertheless, it is convenient to consider the noise output of a narrowband filter as the result of white noise applied to the input. For this purpose, we need to know the relationship of the power in white noise to the noise signal strength. Using the same definition of the variance as in Eqs. (8.16) and (8.18), immediately above, we can write,

$$\tilde{\Phi}(\omega) = \tilde{N}/2, \qquad \tilde{\phi}(\tau) = \tilde{N}\delta(\tau) \tag{8.22}$$

where $\tilde{\Phi}(\omega) = \tilde{N}/2$ is the noise power spectral density and $\tilde{\phi}(\tau)$ is the correlation function of the noise.

8.6 Echo Signals in the Presence of Noise

Now we examine the realistic case in which a scattering object reflects a signal with amplitude components $a_e(t)$ and $b_e(t)$, which we receive in the presence of noise. The composite signal is in the linear superposition of its two components,

$$s_{s+n}(t) = (a_e(t) + x(t)) \cos 2\pi f_c t + (b_e(t) + y(t)) \sin 2\pi f_c t \tag{8.23}$$

Amplitude and phase do not combine linearly, however. These become,

$$A_{s+n} = \sqrt{(a_e(t) + x(t))^2 + (b_e(t) + y(t))^2} \tag{8.24}$$

$$\phi_{s+n} = \text{arc tan}\left(\frac{b_e(t) + y(t)}{a_e(t) + x(t)}\right) \tag{8.25}$$

The question is, "What is the expected statistical behavior of this more complex, combined echo signal and noise waveform?" In order to address this, we first simplify the problem by assuming, for our purposes here, a non-fluctuating target so that $a_e(t) = a_e$ and $b_e(t) = b_e$ are known constants. With this condition, the composite signal envelope becomes,

$$A_{s+n} = \sqrt{(a_e + x(t))^2 + (b_e + y(t))^2} \tag{8.26}$$

at a phase angle of,

$$\phi_{s+n} = \arctan\left(\frac{b_e + y(t)}{a_e + x(t)}\right) \tag{8.27}$$

In this instance, the echo signal amplitude is a constant.

$$A_s = \sqrt{a_e^2 + b_e^2} \tag{8.28}$$

The corresponding pdf for the amplitude of the envelope with constant signal can be calculated in a straightforward manner, although it's a bit tedious to do so. Carrying out the calculation, we obtain,

$$p(A_{s+n}) = \frac{A}{N} \exp\left(\frac{-(A^2 + A_s^2)}{2N}\right) I_0\left(\frac{AA_s}{N}\right) \tag{8.29}$$

which is the *Rice* pdf.[8] Function $I_0(x)$ is a modified Bessel function of the first kind of zero order defined for all x. Here, though, $AA_s/N \geq 0$, as the signal amplitudes are always zero or positive. The modified Bessel function for positive real arguments is shown in Figure 8.16. Observe that $I_0(0)$ has value unity at $x = 0$, and rises steeply from this value for x greater than about 1–2.

The Rice distribution can be visualized as arising from the phasor sum of its signal and noise components, as illustrated in Figure 8.15. Keep in mind that the signal phasor representing the echo signal rotates at the frequency of the transmitted signal or, in the receiver, at an intermediate frequency within the receiver. The noise phasor travels around the origin at the end of the rotating signal phasor.

We check our result in Eq. (8.29) by examining the asymptotic pdf predicted under conditions corresponding to very low and very high values of the SNR $= A_s^2/N$. Beginning with the case in which the signal power is very weak relative to the noise, from Figure 8.16 and Eq. (8.29), we see that,

$$\lim_{A_s \to 0} p(A_{s+n}) = p(A_n) = \frac{A}{N} \exp \left.\frac{-A^2}{2N}\right|_{A_n = A} \tag{8.30}$$

That is, as the echo signal power disappears, the Rician pdf, as mentioned earlier, reduces to the Rayleigh pdf corresponding to noise alone. This can be seen by realizing that the origin of A_n moves to the center of the Argand diagram as A_s vanishes, leading once again to Eq. (8.18). Referring to the expression above and in Figure 8.14, the pdf of noise alone is zero at $A = 0$, rises to a peak located at $A = \sqrt{N}$, and then decays with further increases in A. It's easy to show that the mean value of the output is $A = \sqrt{N}\sqrt{\pi/2}$.

8 Also known both as the *Rician* and the *Ricean*.

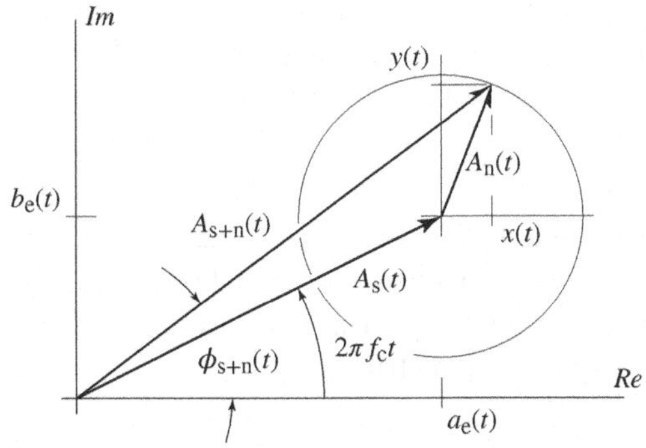

Figure 8.15 Rice phasor. In the complex plane, the Rician signal is the phasor sum of the coherent signal $A_s = a_e \cos 2\pi f_c t + b_b \sin 2\pi f_c t$ and the noise signal $A_n = x(t) + y(t)$. This sum is represented by the phasor $A_{s+n} = A_s + A_n$. Note that the noise phasor rides about the origin at a constant radius on the tip of the signal phasor, which is of constant length. The total signal fluctuates in phase and amplitude with respect to the signal phasor. At the instant shown, the net phase of A_{s+n} leads the signal phasor. But this will change as the noise amplitude and phase vary with time. As the noise phase is uniformly distributed over 2π radians, the tip of the noise vector could be located at any point on the circle indicated. Since the noise amplitude follows a Rayleigh distribution, the noise phasor is very seldom zero, to the total phasor is almost always offset in length and angle from that of the signal.

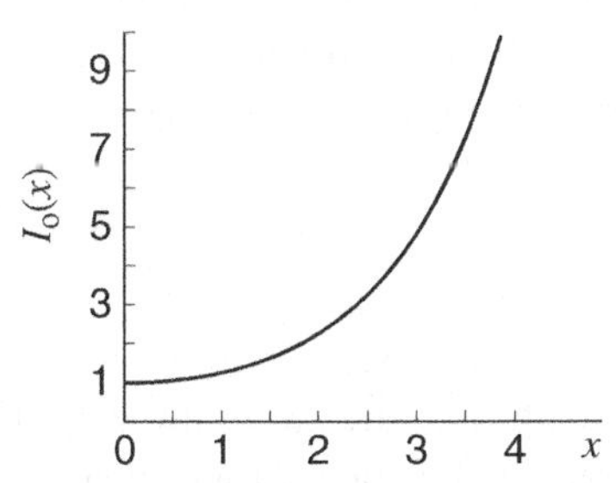

Figure 8.16 Modified Bessel function of the first kind of zero order. $I_0(x)$, $x \geq 0$, superficially resembles an offset parabola or cosh x, with a zero slope and unity intersection of the ordinate. Near the origin, $I_0 \approx 1$. There is no simple expression for the rise for $x >> 1$. The rise with increasing values of the argument is very steep—much greater than that of a parabola, however.

We also ask, what is the behavior of the pdf when the SNR is high? In this instance, $A_s^2/N >> 1$. Using the asymptotic form for $I_0(x)$ and $x = AA_s/N$, we have,

$$\lim_{A_s^2/N \to \infty} p(A) = \frac{1}{\sqrt{2\pi N}} \exp\left(\frac{-(A - A_{sig})^2}{2N}\right) \tag{8.31}$$

which is an "offset" Gaussian pdf with variance N for the total signal, A, with the mode and the mean both located at $A = A_s$. So for SNR >> 1, the practical effect of increasing or decreasing the strength of an already strong echo signal is to shift the Gaussian function along the A axis to the right or left, respectively. This useful result allows relatively simple visualization of the probability distribution function and estimation of signal statistics.

Linear signal detectors were difficult to build in the early days of radar, but the nonlinear properties of a simple diode operating just above its conduction threshold could be exploited to construct "square-law detectors," for which the output signals are proportional to A^2. Square-law detectors are still widely used for radars and communications systems, albeit the implementing circuits are considerably more advanced.

The statistics of square-law detectors are found by simple manipulations of the noise-only and signal-plus-noise waveforms. We retain the Gaussian pdf for the I and Q components of A_n as the independent noise portion of A_{s+n}. For each of the signals, we make the substitution,

$$S = A^2 = (a(t) + x(t))^2 + (b(t) + y(t))^2 \tag{8.32}$$

where we use S to represent the total power in the echo signal plus noise across 1 Ω. The effect of squaring this signal is to reshape the pdf. For noise alone, the new pdf can be found by straightforward but somewhat tedious computation of the new pdf. The result is a special case for χ_ν^2 distribution with $\nu = 2$, which conveniently reduces to the exponential distribution,

$$f_n(S) = \frac{S}{N} \exp\left(\frac{-S}{N}\right) \tag{8.33}$$

where N is the total noise power in the I and Q components of $n(t)$.

The statistics of the power when both signal and noise are present depend on the character of the echo signals, assuming that the noise remains as above. In instances of scattering from point targets or objects with classical geometric shapes, such as planes and spheres, as described in Chapter 10, the echo signal waveform mimics that of the transmitted signals except for the loss in signal strength, the propagation time delay, and a possible Doppler shift at times when the target object is in motion with respect to the radar. In such instances, there is no statistical variation in the received echo waveform, and while the power in both the noise and the signal contribute to an offset of the mean received power, only the noise contributes to the uncertainty in the received power. Under these conditions, the pdf of the power in the echo signal plus noise is given by,

$$f_{cs}(S) = \frac{1}{\sqrt{2\pi N}} \exp\left[-\frac{S - (A_s^2 + N)}{2N^2}\right] \tag{8.34}$$

where the subscript "cs" denotes the presence of the "coherent-signal" echo waveform.

At the other extreme, when the echo signal is randomized by scattering, the pdf of the output signal depends on the relative power in the echo and noise, and on the statistical structure of the signal. When the $\mathrm{SNR} = A_s^2/N >> 1$, and the echo signal fluctuates strongly, the mean-square power is given by $A_s^2 + N^2 = N^2(1 + \mathrm{SNR})$, and the variance by the same. From this,

$$f_{rs}(S) = \frac{1}{\sqrt{2\pi}\sqrt{N(1 + \mathrm{SNR})}} \exp\left[-\frac{\left(S - \sqrt{(A_s^2 + N)}\right)}{2N^2(1 + \mathrm{SNR})}\right] \tag{8.35}$$

where the subscript "rs" denotes the presence of a "random-signal" echo waveform.[9] As the echo signal becomes randomized, the predictability of the initially well-defined echo waveform evolves toward a wholly stochastic function. The basic superposition of the two signals remains; however, i.e., the echo and receiver noise still combine linearly. In the limit as the echo is totally transformed, the combination of the echo and noise becomes a sum of two random processes, in which the echo is one and the receiver noise is the other. As the pdf of the sum is the convolution of the pdfs, we can write down the approximate form by inspection. Recall that when any two distributions are convolved, the means and variances both add. Consequently, the means and variances of the new

9 The form of Eq. (8.35) can be understood by first considering the structure of the Rice signal, Figure 8.15, which is the sum of a deterministic, coherent echo signal and a random noise waveform.

distribution are known directly. We know that random noise has a Gaussian pdf. If we assume that the number of independent scattering contributions to the echo signal is a few more than the I and Q components, and that the phases are random, the echo signal is reasonably approximated by a Gaussian as well. Equation (8.35) is simply a new Gaussian with mean offset and variance calculated from the sum of the echo and noise means and variances.

8.7 Summary

The main points from our discussion of radar detection theory are as follows:

1. "Radar detection" refers to the method of choosing between the hypotheses that a radar echo is or is not present in a received waveform. While there are obvious solutions when the echo signals are strong, perhaps radar designers must address the limits of performance in which the radar echoes are weak, even those approaching the level of the system noise. For this reason, the question of detection performance is almost always a design issue.
2. System noise—apart from interference arising from human activities—is the result of fundamental, discrete, physical processes and cannot be avoided altogether. As a result, the output of radar receivers always contains a stochastic component that must be characterized on the basis of probability theory. We accomplish this to first order by describing the signal in terms of pdfs that allow us to calculate the expected frequency with which the signals fall in specified ranges of their physical units.
3. In many common design problems, the receiver output signal is described by one of a family of pdfs apropos several alternative system implementations.
4. As a practical matter, we most often use the Neyman–Pearson criterion to set the detection threshold, choosing the hypothesis that an echo is present when the receiver output signal exceeds the set detection threshold, and the hypothesis that no echo is present when the receiver output signal falls below this threshold. "Detection" of a radar echo by this scheme can never be accomplished with absolute certainty. At times, some echoes that are present will be missed, while at other times an echo will be deemed present as the result of noise alone exceeding any particular threshold. That is, there will always be some "false alarms" and "missed detections." With sound design, the rates of occurrence of false alarms and missed detections can be controlled and brought to acceptable limits.
5. Radar performance depends on the echo signal strength and the system design, including the level at which we set the probabilities of false alarm for a given probability of detection.

Performance limitations inherent in the detection problem require us to examine the design space carefully and, usually, to trade-off different aspects of system performance, in order to achieve a particular set of system design goals.

9

A Detection Example and Other Considerations

Given the principles of radar detection, we now wish to see how these are applied by use of an example. For this, we can use results from Chapter 8 directly. But in working out an actual problem, it is convenient to develop some additional ideas that are best motivated by example results and some practical considerations.

9.1 A Detection Example

As an example of a detection problem and its analysis, we examine a square-law detector system subject to the Neyman–Pearson criterion. This allows us to set the probabilities for detection and false alarm. Other criteria certainly exist, and result in different settings for the threshold, each with a different set of detection probabilities. For the signal probability distributions we are now using in this example, we know from Chapter 8 that square-law detection results in an exponential pdf for the noise-only case and a noncentral χ^2 pdf when a signal is present. For computational simplicity, we use a Gaussian approximation to the noncentral χ^2 when considering instances in which a signal is present in noise.

For this example, we assume that the radar target radar cross section (RCS) is randomly time varying so that the echo strength is not constant over time. Table 9.2 gives a summary of common assumptions for noise-only and signal plus noise distributions, specifically their mean μ and standard deviation σ for power measurements that go with this assumption. Here, for signal plus noise, we use the high-signal-to-noise ratio (SNR) Gaussian approximation from the right-hand column of the table. Recalling that N denotes the noise power level at the output, and that A_s^2 denotes echo power received in the absence of noise, we can rewrite the expression for σ is a slightly simpler form,

$$\sigma = N\sqrt{1 + \text{SNR}} \tag{9.1}$$

The Rayleigh and offset mean Gaussian distributions are as we pictured previously (Figure 8.14). We now add to the figure of the noise-only and signal-in-noise distributions a notional threshold, and also indicate with shading the area of the pdf for noise corresponding to "false alarms," Figure 9.1. In Figure 9.1 the shaded area in the graph represents the probability of false alarm for this system, that is the probability that we assume an object is present even though only noise is measured.

Our first step in an analysis of the detection performance of a radar is usually to solve for the threshold, S_0, that yields a desired probability of false alarm. For convenience, let the parameters of our system be given in Table 9.1.

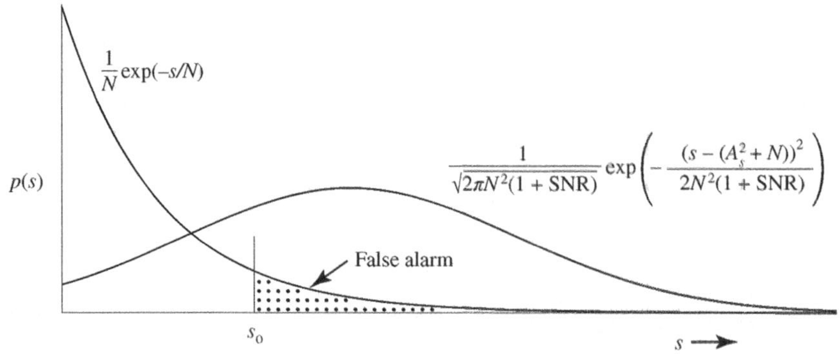

Figure 9.1 PDFs for noise power and signal plus noise power. Also shown is the area corresponding to a false alarm decision for a particular value of the threshold signal, S_0.

Table 9.1 System parameters for our example detection problem.

Parameter	Value
N	1
A_s^2	4
FA rate	0.01

The expected false alarm rate for our design is the integral of the noise pdf from the threshold value S_0 to infinity. Writing this out,

$$P_{\text{FA}}(S_0) = \int_{S_0}^{\infty} \frac{1}{N} \exp\left(-\frac{s}{N}\right) ds = \exp\left(\frac{-S_0}{N}\right) = 0.01 \tag{9.2}$$

Given the simplicity of the exponential pdf, we are able to solve the integral directly to obtain a closed-form expression for S_0

$$S_0 = -N \ \ln(P_{\text{FA}})\Big|_{P_{\text{FA}=0.01}} = 4.6 \tag{9.3}$$

Based on this result, we set our threshold power $S_0 = 4.6$.

Observe that the combination of our required false alarm rate and noise power together uniquely determine the setting of the threshold, S_0. That is, given the requirement that our system produce no more than a specified rate of false alarms, together with the noise performance of our system, we have no choice as to the setting of the detection threshold. But the detection threshold, in combination with other system design choices, also fixes the probability of detection for a specified value of the SNR.

Given the detection threshold, we illustrate this point by next evaluating the probability of detection for a signal of the strength specified in our Table 9.1. Taking the ratio of the signal and noise powers from the table, we find that SNR $= 4:1$, but the statistical properties of the signal power are not given. For now, assume that the echo signal fluctuates randomly, so that the appropriate expressions for mean and standard deviation of the Gaussian approximation for signal plus noise are those from the right-hand column, last line, of Table 9.2. With this and the numerical values from Table 9.1 the appropriate pdf is,

Table 9.2 Multipulse statistics. Averaging received signals over multiple pulses modifies mean and standard deviation, reducing output fluctuations with respect to the mean.

Quantity	Noise	Signal + Noise
Mean, μ_M	$M \cdot N$	$M(A_s^2 + N)$
Standard dev., σ_M	$\sqrt{M} \cdot N$	$\sqrt{M} \cdot N(1 + \text{SNR})^{1/2}$

$$f_{\text{rs}}(s) = \frac{1}{\sqrt{2\pi}\sqrt{N^2(1+\text{SNR})}} \exp\left(-\frac{s^2 - (A_s^2 + N)}{2N^2(1+\text{SNR})}\right) \tag{9.4}$$

The integral under the signal plus noise density function is

$$P_{\text{D}}(S_0) = \int_{S_0}^{\infty} \frac{1}{\sqrt{2\pi N^2(1+\text{SNR})}} \exp\left(-\frac{(s - (A_s^2 + N))^2}{2N^2(1+\text{SNR})}\right) ds \tag{9.5}$$

Or, more compactly, making use of the notation of Table 9.2,

$$P_{\text{D}}(S_0 = 4.6) = \frac{1}{\sqrt{2\pi}\ \sigma} \int_{S_0}^{\infty} \exp\left(-\frac{(s-\mu)^2}{2\sigma^2}\right) dS \tag{9.6}$$

where in this example $\mu = A_s^2 + N = 5$, and $\sigma = N\sqrt{1 + \text{SNR}} = \sqrt{5} \doteq 2.2$.

We must evaluate the integrals in Eqs. (9.5) or (9.6) in order to determine the probability of detection, P_{D}, given the threshold value, S_0, and our other system parameters. In this instance, this cannot be accomplished analytically in closed form, and so must be done numerically. Nevertheless, the definite integral of the Gaussian form is of such great importance in mathematical theory and practice that it has been given a standard form and a name, viz,

$$\begin{aligned} \text{erfc}(x) &= \int_x^{\infty} \frac{2}{\sqrt{\pi}} e^{-x^2}\, dx \\ &= 1 - \int_0^{x} \frac{2}{\sqrt{\pi}} e^{-x^2}\, dx \end{aligned} \tag{9.7}$$

$$\begin{aligned} \text{erf}(x) &= \int_0^{x} \frac{2}{\sqrt{\pi}} e^{-x^2}\, dx \\ &= 1 - \text{erfc}(x) \end{aligned} \tag{9.8}$$

The expression we need is of the form Eq. (9.7), the "erfc(x)," or "error function complement." Its counterpart, Eq. (9.8), the "erf(x)," is the "error function" itself. Accurate values for the $\text{erf}(x)$ and $\text{erfc}(x)$ are obtained readily by direct numerical integration or by use of standard tables for these functions.[1]

1 Tabulations of the error function are found in many texts on probability, and in Abramowitz and Stegun, *Handbook of Mathematical Functions*. These are useful for checking numerical routines, as well as for looking up specific values.

In order to make use of these standard forms, we require some further algebraic manipulation. Rewriting the probability of detection, for example, Eq. (9.6) in terms of the error function we have,

$$P_D(S_0 = 4.6) = \frac{1}{2} - \frac{1}{2}\text{erf}\left(\frac{S_0 - \mu}{\sqrt{2}\sigma}\right)\Bigg|_{\mu=5,\sigma=\sqrt{5}} \tag{9.9}$$

Evaluation of the error function for our example parameters, still with $P_{FA} = 0.01$ gives,

$$P_D(S_0)\Big|_{S_0=4.6} = 0.57 \tag{9.10}$$

For many applications, a detection rate of 57% is not very good, as we will miss almost half the echo signals at the specified SNR. Suppose, instead, we allow a false alarm rate of $P_{FA} = 0.10$, i.e., 10%. Then, repeating the steps above, we find that after this change,

$$S_0 = -N\ln(0.10) = 2.3 \tag{9.11}$$

That is, the detection threshold falls by half with this change in false alarm rate, while at the same time as the probability of detection rises to,

$$P_D(S_0)\Big|_{S_0=2.3} = 0.92 \tag{9.12}$$

So, with the reduced threshold, we expect to detect the great majority of events.[2] The top line in Figure 9.2 illustrates the exponential and Gaussian pdfs corresponding to the given example SNR and P_{FA}, and the corresponding P_D, and threshold S_0, just calculated.

Note that this change in probability of detection comes about solely as a result of accepting an increase in the false alarm rate. We did not change to N or A_s^2, parameters that depend on the *physical construction* of the radar and a given target range and RCS, as contrasted with P_{FA}, a parameter that depends on how *we decide to operate the radar*. With fixed system parameters corresponding to the physical system, the cost for improved detection performance is an increase in false alarm rate. Since these two quantities are interrelated through a common limit, their respective probability integrals—S_0 is a lower limit for calculation of P_D and an upper limit for calculation of P_{FA} (cf. Figure 9.2).

It is very useful for any particular system to understand P_D directly in terms of P_{FA}. Since these two probabilities are both functions of S_0, they can be calculated and plotted parametrically as functions of each other. We accomplish this by holding the SNR of the echo signals constant while allowing the detection threshold to vary. This information can tell us how well we expect our radar to perform for given threshold and any choice of P_{FA}. In other words, for any desired false alarm rate, we can find the corresponding probability of detection. Conversely, for any desired detection probability, we can calculate the false alarm rate. A sample set of such curves is given in Figure 9.3. These curves provide a summary of the radar performance in a detection problem. Collectively, such curves are known as "Receiver Operating Characteristics," or the ROC of the radar.

For any given system, we need to select the optimum operating point for that system. The ROC curves allow us to manually review the cost in terms of false alarms for improving the detection sensitivity. This will be different for different radar designs, and for the same design as adapted for different applications. Once a decision is made with regard to the acceptable false alarm, then the operating threshold can be found from curves or calculations of $P_{FA}(S_0)$.

---

2 Notice that $P_{FA} + P_D \neq 1$. Why should they? There is no reason that they must because they represent different pdfs corresponding to distinct *a priori* circumstances.

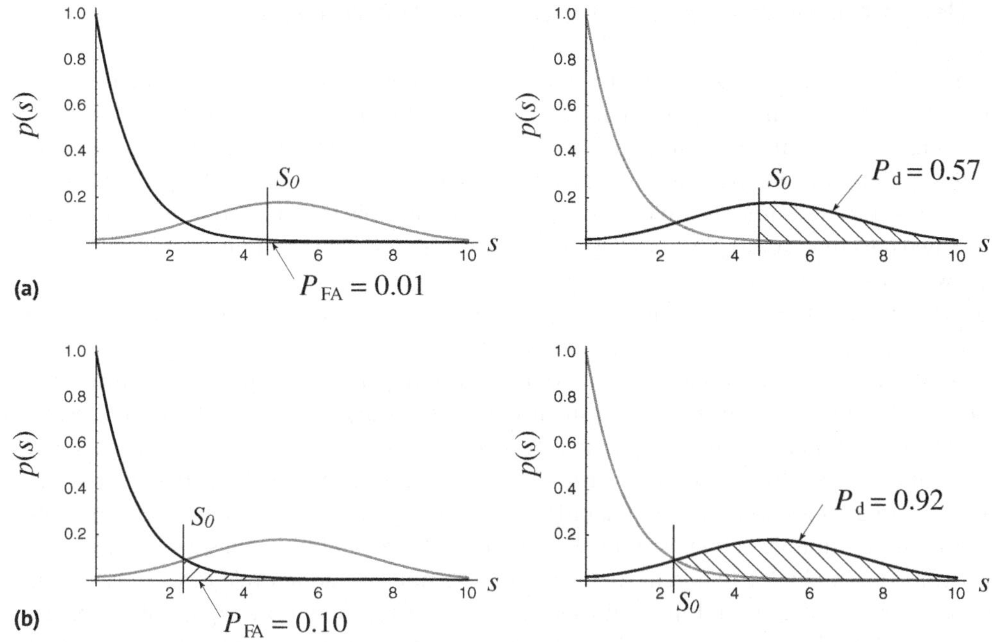

Figure 9.2 Example pdfs for SNR = 4. (a) Exponential noise and offset-Gaussian echo + noise pdfs; S_0 = 4.6. (b) Same as "a" with S_0 = 2.3. Probability of detection is improved markedly by reducing value of S_0, at the cost of a corresponding increase in the probability of false alarms. Tolerance for false alarms in practical systems depends on the application but is usually several orders of magnitude lower than in this example.

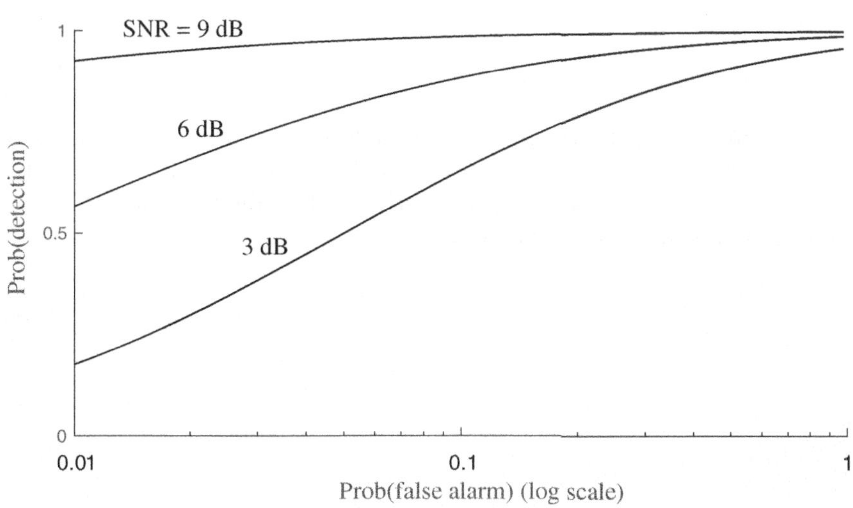

Figure 9.3 Receiver operating characteristic (ROC) curve. Relationship of detection probability to false alarm probability for various SNRs.

9.2 Use of Multiple Pulses and Incoherent Integration

As can be seen from the previous example, sometimes we find that we cannot achieve a very high detection rate while maintaining a low false alarm rate. But this example system relies on the use of a single radar echo to deem the presence or absence of a reflecting object. Often, we can reduce the variance of a measurement—equivalently narrowing the width of the probability density function—if we examine possible echo signals from multiple transmitted pulses. This would make the two pdfs representing the case of an object present and of noise-only more easily separable than when using a single pulse with the same detection threshold.

To get a feel for how this works, consider simple coin flips. For an echo signal present, let "heads," say, represent a detection and "tails" represent no detection.

Because observations of echo signals result from nearly independent samples of statistically varying noise, performing the detection experiment many times and combining the results leads to a more reliable decision than one based on only a single observation, as discussed to this point. Multiple-pulse echoes often—almost always—have more favorable noise statistics than does an observation based on a single transmitted pulse. In radar jargon, the approach of examining possible echoes from multiple echo possibilities to improve the reliability of a measurement is often referred to as "taking looks," wherein each transmission and reception cycle is a "look." In most instances, the processing of multiple pulses amounts to an incoherent averaging of the receiver output signals. That is, the phases are discarded and the amplitudes or powers summed. Mathematically, it is easy to express the multilook signal density functions for both the amplitude and square-law detector cases. The evaluation of the integrals for the linear amplitude detector case is quite complicated, and the results are usually only available numerically, so here we continue our discussion based on use of a square-law detector case where measurements of output power, rather than amplitude, are summed.

For M pulses we form the sum,

$$S = \sum_{i=1}^{M} S_i \tag{9.13}$$

where S_i is the power measured from each echo pulse. We need the pdfs of this sum for noise-alone and for signal-in-noise. The expected μ and σ for our two example cases can be calculated directly, and the results for addition of random variable are particularly simple as the means and variances of uncorrelated RV always add. Thus, here, in particular,

$$\mu_M = \sum_{i=1}^{M} \mu_{s,i} \tag{9.14}$$

and

$$\mathrm{Var}(S_M) = \sum_{i=1}^{M} \mathrm{Var}(S_i) \tag{9.15}$$

or in our summary notation,

$$\sigma_M^2 = \sum_{i=1}^{M} \sigma_i^2 \tag{9.16}$$

Another simplification results from invoking the central limit theorem to approximate the distribution of the sum of many random variables as a Gaussian-distributed random variable. This is a reasonable approximation of the final distribution for $M > \sim 4$. Therefore, we need to know only the mean and variance (or sigma) of the single-look distribution. Here the distributions both become Gaussians but at two different centers with different variances. The multiple-pulse means and standard deviations will be as listed in Table 9.2.

What is the effect taking M looks on our various probabilities, in particular the probability of detection. For a positive quantity, the uncertainty of a measurement is in a sense related to the ratio of its standard deviation to its mean. As this ratio goes to zero, the measurement becomes more and more "exact." Consider this ratio for our noise-only case,

$$\frac{\sigma_M}{\mu_M} = \frac{\sqrt{M} \cdot N}{M \cdot N} \cdot \frac{\sigma_1}{\mu_1} = \frac{1}{\sqrt{M}} \cdot \frac{\sigma_1}{\mu_1} \tag{9.17}$$

In the equation, the subscript "1" has been added to variables on the right-hand side to remind us that these quantities are from a single measurement. This result tells us that the uncertainty of a sum of observations increases more slowly as we increase the number of observations than does the mean. From the formula, relative to the mean of the sum, the relative uncertainty decreases a $1/\sqrt{M}$. Thus, as we "take more looks," we improve our knowledge of the mean, since we can express the mean of a single observation as,

$$< \mu_1 > = \frac{\mu_M}{M} \pm \frac{\sigma_M}{M^{3/2}} \tag{9.18}$$

where the brackets <> indicate the expected mean of an observation. Thus, the more observations we incorporate into our sum, the less the uncertainty in our determination of the mean, μ.

For signal plus noise, the results are similar,

$$\begin{aligned} \frac{\sigma_M}{\mu_M} &= \frac{\sqrt{M} \cdot N(1 + \mathrm{SNR})^{\frac{1}{2}}}{M \cdot (A_{\mathrm{sig}}^2 + N)} \cdot \frac{\sigma_1}{\mu_1} \\ &= \frac{\sqrt{M}}{M} \cdot \frac{N(1 + \mathrm{SNR})^{\frac{1}{2}}}{N(1 + \mathrm{SNR})} \cdot \frac{\sigma_1}{\mu_1} \\ &= \frac{1}{\sqrt{M}\sqrt{1 + \mathrm{SNR}}} \cdot \frac{\sigma_1}{\mu_1} \end{aligned} \tag{9.19}$$

In both cases, we get improvement by $\sqrt{M}$. Note also, we do better for large SNR in the noise plus signal case. If we plot the single-and multilook distributions (Figure 9.4), we can visualize the result of multilook operations as a compression of the spreads of the individual pdfs for noise alone and for signal-in-noise and, for $M > 4$ an obvious drift of event, the exponential pdf toward a bell shape. Of course, from the central limit theorem of probability theory, all pdfs of sums approach a Gaussian form as $M \to \infty$.

Figure 9.5 illustrates the effect of ten summing signals from ten looks for a system with the basic operating parameters in Table 9.1. Comparing Figures 9.1 and 9.5 show the result of summing ten looks is to reduce relative σ for both noise alone and signal plus noise by $1/\sqrt{10} \approx 3.17$. Visually, thresholds §0 = 2.3, 4.6 are now clear of the main portions of both distributions, so that detection performance is greatly improved otherwise unchanged system parameters.

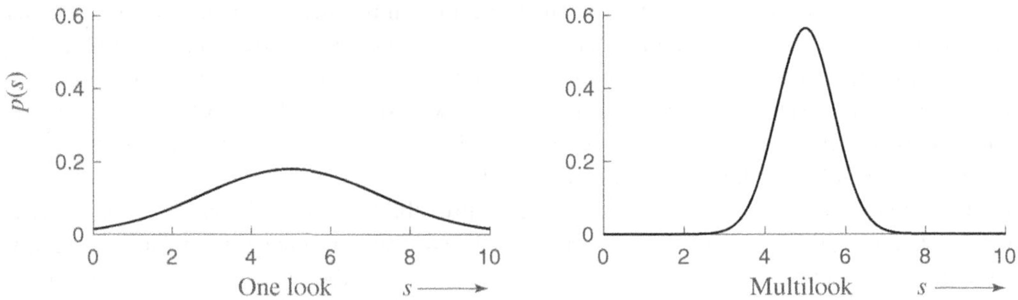

Figure 9.4 Example distributions for single- and multilook distribution pdfs. Adding more observations (looks) better defines the mean of the distribution so that estimation of the signal strength is more accurate. The case pictured is for an SNR of 4, with 1 and 10 looks.

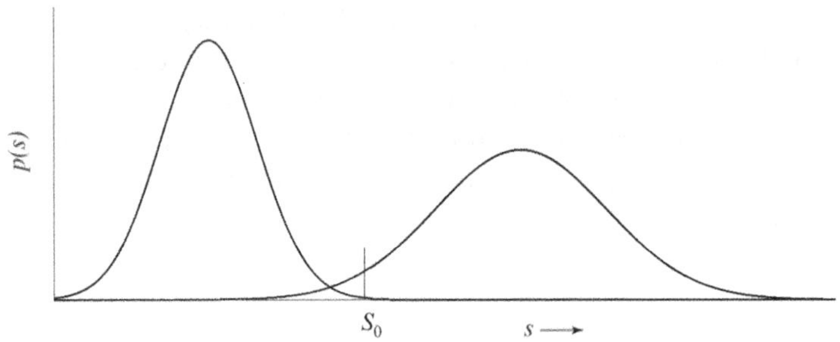

Figure 9.5 Result of ten looks from example system. Figure is derived from system of Figure 9.1 by computing pdfs for ten looks. Using the same threshold as before yields essentially zero false alarm rate. Lowering the threshold would increase the probability of detection while allowing more false alarms.

9.2.1 PDFs for Averaged Observations

General forms for convolution of Gaussian and exponential pdfs are, first for the exponential,

$$p_{\text{exp}}(s) = u(s)e^{-as} \tag{9.20}$$

and

$$p_{\text{exp}_\nu}(s) = u(s)\frac{a^{(\nu+1)}s^{\nu}e^{-as}}{\nu!} \tag{9.21}$$

where $u(t)$ is the unit step or Heaviside function and $\nu = M - 1$ is the number of convolutions. And for the Gaussian,

$$p_{\text{G}}(s) = \frac{1}{\sqrt{2\pi}\ \sigma} \times \exp\left[\frac{-(s-s_0)^2}{2\sigma^2}\right] \tag{9.22}$$

and

$$p_{\text{G}_\nu}(s) = \frac{1}{\sqrt{2\pi(\nu+1)}\ \sigma} \times \exp\left[\frac{-(s-(\nu+1)s_0)^2}{2(\nu+1)\ \sigma^2}\right] \tag{9.23}$$

where s_0 is the offset of the Gaussian mean from the origin.

It is also useful to have the normalized form of the convolved pdfs. By scaling the amplitude and abscissa of the two forms, p_{exp_ν} and p_{G_ν}, we can see the relative effect of the summing, or in this instance, averaging, since we are dividing by the number in the sum, on the relative widths of these pdfs. These forms are,

$$\bar{p}_{\mathrm{exp}_\nu}(s) = u(s)\frac{\nu a^{(\nu+1)}(s\nu)^\nu e^{-a(s\nu)}}{\nu!} \tag{9.24}$$

and

$$\bar{p}_{\mathrm{G}}(s) = \frac{\nu+1}{\sqrt{2\pi(\nu+1)}\,\sigma} \times \exp\left[\frac{-(s(\nu+1)-s_0)^2}{2(\nu+1)\,\sigma^2}\right] \tag{9.25}$$

where the meaning of the variables is the same as above. Equations (9.24) and (9.25) are the forms shown in Figure 9.5.

9.3 Fluctuating Targets: Swerling Models

Up to now, we have considered detection of a known signal-in-noise and the only variability has been in the noise component of the signal. For many targets, the signal itself fluctuates, and we must detect or estimate a "noisy" signal in the presence of other noise inherent in the system.

Why might signals fluctuate? Usually it is due to changes in radar viewing "aspect," or imaging geometry, but physical changes in the object itself can cause fluctuations. Viewing an object from different angles can cause the cross-section to change dramatically. For example, recall the case of the five scattering spheres we considered previously, when we looked at statistics of a collection of spherical scatterers. In that instance, the observed cross-section is highly dependent on viewing angle. Let's say we observe an agricultural area during an aircraft flyby in order to determine the expected crop yield later in the growing season, Figure 9.6. The signal from groups of trees within an imaging element appears to fluctuate since the various glints from tree limbs and the ripening fruit add in a manner similar to the example spheres. That is, the relative phase paths from given points within the resolution element change continuously as we flyby. As a result, the net sum of the signals from the various scattering elements fluctuates approximately according to the Rayleigh distribution.

On the other hand, the object being imaged may itself vary with time, as suggested by Figure 9.7. For example, the RCS of a tree whose branches and leaves form the scattering centers fluctuates as wind moves the limbs of the tree. Similarly, the spatial distribution of wind-driven rain in strong storms is not uniform, and varies with time, as does scattering from large grass crops, such as corn and wheat.

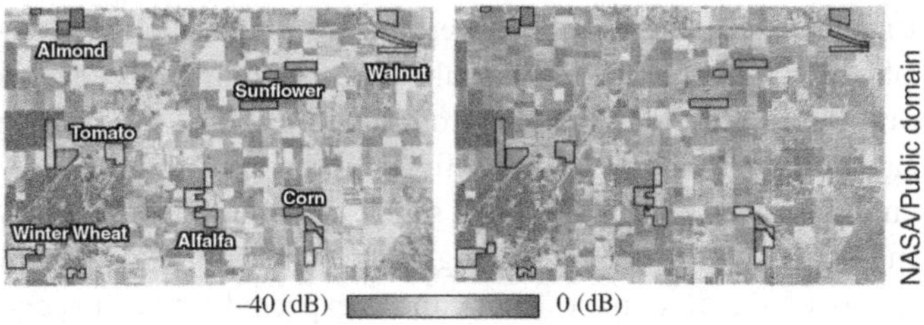

Figure 9.6 Changes in radar backscatter over time. Images acquired on 30 March 2011 (left) and 16 June 2011 (right) of an area near Sacramento, CA, show large changes in average RCS resulting from crop growth. Changes are evident for many types of crops, some of which are labeled in the left-hand image. Several fields with rather large temporal variation are outlined in black.

Figure 9.7 Examples of time-variable RCS. (Left) Wind moves branches and rustles leaves. (Right) Raindrops fall, are stirred by turbulence, and are transported laterally by wind. Such effects introduce a stochastic time-variation into the RCS even though radar is stationary and viewing aspect is constant.

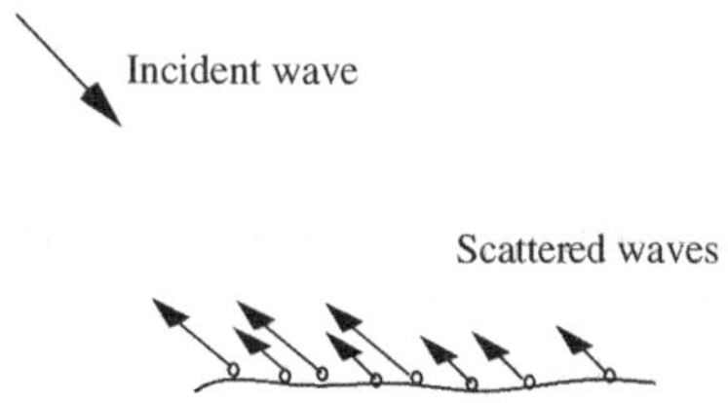

Figure 9.8 Echo signal as a sum of reflected signals. Reflected echo can be viewed as the sum of individual scattered rays from within an illuminated area, or within a processed resolution cell. As with fluctuating target, echoes are generally uncorrelated from pulse to pulse.

The backscatter cross-section of the ground itself also changes due to imaging geometry variations. A reasonable model of the ground is a random set of scattering centers spread over the area within the resolution element of a radar. Each scattering center produces a reflected-wave echo; when these add in phase, the RCS is high, and when the echo phases largely cancel each other, the cross section is small. Also, scattering from ploughed fields is known to be highly depending on the direction to the radar relative to the layout of the rows. If the spacing between scatterers is large or the density of the scatterers varies within the scattering area, then the signal distribution is described by a Rayleigh pdf amplitude (Figure 9.8).

These diverse possibilities lead to many models describing the manner in which radar echo signals fluctuate, each model applicable to a class of scattering conditions. Such models are useful in detailed design of detection and estimation systems. But the number of different types of fluctuation is small, and it is worthwhile then to sort these by some classification rule in order to simplify the number of cases that need to be examined in detection problems. Perhaps the primary scattering target classification scheme in current use is that devised by Swerling (see Bibliography). Swerling examined tbd-number models, of which four are most commonly used. The four cases devolve from the combination of the temporal fluctuation rate and the gross distribution of the relative strengths of the scattering centers. Table 9.3 summarizes the conditions for the four main cases.

The Swerling models are representative cases with a standard set of distributions and time-dependent behavior that engineers may use as a basis for comparing performance among different radar designs. Basically, they define sets of signal statistics that can be expected under a specified set of conditions. These provide specific, well-understood echo signal conditions that can be invoked by engineers in evaluating the performance of a particular design in the design of a radar sensor. Alternatively, the standard models can also be used to standardize conditions for evaluation designs by different engineers. An actual case may or may not easily be categorized as one of these

Table 9.3 Four Swerling classes for fluctuating echo characteristics. Swerling Types I–IV are often used in design specifications of radar performance.

	Target fluctuation rate	
Echo signal	**Slow fluctuation rate**	**Fast fluctuation rate**
	Type I	**Type II**
Component echoes of similar strength	Rayleigh amplitude Exponential power Signal constant for each observation	Rayleigh amplitude Exponential power Signal uncorrelated for each observation
	Type III	**Type IV**
Single echo component dominates	χ^2 (power) χ amplitude Signal constant for each observation	Same as III Signal uncorrelated for each observation

four examples. In a great many instances, however, a given application may be sorted according to the first four Swerling categories. Some examples are:

Type I—Cultural artifacts such as buildings at a great distance, or a surface revisited at slightly different geometries.

Type II—Sensing of Earth's surface or built-up cultural viewed with rapidly changing aspect; also sensing of trees, ocean.

Type III—Sensing of scattering from complex structures, including buildings for which a few surfaces or corners are the dominant source of echo energy; scattering from flocks of birds; scattering from trees with large trunks sensed at low frequencies, and collections of objects such a comets as they approach the sun and begin to break up.

Type IV—Sensing of trees in wind at higher frequencies, scattering from complex moving objects such as moving helicopters.

9.4 Example of a Corner Reflector in a Field

At this point, it is instructive to work through a more complete example. For this, we consider the question of the detectability of a corner reflector located in an agricultural field, by use of a radar with typical system parameters, and ask how well a particular system might perform. To this end, we make use of the concept of the normalized or "specific" RCS, σ^0, defined in Section 15.1. Table 9.4 summarizes the radar system and scattering parameters.

A field of corn has an empirically determined specific RCS σ^0 of −10 dB, i.e., 0.1 m^2/m^2 at a wavelength $\lambda = 2.5$ cm and an incidence angle $\theta_i = 30°$. If the radar antenna pattern projected on the ground at the 30° incidence angle limits the echo to backscattering from an area 10×10 m^2, then what is the RCS associated with this patch of cornfield? In order to address the question, we need to consider several aspects of the problem: (i) The corner reflector must be detected against the background of scattering from cornfield. So the first question is, "What is the RCS background from the cornfield?" (ii) Second, we need to know, "What is the strength of the signal from the

Table 9.4 Example radar system parameters for corner reflector in cornfield.

Parameter		
Transmitted power	100	W
Wavelength	0.025	cm
Antenna gain	46	dB
Beam resolution on ground	100	m^2
Range	400	m
System noise temperature	1000	K
Bandwidth	10	kHz
Corner reflector RCS	50	m^2
Specific RCS of cornfield	0.1	m^2/m^2

corner reflector, σ_{cr}?" (iii) similarly, "What is the contribution to the background signal from the system noise?" and finally, (iv) for the resulting SNR, "What are the probabilities of detection and false alarm for a given choice of system threshold?" This example extends our previous discussion a bit by introducing the background scattering from the cornfield as an additional "noise" source in the problem of finding the corner reflector. In order to detect the corner reflector, we must be able to distinguish its echo from that of the combination of the system noise plus the echo signals from the cornfield.

We determine the cross section of the background by multiplying the specific cross section by the resolution-cell size,

$$\begin{aligned} \sigma &= \sigma^0 \cdot \text{Area of resolution cell} \\ &= 0.1\,(\text{m}^2/\text{m}^2) \times 100\,(\text{m}^2) \\ &= 10\,\text{m}^2 \end{aligned} \tag{9.26}$$

The physical area of the corner reflector producing a 50 m^2 RCS at $\lambda = 0.025$ m is much smaller than that of the resolution element in the cornfield, so we ignore any shadowing effect this reflector has on the natural surface in the resolution element in which it is located. Because the scattering from a particular radar resolution element is random, the phase difference between signals from the corner reflector and the patch of cornfield in which it is located is uniformly distributed over 2π radians. As a consequence, we add the power from these two signals to obtain the expected power.[3]

Applying the radar equation to the above and making use of a dB table to track the various signals from the reflector, the background field, and the noise, we have the results in Tables 9.5 and 9.6.

We now have values P_r and P_n for the strength of the two unwanted signals whose sum forms our total "noise." That is, the "signal-to-noise-plus-clutter" ratio is

$$= \frac{P_{\text{x\_refl}}}{P_n + P_{\text{cornfield}}} \tag{9.27}$$

3 Note also that the reflector has no deterministic relationship on the scale of the wavelength to the other scattering elements in the cornfield. In addition, the precise range between the radar and corner reflector cannot be known to a wavelength. For these several reasons, the phase between the reflector and the composite signal from the field is unknown.

Table 9.5 Power received from corner reflector and background. dB table bookkeeping compares signal levels from corner reflector and from the background cornfield in example radar observation. Calculation is based on Eq. (3.9).

	Components of echo signal		
Radar parameter	**Corner reflector**	**Background**	
P_t	20	20	dBW
G	46	46	
$A = \frac{G\lambda^2}{4\pi}$	3	3	
$(4\pi)^{-2}$	−22	−22	
R^{-4}	−208	−208	
$\sigma = 50, 10$	17	10	
P_r	−144	−151	dBW

Table 9.6 Noise power for cornfield example. Calculation is based on Eq. (3.21).

Radar parameter	Contribution to P_n
$k = 1.38 \times 10^{-23}$	−229 dBW
T_{sys}	30
B	40
P_n	−159 dBW

where P_{x_refl} is the echo power from the corner reflector, P_n is the noise power, and $P_{cornfield}$ is the background echo signal from scattering by 100 m^2 of the cornfield. Expressed in dBW, the denominator of Eq. (9.27) is very close to −150 dBW. Subtracting this from the echo power of the corner reflector, Table 9.5, the result for the signal-to-noise-plus-clutter ratio is 6 dB, or a ratio of 4 : 1. The example of Section 9.1 discusses the detection problem for this SNR and gives the probability of false alarm and detection relationships for this level of echo strength relative to noise, and thus can use our earlier ROC curve to choose our threshold value.

9.5 Summary of Detection: Main Points

1. We can obtain numerical values for the system performance parameters if we know properties of the signals and assume forms for the signal distributions.
2. The trade-off between probability of detection and probability of false alarm is subjective, and can be expressed using the receiver operating characteristic curve.
3. Multiple pulses provide a more reliable detection, and typically increase performance as $\sqrt{M}$.
4. Targets can fluctuate, and there are agreed upon standard models for some cases.

10

Waveforms and Prototypical Systems

The three basic elements discussed previously—transmitters, receivers,[1] and their geometric relationship to scattering objects or "targets"—are combined to form radar systems. We had a brief look at one example in Chapter 2. How many ways can this be done? How do these systems work? What are the relative advantages and disadvantages of the various methods?

The basic types of radar can be categorized by their waveforms as either *continuous wave*, referred to as "CW," or as *modulated*, usually called *pulsed* regardless of the waveform. The geometry can be either *monostatic*, with colocated transmitter and receiver, or *bistatic*, with well-separated transmitter and receiver, as discussed previously. Still, these are not the only possibilities, as it is also possible to combine radar systems in "multi-static" configurations in which more than two sites share transmitting, receiving, and data processing functions. Figure 10.1 illustrates the various possibilities with the extremes of geometries and waveforms located opposite each other.

The full array of types has appeared in practice, and all have important applications. By far, the most important type in terms of installed systems is the monostatic pulsed radar, which has been developed to a high degree. In their simplest realizations, other types sometimes occur in such modest roles—such as automatic door openers—that they may not be recognized as radars. Also, in the past, systems occupying two or more sites and requiring complex signal processing have been very difficult to build. Technological advances mitigate many problems that were impractical to solve previously. For example, use of the Global Positioning System (GPS) or the Russian GLONASS

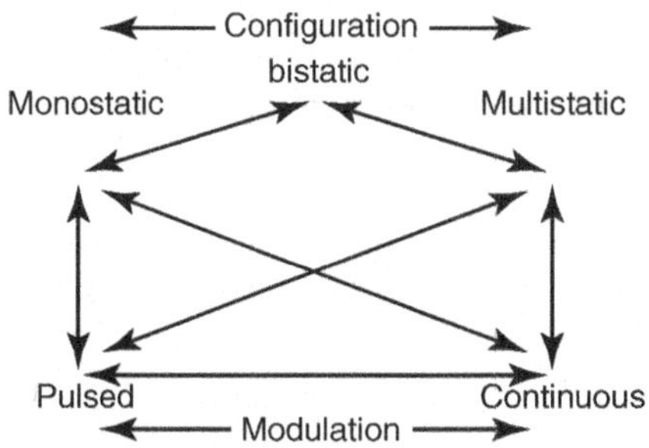

Figure 10.1 Basic radar configurations. Basic radar geometries and waveforms vary continuously between the monostatic, bistatic, and multistatic configurations and continuous wave to complex pulsed waveforms. Multistatic is unusual in practice, but includes a number of important systems, e.g., for remote sensing of Earth's ionosphere and in precision tracking of objects.

1 When used in a general sense, we intend that the terms transmitter and receiver include the necessary antennas.

makes surveys of remotely sited system elements straightforward, while digital networking technology simplifies implementation of wideband communication and data transfer between and among sites. As a result, new uses of the various possible radar configurations continue to be found, so old forms that have fallen out of favor are sometimes revived while new forms may become useful in the future. For this reason, we begin more specific discussions of radar systems by first considering the broad general questions of radar configurations and modulation. In the following we begin with the simplest types of radar, and then work our way up to more complex systems.

10.1 Continuous-wave Radars

10.1.1 Continuous-wave Bistatic

The CW bistatic radar is perhaps the simplest of all types, comprising only a transmitter and a remotely sited receiver. This system relies on detection of signal level changes in the absence of a reference from the transmitter, and is perhaps the most primitive system possible. It could be used to detect perturbations to the propagation path caused by introduction, removal, or other changes—perhaps a change in shape or structure—associated with scattering by a "target" object. A block diagram is given in Figure 10.2, which is very similar to Figure 3.1 used to define RCS. The transmitter emits an *unmodulated* continuous-wave signal which ideally is described by a cosinusoid. The receiver, in its simplest embodiment, requires only an antenna, an amplifier, and one or more filters, followed by a power monitor which we refer to as a "detector." For a fixed geometry, the output of the receiver is an attenuated, delayed version of the transmitted signal. If the transmitted signal is represented by

$$s_t = A_t \cos \omega_0 t \tag{10.1}$$

then, the received signal is just a rescaled and time-delayed version of that transmitted, or,

$$s_r = A_r \cos(\omega_0[t - \tau]) = A_r \cos(\omega_0 t - \phi) \tag{10.2}$$

where $\omega_0 = 2\pi f_0$ is the carrier frequency, τ is the propagation time delay along the scattering path connecting the transmitter and receiver, and ϕ is the corresponding phase delay. The *detector* for sensing the presence of a scattered signal and determining its characteristics follows a filter which isolates the scattered signal in frequency and limits the total associated noise. In a simple system, the detector may be little more than a diode rectifier, a capacitor, and a voltmeter. The presence of

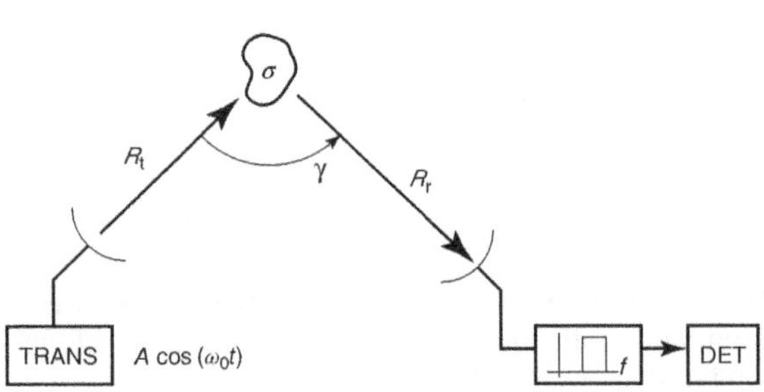

Figure 10.2 CW bistatic radar. This represents the simplest type of radar system. Transmitting and receiving sites are well-separated. Continuous-wave transmitter is used with a simple receiver comprising a bandpass filter and a detector (DET).

a target is noted by reference to the strength of the received signal. In the most basic implementation, the transmitter operates continuously, the antenna positions are fixed, and only incidental communication between transmitter and receiver is needed. The bistatic angle γ and the ranges are the critical geometric parameters.

Simple radars such as this are well-suited to sensing the presence of the scattering object. In addition, some basic measurements such as time of appearance, signal strength, and polarization can provide rudimentary—but sometimes critical—information about the target's scattering characteristics. We already understand generally how this system works from the previous chapters and our discussion of the definition of RCS.

10.1.2 Continuous-wave Bistatic Doppler Radar

The CW bistatic Doppler radar is similar to the simple CW bistatic system just discussed, except that we modify the receiver in order to make the system sensitive to motion of the target. Slightly more complex, this type of system has a number of applications including detection of meteors and their ionized gas trails, measurements of motion in particular directions, and monitoring of satellite orbits. Referring to Figure 10.3, the motion of the target introduces time variability into distances R_r and R_t, so that the propagation time delay between transmitter and receiver now changes with time, thereby, Doppler shifting the frequency of the received signal with respect to that of the transmitter, f_0. From the classical theory of Doppler shift, the change in received frequency is,

$$\Delta f = f_{\text{Dop}} = \frac{d}{dt}\left[\frac{R_t + R_r}{\lambda}\right] = \frac{\mathbf{v}\cdot(\mathbf{u}_{\text{scat}} - \mathbf{u}_{\text{inc}})}{\lambda} \tag{10.3}$$

or in terms of the velocity of the target and the wave vectors,

$$f_{\text{Dop}} = \frac{\mathbf{v}\cdot(\mathbf{k}_{\text{scat}} - \mathbf{k}_{\text{inc}})}{2\pi} \tag{10.4}$$

The actual received frequency, called the *antenna frequency*, is $f_r = f_0 + f_{\text{Dop}}$.

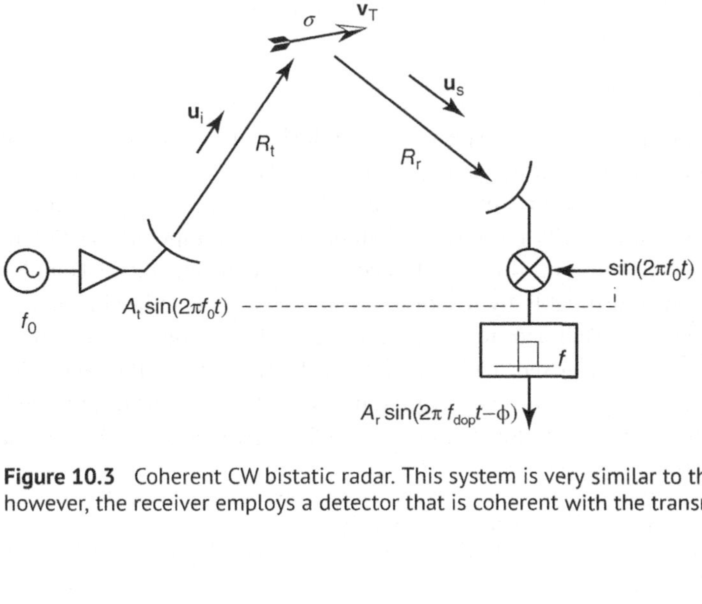

Figure 10.3 Coherent CW bistatic radar. This system is very similar to that of Figure 10.2. In this instance, however, the receiver employs a detector that is coherent with the transmitter.

Measurement of the Doppler frequency requires use of a frequency reference source, or *frequency standard*, at the receiving site. There are many ways to accomplish this. One way in which the frequency change can be measured is to multiply the received signal by a reference signal,[2]

$$s_{\mathrm{ref}} = \cos(2\pi f_0 t) \tag{10.5}$$

followed by a low-pass filter (LPF) to obtain a signal at the difference frequency, which is f_{Dop}. In Figure 10.3 the choice of transmitter signal is, for variety,

$$s_{\mathrm{t}} = A_{\mathrm{t}} \cos(2\pi f_0 t) \tag{10.6}$$

with the result that the echo at the receiver antenna becomes,

$$s_{\mathrm{r}} = A_{\mathrm{r}} \cos(2\pi(f_0 + f_{\mathrm{Dop}})t - \phi) \tag{10.7}$$

At the output of the mixer,

$$s_{\mathrm{out}} = s_{\mathrm{r}} \cdot s_{\mathrm{ref}} = \frac{A_{\mathrm{r}}}{2} \cos[(2\pi f_{\mathrm{Dop}} t - \phi] - \frac{A_{\mathrm{r}}}{2} \cos[(2\pi(2f_0 + f_{\mathrm{Dop}})t - \phi] \tag{10.8}$$

The second term at twice the transmitter frequency is eliminated by the LPF, so that only the low-frequency, or *baseband*, term remains.

In practice, this approach is a little too simple to be effective, because interpretation of the target velocity requires knowledge of the *sign* of the Doppler shift or of the absolute phase along the path in order to distinguish between a lengthening or shortening path, $R_{\mathrm{t}} + R_{\mathrm{r}}$, neither of which are available in the usual observation. Nevertheless, the example illustrates the principle of the system; we discuss the practical issues later.

In addition, there is another, implicit assumption embedded in Figure 10.3 that deserves mention. Generation of the reference signal, Eq. (10.5), requires information at the receiver as to the precise time base in use at the transmitter. This can be accomplished by use of a common frequency reference such as the GPS satellite system, for example, by use of a direct connection between the transmitter and receiver sites, or by use of synchronized oscillators. Systems that match or share time bases between the transmitter and receiver are said to be *coherent* in the sense that precise frequency and often phase relationships are established among the various system elements.

10.1.3 CW Monostatic Doppler Radar

Conceptually, the CW monostatic Doppler radar is a special case of the bistatic version. Colocating the transmitter and receiver offers considerable system simplification. However, in that it eliminates the practical need to exchange information relating time and frequency between separated transmitting and receiving sites. Variants of CW monostatic systems are in widespread use. Often, the purpose of such a system is to detect motion, which can be accomplished by sensing the Doppler shift.[3] This can be achieved readily and reliably by use of a portion of the transmitted signal as the frequency reference for the receiver. We illustrate the monostatic configuration in Figure 10.4, which shows the signal from the antenna mixed with a sample of the transmitter reference signal in order to observe the Doppler-shift in the received signal. In this system, signals from targets with changing range appear at the output of the low-pass filter at a frequency of f_{Dop}. The special

---

2 In the vernacular of the radio and radar systems engineer, the process of *mixing* two signals refers to the multiplication employed in heterodyne systems. The circuit for accomplishing this is referred to as a *mixer*.

3 Many automatic door openers operate on a similar principle of Doppler detection albeit using acoustic rather than radio techniques.

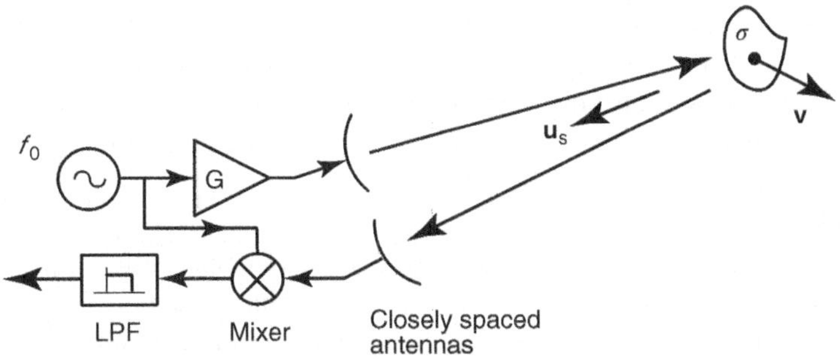

Figure 10.4 Monostatic CW radar. With colocated transmitter and receiver it becomes trivial to provide a coherent reference between transmitter and receiver. Often mixing of the transmitted and received signals is accomplished by leakage between the two antennas.

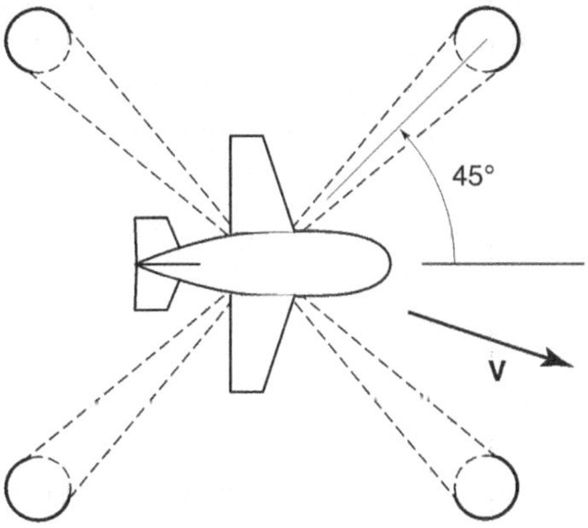

Figure 10.5 Doppler navigation system. Use of the monostatic CW system, Figure 10.4, with multiple beams provides a method for sensing velocity over the ground. Note that, as a result of winds, the velocity over the ground, $\mathbf{v}$, is not necessarily aligned with the axis of the aircraft.

case of a signal with $f_{\text{Dop}} = 0$ is shifted to zero frequency, i.e., the output of the LPF is an unvarying constant, "DC," value for echoes from targets that are at constant range with respect to the radar. Equation 10.4 for the bistatic Doppler shift can be reduced to the monostatic case by setting $\mathbf{u}_{\text{inc}} = -\mathbf{u}_{\text{scat}}$. Then

$$f_{\text{Dop}} = \frac{2}{\lambda}\mathbf{v} \cdot \mathbf{u}_{\text{scat}} \tag{10.9}$$

A few examples of applications of CW radars can provide a feel for what can be accomplished with such systems.

10.1.3.1 Doppler Navigation Radar

Doppler radar can be used to establish the path of an aircraft or other vehicle over the ground by using the Doppler effect to sense the velocity relative to the horizontal. Integration of the measured velocity then gives the path. Alternatively, the measured velocity can be transferred to another system for use as part of a comprehensive navigation solution. Figure 10.5 indicates one way in which the velocity measurement is accomplished. In a typical system, four beams are established at significant angles to the vertical; each beam measures the component of velocity projected onto the radius from the aircraft that lies along the beam axis. The difference between velocities sensed by two beams in the same plane, corrected for the angle of the beam relative to the surface gives the horizontal velocity in the plane containing the beams. The two components of **v** over the ground obtained from signals in two orthogonal planes are then combined to obtain the total motion.

10.1.3.2 Satellite Detection Radar

The problem of detecting and monitoring of objects in Earth orbit is one of considerable importance. The laws of motion are such that once an object is detected and its orbit calculated, its path can be predicted for some time into the future. Subsequently, known objects can be monitored by telescopes or precision tracking radars. But how are new objects to be detected and located the first time, and how are objects to be relocated after changes in the orbit as the result of spacecraft maneuvers or of collisions with space debris?

One solution to this problem is the use of a bistatic continuous-wave system set up for this purpose. The geometry is suggested by Figure 10.6. A broad antenna beam is oriented east-to-west, so that any satellite must pass through it at some point in its orbit; this will occur at least twice per day for low-altitude Earth orbiters. The configuration can be either monostatic or bistatic. A second, overlapping beam is used at the receiver. The passage of a satellite through the beam is indicated by the presence at the receiver of a Doppler-shifted signal consistent with orbital motion at the receiver.[4] An inexpensive variant of this system "hitchhikes" on existing transmitters built for other purposes. Television and FM radio stations usually radiate carrier signals with near state-of-the-art frequency stability. Some research projects have deployed arrays of inexpensive receivers to detect the echoes of these carriers from meteors, as discussed earlier.

10.1.3.3 Forward-scatter Fences

A similar example with a radically different application is the "forward-scatter fence," sketched in Figure 10.7. Such systems are similar to that used in the original Taylor and Young experiment

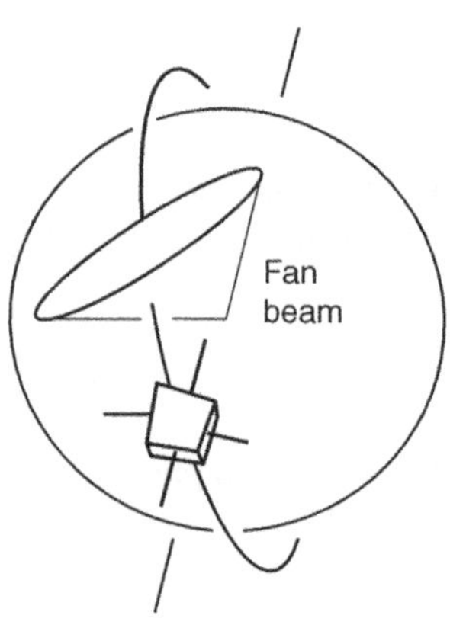

Figure 10.6 A satellite fence. A CW bistatic system with fan beams can detect all satellites within a few orbits. Such a fence is best constructed using multiple fan beams with spaced transmitters and receivers, so that the typical system would be either bistatic or multistatic (see Section 10.1.3.3).

4 Note citations in Willis, N. J., *Bistatic Radar*, esp. 2.2.5, pp. 38–ff., Artech House, Boston (1991).

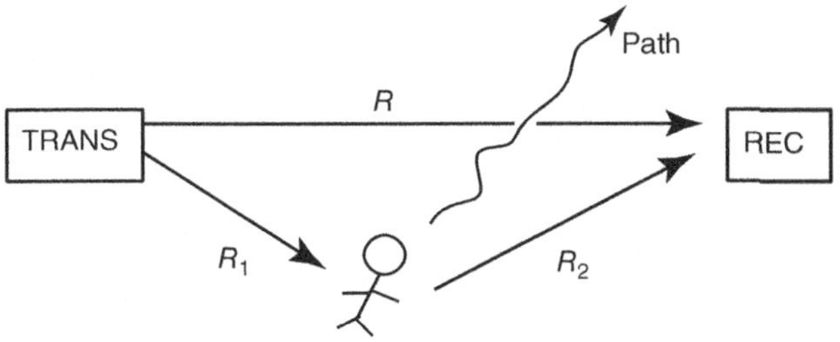

Figure 10.7 A forward-scatter fence. Such systems provide a sensitive means of detecting objects or animals crossing the fence.

at NRL. In this instance, the receiver will observe the variation in signal strength, or "beat" signals associated with interference between the direct, R_d, and variable scattered paths, $R_t + R_r$. Such a system can be acoustic or electromagnetic. Owing to the wave nature of electromagnetic and acoustic propagation, the "fence" actually fills a volume approximately centered on the geometric path. Forward-scatter fences are employed, as the figure suggests, as alarm systems for detection of intruders. At typical wavelengths the volume is too large to evade simply by stepping across.

A purely remote sensing version of the forward-scatter fence is hitchhiking on very low frequency transmissions from large military installations to study the effect of lightning strokes traveling between cloud *tops* and the upper atmosphere on Earth's ionosphere. Discovered only in the late 1990s, this phenomenon is thought to be a basic process in the connection between Earth's external environment and weather. In this instance, the system detects changes in amplitude and phase of the received signal that result from lightning-current induced perturbations to the ionospheric layer.

10.2 Modulated Radars

10.2.1 Monostatic Pulsed Radars

Pulsed radars operate by emitting short bursts of energy and detecting discrete, time-limited echoes scattered by objects of interest. In terms of numbers deployed and generality of application, these are the most important type of radar systems. Pulsed systems represent a substantial increase in complexity over simple CW systems because of the need for a modulator to control the transmitted waveform and, in the common monostatic system employing a single antenna for both transmission and reception, the need to switch the antenna rapidly and repeatedly between transmitter and receiver. Protection of the receiver from overload by the transmitter is also an important issue.

The geometry of a simple emitted pulse waveform and its echo is shown in Figure 10.8. In order to understand the sequence of pulse emission, target-wave interaction, and occurrence of the return echo, consider the result of emitting a sequence of pulses each of duration t_p in the presence of two targets, T_1 and T_2, located at ranges R_1 and R_2, respectively. The transmitted pulse emerges from the antenna in the form of a spherical wave front expanding at the speed of light, c_0.[5] The "thickness" of the wave front is $c_0 t_p$, as indicated in Figure 10.8. As the pulse passes each of the two targets in turn, some of the energy in the wave is backscattered toward the radar, where for simplicity we treat

5 More aptly, as a segment of a spherical wave surface, in the case of strongly beamed radiation.

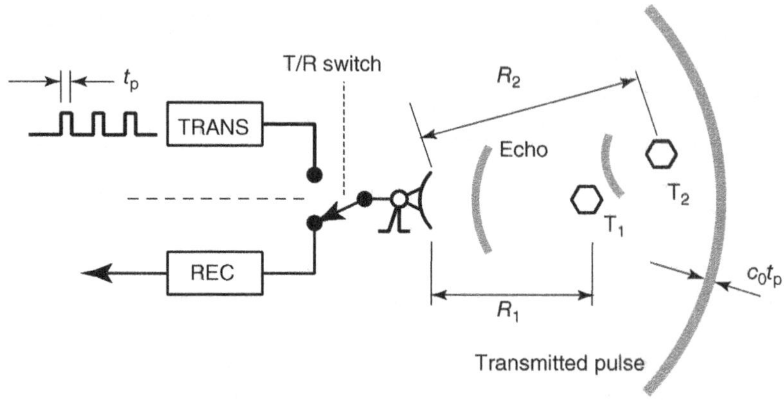

Figure 10.8 Pulsed radar geometry. A simple pulse radar with transmit/receive (T/R) switch operates by launching one or a series of pulses from a single antenna. The pulse waveform moves outward in a spherical shell at the speed of light. Echoes from objects T_1 and T_2 become centers of new shells expanding about the objects location.

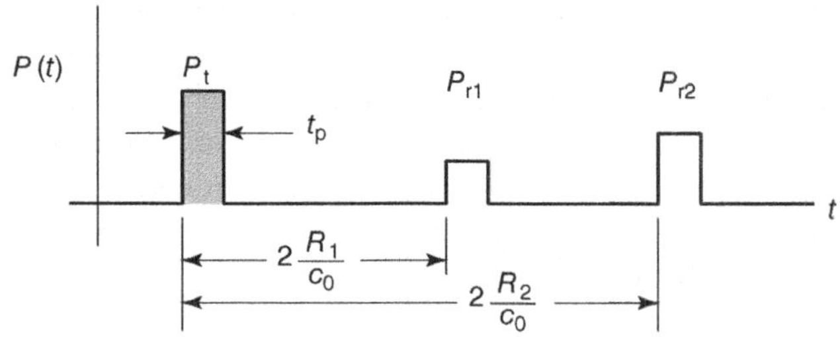

Figure 10.9 Pulse radar timing. The timing diagram indicates the received sequence and relative arrival times of the echoes; pulse symbols represent the transmitted (shaded) and received (open) power. Actual transmitted pulse waveform typically is a gated cosine at some specified frequency resembling that sketched in Figure 10.10.

the targets as scattering points. This scattered echo energy also expands in spherical wave fronts at speed c_0 and with thickness c_0t_p, but this time centered on the targets.

Figure 10.9 indicates the sequence of events at the radar. First, the transmitted pulse of power P_{trans} is radiated for time t_p. Applying our basic concepts from Section 2.3.1, echoes from the two targets are received after a delay of $2R/c_0$ from the time the transmitter pulse was radiated. Note that this time represents the delay between radiation of the *leading edge* of the transmitted pulse and reception of the *leading edge* of the received pulse, as indicated for the two targets in Figure 10.9. The sequence shown is repeated for each transmitted pulse.

10.2.2 Scanning Beam Pulse Systems

Typically, pulsed systems employ antennas with scanning beams, in the form of either large electromechanical systems or electronically steered arrays.[6] These antennas successively direct the

---

6 For example, the NASA Deep Space Network 70 m antennas are roughly the height of a 25–30-story office building; yet, much of the structure must be steered over a hemisphere to an accuracy of better than λ/D. At the highest frequency of operation $\lambda \approx 0.01$ m, the beam is sufficiently narrow as to require angular control of the

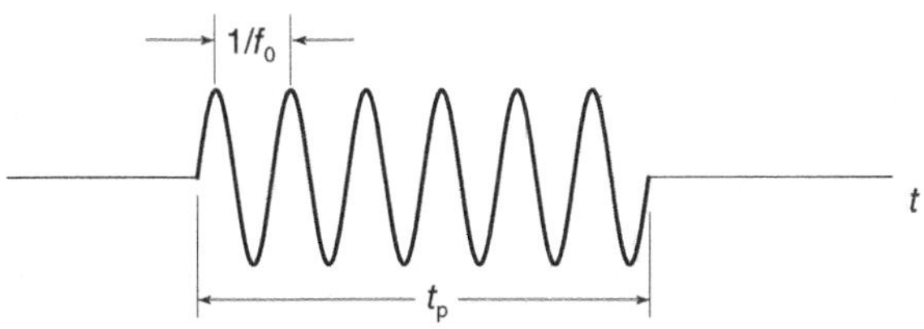

Figure 10.10 Simple pulse waveform. The waveform shown comprises six cycles of a sine function at frequency f_0. As shown, the signal begins and ends precisely at the zero crossing of the cosine—this is by no means required. In practice the envelope of the pulse is carefully controlled to shape the radiated spectrum.

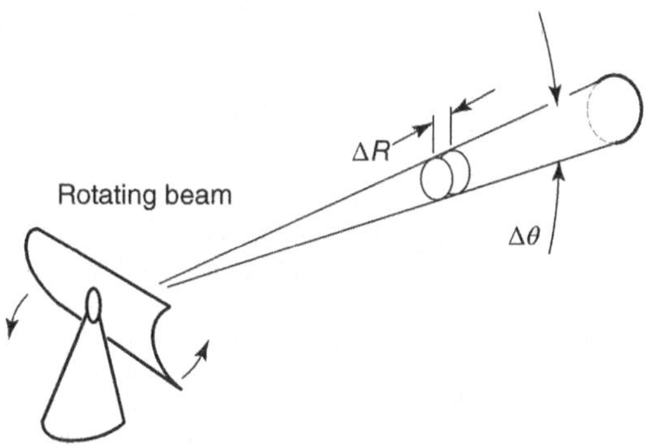

Figure 10.11 Scanning beam monostatic radar. The sketch depicts a typical mechanically scanned search radar; the search volume is divided into resolution elements defined by ΔR, $\Delta\theta$. Most such radars operate in a pulse mode.

radiated energy and the receiving beam through different, but usually adjacent, volumes of space. In this way, a single antenna can sense objects over a wide range of distances and directions, as suggested by Figure 10.11. Obviously, such a system maps target location directly into polar coordinates R, θ, ϕ.

10.2.3 Range Resolution of Pulsed Radars

The term *range resolution* of a radar refers to the ability of a radar system to distinguish two closely spaced "point scatterers" in the range direction. That is, if our targets T_1 and T_2 in Figure 10.8 were transformed into two-dimensional objects with zero thickness along the range direction, what minimum range separation is required in order that they could be distinguished from each other based on study of the radar output. For our simple system example, this can be related to the "thickness" of the radar pulse in flight, $c_0 t_p$. One way to view this is that the time required for the transmitted pulse to pass over a zero-thickness scatterer is just t_p, so the duration of the echo is the same. Upon receiving an echo from a vanishingly thin object, then, the actual change in range of the *leading edge of the pulse* is $c_0 t_p$, but taking the round trip into account, the effective thickness of the pulse in range is only $c_0 t_p/2$. Therefore, the echo duration is equivalent to that of a vanishingly thin pulse interacting with an object that has thickness $c_0 t_p/2$, the same as before. Use of a "thick" pulse produces the appearance of a target that is at least as thick. So two objects that are

pointing at the level of about 10^{-4} radians. Typical radar systems must be able to change direction quickly, however, which requires a smaller structure of phased-array technology.

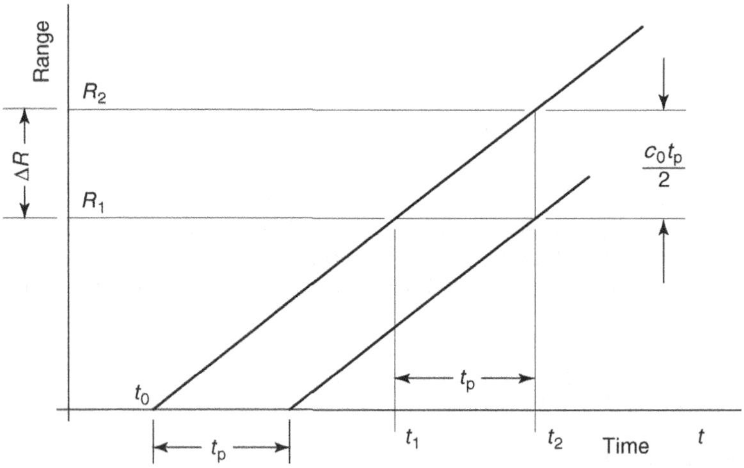

Figure 10.12 Range resolution of monostatic radar. Range of a scattering point corresponding to a backscattered echo received as a function of time after a radar transmission is plotted as lines with slope $c_0/2$. Two specific ranges are separated by $\Delta R = c_0 t_p/2$. For radar transmission emitted at time t_0 echoes from point targets at ranges R_1, R_2 are observed at times t_1, t_2, respectively. Arrival times from an extended radar pulse are determined by considering lines passing through times t_1 and $t_1 + t_p$. From the range axis, an echo observed at time t_2 corresponds to ranges R_1, R_2, and all range points in between, i.e., within ΔR. Similarly, echo of pulse of duration t_p from point range R_1 is extended in time over the interval t_p, between t_1 and t_2.

separated physically by less than about $c_0 t_p/2$ in the range direction cannot be distinguished on the basis of the radar echo.

An alternative, more analytical, way of looking at this question is illustrated in Figure 10.12, which plots radar range versus the arrival times of echoes from objects at ranges R_1 and R_2 for signals launched at times t_0 and $t_0 + t_p$. Echoes of a signal launched at time t_0 are observed at times t_1 and t_2, respectively, corresponding to the range difference ΔR. A signal launched at time $t_0 + t_p$ results in an echo from range R_1 at time t_2. All signals launched in the time interval $t_0 < t < t_0 + t_p$ produce simultaneous echoes at time t_2 from some range in the interval $R_1 < R < R_2$, showing the inability to distinguish between echoes from a span of ranges given by $c_0 t_p/2$.

10.3 Range Ambiguity in Pulse Radars

10.3.1 Geometry and Timing of Range Measurements

Although conceptually simple, the desire to operate a pulsed radar so as to continuously monitor targets leads to an interesting but fundamental complication. This arises when a sequence of pulses is radiated so rapidly that more than one pulse is in flight within the detectable range of targets. This new situation is depicted in Figure 10.13, which is similar to Figure 10.8. The difference between the two figures is that Figure 10.13 shows three transmitted pulses and a single target, as compared with the single transmitted pulse and two targets of Figure 10.8. In Figure 10.13, echo E_0 is closest to the receiver, while the last pulse emitted, pulse P_3, has not yet reached the target. Echo E_{-1}, from the transmitted pulse preceding pulse P (pulse P_{-1}, not shown), has already been received.

The timing diagram corresponding to the multiple pulse geometry is given in Figure 10.14. Echo E_{-1} is the received pulse closest in receive time to transmitted pulse P_2; Echo E_2 will not arrive at the receiver until after two more pulses are emitted, i.e., after pulse P_4 is transmitted (not shown).

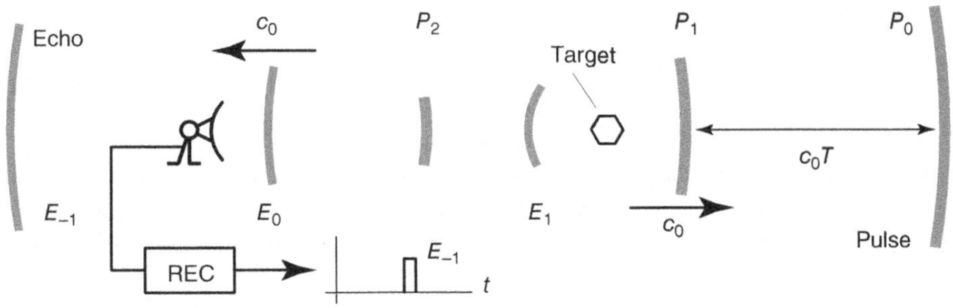

Figure 10.13 Multiple pulse geometry. This figure depicts a more general situation than that of Figure 10.8. Here, for a single representative target, there are multiple transmitted pulses in flight. The target is located beyond the range corresponding to the spatial distance between three adjacent pulses in flight (see text).

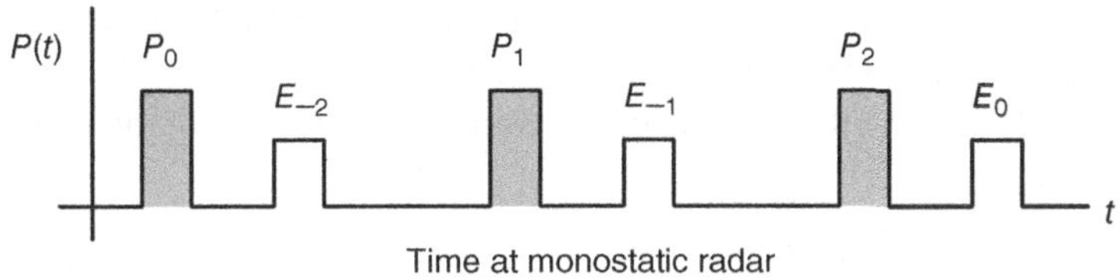

Figure 10.14 Multiple pulse timing. The timing diagram indicates that an echo from a particular transmitted pulse, as depicted in Figure 10.13 is not received until two additional pulses have been emitted (see Figure 10.13, text).

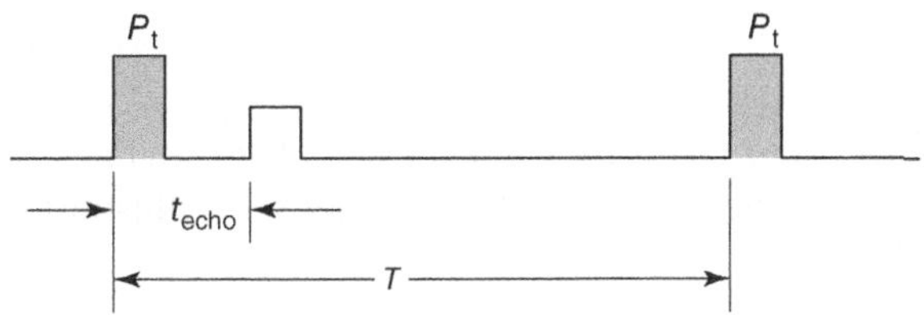

Figure 10.15 Unambiguous flight time of a pulse. The unambiguous flight time of a pulse is the interpulse interval. Any target at a range $R < R_u = c_0T/2$ has an echo time, t_{echo} corresponding to its true range (see text).

Clearly, the one-to-one correspondence between the transmitted pulse and the echoes from a given target, which is clear from the diagram of Figure 10.13, is ambiguous at the receiver because there is no label on the arriving pulses with which to sort them out. As a consequence, the range, as determined from the time of transmission to the time of reception of the *next* received pulse, is not correct for the geometry depicted because the spatial separation between successive transmitted pulses is less than the round trip distance from the radar to the target object and back.

Figure 10.15 illustrates the relationship between the flight time, t_{echo}, and the interpulse interval, T, for the emitted pulses. In the figure, the echo timing will be unambiguous when $t_{echo} < T$, where for now, we have ignored the periods of overlap between the emitted and received pulses. With this, we can write the unambiguous flight time t_u as,

$$t_u \leq T \tag{10.10}$$

to which there is a corresponding unambiguous range R_u,

$$R_\mathrm{u} = \frac{c_0 T}{2} \tag{10.11}$$

Range R_u is the maximum target distance for which there can be no confusion.

The *apparent range*, R_app, is the range inferred from the time between a pulse transmission and that of the next arriving pulse. The apparent range is different from the true range when the true range exceeds the unambiguous range. The two range measures are connected by a simple relationship,

$$R_\mathrm{app} = R_\mathrm{target} - n \cdot R_u \tag{10.12}$$

$$R_\mathrm{app} = R_\mathrm{target} - n \cdot \frac{c_0 T}{2} \tag{10.13}$$

where integer n corresponds to the ambiguity in range. The apparent range is always the true range minus an integer multiple of the unambiguous range.[7]

The *pulse repetition frequency* or PRF is the rate of pulse transmission, or the inverse of the inter-pulse interval,

$$\mathrm{PRF} = f_\mathrm{PRF} = \frac{1}{T} \tag{10.14}$$

Expressing the unambiguous and apparent ranges in terms of the PRF,

$$R_\mathrm{u} = \frac{c_0}{2 f_\mathrm{PRF}} \tag{10.15}$$

$$R_\mathrm{app} = R_\mathrm{target} - \frac{n c_0}{2 f_\mathrm{PRF}} \tag{10.16}$$

That is, the PRF and the unambiguous range are inversely related, while the offset between the true range and the apparent range is a multiple of f_PRF^{-1}. The following table provides some illustrative numerical values for the unambiguous range.

PRF	R_u
0.1 kHz	1500 km
1 kHz	150 km
10 kHz	15 km
1 MHz	150 m

(10.17)

10.3.2 Resolution of Ambiguous Range

Given the relationship between apparent range, R_app, and the PRF, Eq. (10.16), how might we resolve the ambiguity? Considering Eq. (10.16), there are two unknowns, R_target and n, since for the remaining two variables (i) the PRF is controlled at the radar and (ii) R_app is observed. We can

7 The phenomena of range ambiguity explain the occurrence of "mysterious" radar "sightings" of invisible targets. The invisible target is the echo of an actual target at a range $R_\mathrm{T} > R_\mathrm{amb}$. A search for the target at the apparent range would be directed to an incorrect target location.

generate a second equation apropos the range by differentiation of Eq. (10.13), which shows that the variation of R_{app} is directly proportional to the number of ambiguity, n,

$$\frac{\partial R_{app}}{\partial T} = 0 - \frac{n\,c_0}{2} = -\frac{n\,c_0}{2} \tag{10.18}$$

This suggests that if we change $T = 1/f_{PRF}$ slightly to $T \to T + \delta T$, then $R_{app} \to R_{app} + \delta R_{app}$, thus providing a relationship between n and the changes δR_{app}, δT. Expressing δT in terms of a change in the PRF,

$$\delta R_{app} = -\frac{nc_0}{2}\left(\frac{1}{f_{PRF_2}} - \frac{1}{f_{PRF_1}}\right) \tag{10.19}$$

This is a useful result because PRF_1 and PRF_2 are set at the radar and δR_{app} is the measurable result of the change. In effect, by using the observations for the two values of PRF we provide ourselves with two equations with which to solve for the two unknowns, n and R_{target}. Solving Eq. (10.19) for n,

$$n = \frac{2}{c_0}\delta R_{app}\left(\frac{1}{f_{PRF_1}} - \frac{1}{f_{PRF_2}}\right)^{-1} \tag{10.20}$$

while R_{target} follows from Eq. (10.12) or (10.16).

As a specific example of how this might be implemented, choose the second PRF to be an integer ratio, less than unity, of the first,

$$f_{PRF_2} = \frac{N}{M} f_{PRF_1}, \quad \frac{N}{M} \langle 1 \text{ for } \delta T \rangle\, 0 \tag{10.21}$$

Then, the associated change in the apparent range is,

$$\delta R = -\frac{nc_0}{2}\frac{1}{f_{PRF_1}}\left(\frac{M-N}{N}\right) \tag{10.22}$$

For the particular case where M is obtained by incrementing N by one,

$$M \to N + 1, \qquad \delta R = -\frac{nc_0}{2}\frac{T}{N} = -\frac{n}{N}R_u \tag{10.23}$$

This leads to the relationship between n and δR_{app} shown in Table 10.1,

In this example, the interpulse period is always increased. Thus in this scheme, a target located originally at an unambiguous range remains so!

Table 10.1 Apparent range vs. ambiguity number. For a known change in PRF, apparent range is systematically related to the ambiguity number.

n	δR_{app}
0	0
1	$-D\frac{c_0}{2}\delta T$
2	$-2D\frac{c_0}{2}\delta T$
⋮	⋮

10.3.2.1 Example with Two Targets

Figure 10.16 provides an example using two targets in which the PRF is increased between the upper and lower panels. Comparing the two cases, the echo time delays, as measured from the initiating transmitter pulse, are the same in the two sketches. The difference in the two panels is that the interpulse interval between two successive transmitter pulses is less in the lower panel than in the upper panel. Because target T_1 is at an unambiguous range for both PRFs employed, i.e., $n = 0$ for the first target, the apparent range of T_1 is the same in the two panels. Target T_2, however, lies beyond the ambiguous range, R_u, for this system, the ambiguity for T_2 is $n = 1$. Therefore, the apparent range of T_2 changes—it increases in this example—when the PRF is increased.

10.3.2.2 Implications of Range Ambiguities

What do we lose if we choose PRFs such that the target range is ambiguous? Nothing—provided that (i) we have sufficient ancillary information to resolve the ambiguity, or (ii) our system allows for a solution to target range ambiguity, and (iii) we are not concerned with the targets within the blind range interval. Under such circumstances, we may not care at all if $R_{\text{target}} > R_u$. At the same time, there can be advantages to using a small R_u, since small values of R_u imply the use of a high PRF, which increases the average power illuminating the target object.

10.3.2.3 Blind Ranges

There are also range intervals at which objects cannot be detected by a monostatic pulsed system. These are referred to as *blind ranges*. Blind ranges result from the dual use of a single antenna for transmission and reception in a monostatic system, which inherently precludes reception of echoes during transmission intervals. Quantifying this, blind ranges occur during the intervals when the transmitter is operating (cf. Figure 10.9),

$$\Delta R_{\text{blind}} = \left(R_u, R_u - \frac{c_0 t_p}{2} \right) \tag{10.24}$$

since during these intervals the transmitter is turned on and the receiver is "blinded" by radiation from the transmitted pulse, which is always much stronger than an echo from a distant target.

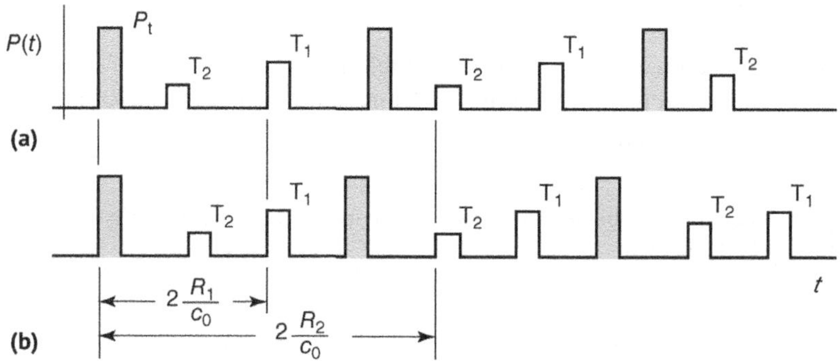

Figure 10.16 Resolving ambiguous range. (a) Example timing diagram in which target T_1 is within the unambiguous range, while target T_2 lies just past the unambiguous range. (b) Result of increasing the PRF with respect to that of the upper panel; the apparent range of the unambiguous target, T_1 remains the same, while that of T_2 increases.

In most designs, the receiver is disconnected from the antenna in order to protect it from the transmitter pulse.[8] Blind ranges can be eliminated by systematically varying the PRF.

How is this different from a continuous-wave monostatic system?

10.3.3 Example of a Radar "Looking" at the Moon

Our moon is certainly a common radar "target." The moon has been studied by powerful ground-based radars for many years, first in order to understand its characteristics as a radio relay reflector, and later to study and understand its surface properties, to determine its precise shape, and to refine its orbit for geophysical and astronomical purposes.[9] With present technology, echoes from the lunar surface can be obtained readily with amateur equipment. For some powerful communication and radar systems, the moon is considered more of a nuisance than an interesting object since it also can act as a source of unwanted echoes arising from range ambiguities beyond R_u.[10]

For radar studies of the moon, in many instances it is desirable to have no more than one pulse enveloping the moon at any one time, setting an upper bound on the PRF. If we choose, for example,

$$f_{\text{PRF}} = 15 \text{ Hz} \tag{10.25}$$

the corresponding pulse separation in space is 20,000 km, which is considerably greater than the 3476 km diameter of the moon. At the same time, the distance between the centers of Earth and the moon is approximately 400,000 km, which is 40 times greater than the unambiguous range at a PRF of 15 Hz. The distance between a radar on Earth's surface and a point on the Earth side of the lunar surface is less than this by up to about 8000 km, so we can have 39 pulses in flight. The geometry is given in Figure 10.17.

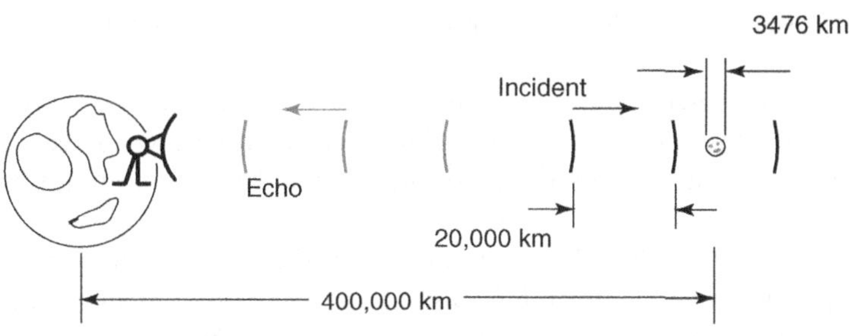

Figure 10.17 Radar looking at the moon. Figure is sketched for a PRF of 15 kHz, corresponding to a 20,000 km pulse separation. The pulse separation is such that only a single pulse intercepts the moon at any one time, so the "target" is unambiguously resolved in range (and as it happens, in Doppler as well). The range to the moon itself is highly ambiguous, however. Because the distance from Earth to Moon is known to centimeters, the range ambiguity is readily resolved on the basis of *a priori* information and thus is of no consequence.

---

8 For typical high power transmitters, during transmissions the receiver must be carefully isolated from the antenna in order to prevent destructive overload from exposure to high levels of transmitted energy. Consider that for a 1 MW transmitter, even tbd dB of isolation might not be sufficient.

9 The Moon is receding from the Earth at a rate of about 0.01 m per year. This is known from LIDAR measurements utilizing reflections from optical corner reflectors placed on the moon by Apollo astronauts.

10 On the other hand, radio amateurs find challenge and pleasure in establishing two-way communications among widely separated parts of the world by the use of "moon bounce" techniques.

Does the existence of this large number of pulses along the path between Earth and Moon matter? Because we know the actual distance to the moon to about 1 cm from light detection and ranging (LIDAR) measurements, a range ambiguity of 10^7 m is not an issue.[11] That is, we can always use an accurate lunar ephemeris to resolve unambiguously the time delay of an echo relative to the particular transmitter pulse that produced it.

The 15 Hz PRF used in this example is unusually low. Recall that we chose it in order that no more than a single pulse would envelop the moon at any time. In many cases, the question of echoes arriving simultaneously from multiple pulses is not a consideration as object sizes are small compared with permissible unambiguous ranges. Nevertheless, limits imposed by range ambiguities can be an extremely important issue, as we shall see when we examine the fundamental questions of radar resolution.

10.4 Comparison of Simple Pulse and Continuous-wave Radars

We have seen that we can measure a component of velocity very accurately with a CW radar, although such a system provides no information regarding range. On the other hand, we can measure range accurately with a single radar pulse and the time of arrival of its echo, which is insensitive to target velocity. A simple *pulse train* is inherently ambiguous in range, however. The ambiguity can be resolved provided that we have some freedom to manipulate the waveforms. In a sense, the use of the pulse train to measure range is similar to the use of a ruler without major divisions. One end of the ruler is at the target object, while the other extends to the radar. You can see the divisions at the radar, but you cannot count the divisions along the ruler to determine the distance to the scattering object. One does know, however, the precise separation distance between the divisions of this unmarked scale. Changing the PRF adds a second ruler with a different, known scale which is also precisely. As we have seen, this is adequate to find the true range, provided the scales are chosen carefully.

It is interesting that this method of resolving the range ambiguity by changing the scale of the ruler is implemented by changing the *frequency* of repetition of the pulses. That is, we increased somewhat the overall bandwidth and structural complexity of the sensing data by selecting two distinct choices of the PRF.

Comparison of the CW radar with the simple pulse train radar raises the extremely important question of how to measure range and range-rate simultaneously: Can we measure velocity accurately with a pulsed, "ranging" radar? Can we measure range accurately with a "CW" radar? Generalizing, What are the underlying principles and limitations in the choice of radar waveforms insofar as optimum measurements of range and velocity are concerned?

10.5 Velocity Measurement with a Pulse Radar

10.5.1 Randomly Gated Pulse Train

Measurement of target range R by determining the flight time of a simple waveform is a common motivation for use of pulse radars. An obvious question is, "Can we simultaneously determine

---

11 LIDAR is a form of radar based on LASER transmitters and telescopic "antennas." The range to the moon and its recession from the Earth have been measured using a system constructed at the Lick Observatory. Bright optical echoes are obtained from an array of corner reflectors left at their landing site by the Apollo 11 astronauts for this purpose.

the range-rate $\dot{R}$ on the basis of the Doppler effect?" If so,"How accurately can we make such a measurement?"

Consider a single transmitted pulse,

$$s(t)_{\text{pulse}} = \sqcap\left(\frac{t}{t_{\text{p}}}\right)\cos(2\pi f_c t + \phi) \tag{10.26}$$

where the frequency is now denoted f_c for "carrier," and the phase ϕ is random with respect to the start of the pulse window $\sqcap\left(\frac{t}{t_{\text{p}}}\right)$, where the pulse function is defined by,

$$\sqcap\left(\frac{t}{t_{\text{p}}}\right) = \begin{cases} & |t| \leq \frac{t_{\text{p}}}{2} \\ & |t| > \frac{t_{\text{p}}}{2} \end{cases} \tag{10.27}$$

A sample waveform is sketched in Figure 10.18. The duration of the pulse is t_{p}, as before. Note, however, that because the starting phase of the pulse is random, as drawing of a second realization will generally have a different starting phase, rather than the approximately ϕ =190° of this example. In the idealized case, the amplitude is constant, as shown.

The dilemma of using this simple pulse for the measurement of both range and range-rate is embodied in the well-known uncertainty relationship governing pulses—the more localized the pulse is in time, the more widely spread it becomes in frequency, and vice versa. The one-sided power spectrum of the pulse described by Eq. (10.27) is given in Eq. (10.29) and sketched in Figure 10.19.

$$|S(f)_{\text{pulse}}|^2 = \frac{1}{2}\text{sinc}^2[t_{\text{p}}(f - f_c)] \tag{10.28}$$

where $\text{sinc}(x) = \sin(\pi x)/(\pi x)$. The uncertainty relationship appears in the inverse scaling between the pulse duration and the width of the pulse spectrum. For example, a very short pulse with small t_{p}

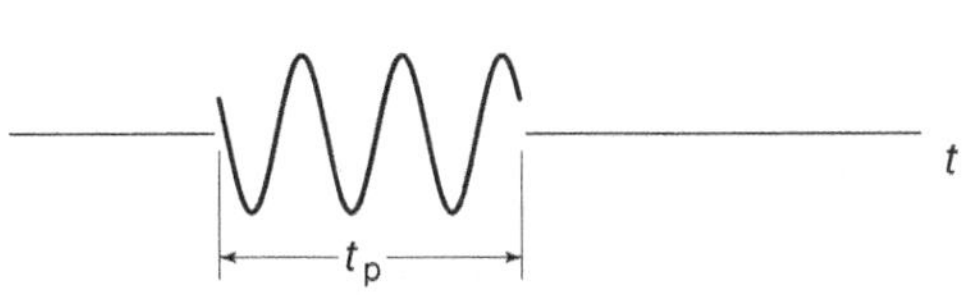

Figure 10.18 Randomly gated pulse waveform. Example cosinusoidal pulse signal of duration t_{p} with random initial phase.

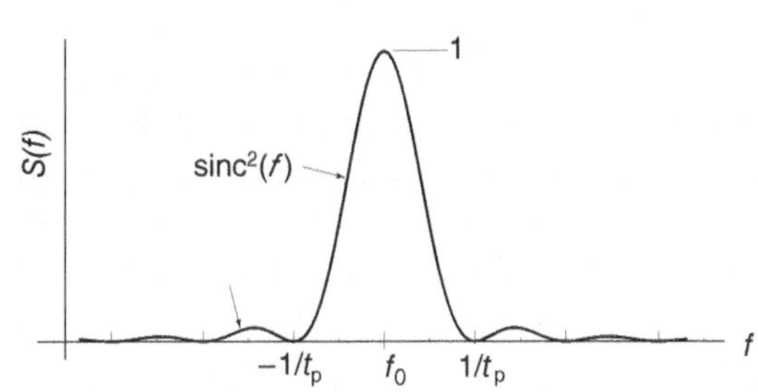

Figure 10.19 Power spectrum of randomly gated pulse. Power spectrum depends only on the carrier frequency f_0, and is independent of the phase at which the gate is applied. Zeroes of the spectrum are spaced by units of $1/t_{\text{p}}$ from the center frequency.

is well localized in time but is widely dispersed in frequency, in the sense that its Fourier transform occupies a wide effective range of frequencies $\propto 1/t_\mathrm{p}$.

We also need to consider the spectrum of a sequence of pulses. For simplicity, assume that the pulses are regularly spaced and separated in time by T. For now, though, let the starting phase of the carrier oscillation be uncontrolled. We can describe such a pulse train by,

$$s_{\mathrm{noncoh}}(t) = \sum_{n-1}^{N} \sqcap\left(\frac{t-nT}{t_\mathrm{p}}\right)\cos(2\pi f_0(t-nT)+\phi_n) \tag{10.29}$$

where ϕ_n is a random phase for each pulse in the burst. We include the time offset nT because we do not require information from any of the other transmit pulses. The structure of such a signal is illustrated in Figure 10.20. The power spectrum of this randomly-phased pulse train is the same form as that of Eq. (10.29), Figure 10.19, except that the strength of the spectrum is increased by the number of pulses in the train. That is,

$$\left\langle |S(f)_{\mathrm{noncoh}}|^2\right\rangle = \frac{Nt_\mathrm{p}}{4}\mathrm{sinc}^2[t_\mathrm{p}(f-f_\mathrm{c})] \tag{10.30}$$

where the "⟨⟩" indicate the statistical expected value of the enclosed spectrum.

The range resolution achieved depends directly on the pulse duration, t_p; shortening t_p improves the resolution since the uncertainty in round trip time-of-flight for the pulse becomes less as the pulse becomes more compact and the arrival time at the receiver becomes more certain. Using the pulse width as a rough measure of the arrival time uncertainty, as in Section 10.2.3,

$$\delta R \simeq \frac{c_0}{2}\, t_\mathrm{p} \tag{10.31}$$

For a 1 μsec pulse,

$$t_\mathrm{p} \simeq 10^{-6} \Rightarrow \delta R \simeq 150 \text{ m} \tag{10.32}$$

The bandwidth of such a pulse is, however,

$$B_\mathrm{p} \simeq \frac{1}{t_\mathrm{p}} = \frac{1}{10^6} = 10^6 \text{ Hz} = 1 \text{ MHz} \tag{10.33}$$

Clearly, a requirement for fine range resolution can lead to the use of large bandwidths.

Now consider the use of such a pulse to measure the speed of an aircraft for other object traveling directly toward a monostatic radar at a speed of 115 m s^{-1} ($\simeq$ 250 kts) utilizing a wavelength of 3 cm. The Doppler shift is $f_{\mathrm{Dop}} \simeq (2 \cdot 115)/0.03 \simeq 7.7\,\mathrm{kHz} \simeq 0.01\, B_{\mathrm{pulse}}$. Backscatter from an object traveling at this rather high speed produces a shift in frequency that is only about one percent of the pulse bandwidth. While it would be fairly straightforward to distinguish between a stationary object and one moving at the assumed radial velocity, a fine measurement of velocity on the basis of the spectral shift would be very difficult. One approach would be direct comparison of the spectra of the transmitted and received pulses, in which case measurement of a Doppler frequency shift would require that we distinguish between two broad functions which are shifted

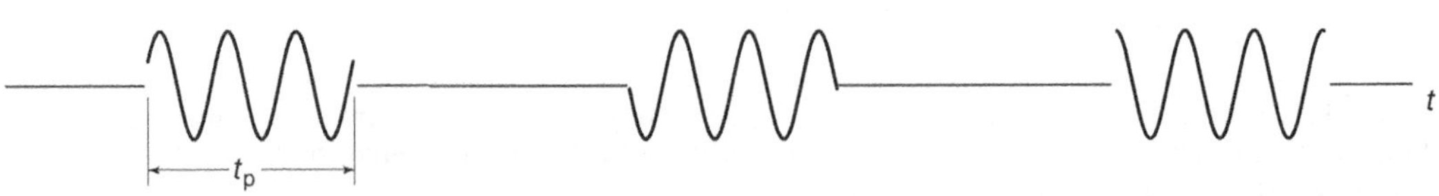

Figure 10.20 Non-coherent pulse train. Three-pulse train made up of randomly phased pulses, such as that in Figure 10.18. Note the different initial phases of the three pulses. This reflects the randomness symbolized by the ϕ_n in Eq. (10.30).

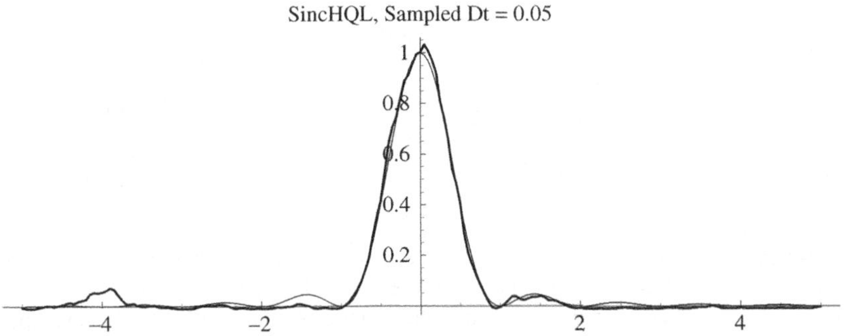

Figure 10.21 Power spectra of a pulse and a pulse with noise. Comparison of power spectrum of a pulse with noise with the underlying noiseless pulse illustrates difficulties in accurate determination of the frequency of the underlying pulse with noise is present. Peak of noisy pulse is shifted to higher frequencies by 10% of the pulse bandwidth, while computation of centroid results in about a 2% shift to the left. Note that deviation of noisy spectrum from ideal is typically much less than about 10% in power; noisy spectrum has been shifted downwards by average level of noise to facilitate comparison in vicinity of main peak. Spectrum in the figure is centered at f_0, while units of frequency are $1/t_p$.

only slightly with respect to one another. Figure 10.21 indicates the problem. While such comparisons can be carried out to infinite precision with noiseless signals of ideal shape, they are severely compromised by the presence of even small amounts of noise and system waveform distortion in either the transmitter or receiver. In the example of Figure 10.21 determination of the location of the noisy spectrum in frequency based on use of the peak spectral power is in error by about +10%, while use of the centroid produces and a negative error of about 2%. This illustrates the difficulty of obtaining a measurement of the center frequency that is more accurate than a few percent of the pulse bandwidth when even a modest amount of noise is present.

Alternatively, we might try to determine the echo frequency by counting zero crossings of the waveform rather than by attempting to measure spectral shift in the frequency domain. Again, this works well at high values of signal-to-noise ratio, but becomes imprecise for determining frequency when noise is added. Again, the limitation is a result of the limited temporal sample of the waveform associated with the use of a short pulse.

A third alternative is to use different sequential range measurements. For this calculation,

$$v_{\text{radial}} = \frac{R_{\text{app}}(t_2) - R_{\text{app}}(t_1)}{t_2 - t_1} \tag{10.34}$$

This method is commonly employed in simple radars, but usually is the least accurate option, since the velocity error is roughly $2\,\delta R_{\text{app}}/(t_2 - t_1)$, and δR_{app} in our present example is ~100 m. Because this range error is of the order of the velocity we wish to measure, we would need to observe for about 20 seconds to obtain an ≈ 10% error.

10.5.2 Coherent Pulse Train

Yet another alternative is the use of a *coherent pulse train*, which is a sequence of pulses similar to that of Eq. (10.30) but with one critical difference. The sequence is said to be "coherent" when the phase of the underlying carrier frequency within the burst is controlled. One method for accomplishing this is to emit sections of a cosinusoidal waveform, such as,

$$s_{\text{coh}}(t) = \sum_{n=1}^{N} \sqcap\left(\frac{t - nT}{t_p}\right)\cos(2\pi f_c(t) + \phi) \tag{10.35}$$

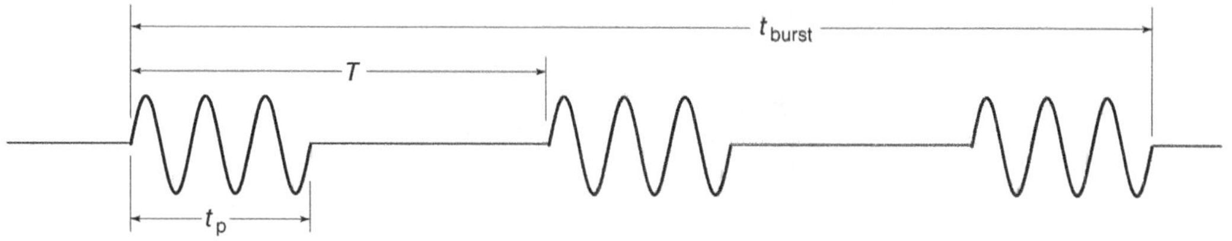

Figure 10.22 Coherent pulse train. A coherent pulse train (only the first two and the last pulses are shown) preserves a deterministic phase relationship among the pulses in the burst. Here $\phi = 0$, $t_p = 3f_c^{-1}$, and $T = mf_c^{-1}$, where m is an integer, $t_{burst} = Nt_p$, and the other variables are those in Eq. (10.36). In this example the starting phase and ending phase for each pulse is the same; this need not be the case, however (see text).

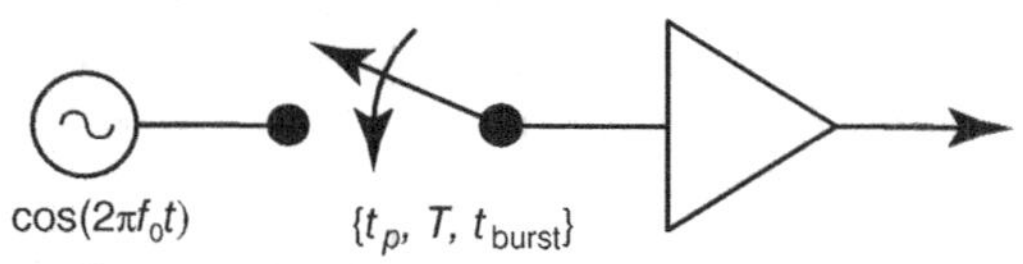

Figure 10.23 A coherent pulse train generator. This example preserves the continuous phase of the underlying cosinusoid.

where n is an integer and $T > t_p$ is the inter-pulse period, as before, and the duration of the burst is $t_{burst} = NT$. For this waveform the phase of $\cos(2\pi f_c t + \phi)$ is the same as that of a continuous carrier viewed through time "windows" represented by the replicated gate function. An example of a coherent pulse burst in which the pulse duration and the interpulse period correspond to integer numbers of cycles of the carrier frequency is shown in Figure 10.22. The key idea is that the phase of each individual pulse is deterministic and known for use in the detection process. With that in mind, it is not necessary that t_p, T, and f_c be rationally proportional, as in the example. Only that the relationships be known. Similarly, it is not necessary that the phase of each pulse be the same or that the frequency be constant—they need only be controlled in some deterministic manner.

10.5.3 A Simple Coherent Pulse Train Generator

Coherent pulse bursts corresponding to Eq. (10.36) with an arbitrary but fixed value of *phi* are easily generated with an oscillator and switch. The oscillator runs continuously in order that the underlying coherence of the waveform is maintained. A simplified diagram is shown in Figure 10.23. In practice, the source oscillator is also used as the time reference for determining t_p, T, and t_{burst}.

It is important to avoid implementations that do not maintain coherence, for example by switching the oscillator power, as in Figure 10.24. Use of this scheme will result in the non-coherent pulse train of Eq. (10.30), where ϕ_n will take on the random start up phase of the oscillator for each pulse window. In this instance, the randomness of the carrier phase among the pulses makes it impossible to predict the details of the resulting waveform, in contrast to the situation in the waveform given in Eq. (10.36). As we shall see, this is a critical distinction in the construction and use of radar waveforms.

10.5.4 Power Spectrum of Coherent Pulse Train

The power spectrum of a coherent pulse train is quite different from that of a single pulse or a train of randomly phased pulses. An example showing the main characteristics is displayed in Figure 10.25. The features in the spectrum are related directly to the parameters of the coherent pulse train as described in Eq. (10.36) and summarized in Table 10.2.

The form of the coherent spectrum is given by,

$$|S(f)_{\text{coh}}|^2 = \frac{t_{\text{p}}}{4}\,\text{sinc}^2\left[t_{\text{p}}(f - f_{\text{c}})\right]\left(\frac{\sin(\pi NTf)}{\sin(\pi Tf)}\right)^2 \qquad (10.36)$$

As $T > t_{\text{p}}$, and usually $N \gg 1$, the most rapid variation with frequency occurs in the numerator of the ratio of sines, which behaves much like a "sinc" function with values of unity at the zeros of the denominator $\sin(\pi Tf)$. This behavior controls the spacing of the spectral lines, which occur

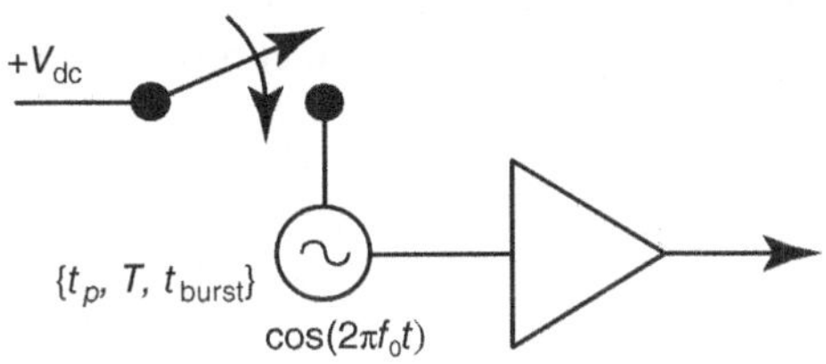

Figure 10.24 Non-coherent, switched oscillator pulse burst generator. Since a typical oscillator can start with random phase, the phase of the generated pulses is uncontrolled.

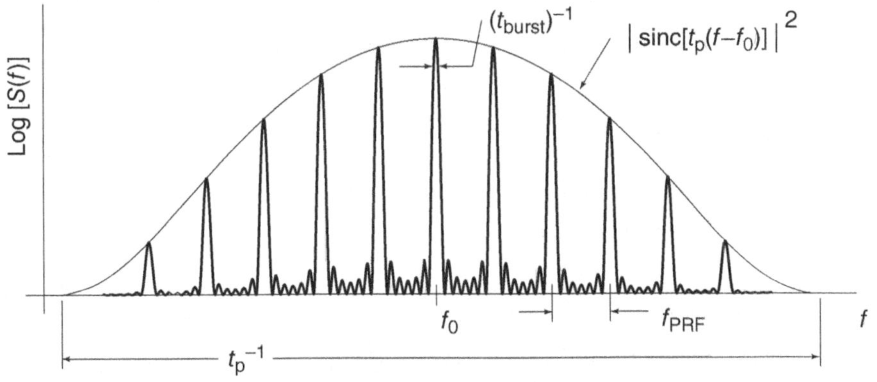

Figure 10.25 Power spectrum of coherent pulse train. The spectrum is a comb of fine lines, within an envelope of the individual pulse spectrum (see text, Table 10.2).

Table 10.2 Spectral characteristics of a coherent pulse train.

Feature	Scale	Controlling parameter
Fine line width	$\simeq D\dfrac{2}{t_{\text{burst}}}$	Burst duration
Fine line spacing	$= D\dfrac{1}{T}$	PRF
Width of envelope	$\simeq D\dfrac{2}{t_{\text{p}}}$	Individual pulse duration

at values of f such that $fT = 0, \pm 1, \pm 2, \ldots$ On the other hand, the multiplicative $\mathrm{sinc}^2[t_p(f - f_c)]$, varies relatively slowly with respect to either the numerator or denominator of the sine fraction. The fine line width, $1/NT$, determines the precision to which the signal frequency can be measured. This value depends only on the *total duration of the pulse burst*, $t_p = NT$, and is independent of the individual pulse length, t_p, on which time localization or range measurements depend. This implies that the frequency resolution—the measurement of the Doppler shift for determining radial velocity—can be independent of the range accuracy obtained from a simultaneous measurement of echo time delay! This result is subject to both range and frequency ambiguities, however. The following example illustrates how this works.

10.5.4.1 Example Spectrum of a Coherent Pulse Burst

The previous section develops the general formula for a coherent pulse burst. We now look at a specific example of an echo from our 115 m s^{-1} target. We assume that the target is approaching the radar so that the radial component of velocity actually is negative ($\upsilon_{\text{radial}} = -115$ m s^{-1}). The decreasing range results in a positive Doppler shift, however. Suppose we were to choose a waveform described by Eq. (10.36) with the parameters,

$$\begin{aligned} t_{\text{burst}} &\simeq 10^{-2}\ \text{s} \\ T &\simeq 10^{-4}\ \text{s} \\ t_p &\simeq 10^{-6}\ \text{s} \end{aligned} \tag{10.37}$$

These parameters represent a pulse burst containing exactly 100 individual pulse intervals.

Consider then that the spectral line width is,

$$\delta f \simeq t_{\text{burst}}^{-1} = 100\ \text{Hz} \tag{10.38}$$

while for the line spacing,

$$\Delta f = T^{-1} = \ \text{PRF} \ = 10\ \text{kHz} \tag{10.39}$$

Equation (10.41) provides an explicit formula for the fine line width in terms of the number of pulses in the burst and the PRF. Because $T \cdot f_{\text{PRF}} = 1$,

$$\delta f = \frac{T \cdot f_{\text{PRF}}}{t_{\text{burst}}} = \frac{f_{\text{PRF}}}{\text{pulses/burst}}\ \text{Hz} \tag{10.40}$$

As in Table 10.2, δf is independent of the individual pulse width, t_p.

The central portion of the example spectrum is shown in Figure 10.26. In the transmitted spectrum, the spectral lines are perfectly aligned with 10 kHz intervals on either side of the carrier frequency f_c. The received spectrum is similar, except that the spectral lines are all shifted to the right by f_{Dop}, corresponding to a positive Doppler effect. Considering that the fine line widths are $\delta f \simeq 100$ Hz, the Doppler shift can be measured easily to about $2\delta f \simeq 200$ Hz or better—provided that in an experimental situation it is possible to determine which are the corresponding lines in the transmitted and received spectra. The ambiguity in frequency measurement here is the spacing between the spectral lines, Eq. (10.40), or

$$f_{\text{amb}} = 10\ \text{kHz} \tag{10.41}$$

In our example $f_{\text{Dop}} < \text{PRF}$, there is no folding, based on our *a priori* knowledge of the general range of speeds. This need not be the case in a real situation.

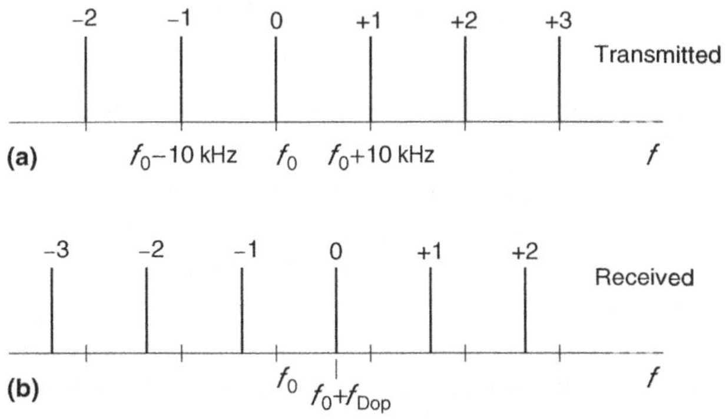

Figure 10.26 Central region of example coherent pulse spectrum. Panels compare power spectra of a transmitted (a) and Doppler shifted coherent pulse burst (b). Line spacing is that of the PRF = 10 kHz. For more than a few pulses in the burst, the width of the burst envelope, which is governed by the length of an individual pulse, t_p, is very broad compared with the width of the spectral lines within the burst, as suggested by the figure. This makes it impractical to identify unambiguously small shifts in frequency on the basis of the line amplitudes.

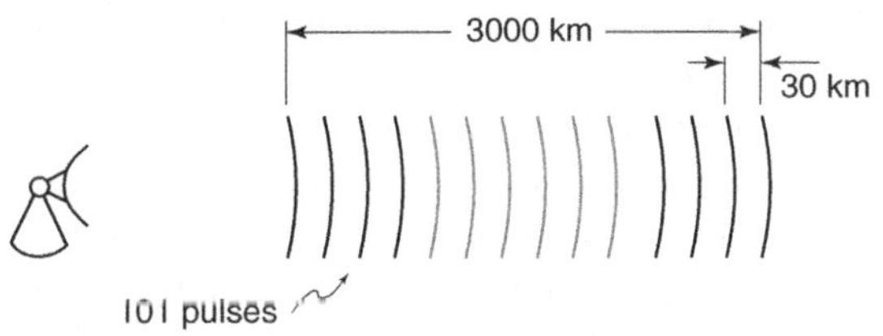

Figure 10.27 One hundred one pulses in space. Figure corresponds to parameters of Figure 10.26 except that one additional pulse has been added so that there are 100 intervals between first and last pulses emitted. Targets separated by less than one half the pulse burst length, 15,000 km in range, will have ambiguous arrival times, mirroring the frequency ambiguity arising from the multiple lines in the frequency domain.

For a coherent pulse train analyzed in the frequency domain, this situation exactly parallels the problem of range ambiguity discussed previously in the time domain. For a target moving at a constant speed the frequency ambiguity can be resolved by varying the PRF, a manner similar to that used to sort out range ambiguity in Section 10.3.2. Since the line spacing is the same as the PFR, changing the PRF expands or compresses the line pattern, so that a measurement of that change in apparent Doppler shift permits us to resolve the ambiguity.

Figure 10.27 illustrates the dimensions of this wavetrain in space. (It is always a good idea to have this type of picture in mind when beginning an analysis of a radar signal processing problem.) The appropriate dimensions in this picture are: (i) the individual pulse spacing, $c_0T = 30$ km, (ii) the ambiguous range is one half the spatial separation of the individual pulses,

$$R_u = \frac{c_0}{2}T = \frac{c_0}{2PRF} = 15 \text{ km} \tag{10.42}$$

while (iii) the length of the pulse burst is $c_0\, t_{burst} = 3000$ km. Keep in mind that the folding frequency in the frequency domain and the unambiguous range in the time domain cannot be chosen independently, since it is easily shown that,

$$R_u f_{amb} = \frac{c_0}{2} \tag{10.43}$$

It is interesting to compare this example with that of the single pulse. For the parameters employed,

i. The Doppler shift of 7.7 kHz was easily measured to 0.1 kHz with a single burst of 10 ms, whereas this is not possible with a single pulse;
ii. It can be difficult to distinguish $f_{\text{Dop}}$from $f_{\text{Dop}} \pm f_{\text{PRF}}$, as the Doppler shift is ambiguous in a measurement using the pulse train, whereas for the simple pulse the measurement is unambiguous, albeit relatively imprecise. In visualizing this, keep in mind that the envelope of the spectrum is nearly flat since $T/t_{\text{burst}} \gg 1$.
iii. Range is ambiguous within the pulses in the burst, but is unambiguous for the simple pulse;
iv. Reducing the range ambiguity within the pulse train increases the Doppler ambiguity, and vice versa.

Returning to our questions from the beginning of this section, obviously a pulsed system can be used to determine not only range, but Doppler shift as well. Simple pulse waveforms are limited by the uncertainty principle from Fourier theory, however, which tells us that shortening the time duration of a waveform spreads its energy in frequency and vice versa. Thus, use of a simple "short" pulse optimized for range resolution has a commitment limit on how well it can be used to determine range rate from the Doppler frequency shift, while use of a simple "long" pulse optimized for frequency resolution has limited utility for determination of range.

The limitations on the simultaneous determination of range and range rate can be bypassed by the use of trains of pulses or pulse bursts. Extension of the waveform in time introduces a series of discrete lines into the frequency spectrum. The width of the spectral lines is independent of the duration of the individual pulses in the pulse burst. Combining short pulse lengths with a long duration of the overall pulse burst provides us with the ability to simultaneously control the range and frequency resolution, albeit at the cost of introducing range and Doppler frequency ambiguities.

10.6 Frequency-modulated, Continuous-wave Radar

Having examined the question of using a pulsed radar to measure Doppler shift effectively, we now ask, "Can we use a 'CW' radar measure range?" There is an answer, of course, but it is obscured by semantics. Strictly speaking, a *continuous-wave* radar emitting a cosinusoidal waveform such as we discussed in Section 10.1.1 cannot be used to determine range. A *continuously transmitting* radar can achieve this feat, however, even if the transmitted power is constant. In the latter instance, the argument of the cosine waveform is modulated, thereby conferring some pulse-like characteristics on the "CW" signal.

The most common form of a modulated CW system is the *frequency-modulated, continuous-wave* design, or FM-CW system. A simplified block diagram is shown in Figure 10.28.[12] In this system, the transmitter operates continuously, but the radiated frequency varies linearly with time, or *sweeps*, between two fixed limits. The frequency is then immediately reset to the starting value and the sweep repeated without delay. The solid line segments in Figure 10.29a illustrate this choice for the time variation of frequency of such a waveform. Referring to the figure, the variation of transmitted frequency with time forms a saw-tooth pattern, where the frequency is discontinuous

12 This is essentially the system discussed in Section 10.1.3, with the addition of the swept-frequency modulation.

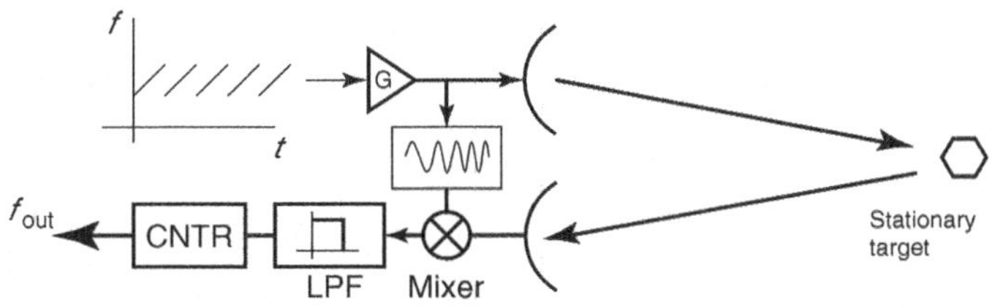

Figure 10.28 FM-CW radar block diagram. In this system, the transmitter operates continuously while the transmitter frequency is swept linearly between fixed limits over time. Conceptually, f_{out} is determined by counting the number of zero crossing per unit time of the waveform at the LPF output. Alternatively, f_{out} can be determined equally well by spectral analysis of the LPF output (not shown).

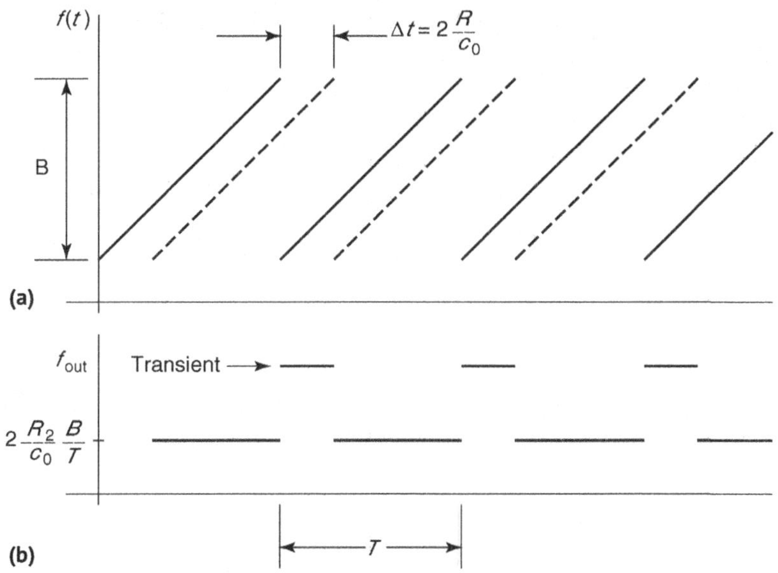

Figure 10.29 FM-CW time-frequency variation. (a) Transmitted signal linearly sweeps in frequency over a bandwidth B (solid lines); echo signals have a similar variation delayed by t_e (dashed lines). (b) Difference in frequency between overlapping traces of transmitted and received frequencies are proportional to range.

at the point between the end of one sweep and the beginning of the next. Upon reflection from a stationary point target, Figure 10.28, the frequency of the echo has the same frequency variation as the transmitter, although the waveform is delayed in time by the round trip flight time required for the pulse to travel to the target and return to the radar, as shown by the dashed line in Figure 10.29a. In the receiver, the combination of the mixer and LPF selects the signal at the difference frequency between the transmitter and receiver, which is represented by the vertical distance between solid and dashed lines in panel (a) of Figure 10.29; the difference frequency is shown as f_{out} in the lower panel (b) of the figure. After eliminating the transients at the ends of the sweeps from the counter input the output frequency measured by the counter is proportional to range. In order to see this, note that the value of f_{out} can be calculated from the bandwidth spanned by the sweep, B, and the time required to complete the frequency sweep, T, as shown in Figure 10.29, where the rate of change of transmitter frequency is $df/dt = B/T$,

$$f_{\text{out}} = f_{\text{trans}} - f_{\text{rec}} = t_{\text{echo}} \frac{df}{dt} \tag{10.44}$$

$$= t_{\text{echo}} \frac{B}{T} = \frac{2R}{c_0} \frac{B}{T} \tag{10.45}$$

$$= \frac{R\,B}{R_{\text{u}}} \tag{10.46}$$

It follows that,

$$R = \frac{R_{\text{u}}}{B} f_{\text{out}}, \qquad \delta R \sim \frac{R_{\text{u}}}{B} \delta f_{\text{out}} \tag{10.47}$$

where δf_{out} is the uncertainty or error in the determination of f_{out}. That is, the addition of the FM sweep to the CW signal transforms the time delay associated with the echo signal from a highly ambiguous phase shift into an observable frequency difference.

Although at first glance it seems more complex, in its own way, the FM-CW waveform is as simple as the basic pulse train. In fact, it appears in many radar systems as the solution to particular problems in signal design. One application in which it appears in a particularly natural way is that of the so-called "synthetic aperture radar" mapping systems. We examine the mapping radar problem in detail later.

10.6.1 Example Sweep Parameters

Consider a particular choice of parameters,

$$B = 10^6 \text{ Hz}, \qquad T = 1 \text{ s} \tag{10.48}$$

For these values,

$$\delta R_{\text{error}} \sim \frac{150,000 \text{ km}}{10^6 \text{ Hz}} \cdot \delta f_{\text{error}} \text{ Hz} = 150 \cdot \delta f_{\text{out}} \text{ m} \tag{10.49}$$

Since we can readily measure frequency to an accuracy of $\delta f_{\text{out}} \simeq 1$ Hz in 1 s, the range uncertainty for observations over a single sweep are approximately $\delta R_{\text{error}} \lesssim 150$ m for this choice of bandwidth, an accuracy similar to that of a pulsed radar employing signals with $t_{\text{p}} \simeq 1$ μs. Note that earlier we saw that $B \simeq 10^6$ Hz in a pulse system gives comparable range accuracy, so in these examples, the accuracy of the FM-CW and pulse systems are comparable.

10.6.2 Aircraft Altimeter Example

An aircraft altimeter represents a typical application of an FM-CW. Radar altimeters are used to determine the aircraft height above the ground in the landing phase, typically from a one or two thousand meter altitude to the point at which the wheels touch the runway. A common arrangement is shown in Figure 10.30. Here the transmitting and receiving antennas are located separately, one under each wing, while the electronic elements are packaged together. Although the transmitter and receiver are separated, this is essentially a monostatic configuration chosen to improve the isolation between the transmitter and receiver. The need to operate to near-zero range makes implementation of a pulse system difficult in this application because the round-trip flight time of the radar signal becomes extremely short as the aircraft approaches the ground. Hence, it is difficult to carry out the necessary transmitter/receiver switching and to handle the blind range when at low altitudes.

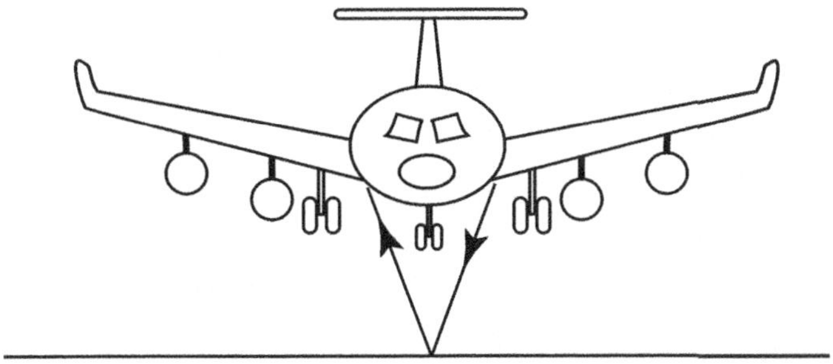

Figure 10.30 Landing radar example. Aircraft altimeters often employ a swept frequency technique since such radars can operate to zero range. Transmitting and receiving antennas are often separated to improve isolation between the two; here one antenna is located under each wing. For practical use in landing, such radars must work from altitudes of ≃ 10,000 m to essentially zero.

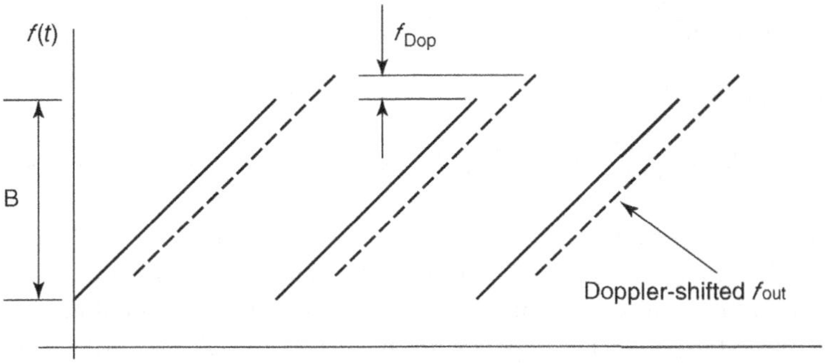

Figure 10.31 Range-rate effects on FM-CW systems. Doppler effect introduces a secondary frequency shift, which is interpreted as an additional incremental range in FM-CW systems. In the figure, the Doppler shift appears as a vertical displacement parallel to the ordinate. Compare locations of dashed lines here and in Figure 10.29.

10.6.3 Range-rate Effects in FM-CW Systems

Consider instances in which the range is changing with time so that the range rate is not zero, i.e., $\dot{R} \neq 0$, as in the aircraft landing example. In this instance, the transmitted and received frequency plot must be modified by the Doppler contribution to the received signal. What is the effect of varying range on an FM-CW system? In Figure 10.31 the sawtooth waveform of Figure 10.29 representing the received signal frequency is shifted upwards by an amount representing a positive Doppler shift. This is the direction of change associated with our aircraft altimeter example, as the aircraft approaches the ground. Assuming for the moment that the target is within R_u of the radar, the frequency of the output signal to the counter is now given by,

$$f_{\text{out}} = \left(\frac{2R}{c_0} \frac{B}{T} - f_{\text{Dop}} \right) = \frac{2RB}{c_0 T} - 2\frac{\mathbf{v} \cdot \mathbf{u}_{\text{scat}}}{\lambda} \tag{10.50}$$

where the new term gives the effect of the rate of change in aircraft altitude. The Doppler shift introduces a range error because the apparent range in this system

$$R_{\text{app}} = f_{\text{out}} \frac{R_u}{B} \tag{10.51}$$

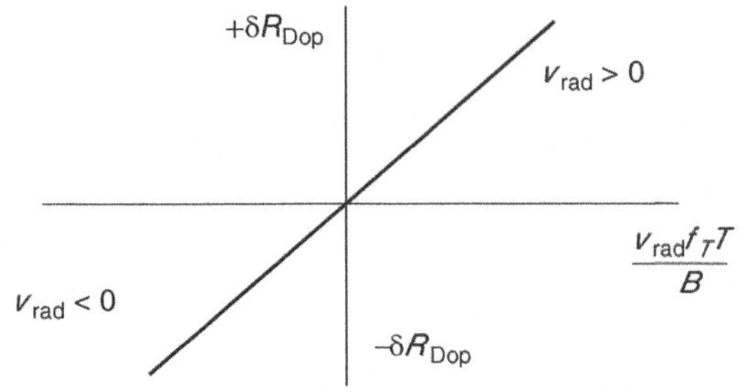

Figure 10.32 FM-CW Doppler range error. The Doppler range error is proportional to radial velocity, mean transmitted frequency, and the inverse slope of the frequency sweep.

includes a variable range-rate effect, f_{Dop}, in addition to the range time-delay,

$$R_{\text{app}} = R_{\text{target}} - f_{\text{Dop}}\frac{R_u}{B} - nR_{\text{u}} = R_{\text{target}} - nR_{\text{u}} + \delta R_{\text{Dop}} \tag{10.52}$$

Obviously, the error due to the Doppler effect is,

$$\delta R_{\text{Dop}} = -f_{\text{Dop}}\frac{R_u}{B} = \left(f_{\text{trans}} \cdot \frac{T}{B}\right) v_{\text{radial}} \tag{10.53}$$

Equation (10.54) tells us that in such systems the measurement of range and velocity are intrinsically coupled in a *range-Doppler ambiguity*. In this situation, a particular range measurement, R_{app}, corresponds to many different possible combinations of actual range and Doppler shift. The sensitivity of the range error to the target motion depends on the sweep rate $df/dt = B/T$, and is decreased by increasing the rate of change of frequency. Unlike the range ambiguity associated with the repetitive transmitter waveform, the ambiguity in this instance is a continuum in which range error and radial target velocity are in proportion to each other, Eq. (10.54).

10.6.4 Range-Doppler Ambiguity in FM-CW Systems

From the previous section we see that the FM-CW system has two ambiguities: (i) the range-Doppler ambiguity that arises from range-rate effects; and (ii) range time-delay ambiguity associated with confusion of the echo sequence relative to the pattern of transmitter frequency sweeps. Note that these differ in a fundamental manner in that the first, the FM-CW range-Doppler ambiguity, varies continuously with the Doppler shift, whereas the time-delay ambiguity results in discrete errors that are multiples of $R_{\text{u}} = c_0 T/2$.

We can plot the results in terms of the range error δR_{dop} versus the radial velocity, or in terms of the radial velocity versus the apparent range. The behavior is sketched in Figures 10.32 and 10.33. Figure 10.32 shows the ambiguity between range and velocity expressed in terms of the Doppler-range error, Eq. (10.54). This figure illustrates that one can not distinguish between an increment in range of size δR_{Dop} and a velocity of $v_{\text{radial}}\, f_{\text{trans}}\, T/B$. Figure 10.33 shows how the ambiguity maps into the apparent range, including the range ambiguity associated with the repetitious nature of the frequency sweep. The latter ambiguity is the same as that analyzed previously for the pulse radar with the range-Doppler ambiguity superimposed. The seriousness of the range-Doppler ambiguity in FM-CW systems depends on the combination of the system parameters and the application. We illustrate this with our aircraft altimeter example.

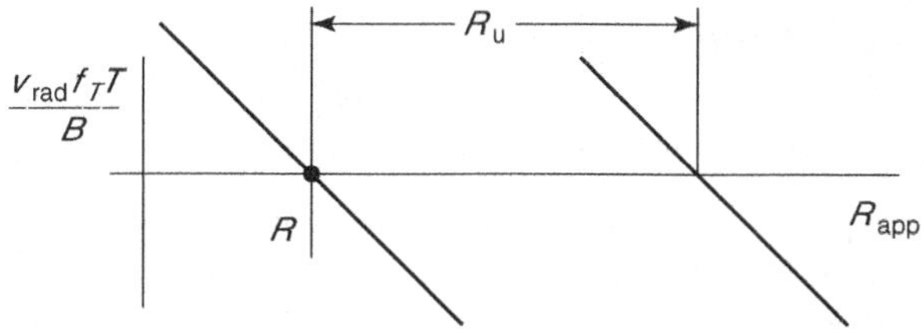

Figure 10.33 Normalized radial velocity vs. apparent range. FM-CW systems also exhibit range ambiguities, as illustrated.

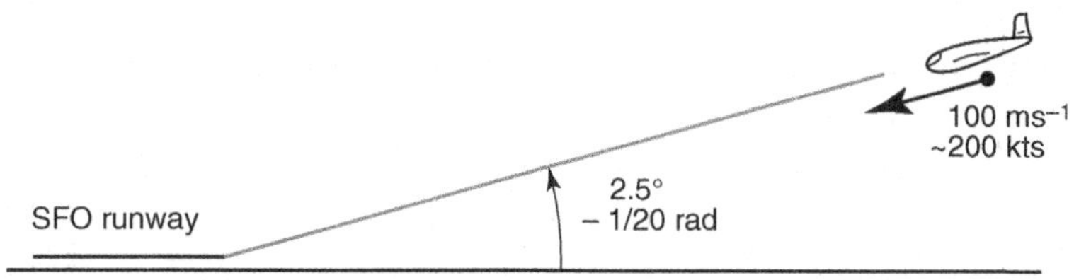

Figure 10.34 Aircraft approach path. The typical approach geometry shown can be used to estimate the magnitude of the Doppler range error in an aircraft altimeter.

10.6.4.1 Range-Doppler Ambiguity in Aircraft Radar Altimeter

Realistic parameters for an altimeter might be,

$$
\begin{aligned}
f_{\text{trans}} &\simeq 4.9 \text{ GHz} \approx 5 \times 10^9 \\
B &\simeq 140 \text{ MHz} \approx 10^8 \\
T &\simeq \frac{1}{160} \text{ sec} \approx 10^{-2} \\
\delta f &\simeq 10^2 \text{ Hz (C-band)}
\end{aligned}
$$

from which we calculate the following altimeter performance characteristics,

$$R_u \simeq \frac{T \cdot c_0}{2} \simeq 1.5 \times 10^8 \cdot 10^{-2} = 1.5 \times 10^6 \text{ m} \tag{10.54}$$

$$\delta R_{\delta f} \simeq \frac{R_u}{B} \delta f \simeq \frac{1.5 \times 10^6}{10^8} \cdot 10^2 \simeq 1 \text{ m (!)} \tag{10.55}$$

So we have a nominal $\delta R \approx 1$ m accuracy with an $R_{\mathrm{u}} \approx 1500$ km range ambiguity. The top of Earth's atmosphere is at roughly 90 km altitude, so the range ambiguity is more than adequate for unambiguous operation from aircraft.

What is, δR_{Dop}, the range error due to the Doppler effect associated with a changing range? In order to estimate this we need to know the typical rate of descent. Referring to Figure 10.34, we estimate the vertical component of the velocity as,

$$v_{\text{rad}} = v_{\text{vertical}} \simeq \frac{-100}{20} \simeq -5 \text{ m s}^{-1} \tag{10.56}$$

The range error due to closing velocity is, Eq. (10.54),

$$\delta R_{\text{Dop}} = f_{\text{trans}}\, v_{\text{rad}}\, \frac{T}{B} \simeq -\frac{5 \times 10^9 \cdot 5 \cdot 10^{-2}}{10^8} \simeq -2.5 \text{ m} \tag{10.57}$$

if uncompensated. This is a constant bias if v_{rad} is constant.

10.7 Inferences from Properties of Simple Signals

Although we have examined only a few simple cases, some important points have emerged:

i. At the extremes, sine waves and simple pulses can give essentially arbitrary precision and accuracy in velocity and range, respectively, but are completely unsuited for use if the measurement goals are interchanged with the two signal types;
ii. FM-CW can be used to measure range, but with an ambiguity between range and velocity, in addition to that associated with repetition of the waveform;
iii. Pulse bursts can be used to measure velocity, but also with an ambiguity in range and velocity.

With regard to ambiguity, the examples of this chapter are special cases of a very general result that we investigate further in chapters A and B.

11

Range-Doppler Imaging

We have seen that if we choose our modulation properly, we can measure both the range and the velocity of objects. This technique can be extended to permit entire image maps to be derived from radar echoes, provided we choose the proper imaging geometry and signal processing methods. We refer to radar systems capable of providing two-dimensional (2D) maps of surface reflectivity as *imaging radars*, as they are capable of displaying the radar reflectivity in an image format. An imaging radar is usually characterized by rather fine resolution, that is, the ability to localize objects to a fine grid, and the ability to derive the locations of many objects nearly simultaneously. These images look very much like pictures of the surface, except that they depict variations in the ability of a surface to reflect radar signals instead of visible light.

In this chapter, we present the problem of mapping the reflectivity of planetary bodies in the solar system. This is but one application of a more general technique called range-Doppler imaging. However, it is a fairly easy to understand example as the geometries of imaging and of the object under study are rather straightforward. As may be deduced from the name of the method, images are produced by estimating both the range delay and the Doppler shift of echoes from each point on the surface of the object under study. Range-Doppler imaging finds application in diverse areas, such as mapping the Earth's surface, imaging ships at sea, and for Doppler ultrasound devices in medicine.

11.1 Planetary Imaging

One of the earliest applications of imaging radar was to map the surfaces of bodies in the solar system. In 1946, the first radar echoes from the moon were recorded, so that this moment may be considered the birth of planetary radar astronomy. This experiment was a simple detection, with imaging yet to come.

We start by examining how we can extend a simple detection experiment to an imaging experiment. First, we review our range and Doppler (velocity) measurements. We will see that most points on a rotating surface have a distinctive signature when viewed in terms of range delay and Doppler frequency, apart from a very important pairwise identification of surface points called the north/south ambiguity. Analysis of the range and Doppler frequency measurements will allow us to form high resolution maps of surface reflectivity.

11.1.1 Range Rings

We begin by determining the set of points on a planetary surface that can return radar echoes at the same time when illuminated by a radar transmitter and antenna. Here we assume that the radar transmits a finite-length pulse of energy at an object. Because the range resolution is finite and hence nonzero, points at nearly the same range effectively all return echoes simultaneously. In fact, the interval of range distances returning echoes at the same time may be taken as the range resolution of the radar.

Consider an experiment in which we transmit a single pulse at an object in the solar system, the planet Venus, as in Figure 11.1.

Now we examine how the received echo power as a function of time relates to the surface reflectivity of Venus. The point on Venus closest to the Earth will return energy first, while each succeeding point on the surface is a little further from Earth and has slightly more delay (see Figures 11.2 and 11.3). While only points on a single circle are truly at equal distances from the radar instrument, and thus have equal delay, finite range resolution forces us to treat all points in an annulus about the circle as being at equal range. Thus, the resolution determines how many distinct rings may be placed on the surface. In other words, finer range resolution leads to more distinct rings, and a more exact location of scattering points on the surface. These resolution rings are often referred to as range "bins," as they represent collections of signal energy about some specific range delay.

As can be seen in Figures 11.2 and 11.3, each step in time of the received radar echo is going to correspond to a slightly different piece of the surface. If we plot the echo received as a function of time, we might see a power vs. time plot as shown in Figure 11.4. We now need to relate the scattered intensity at the various time delays of the reflected signal to the location of the pulse across Venus.

Now we can ask how much the echo is spread out in time. As can be inferred from the imaging geometry, the amount the pulse is stretched to receive depends on the size of the planet. Suppose, as in the case of Venus, the planet's radius is 6000 km. The maximum delay relative to the time when the echo is first received by the radar will be the usual $\frac{2R}{c}$, or about 0.04 s (Figure 11.5).

Figure 11.1 A planetary imaging experiment. A powerful radar on Earth broadcasts a signal toward Venus, and listens for the echo.

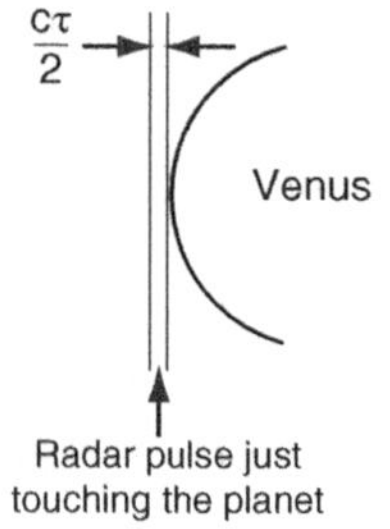

Figure 11.2 Pulse from Earth first arrives at Venus. The picture depicts the moment when the pulse first interacts with Venus' surface.

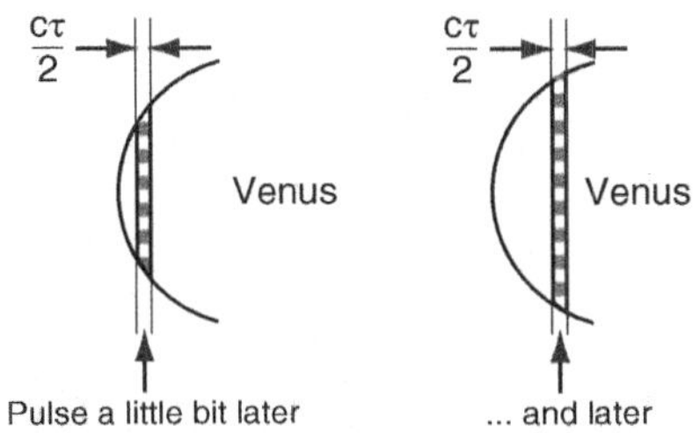

Figure 11.3 Pulse propagates across Venus. The same pulse from Figure 11.2 seen as it moves forward in time. Each point in the cross-hatched region returns energy to the radar at about the same time. The spatial extent of the pulse in the range direction is equal to the radar resolution, as determined from the pulse length.

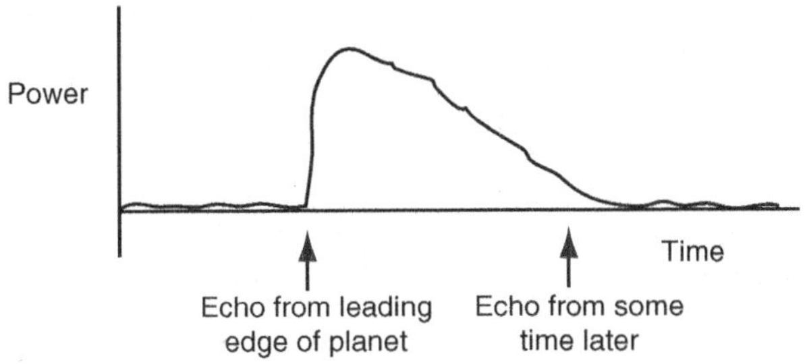

Figure 11.4 Received echo power as a function of time. The echo from the front of the planet is seen first, with energy reflected from other points on the surface arriving later as these points are farther from the Earth.

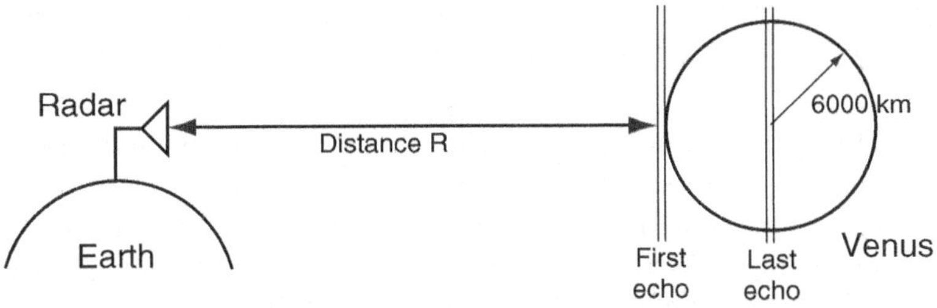

Figure 11.5 Imaging geometry showing spatial location of pulse for generation of first echo and last echo from Venus. The pulse first touches the closest point on the planet's surface to Earth, and then propagates over the surface until the poles are imaged. No echo is returned from the back side of the planet.

Energy from the last echo will be received at

$$
\begin{aligned}
t_{\text{last}} = \frac{2 \cdot (R + 6000\,\text{km})}{c} &= t_{\text{first}} + \frac{2 \cdot 6000\,\text{km}}{c} \\
&= t_{\text{first}} + 0.04\,\text{s}
\end{aligned}
\tag{11.1}
$$

Therefore, if we plot the received power as a function of time $t - t_{\text{first}}$ from 0 to 0.04 s, we can map the integrated reflectivity of each range ring (through the radar equation) on the planet (Figure 11.6).

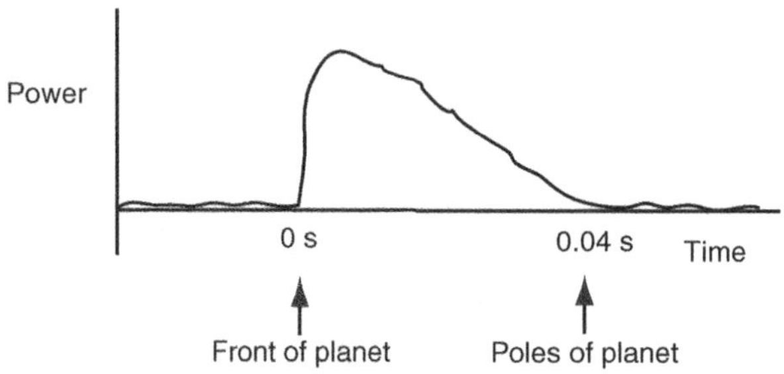

Figure 11.6 Received power as a function of time for Venus imaging geometry. The total echo time will be about 0.04 s for a planet with a radius of 6000 km.

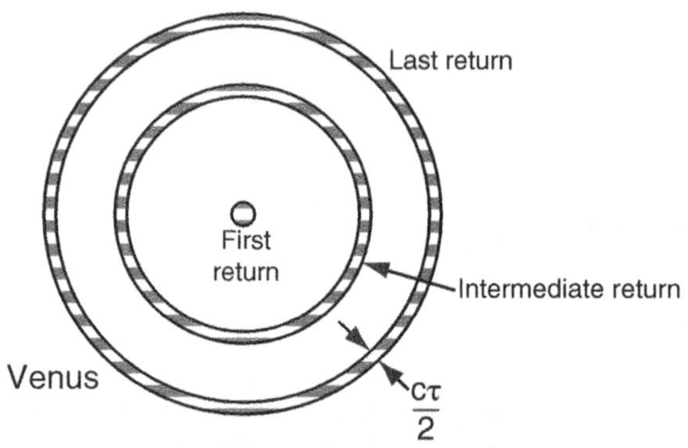

Figure 11.7 Face-on view of Venus showing the location of points contributing to observed scattering for each instant of time. Three instances are shown. The innermost circle is the first return, when the pulse just arrives at Venus. The outermost ring shows the area contributing when the pulse finally reaches the poles of the planet. The intermediate ring shows a sample range ring for a time between the two extremes.

We can visualize this also by changing our drawing to examine the planetary disk, with the direction to the planet now into the page, as in Figure 11.7. This depiction makes it clearer that at any given time energy is received from a single annular ring on the surface. Thus, plotting echo power as a function of time results in our ability to discriminate signals into range rings. Range ring analysis gives only one dimension of variation, while our desired image is 2D. Hence, range analysis alone does not result in an optimal image of the surface, but it is a start.

11.1.2 Doppler Bins

Range analysis provides a means to discriminate echoes from different annular rings on a planetary surface. It remains to be able to separate echoes at different locations along these rings. In this section we show that measuring the Doppler frequencies of the echoes, in addition to the range delays, allow this separation. The variation of Doppler frequency from points at different locations on the surface result from the different apparent velocities of the points. As we already know, objects at different relative velocities impart different Doppler shifts to their radar echoes.

Varying velocities on a planet result from any rotational motion present. Hence, since many planetary bodies spin, we might expect the reflected signals to be shifted in Doppler frequency. If we can relate relative Doppler velocities to location on the planet, we can map the location of echoes by measuring Doppler frequency shifts. To do so, consider a planet spinning with a spin vector at right angles to the radar look direction, as shown in Figure 11.8.

If, instead, we view the spinning planet from above, we visualize the point motions as in Figure 11.9. This view makes it easier to derive the line of sight velocity for any location on the planet.

It can be seen in Figure 11.9 that points on the planet's surface on the top half of the diagram will be moving toward the radar, while points on the bottom will be moving away. Visualizing the planet motion thus, we can now calculate the magnitude of the Doppler shift for each point on the planet's surface. For a rotation rate of F rotations per second, and radius r, we use the coordinates as defined in Figure 11.10 to define a velocity vector for each point on the surface. Note that for a spinning planet, any point has a velocity vector perpendicular to and proportional to its radius vector.

Since the rotational rate and planet's radius are known, we calculate the magnitude of the velocity —V— of a surface point as the distance traveled around the circle divided by the time it takes to complete one rotation, which is given by

$$|\mathrm{v}| = \frac{2\pi r_p}{\mathrm{P}} = 2\pi r_p \mathrm{F} \tag{11.2}$$

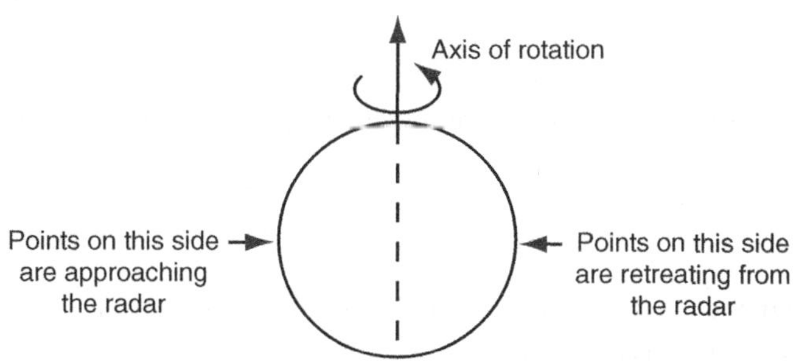

Figure 11.8 Spinning planet illuminated by a radar. If the planet's spin axis is in the page and vertical, with a right-handed spin direction, then for a radar at a location in front of the page, points on the right-hand hemisphere are retreating from the radar. Conversely, points in the left hand hemisphere approach the radar. The point velocities would be in the opposite direction for a planet with a left handed spin vector.

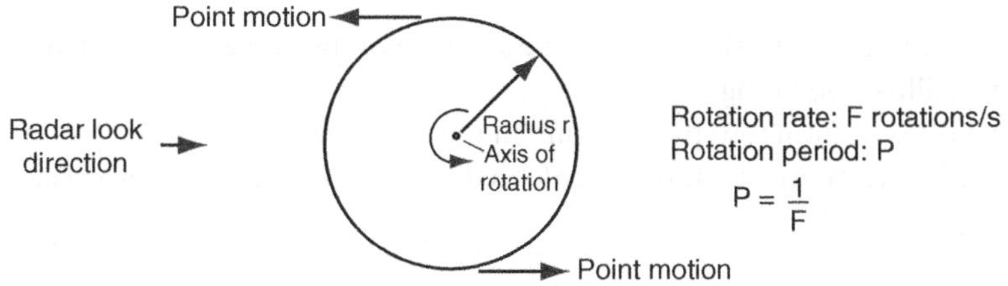

Figure 11.9 View from above spinning planet. With the planet's spin axis out of the page and right-handed, points on the upper hemisphere are approaching the radar, while points in the lower hemisphere retreat from the radar. The actual velocities of the points are related to the rotational rate of the planet, or, equivalently, its rotational period (see Figure 11.10).

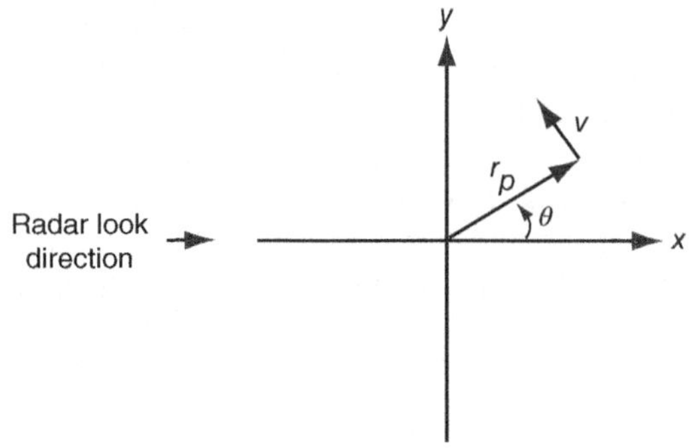

Figure 11.10 Geometry for computing the velocity vectors for each point on a spinning planet. The planet is again viewed from above. A point at radius r_p has a velocity υ at right angles to the radius vector.

Application of a little geometry and analysis identifies the x- and y-components of the velocity vector, and we obtain

$$\begin{aligned} \upsilon_x &= -\frac{2\pi r_p}{\mathrm{P}} \sin\theta \\ \upsilon_y &= \frac{2\pi r_p}{\mathrm{P}} \cos\theta \end{aligned} \tag{11.3}$$

Continuing, we obtain the magnitude of the line of sight Doppler velocity from the dot product of V with the look direction, here in the x-direction. Thus, the observed Doppler shift in frequency will be

$$f_{\mathrm{Dop}} = \frac{2}{\lambda} \cdot \frac{2\pi r_p}{\mathrm{P}} \sin\theta \tag{11.4}$$

We must next relate this Doppler frequency to the position of a point on the planet's surface, and so decompose the position r, θ, z to Cartesian x, y, z coordinates, as

$$\begin{aligned} r_x &= r_p \cos\theta \\ r_y &= r_p \sin\theta \\ r_z &= z \end{aligned} \tag{11.5}$$

The z coordinate in Eq. (11.5) is the position above the equator for a point p, and whose distance from the spin axis is r_p.

It is important here to make sure we understand the geometry. We plot the planet in a front view as seen by the radar, which we illustrate in Figure 11.11.

The point p is located at a radius r_p from the spin axis, and in this view we only see its projection in the y-direction. From this we have another condition, enforced by all points being on the surface on the sphere:

$$r_x^2 + r_y^2 + r_z^2 = r^2 \tag{11.6}$$

or

$$r_p^2 + r_z^2 = r^2 \tag{11.7}$$

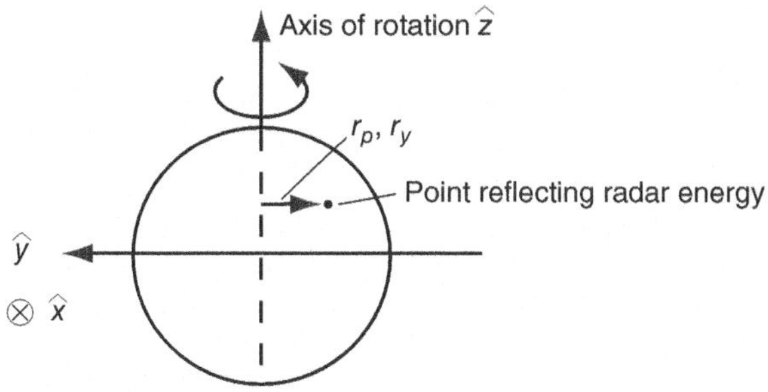

Figure 11.11 Plane of sky geometry for calculating the Doppler shift of a point on the planet's surface. A scattering point p is located at a radius r_p from the spin axis. In this view it is important to remember that we only observe the projection of the full velocity vector in the y-direction.

Hence,

$$\begin{aligned} r_x &= \sqrt{r^2 - r_z^2}\cos\theta \\ r_y &= \sqrt{r^2 - r_z^2}\sin\theta \\ r_z &= z \end{aligned} \tag{11.8}$$

The Doppler shift, from the above, is

$$f_{\text{Dop}} = \frac{2}{\lambda} \cdot \frac{2\pi}{\text{P}}\sqrt{r^2 - r_z^2}\sin\theta \tag{11.9}$$

while the (y, z) location of the point is

$$\begin{aligned} r_y &= \sqrt{r^2 - r_z^2}\sin\theta \\ r_z &= z \end{aligned} \tag{11.10}$$

Therefore, the Doppler shift of a point on the surface at location r_y, r_z follows from Eqs. (11.9) and (11.10), or, equivalently from Eqs. (11.4) and (11.5), is

$$f_{\text{Dop}} = \frac{2}{\lambda} \cdot \frac{2\pi}{\text{P}} r_y \tag{11.11}$$

The result of this manipulation is perhaps somewhat surprising in its simplicity. The Doppler shift is independent of z, and depends only linearly on the projected position of the point off the spin axis. So, cells of constant Doppler shift are vertical lines, and therefore, Doppler bins appear as vertical stripes on the planet (see Figure 11.12). The concept of Doppler bins is entirely analogous to that of the range bins described in the previous section, so that it is evident that since we associate some resolution with our estimates of Doppler frequency, the interpretation of Doppler frequency with position on the planet leads to a finite resolution in position as well.

From this analysis, we see that if we use a continuous wave radar to look at the surface of the planet, the observed Doppler shift separates the return into strips rather than rings. This is a different geometrical filtering of the data than we saw with the range measurements. But it still smears the results over large areas on the surface, making it difficult to isolate the returned power from a particular point on the planet. We shall see next that by combining both range delay and Doppler frequency observations, we can resolve the 2D pattern of radar reflectivity on the surface.

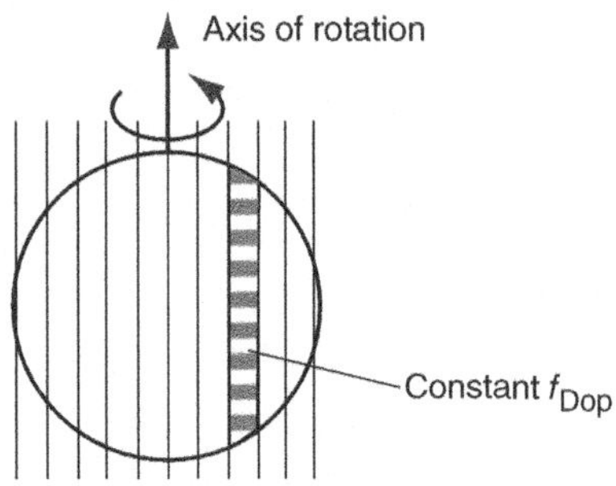

Figure 11.12 Loci of constant Doppler shift form vertical lines on the face of the planet. Hence for a given Doppler range, denoted a Doppler bin, the planet is divided into a series of stripes. The total energy reflected from within each stripe is recorded as the value of the power in the appropriate Doppler bin.

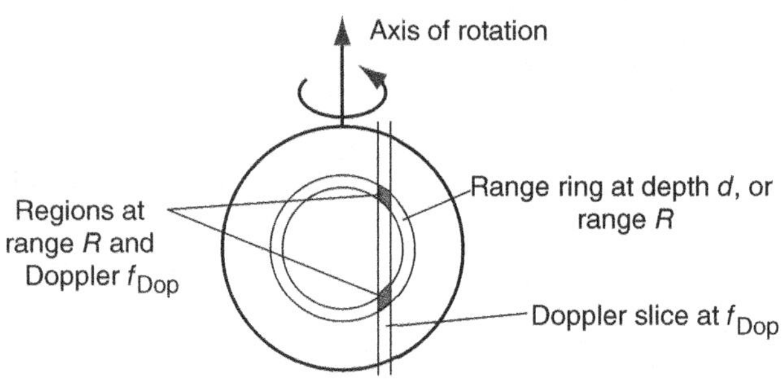

Figure 11.13 Imaging the planet surface by combining the range and Doppler measurements. Points on the surface at constant range from the range rings, and points of constant Doppler shift form vertical lines on the face of the planet. By determining both quantities, it is possible to identify the return at a particular range, Doppler pair with two distinct spots on the surface of the planet. The inability to distinguish between the spots in the upper and lower hemispheres is called the north–south ambiguity of a radar range-Doppler image.

11.1.3 Imaging: Combining Range and Doppler

If our goal is to create a high-resolution image of radar reflectivity of a planetary surface, we want to be able to resolve the radar echoes returned from small patches on the surface. While either range rings or Doppler slices are helpful in isolating the return from different parts of the projected surface, each still covers a large part of the ground. Range rings do not resolve in the angular direction from the planet's apparent center, while Doppler slices do not separate echoes in the vertical direction at all. We can do a better job of creating an image if we use both range and Doppler estimates at the same time.

Suppose we could examine both the range and the Doppler properties of the radar echoes all at once. We could then decompose the echo from the planet as in Figure 11.13, producing unique estimates from much smaller regions on the surface. In this imaging approach, there are still two fairly small spots on the surface with the same range R and frequency f_{Dop}. This is an example of the north/south ambiguity that plagues range-Doppler imaging, but even in its presence we would be identifying the radar echoes from very small parts of the surface.

Hence, if we transmit a pulse and measure its Doppler shift and range, we can produce a high-resolution image of the surface. Range tells us what range ring we are on, while the Doppler tells us how far left or right of the axis is the point whose echo we have recorded. If we can measure both delay and frequency very accurately, we can produce a high-quality image.

11.1.4 North–South Ambiguity

One problem with the range-Doppler algorithm is that for each combination of delay and frequency there are two points contributing to the total echo. An example was obvious in the previous section on planetary imaging, and it is this famous example that led to the nomenclature of north–south in describing the ambiguity. Any points that are located in the south show up in both the north and south hemispheres of the image; the same is true for points in the north. In fact, any range-Doppler imaging algorithm suffers from forms of this ambiguity, in that in many cases there can be several points on an object that have the same range from radar to object, and the same projected Doppler velocity. In these instances, it will always be impossible to uniquely determine a reflectivity map for the object.

There are a few methods that have been used to avoid the ambiguity problem. Each involves altering either the imaging geometry or the design of the radar so that only one possible range-Doppler pair is observable at any given time, or else the system is augmented by additional measurements that can resolve the ambiguity. For example, one approach is to change the radar antenna and wavelength so that only one hemisphere is illuminated at any time. Then the echo will be received only from the north or the south, and the images may be readily interpreted without the uncertainty that the ambiguity provides. Another method that has been used fairly often is to use an array of antennas as an interferometer, which puts a positive interferometer phase on one hemisphere and a negative phase on the other. Then the returns can be separated by additional signal processing that estimates the interferometer phase as well as range delay and Doppler frequency.

Unfortunately, for most targets in the solar system, neither of these work well due to physical limitations on antenna or array size and location. The targets are simply too far away to form fine enough antenna beams to illuminate only one hemisphere, and radar interferometers appropriate for planetary applications are rare. The north–south ambiguity remains a difficult problem.

11.2 Measuring Doppler Shifts

We have seen that we require simultaneous measurement of range delay and Doppler frequency to produce high-quality images of a planetary surface, or indeed of any object we wish to image with range-Doppler methods. We know that we can measure range by transmitting short radar pulses and timing the echo return, and the shorter the pulse the better range resolution we will achieve. We also know that we can measure Doppler by transmitting continuous wave signals and mixing the returned echo with the transmitted waveform. This requires the transmission of very long radar pulses, which is at odds with our previous desire to use many short pulses to estimate the range of an object. So the problem in practice is how to measure both at once. We addressed a similar problem earlier by looking at waveform design, and found that there were certain waveforms that lent themselves to simultaneous estimation of delay and frequency. Now we shall see how we can implement this type of waveform and its analysis in a real system.

11.2.1 Doppler-phase Relationship

We will find it useful here to begin with the relationship between the phase of a radar echo and its Doppler frequency. Suppose we have an object moving toward the radar at velocity v, and we illuminate it with a series of very narrow pulses, say a few cycles at a high carrier frequency, as in Figure 11.14. This was the case we described earlier of a radar speed gun for sports activities, such as the measurement of the speed of a fastball thrown by a baseball pitcher.

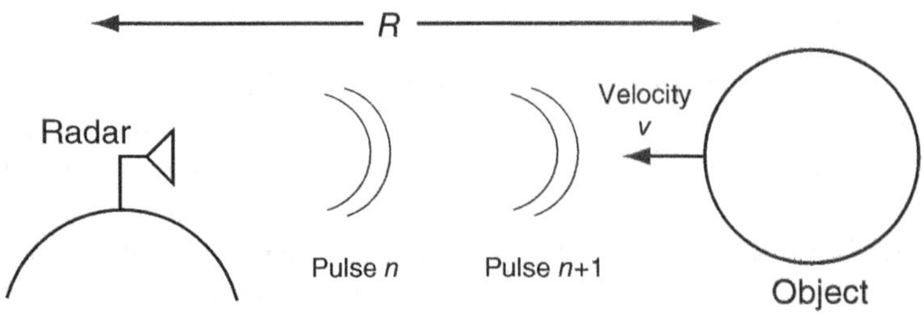

Figure 11.14 Imaging an object with a pulsed radar system. Multiple pulses are transmitted in sequence from the radar transmitter, and each generates an echo from the object's surface.

If the object is located a distance R from the radar, and the transmitted signal is

$$S_{\text{trans}}(t) = \sqcap\left(\frac{t}{t_p}\right) e^{-j2\pi f t} \tag{11.12}$$

then the reflected echo is delayed in time by the round-trip radar delay $\frac{2\text{R}}{\text{c}}$, and is observed at an amplitude dependent on the reflectivity of the surface and by spherical wave spreading. We shall ignore the amplitude for now, and so the received signal can be modeled as

$$S_{\text{rec}}(t) = \sqcap\left(\frac{t - \frac{2R}{c}}{t_p}\right) e^{-j2\pi f \frac{t-2R}{c}} \tag{11.13}$$

In our receiver, we can remove the carrier and generate a signal based solely on the object by mixing the received signal with the complex conjugate of the transmitted waveform at an arbitrary phase ϕ_0:

$$\begin{aligned} S_{\text{mixed}}(t) &= \sqcap\left(\frac{t - \frac{2R}{c}}{t_p}\right) e^{-j2\pi f \frac{t-2R}{c}} \cdot e^{+j(2\pi f t + \phi_0)} \\ &= \sqcap\left(\frac{t - \frac{2R}{c}}{t_p}\right) e^{j2\pi f \frac{2R}{c}} \cdot e^{j\phi_0} \end{aligned} \tag{11.14}$$

What is the phase of this signal over the window? Alternatively, what is the complex received signal over this time?

$$\begin{aligned} s(t) &= e^{j2\pi f \frac{2R}{c}} \cdot e^{j\phi_0} \\ &= e^{j2\pi \frac{R}{\lambda}} \cdot e^{j\phi_0} \end{aligned} \tag{11.15}$$

The radar transmitter now emits a second pulse toward the object. For this next pulse, we will receive a similar signal but with the range decreased slightly, because R is slightly less. So if the next pulse follows at time $\text{T} = \frac{1}{\text{PRF}}$ later, it will be given by

$$s(t) = e^{j\frac{4\pi}{\lambda}(R - vT)} \cdot e^{j\phi_0} \tag{11.16}$$

We can form a sequence of observations in which each echo is similar in amplitude to its predecessors, but the phase gets progressively less for each pulse due to the decrease in distance. In other words, the propagation phase decreases as the object approaches the radar. Here we have assumed that the pulse length τ is so short there is no appreciable motion of the object during that time.

We can form a sequence of measurements $s_k(t)$, where each measurement is the mixed-down complex signal from successive pulses, or

$$\begin{aligned}
s_0(t) &= e^{j\frac{4\pi}{\lambda}(R)} \cdot e^{j\phi_0} \\
s_1(t) &= e^{j\frac{4\pi}{\lambda}(R-vT)} \cdot e^{j\phi_0} \\
s_2(t) &= e^{j\frac{4\pi}{\lambda}(R-2vT)} \cdot e^{j\phi_0} \\
&\vdots \\
s_i(t) &= e^{j\frac{4\pi}{\lambda}(R-ivT)} \cdot e^{j\phi_0}
\end{aligned} \tag{11.17}$$

Hence, each successive pulse has a phase changing by $vT \cdot 4(\pi/\lambda)$, with each successive sample spaced in time by T. We can therefore view this as a sampled waveform, forming a signal with linear phase change. The slope of this linear phase change is proportional to the line of sight velocity of the object. Since linear phase charge equates to a constant frequency, we can associate a frequency with this signal. We can compare the frequency observed in this way with the conventional Doppler frequency shift, which we know how to relate to the position of the scattering point on the object. Consider just the phases of s_i:

$$\phi_i = \phi_0 + 4\pi\frac{R}{\lambda} - i\frac{4\pi}{\lambda}vT \tag{11.18}$$

Now, from the definition of frequency

$$f = \frac{1}{2\pi}\frac{d\phi}{dt} \tag{11.19}$$

the effective frequency of this signal is thus

$$\begin{aligned}
f_{\text{eff}} &= \frac{1}{2\pi}\frac{d\phi}{dt} = -\frac{1}{2\pi}\frac{4\pi}{\lambda}v \\
&= -\frac{2}{\lambda}v
\end{aligned} \tag{11.20}$$

This, of course, is just the Doppler shift of a target moving toward us at velocity v. Hence, by using a pulsed waveform we can measure the range and Doppler shift of an echo simultaneously. This will allow us to find from an image of the surface of the object under study.

11.3 Imaging Algorithm

We are finally ready to discuss how to convert the received echoes from a planetary surface to an image. Remember that we want to decompose the echoes into range and Doppler estimates, which will map into points on the observed surface of the planet. For simplicity, suppose that we have only one very bright spot on the planet face. We illuminate the entire planet with a sequence of short sinusoidal pulses. The echo we receive in time then looks like the plot in Figure 11.15.

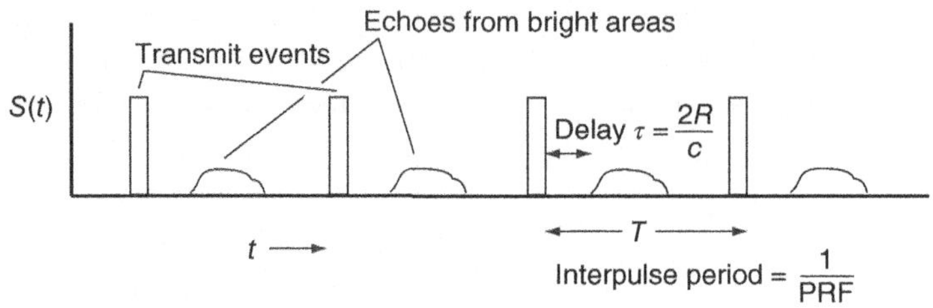

Figure 11.15 Received echo $s(t)$ as a function of time for pulse-illuminated planet. Each transmit event is denoted by a rectangle at the time of transmission, with the echo following after a delay related to radar range. There will be one echo for each transmit pulse, separated in time by the interpulse period, or PRF.

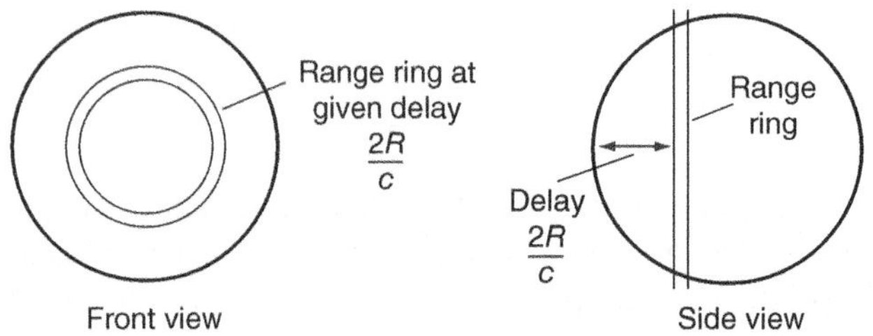

Figure 11.16 Two views of the range ring corresponding to the bright echo observed at a given delay time, per Figure 11.15. The delay may be obtained from the radius of the range ring (left) or from its position with respect to the closest point of the planet (right).

Examination of the received waveform shows the delay, corresponding to a range ring, that contains the bright area. The delay is measured from the time of the transmit event to the time when the echo appears. This allows us to associate the echo with just one range ring. In this experiment, we might visualize the return as in Figure 11.16. Here the single range ring that contains the bright point can be mapped to a fixed position on the planet.

At this point, we can see in what range ring the bright area lies, but we cannot localize it further. As we argued above, we also need to add the Doppler information as to the left–right location of the brightly scattering point. Rather than present our time series as a single function of time, it will be more useful to stack every received pulse next to the previous pulse and form a 2D matrix of received signal strengths. We might plot the results of the echo reception then as in Figure 11.17. The row number of each entry in the matrix corresponds to time delays, while the column depends on the transmitted pulse number.

This matrix depicts the range delays we need to localize the echo to range rings, and we yet have to estimate the Doppler frequency of the echoes. We know that we can estimate the Doppler shift of the bright area by looking at the phase shift of the signal from pulse to pulse, or, equivalently, measuring the frequency of the time series obtained from pulse to pulse signals at the given range delay. The conversion of a sequence of complex values to frequency is readily done using a Fourier transform, so the next step in the algorithm is to calculate the Fourier transform across the radar pulses at each time delay. Thus, if we visualize the data as stored in a matrix as in Figure 11.18, we can readily determine both the range delay and the Doppler frequency of a given part of the echo history.

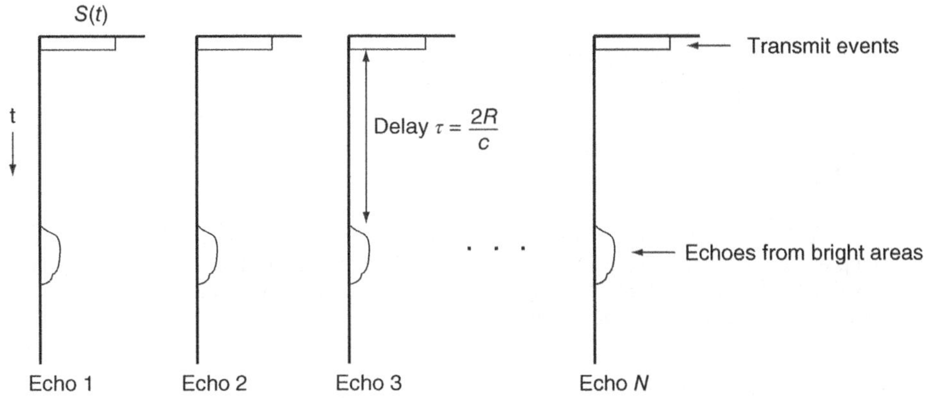

Figure 11.17 Another way to visualize the radar return as a function of time depicted in Figure 11.15. Here the delay from the transmit event to the echo is shown vertically, and the echo from each transmit event is stacked next to that of the previous pulse.

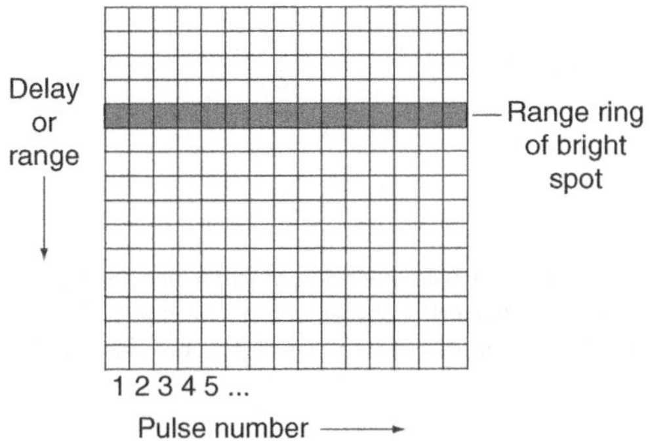

Figure 11.18 Matrix representation of the radar return as a function of time. Here we represent the radar echo at each point in time, and from each pulse, as a 2D grid of numbers. This is simply another way of visualizing the signal depicted in either Figure 11.15 or Figure 11.17. This representation maps easily to a set of numbers we can analyze with a digital computer.

Therefore, we calculate the Fourier transform (or FFT) across the columns in the matrix, and thus estimate the Doppler frequency of the bright spot. Hence, we know both range and Doppler, so our face-on view of the planet becomes as in Figure 11.19. From this view it is clear that the return can only arise from two distinct regions on the planet. However, the north/south ambiguity remains and makes interpretation of the resulting map difficult.

In summary, by looking at the range delay and Doppler spectrum of the echo, we can localize the bright spot to one of two small areas on the planet. This dual solution is simply the north–south ambiguity we discussed previously.

If instead of a single bright spot, we have a complete surface to image, we can model it as a superposition of scatterers at different spots on the planet. We apply the approach of putting the data in a delay/pulse matrix and then decomposing the time series corresponding to each range ring through a Fourier transform. This gives us a unique number for every possible combination of range and Doppler, which we can visualize as an image of the reflectivity of the surface. The resulting map is depicted pictorially in Figure 11.20.

All that remains is to put these echoes in the proper spots on the map.

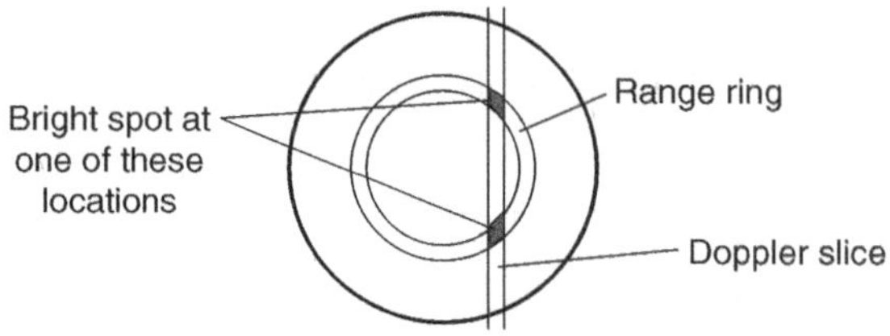

Figure 11.19 Identification of the two spots on the surface of the planet that can a yield an echo at a given range delay and Doppler frequency. It is impossible to determine which of the two locations is responsible for echo, however, without other information, leading to the north/south ambiguity that is present in all range/Doppler images acquired using this imaging geometry.

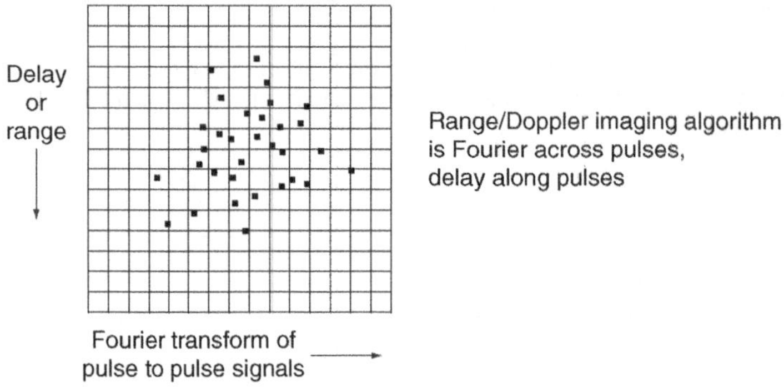

Figure 11.20 Matrix obtained by calculating the Fourier transform across the series of radar echoes for each possible range delay time. Each value in the matrix corresponds to the power in a particular range delay/Doppler frequency pair. Hence, each number corresponds to a unique north/south pair of locations on the planet's surface.

11.4 Mapping Range/Doppler to Location on the Planet

The final task after producing an unambiguous range/Doppler image of the planet is to map the range/Doppler grid onto a useful coordinate system, such as latitude and longitude. To simplify this mapping, for now we ignore planetary rotation and define coordinates as in Figure 11.21.

We need to equate our range/Doppler values to latitude and longitude values on the surface. Note that if we know the vector from the center of the planet to a point on the surface $\mathbf{r} = (r_x, r_y, r_z)$ and if we choose a coordinate system as before (see Figure 11.22), then

$$\begin{aligned} r_z &= a\sin(\text{lat}) \\ r_x &= -a\cos(\text{lat})\cos(\text{long}) \\ r_y &= -a\cos(\text{lat})\sin(\text{long}) \end{aligned} \tag{11.21}$$

This is a bit different than what we usually see but is driven by our Cartesian coordinate choice (x, y, z), as shown in Figure 11.22.

Recall from before that we could relate r_y to Doppler by

$$f_{\text{Dop}} = \frac{4\pi}{\lambda \text{P}} r_y \tag{11.22}$$

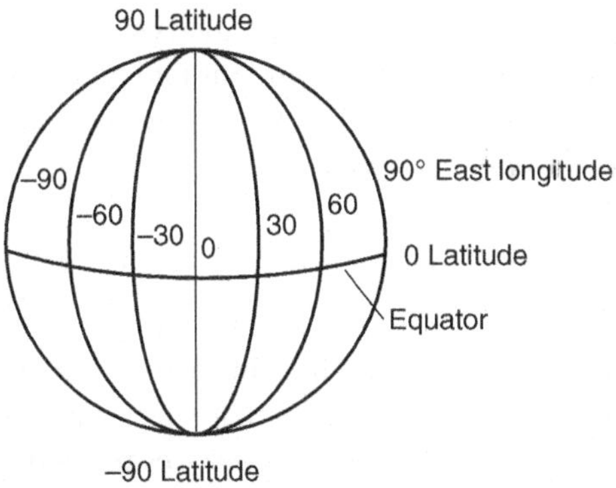

Figure 11.21 Planetary coordinate system referenced to sub-Earth point. We can define a set of coordinates with the origin at the closest point on the planet in terms of latitude and longitude values. 90° latitude refers to the north pole, and −90° the south pole. The equator lies halfway in between. The 0° meridian passes through the sub-Earth point, with meridional circles forming lines of constant longitude.

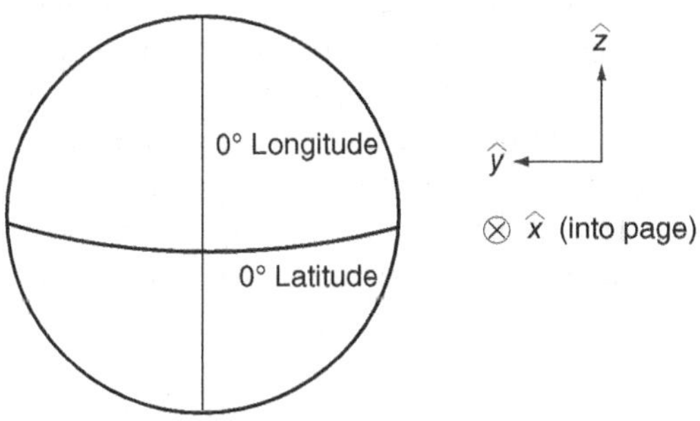

Figure 11.22 Planetary coordinate system for relating range-Doppler measurements to location on the planet surface.

and delay is related to r_x by

$$\text{delay} = a + r_x \tag{11.23}$$

so that

$$\begin{aligned} f_{\text{Dop}} &= \frac{4\pi}{\lambda \text{P}} a \cos(\text{lat}) \sin(\text{long}) \\ \text{delay} &= a - a\cos(\text{lat})\cos(\text{long}) \end{aligned} \tag{11.24}$$

where delay is in meters, and f_{Dop} is in Hertz.

Hence

$$\cos(\text{lat})\cos(\text{long}) = \frac{a - \text{delay}}{a} \tag{11.25}$$

and

$$\cos(\text{lat})\sin(\text{long}) = -\frac{F_{\text{Dop}}\lambda \text{P}}{4\pi a} \tag{11.26}$$

Thus, we can get the longitude unambiguously and get a solution for $\cos^2(\text{lat})$, which we can now see is just another way to look at the north–south ambiguity. There are two solutions for latitude in this equation, both a positive and negative one, which is what we saw when we plotted the areas contributing to an echo at a single pair of range delay and Doppler frequency.

11.5 Sampling Effects

Apart from the north–south ambiguity, we have seen that range-Doppler imaging can be used to create high-resolution maps of surfaces. But there are geometrical constraints that affect the usefulness of this imaging algorithm. An important issue arises in the imaging of surfaces using the range-Doppler method related to the allowed values of range and Doppler that may be determined in any given experiment. Since we are dealing with a sampled system due to the multiple pulse nature of the experiment, we must consider whether the values in the matrix we have calculated are properly sampled, and that no aliasing occurs.

Because we are using a pulsed waveform, we are in effect sampling the Doppler spectrum at a finite rate. As you know from experience, this is going to put aliasing constraints on our signal analysis. In the case of range/Doppler imaging, sampling influences both the range and Doppler measurements. What are these constraints?

Lets begin with the range constraint. We want to be able to measure the range to each point on the surface unambiguously. Therefore, the interpulse period must be large enough to fit the entire echo between pulses. The situation is depicted graphically in Figure 11.23.

The minimum interpulse period must be long enough to fit ranges up to the radius of the planet, a:

$$T_{\min} \geq \frac{2a}{c} \tag{11.27}$$

where T is the interpulse period, and by definition is 1/PRF. This defines our maximum pulse rate, or sample rate. We now need to use this signal to sample the Doppler spectrum of the echo. Now we must ask if a PRF of this rate is sufficient to properly sample the Doppler spectrum according to the Nyquist criterion. Hence, first we must calculate the width of the Doppler spectrum. This is easily seen by visualizing the Doppler velocities as in Figure 11.24.

From here we see that the maximum Doppler velocity is

$$v_{\max} = \frac{2\pi a}{P} \tag{11.28}$$

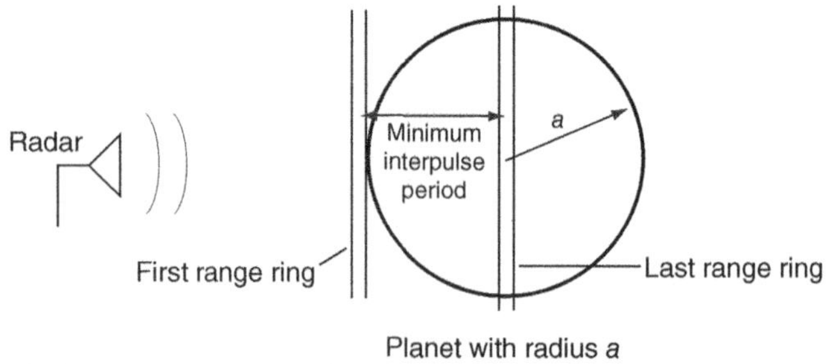

Figure 11.23 Minimum interpulse period for imaging the full disk of a planet. The pulse rate must be chosen so that only one pulse is interacting with the planet's surface at any point in time, otherwise we would have data aliased in position. This leads to a minimum time between pulses, or, equivalently, a maximum pulse rate.

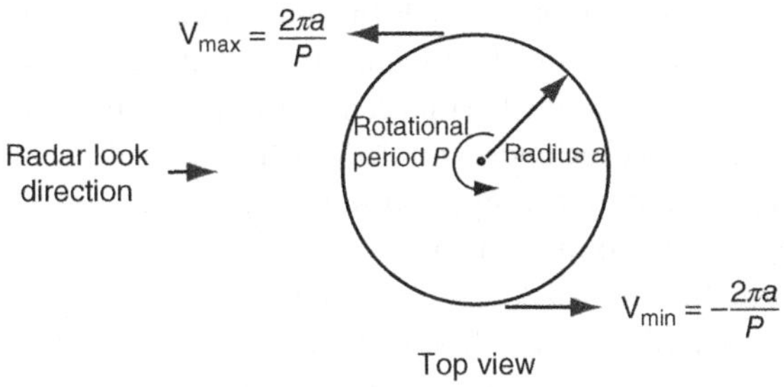

Figure 11.24 Maximum Doppler velocities for our radar echoes. The maximum velocities are seen at the extrema of the planetary disk, where the projected velocities are greatest. the actual Doppler shifts seen depend on the rotational period *P* and the planet radius *a*.

and

$$\upsilon_{\min} = -\frac{2\pi a}{\mathrm{P}} \tag{11.29}$$

from which

$$\begin{aligned} f_{\mathrm{Dop,max}} &= \frac{2}{\lambda}\frac{2\pi a}{\mathrm{P}} = \frac{4\pi a}{\lambda \mathrm{P}} \\ f_{\mathrm{Dop,min}} &= -\frac{4\pi a}{\lambda \mathrm{P}} \end{aligned} \tag{11.30}$$

so the total Doppler bandwidth is

$$\mathrm{BW}_{\mathrm{Dop}} = \frac{8\pi a}{\lambda \mathrm{P}} \tag{11.31}$$

Thus, the minimum PRF becomes

$$\mathrm{PRF}_{\min} \geq \frac{8\pi a}{\lambda \mathrm{P}} \tag{11.32}$$

or

$$T_{\max} \leq \frac{\lambda \mathrm{P}}{8\pi a} \tag{11.33}$$

This gives us an upper bound on *T*, while range gave us a lower bound on *T* to avoid aliasing the range estimate. Summarizing both of these results in a single equation, the permissible range of *T* is

$$\frac{2a}{c} \leq \mathrm{T} \leq \frac{\lambda \mathrm{P}}{8\pi a} \tag{11.34}$$

So if we went to image properly, we need to select *T* to fit in this range.

A bad situation results if we choose *T* outside this range. We will either not see all ranges unambiguously, or we will undersample the Doppler spectrum. This can easily happen in practice if the wavelength is not properly chosen. For example, suppose we squeeze the range of allowable *T* to zero, that is, we let

$$\frac{2a}{c} = \frac{\lambda \mathrm{P}}{8\pi a} \tag{11.35}$$

or, in terms of wavelength λ,

$$\lambda = \frac{16\pi a^2}{c\,\mathrm{P}} \tag{11.36}$$

If you examine Eq. (11.34) above, you will see that as long as λ exceeds the limit $16\pi a^2 cP$, we can image the surface unambiguously in both range delay and Doppler frequency, while if λ is smaller than this limit, we cannot. Either the range will be undersampled in that echoes from several ranges are received simultaneously, or the Doppler spectrum will be undersampled and aliasing will occur. In either of these situations, we cannot form a proper map relating radar reflectivity to position on the surface, no matter how we design the rest of the radar instrument.

The above limits hold whenever we illuminate the entire disk of the planet. If our antenna beam is narrow enough so that only a portion of the projected surface is illuminated, then the appropriate limits on interpulse period are determined by the portion of the surface that is visible by the radar, within its illumination pattern.

In planetary radar literature, if the conditions above are violated, the planet is said to be *over-spread* at wavelength λ. If imaging is feasible, with the sampling conditions satisfied, the target is said to be *underspread*.

Let us examine some examples of range-Doppler imaging to see what the wavelength limits might be in practice. Specifically, what are the limits for some planetary bodies?

Venus: Here $a = 6052$ km, $P = 243$ days. Then

$$\lambda_{\text{critical}} = \frac{16\pi a^2}{c\,P} = 29.2\,\text{cm} \tag{11.37}$$

Mars: For Mars, $a = 3390$ km, $P = 24.6$ h $= 1.025$ days

$$\lambda_{\text{critical}} = \frac{16\pi a^2}{c\,P} = 21.7\,\text{m!} \tag{11.38}$$

Mercury: For Mercury, $a = 2440$ km, $P = 58.6$ days

$$\lambda_{\text{critical}} = \frac{16\pi a^2}{c\,P} = 19\,\text{cm} \tag{11.39}$$

12

Swath-imaging Radars

When a radar system's primary output product is a two-dimensional (2D) map, or image, of the ground, that system is called an *imaging radar*. Such radars are often, but not always, relatively high-resolution instruments that produce a picture of the RCS distribution on the surface. That is, they create an image of the specific cross section, σ^0, averaged or smoothed over the resolution area on the mean surface. In this chapter, we discuss the viewing geometry of imaging radar, methods to generate high-resolution image products, design approaches, and performance analysis. An important class of imaging radar is *synthetic aperture radar*, a term which some loosely use as synonymous with imaging radar. However, synthetic aperture radar, or SAR, refers specifically to one of the various techniques available for obtaining high resolution in a surface image. In this chapter, we are concerned with radars that are capable of mapping the surface of Earth, or another body, from a platform that moves along a flight track in order to cover large ground areas. The platform may be an aircraft, or a satellite in orbit above the body of interest. Techniques introduced here are readily extended to several different geometries.

An imaging radar systematically measures the RCS of every resolution element on the surface. While the most obvious approach might be to direct a narrow beam antenna downward and make a RCS observation over all points of interest, for high-resolution systems the number of pulses required, the need to scan the surface by manipulation of the antenna beam, and the signal processing requirements associated with the large number of pulses needed quickly render such an approach impractical. We find it much easier to transmit pulses only occasionally, and then analyze the radar echoes as a means of separating the portions of the echoes that arise from different points on the ground. We have seen previously that we can separate echoes from objects at different distances within a radar echo because each distance corresponds to a different echo time delay, and we will make use of that principle here. We also have seen that we can separate echoes from objects in slightly different directions on the basis of Doppler frequency shift, and we will use this method also, among others. Overall, the goal is to discriminate among echoes from various different scattering locations contained in the region of radar illumination.

12.1 Imaging Geometry and Terminology

Most imaging radars operate by systematically scanning a *swath* along a flight track, where "swath" refers to the strip of illuminated surface that falls within the radar antenna illumination pattern. This mode of operation is thus often referred to as strip-mapping, because the output maps describe strips along the ground. Alternatively, it is possible to maintain the radar illumination on a smaller region on the surface as the radar flies overhead, increasing the available time during which the

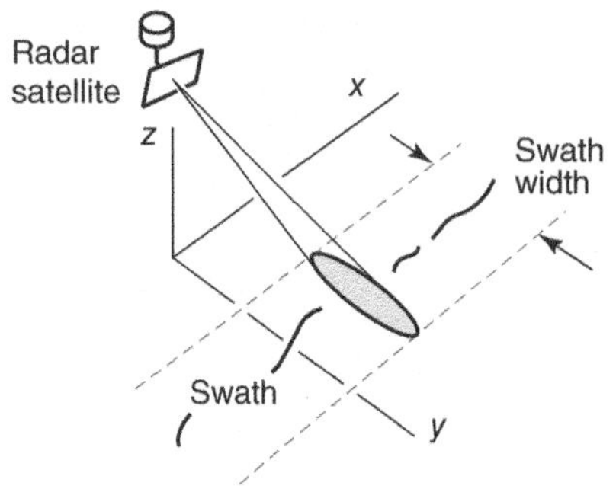

Figure 12.1 Imaging radar geometry. Side-looking radar carried by satellite, or aircraft (not shown) located at altitude z. "Across-track" or range-direction is y. "Along-track" or azimuth direction is x. For purposes of illustration flight path is linear and ground is flat.

system can observe the area of interest. Constraining the illumination to a restricted size area on the ground in such a maneuver is referred to as a *spotlight* mode of operation.

The generic imaging geometry and coordinates are shown in Figure 12.1, which shows a radar system illuminating a swath off to one side of the flight track projected on the ground. A coordinate system centered on the nadir point is shown, where the x-direction is called the along-track, or azimuth, direction, the y-direction is called the across-track, or ground range, direction, and the z-direction is the altitude of the platform, with the corresponding unit vectors denoted by carets ($\hat{x}$, $\hat{y}$, $\hat{z}$), respectively. This nomenclature derives from the projection of unit vectors in the image plane into coordinates defined by the radar sensor motion.

As we shall see below, as a radar is used to illuminate the area more directly beneath the sensor, or nadir point, it becomes increasingly difficult to generate high resolution images of the surface. For this reason, a radar is usually designed to "look" to the side of the flight path, away from the exact nadir direction, as shown in Figure 12.1. This choice led to the term side-looking radar, or SLR, as it describes the basic features of the geometry. In addition, the radar may be aimed slightly ahead or behind the platform carrying the radar, with the angle forward or backward denoting the squint angle. Because it simplifies the radar processing, most often an imaging radar is configured to illuminate the ground at near-right-angles to the flight track, or with a squint angle of 0°. We begin our analysis of SLR imagers with this configuration.

12.2 Side-looking Geometry Over a Flat Earth

Consider a view of a radar imaging geometry as described above, but with the plane of the page perpendicular to the direction of motion and defined to contain the antenna, the sub-radar point, and a point to be imaged, as in Figure 12.2.

The distance, or range, from the radar to the imaged point is related to across-track, or "ground range," distance and altitude through Pythagoras' theorem, if Earth is assumed to be locally flat. Using a curved Earth gives a somewhat more complicated relation, but is not needed for our analysis at this point. We call the angle between the vector from radar to ground point, or look vector, and the vector to the nadir point the look angle, θ, of the observation. Its complement is the depression angle, θ_d, which is the angle that the radar "looks" below the horizon, Figure 12.2.

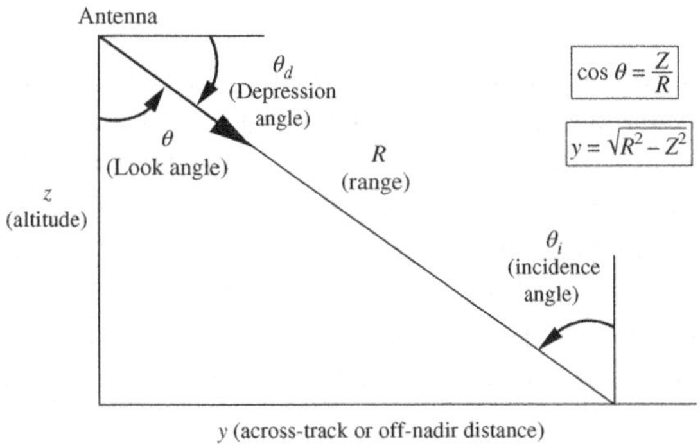

Figure 12.2 Details of side-looking geometry. Angle of incidence, θ_i, is measured from local vertical at the scattering point. Depression angle, θ_d, and grazing angle, $\gamma = \pi - \theta_i$, are measured from ray relative to local horizontal radar and at scattering point, respectively. Look angle, θ, is measured from local nadir at radar. Figure is drawn relative to a flat Earth, but definitions of angles relative to local ray directions and verticals apply to case of a curved Earth as well.

The imaging geometry is defined by the look angle, while the properties of the surface are more closely related to the angle between the look vector and a unit outward normal from the surface, which we refer to as the incidence angle, θ_i. To the extent that we can neglect the curvature of the Earth the depression and incidence angles are the same, $\theta = \theta_i$.

In later chapters, we also find it convenient to relate the look angle to the altitude of the sensor,

$$\theta = \cos^{-1}\frac{z}{R} \tag{12.1}$$

Note that objects located along the surface ahead of and behind the sub-radar point have radar echoes that occur later in time than locations directly opposite the radar platform. This occurs because the slant range to objects always increases with increasing ground distance. Thus, if we transmit a single radar pulse, measuring its echo intensity as a function of time allows us to determine the radar cross section as a function of time, or equivalently, range. To relate the cross section as a function of range to the location of the point on the ground, we use the dependence of ground range y on range R,

$$y = +\sqrt{R^2 - z^2} \tag{12.2}$$

Given an observation of echo intensity vs. time at a given position along the flight track, Eq. (12.2) allows us to determine the reflectivity at any point along the projection of the antenna look direction on the surface.

More generally, note that Eq. (12.2) is the equation of a circle or radius R centered on the radar. This circle arises from the intersection of the sphere of constant range at any fixed echo time simultaneously with the plane of the surface and the plane perpendicular to the flight path through the location of the radar. But there are two such locations, one on each side of the radar track, corresponding to the $\pm$ signs in the evaluation of the radical. It is implicit in the calculation here that the side-looking antenna effectively eliminates echoes from the opposite side of the radar track, hence permitting us to choose the positive sign for the radical in Eq. (12.2), consistent with Figure 12.1. In practice this can never be achieved perfectly, so that the right/left ambiguity inherent in solution for the ground range is important, especially when relatively strong scattering areas relocated on the opposite side of the track from the main beam of the antenna.

12.2.1 "Squint" Geometry

The radar system may not be pointed exactly at right angles to the flight track at all times. Unanticipated motion of the platform, or selecting a curved trajectory, may result in the antenna beam axis, or "boresight," inadvertently moving ahead or behind of the perpendicular. Moreover, the beamwidth of the antenna in the along track direction is not infinitesimal, so that in actuality a small patch of ground is illuminated at any time, as suggested in Figure 12.1. Hence, even when the radar antenna is pointed 90° from the flight path direction, only a portion of the illuminated swath will be exactly at a right angle from the flight track. Every other point will be at some other angle, which we call the squint angle, ϕ, defined with respect to the true side-looking direction (Figure 12.3).

As the radar moves over its track, it will receive echoes from all the points on the ground within its illumination pattern. In order to form an image of the surface, we must distinguish among these points. For this purpose, we add the Doppler effect to our side-looking imaging toolbox. Consider the Doppler shift of the echo reflected from a given point on the surface. As we have seen, the Doppler shift of a radar echo is the component of the relative velocity of the radar and the scatterer in the line of sight direction to the radar,

$$f_{\text{Dop}} = \frac{2}{\lambda} \cdot (\mathbf{v} \cdot \mathbf{u}) \tag{12.3}$$

where $\mathbf{u}$ is a unit vector from the radar to the point on the surface and $\mathbf{v}$ is the relative velocity of the radar over a stationary surface. The dot product in Eq. (12.3) represents the equation of a cone with axis in the direction $\mathbf{v}$ and half-angle $\alpha = \cos^{-1}((\mathbf{v}/v) \cdot \mathbf{u})$. The locus of points with the same Doppler shift is the intersection of this cone with the imaged surface. For a flat-Earth this locus is a hyperbola. At a constant range, the unit vector will be different for each point, hence its Doppler shift will be different from that of its neighbors. We can restate Eq. (12.3) in terms of the imaging geometry as described by the look and squint angles, θ, ϕ, under the assumption that the velocity

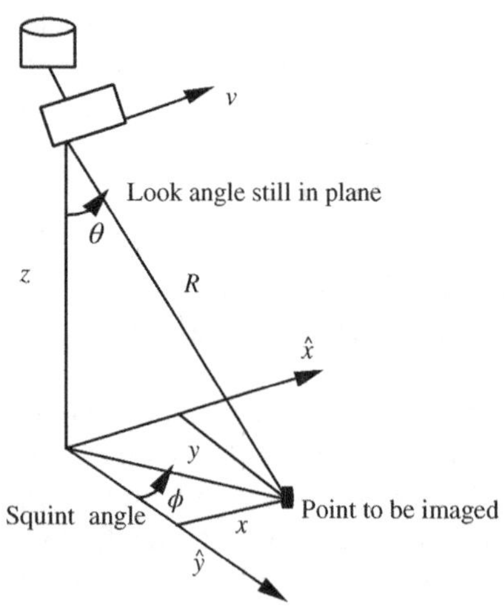

Figure 12.3 Squint angle definition. Squint angle indicates the angular location of the point for interest. This angle, ϕ, is the rotation of a vertical plane containing the z-axis containing the point of interest. That is, the squint angle is the usual azimuthal angle of a spherical coordinate system.

vector $\mathbf{v}$ is entirely in the $\hat{x}$ direction. Under this assumption, $\mathbf{v} \cdot \mathbf{u}$ is the velocity v multiplied by the $\hat{x}$ component of $\mathbf{u}$. The vector from the radar to the point is,

$$R_x = \sqrt{R^2 - z^2} \sin \phi$$
$$R_y = \sqrt{R^2 - z^2} \cos \phi$$
$$R_z = z$$

Noting then that,

$$u_x = \frac{\sqrt{R^2 - z^2}}{R} \sin \phi \tag{12.4}$$

and also that

$$\sin \theta = \frac{\sqrt{R^2 - z^2}}{R} \tag{12.5}$$

we find the following relation for Doppler shift as a function of location specified by the look and squint angles,

$$f_{\text{Dop}} = \frac{2v}{\lambda} \sin \theta \sin \phi \tag{12.6}$$

12.3 How to Make an Image

In the previous sections, we related along the surface from the sub-radar point to radar range, which we determine from the time delay, and position forward or backward from the perpendicular to the flight track by squint angle, which we determine from the Doppler shift. Observe that the location of every point on the ground is specified uniquely by its range from the radar and its Doppler frequency. Hence, for every point in the distribution of the echo signal in time and frequency, we can associate the echo strength at that point with a discrete location on the surface. By including the factors in the radar equation, we can then assign a radar reflectivity to each location on the surface.

In practice, it is straight-forward to gate the echo in the time domain in order to obtain the reflected power as a function of time, and hence range. We receive echoes from many points on the ground nearly simultaneously; however, and it is necessary to separate each of these in frequency to resolve the surface in the along-track direction. Thus, to form the image, we decompose the total received signal according to time delay and Doppler frequency, and assign the resulting power to the appropriate bin in the image array. Our ability to discriminate closely spaced bins in the output image determines the *resolution* of the radar system. The spacing of the bins in range depends on our ability to distinguish variations in time delay, and the spacing in azimuth depends on our ability to separate Doppler frequencies.

How to accomplish the measurement of the echo distribution? From our matched filter discussion, recall the structure of a filter spanning delay-Doppler space, which we sketch in simplified form in Figure 12.4. In this figure, each box across represents one additional unit of time delay, while each box down represents an additional unit of frequency shift. The squared amplitudes at the output of each of these matched filters are proportional to the surface RCS, σ^0, where the spacing in time, the spacing in frequency, and the time-frequency spread of the ambiguity function determine the resolution of the radar system. This 2D grid covers the surface, albeit in a distorted geometry. But using Eqs. (12.2) and (12.6) we can remap this set of intensity measurements into a cartigraphically accurate coordinate system.

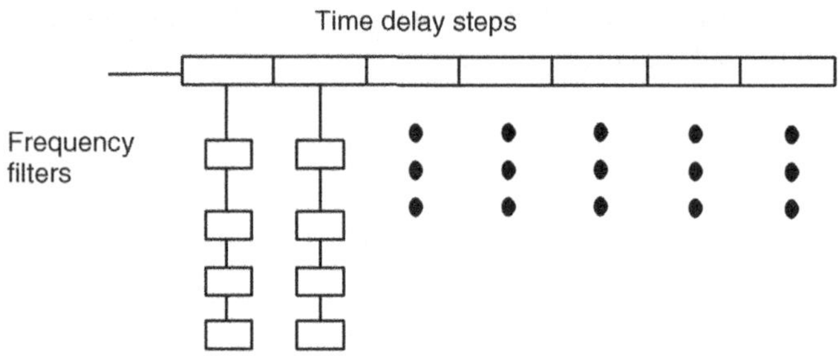

Figure 12.4 Matched filter grid. Processor for SLR takes a form similar to that of the matched filter array in our discussion of the ambiguity function, Appendix A. Here, though, the sorting is a little different. Each time delay step refers to a successive pulse interval, so that each vertical column represents the sweep of a radar pulse. Hence, horizontal time interval represents a different transmission/reception point along the flight path of the radar. Because the speed of light is much greater than the practical speed of the radar platform, the essentially rectangular grid shown results. Data from a sequence of pulses are sorted according the range delay, and the results applied to the filter banks.

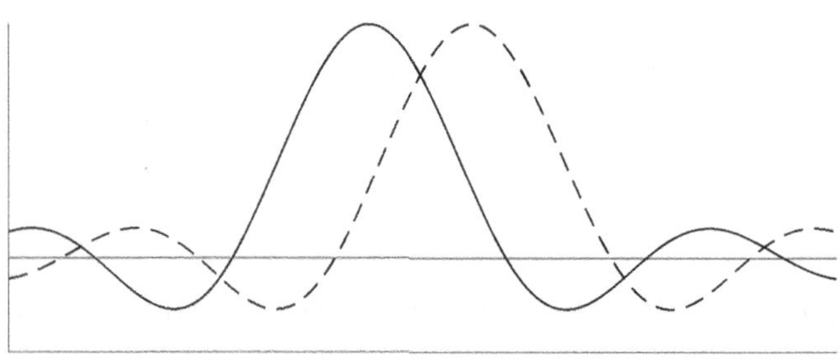

Figure 12.5 Two point target responses. When the separation of the two impulse responses is less than the resolution of the radar, the two echoes are indistinguishable from a single, slightly broader object. We can define resolution by this criterion—it is the minimum separation distance at which we can confidently identify two distinct objects.

12.4 Range Resolution in Imaging Radars

The ability of a radar system to distinguish, or *resolve*, closely-spaced objects is a key descriptor of system performance. Typically, an imaging radar system is designed to achieve similar resolution in both dimensions of its 2D image, but not always. Thus, it is common practice to specify the resolution in each dimension independently. An approximate value for resolution, which we will use here, is the peak-to-null widths and area of the radar in τ, f_{Dop} space. This characteristic is the *impulse response* of the radar. In order to understand how this affects the image resolution in terms of resolution along the surface, we need to do a little more. The ability to discriminate echoes in range is related to the transmitted pulse length, which determines the range impulse response (Figure 12.5). For now assume a transmitted pulse consisting of a cosine burst,

$$s(t) = S \,\text{rect}\left(\frac{t - t_0/2}{T}\right) \cos \omega_0 t \tag{12.7}$$

Since the pulse length is T, the range resolution is,

$$\rho_{\text{range}} = \frac{cT}{2} \tag{12.8}$$

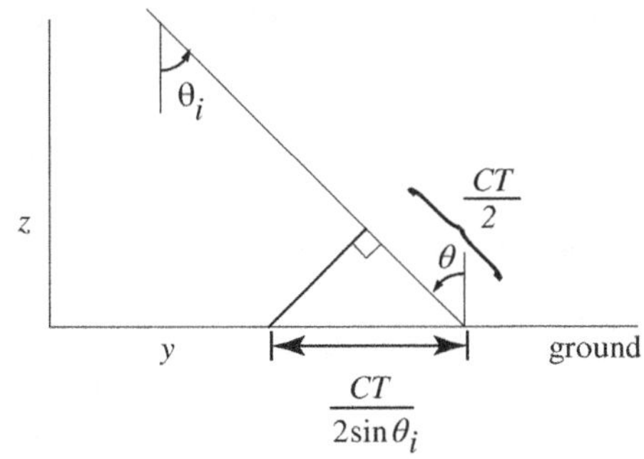

Figure 12.6 Slant range vs. ground range resolution. Resolution in the ground (y) direction is the projection of the slant range resolution element onto the surface.

The ground resolution corresponding to this pulse length is greater than this. However,

$$\rho_{\text{gnd}} = \frac{cT}{2\sin\theta_i} \tag{12.9}$$

because the projection onto the ground geometrically increases the distance corresponding to the beginning and end of the pulse (Figure 12.6). Often we will refer to the range direction as the *slant range* direction in order to emphasize that we are referring to the distance measured as the radar pulse-crow flies.

12.5 Azimuth Resolution of Real Aperture Imaging Radars

Perhaps the simplest method of implementing an imaging radar is to set the azimuth resolution not by a bank of Doppler filters, as in Section 12.3, but rather by the limits of the azimuth direction antenna pattern. In this approach, they transmit a pulse and record its echo as a function of time in order to scan the surface in the range dimension, while letting the physical extent of antenna beamwidth in the azimuthal direction determine the along-track resolution. Such a configuration is called a "real" aperture radar, because the actual, or physical, aperture of the antenna in the along-track direction sets the resolution. The azimuth impulse response will be a sinc function in angle with width λ/D_{az} for a uniformly illuminated physical antenna aperture of length D_{az}. Thus the azimuth of along-track ground resolution for a point at range R is,

$$\rho_{\text{az}} = \frac{R\lambda}{D_{\text{az}}} \tag{12.10}$$

Since this value is specified in a direction parallel to the flight track, there is not a corresponding $1/\sin\theta$ dilution, as occurs in the range case.

We can now derive the radar equation for a real aperture radar system. This approach is similar to that in the radar equation chapter previously, except that now we are illuminating an extended surface and must consider the specific cross-section σ^0 and the resolution rather than the RCS of discrete objects, σ. Thus,

$$P_r = \frac{P_t G_t}{4\pi R^2} \times \Delta A_{\text{surf}} \cdot \sigma^0 \times \frac{A_r}{4\pi R^2} \tag{12.11}$$

where ΔA_{surf} is the illuminated surface area. What is this area for our radar? It is equal to the resolved area on the ground, or,

$$\Delta A_{\text{surf}} = \rho_{\text{gnd}} \times \rho_{\text{az}} \tag{12.12}$$

For a rectangular resolution element, this is just the product of the ground range and azimuth resolutions. Substituting from Eqs. (12.9) and (12.10),

$$\Delta A_{\text{surf}} = \frac{cT}{2\sin\theta_i} \times \frac{R\lambda}{D_{\text{az}}} \tag{12.13}$$

where D_{az} is the length of the antenna in the azimuth direction. Now using Eq. (12.13) in the radar equation above (Eq. 12.11)

$$P_{\text{r}} = \frac{P_t G_t}{4\pi R^2} \times \frac{cT}{2\sin\theta_i} \cdot \frac{R\lambda}{D_{\text{az}}} \cdot \sigma^0 \times \frac{A_r}{4\pi R^2} \tag{12.14}$$

Notice that the denominator has a factor R^4 while the numerator has a factor R, thus the received power varies as R^3 rather than R^4. Hence, the SNR for the imaged surface is less sensitive to changes in range for this radar than in our previous case. Equation (12.14) follows because the resolution cell size increases with range (Eqs. 12.10 and 12.13), increasing its contribution to the reflected power. At the same time the echo signal intensity from a given scatterer falls off as the fourth power of range, as before, so that SNR still decreases as range increases. It simply decreases at a rate that is one power of range less rapid.

12.6 Swath Width

For most imaging applications, the total area observed by the radar in a given time is an important design consideration. Therefore, we often must constrain the coverage of a strip mapping radar design as a fundamental requirement that the system must meet. We have already defined the width of radar swath as that swept out by the antenna along the ground. If we know the antenna size, we can approximate the swath width as the region illuminated by the range antenna aperture, again projecting an angular beamwidth onto the surface,

$$W_{\text{swath}} = \frac{R\lambda}{D_{\text{range}}\cos\theta_i} \tag{12.15}$$

where D_{range} is the range aperture of the radar antenna. We now have a $1/\cos\theta_i$ dilution factor rather than $1/\sin\theta_i$, because we are now interested in the projection of the beamwidth, which is a distance perpendicular to the look direction, onto the surface. In other words, this is the distance along the ground between the intersections of the two sides of the antenna pattern. Note that to increase swath we would *decrease* D_{range} while to increase azimuth resolution we would *increase* D_{az}. For this reason radar antennas are often long and narrow assemblies or apertures in order to generate a beam which is wide across-track and narrow along-track.

In calculating the swath width in Eq. (12.15), the antenna beam is treated in an idealized manner, as though transmitted and received energy were confined perfectly to the angular interval λ/D_{range}, and that the illumination and response to signals outside that ideal beam were zero. It is important to recognize that such is certainly not the case. As pointed out above (Section 12.2), complete design analyses must consider contributions from sensed areas outside of the primary design swath.

12.7 Scattering from a Surface

The above equations for SNR give a value that can be calculated if we know the parameters of the radar system and also the properties of the surface, which we express as the normalized cross section σ^0. But σ^0 represents the mean cross section for a given surface type, while the actual return measured by a radar making a single measurement of the surface is a statistically varying quantity. If only a small number of independent scatterers is present in the resolution element, the variation in σ^0 may be similar to the example variation in cross section from the five spheres given in Section 5.4. If the imaged surface contains a great many scatterers with no dominating scattering centers, the statistics of the echo from each resolution element will be closer to those of Section 5.5, which are the same as for the previously defined Swerling Type II target. That is, the amplitude will be Rayleigh distributed, and the power exponentially distributed with a mean equal to the normalized cross section. Thus, backscattering from a uniform surface will be observed to be highly variable in brightness with a well-determined mean. This behavior is sometimes referred to as "speckle" by analogy with coherent optical illumination.

Consider a case of Type II scattering, in which we have a radar system that operates with a power law detector. The Rayleigh distribution of cross sections is determined by a single parameter, the mean power, from which the average specific cross section, σ^0, follows, and which can be estimated only by examining the power from many resolution elements. To estimate the distribution, then, we can simply average the returns from many pulses or range resolution elements to find the mean power. Of course, this is done at the expense of spatial resolution, and represents a fundamental trade-off in system design. Similar balancing of resolution and accuracy in intensity is another needed evaluation for any radar design, and will differ according to the application.

12.8 A Simple System Design

We have now introduced most of the fundamental principles of imaging radars, and are now in a position to design a radar that can produce images. We can work an illustrative example to develop an implementation approach and analyze its performance. For this purpose, suppose that our requirements are those listed in Table 12.1.

Many, many different designs are possible! We will just choose one. One thing that we would like to show is how some design parameters are more firmly fixed by the requirements than are others. In fact, some requirements can be said to control the overall design, and this is true to a large extent here. In order to see how this works, lets's start with the coverage requirement. Recalling that the aircraft velocity 250 m s^{-1}, the coverage requirement of 75 km^2 min^{-1} determines the swath width, quite independently of any other considerations,

$$W_{\text{swath}}\ \text{m s}^{-1} = 75 \times 10^6\ \text{m min}^{-1} \tag{12.16}$$

Therefore,

$$W_{\text{swath}} = \frac{75 \times 10^6}{60 \times 250} = 5000\ \text{m} \tag{12.17}$$

Table 12.1 Design requirements for a simple side-looking radar system.

1. Image the surface at a ground resolution of 100 × 100 m
2. Estimate σ^0 with a standard deviation of no more than 30% from target fluctuation effects alone
3. Achieve a single-look SNR of 10 dB for $\sigma^0 = -20$ dB
4. Map 75 km^2 min^{-1}
5. As a radar platform, use a 737 aircraft flying at 250 m s^{-1}
6. Operate at 3 cm-λ

The swath width in turn fixes the vertical, D_{range}, dimension of the radar antenna. This allows us the second step of calculating the antenna width D_{range}. Here, since it was not otherwise specified, we will assume a look angle of 30°, with the results that,

$$W_{\text{swath}} = \frac{R\lambda}{D_{\text{range}} \cos 30^\circ} \tag{12.18}$$

$$\Rightarrow R = \frac{5000 \times D_{\text{range}} \cos 30^\circ}{\lambda} \tag{12.19}$$

Now, we have just encountered our first trade-off, which is R vs. D_{range}, in that we only have a relative requirement for these quantities and selection of their optimal values depends on other factors. For the moment, we resolve this by assuming that the aircraft flies at height of 10 km (33,000 ft), a reasonable altitude for most passenger-type jet aircraft. Under this assumption

$$D_{\text{range}} = \frac{R\lambda}{W_{\text{swath}} \cdot \cos 30^\circ} = \frac{11,547 \cdot 0.03}{5000 \times 0.866} = 0.08 \text{ m} \tag{12.20}$$

where we have used the relation $R = z/\cos\theta$ to obtain the range to the swath center given the look angle and the aircraft altitude.

Our specifications call for a final resolution of 100 × 100 m. But we have an additional specification on fluctuation of the power from speckle, which we recall is related to the number of "looks" or independent measurements of the surface that we obtain. For each look we know that, $\mu = \sigma$ so for M looks, $\mu = Mx$, x being the average scattering cross section, and $\sigma^2 = M \cdot x^2$ or $\sigma = \sqrt{M}\, x$. Our requirement is for no more than a 30% variation. Consequently, we see that or $\sigma/\mu \leq 0.3$, which can be satisfied with $M \approx 11$. Thus our 100 × 100 m output resolution element must be divided into 11 individual measurements of echo intensity which are summed to obtain the desired accuracies in both spatial area and in the variation in intensity.

For now, we assume that we obtain our 11 independent "looks" by averaging in the range direction. That is, a design for finer range resolution than is finally required, and average range pixels to reduce the statistical fluctuations. This approach is seldom used in practice because it greatly increases the needed signal bandwidth required. Proceeding, the single-look range resolution requirement becomes,

$$\frac{100 \text{ m}}{11} = 9.1 \text{ m} \tag{12.21}$$

in the plane of the image, or 9.1 × sin 30° = 4.55 m in slant range. Given the slant range resolution, choose the pulse length in the usual manner using,

$$\rho_{\text{range}} = \frac{c \cdot T}{2} \tag{12.22}$$

Or, substituting our known parameters,

$$T = \frac{2\rho_{\text{range}}}{c} = \frac{2 \cdot 4.55}{3 \times 10^8} = 30.3 \text{ ns} \tag{12.23}$$

Thus, from our choice to use increased range resolution to obtain the required statistical stability in the image products, the transmitted pulse can not be longer in duration than 30.3 ns.

Now that the range parameter is set, we now examine how to achieve the needed azimuth resolution. Here our requirement is to obtain 100 m discrimination between two points along track. Equation (12.10) in our discussion of real aperture systems gives the projection of the azimuth antenna pattern onto the ground surface. Applying this here,

$$D_{\text{az}} = \frac{R\lambda}{\rho_{\text{az}}} = \frac{11,547 \times 0.03}{100} = 3.46 \text{ m} \tag{12.24}$$

Thus, our antenna dimensions are 0.08 by 3.46 m in the range and azimuth related dimensions, respectively. Given our assumed flight altitude and choice of method for achieving the required statistics, these antenna dimensions are then set entirely by the specified wavelength, mapping rate, and resolution.

With the antenna size set, we have the information needed to consider the SNR requirement. For a single pulse, we use our radar equation in the form for the use of a real aperture, Eq. (12.11), above, which we repeat here for convenience,

$$P_r = \frac{P_t G_t}{4\pi R^2} \times \delta A_{\mathrm{look}} \times \sigma^0 \times \frac{A_\mathrm{r}}{4\pi R^2} \tag{12.25}$$

where δA_{look} is the area of the surface sampled by a single pulse. The noise power is again described by,

$$P_\mathrm{n} = kT_{\mathrm{sys}}B \tag{12.26}$$

From the work on image resolution, we either know or have information sufficient to calculate all the parameters in Eqs. (12.25) and (12.26), except for the antenna efficiency, η_{eff}, P_t, and T_{sys}. The efficiency η_{eff} is a matter of antenna design; we are free to choose P_t and T_{sys} from a wide range of possibilities based on practical considerations, one of which is the cost.

What are the other values? Beginning with the bandwidth as the reciprocal of the transmitted pulse length,

$$B \approx \frac{1}{T} = \frac{1}{30.3\ \mathrm{ns}} = 33\ \mathrm{MHz} \tag{12.27}$$

For the antenna gain, here, we assume an approximate value, $\eta_{\mathrm{eff}} = 50\%$, although this is for convenience here. If this were an actual design, an antenna engineer could provide an accurate number for the η_{eff}.

$$G_\mathrm{t} = \frac{4\pi A}{\lambda^2} \times \eta_{\mathrm{eff}} = \frac{4\pi \times 0.14 \times 3.46 \times 0.5}{(0.03)^2}$$
$$= 1932$$

Continuing with the radar equation, the area illuminated by a single pulse, δA_{look}, is determined by the combination of the pulse width and the real aperture azimuth beamwidth,

$$\delta A_{\mathrm{look}} = 100 \times 9.1\ \mathrm{m}^2 = 910\ \mathrm{m}^2$$

Recalling the σ^0 and antenna area parameters we need to consider from our initial and derived specifications,

$$\sigma^0 = -20\ \mathrm{dB} \Rightarrow 0.01\ \mathrm{m}^2/\mathrm{m}^2$$
$$A_\mathrm{r} = 0.08 \times 3.46 \times 0.5 = 0.14\ \mathrm{m}^2$$

We are now left with only with P_t and T_{sys} undetermined. Using the requirement for a single-look SNR of 10:1, combine Eqs. (12.25) and (12.26) to obtain,

$$\frac{\mathrm{Pr}}{P_\mathrm{n}} = \frac{P_\mathrm{t} G_\mathrm{t}}{4\pi R^2} \times \delta A_{\mathrm{look}} \times \sigma^0 \times \frac{A_\mathrm{r}}{4\pi R^2} \times \frac{1}{kT_{\mathrm{sys}}B} = 10 \tag{12.28}$$

Rearranging terms to isolate the remaining unknowns of transmitter power and system temperature, we obtain.

$$\frac{P_t}{T_{sys}} = \frac{10 \times 4\pi R^2 \times 4\pi R^2 \times k\, B}{G_t \times \delta A_{look} \times \sigma^0 \times A_r} \tag{12.29}$$

$$= \frac{10 \times 4\pi \times 11,547^2 \times 4\pi \times 11,547^2 \times 1.38 \times 10^{-23} \times 33 \times 10^6}{1932 \times 910 \times 0.01 \times 0.14} \tag{12.30}$$

$$\approx 5.2 \tag{12.31}$$

This gives us the ratio of transmit power in watts to system temperature in kelvins. We are free to select any values for these two parameters as long as they meet or exceed the ratio of 5.2 that we have just determined. For example, if we select a receiver with $T_{sys} = 1000$ K, then the required peak transmitter power required to drive the pulse is $P_t = 5200$ W. On the other hand, access to a lower noise temperature receiver available with, say, $T_{sys} = 100$ K, leads to a transmitter with a peak power of only $P_t = 520$ W. For this application, we might choose the less expensive receiver and higher power transmitter—or vice versa—depending on other considerations not related directly to the radar performance. These might include such factors as cost, available power, packaging size constraints or the lack thereof, and plans for any future performance upgrades. With the choices made above, however, final engineering specifications are summarized in Table 12.2. These specifications tell an engineer how to build the hardware. Note that at this stage the hardware engineer requires little understanding of the original problem. But the system engineer must understand both!

Table 12.2 Final engineering specifications for example imaging radar.

Platform: 737 aircraft flying at 10 km altitude.
Antenna: 3.64 m along track by 0.08 m wide, side-looking antenna beam center 30° off nadir.
Transmitter: 5200 w, pulse length 30.3 ns, repetition rate once every 100 m or 2.5 pulses per second.
Receiver: 1000 K noise temperature.
Processor: 11 looks in range, none in azimuth.

13

Range and Azimuth Processing

13.1 Range Processing

13.1.1 Fine Resolution vs. High SNR

While it was not too difficult to design a radar capable of 100 m resolution from an aircraft platform using real aperture techniques, for many uses we require much greater accuracies and resolutions. We were able to achieve fairly fine (9.1 m) ground resolution in the range direction by transmitting a very short (30.3 ns) pulse, but even for this modest system we required a transmitter capable of delivering 5200 W into that short pulse. If we needed finer resolution, or required a platform at a greater range, we would quickly find that the transmit powers necessary for typical SNR specifications grow quickly beyond what is available today. In addition, the requirement to put very high peak powers into short pulses can lead to breakdown of cables, connectors, or the antenna itself, as the high voltages required produce arcs or other undesired effects. The competing requirements for high resolution, which implies short transmit pulses, and high transmitted energy, which implies long transmit pulses, make implementation of high-performance radar systems difficult. In this chapter we will investigate range coding, which will allow us to design long, low peak power, transmit waveforms supporting fine spatial resolution, effectively decoupling the range resolution and SNR specifications.

13.1.2 Range Coding

We'll begin by improving the across-track, or range, resolution. So far we have relied on a physically short pulse to discriminate in range, where $\rho_{\text{range}} = \frac{cT}{2}$. Here the receiving system functions as a simple filter of bandwidth $\mathrm{B} = \frac{1}{T}$. If instead we employ a matched filter in the processor, the impulse response of the system may be written

$$g(t) = \int_{-\infty}^{\infty} s(\tau)s^*(\tau - t)d\tau \tag{13.1}$$

where $s(t)$ is the transmitted waveform and we have neglected the change in amplitude of the signal due to propagation. We can recognize this as the ambiguity function with no Doppler frequency shift. From the convolution theorem,

$$g(t) = \int_{-\infty}^{\infty} |S(f)|^2 \exp(-j2\pi f t)df \tag{13.2}$$

From this expression we recognize that the width of $g(t)$, which is the impulse response of the system, depends on the spectrum of $s(t)$, rather than its time domain extent T. Thus we can generate resolution using any signal with a large bandwidth, rather than requiring very short pulse times, if we use matched filters in the processor subsystem. In particular consider the case where $S(f)$ is unity over some bandwidth and the phase is arbitrary. Then

$$g(t) = \int_{f_c-B/2}^{f_c+B/2} exp(-j2\pi ft)df \tag{13.3}$$

$$= Bexp(-j2\pi f_c t) \cdot \frac{\sin \pi Bt}{\pi Bt} \tag{13.4}$$

We are usually interested in the magnitude of the impulse response, so we can neglect the leading phase term as regards resolution of the radar system. The impulse response is seen here to be a sinc function with width $1/B$, again independent of the pulse length in time and only dependent on bandwidth. We then can calculate the impulse response, and hence the resolution, as a function of bandwidth B, leading to

$$\rho_{\text{range}} \simeq \frac{1}{B} \cdot \frac{c}{2} = \frac{c}{2B} \tag{13.5}$$

Thus using a matched filter would permit us to transmit physically long pulses yet achieve fine resolution if the bandwidth is large. Since the long pulse is "compressed" into a shorter pulse, this matched filtering operation is sometimes called *pulse compression*. The degree of compression may be expressed as the ratio of the two lengths, the original transmitted pulse length and the resolution of the impulse response after the matched filtering operation:

$$\text{compression ratio} = \frac{\frac{cT}{2}}{\frac{c}{2B}} = T \cdot B \tag{13.6}$$

This ratio is often denoted as the "time-bandwidth" product of the waveform, as it is equal to the multiplication of those two quantities. A signal with a high time-bandwidth product will afford a great deal of pulse compression, and a signal with a small time-bandwidth product very little. So, if we need to transmit a certain amount of energy to observe an object in naturally occurring noise, we can spread that energy out over a long pulse to keep the peak transmit power down, as long as we use a waveform with sufficient bandwidth to result in our specified resolution.

Note that the time-bandwidth product of a simple uncoded pulse is unity: $T \cdot 1/T = 1$.

Let us contrast the two waveforms again, comparing properties of the coded and uncoded approaches to waveform design (Figure 13.1). Note that the impulse response from the cosine pulse is strictly limited in time, and requires very little complexity in the data processor. The high bandwidth signal has sidelobes that ring for a long time, and requires a matched filter processing operation. Summarizing,

Cosine pulse	High-bandwidth signal
No sidelobes	Sidelobes
Wide	Narrow
Easy processor	Complicated processor

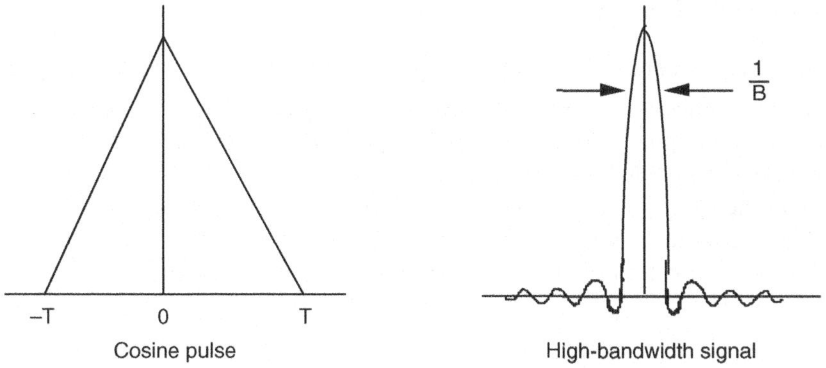

Figure 13.1 Cosine and high-bandwidth envelope impulse responses, etc.

13.1.3 The Chirp Waveform

The previous discussion holds for any transmitted waveform, although the sinc impulse response follows only for waveforms of uniform spectral density over some specified bandwidth. Many waveforms are possible, and in use today. One very common high bandwidth signal is the linear fm, or "chirp," signal, which is a signal with a linear time vs. frequency relationship. This waveform does have the uniform spectral density characteristic, so that we would expect its impulse response after matched filtering to have the sinc form. Since phase is the integral of angular frequency, we can express a chirp waveform as

$$s(t) = \cos\left[2\pi f_c t + 2\pi\alpha\frac{t^2}{2}\right], \quad |t| \le \frac{T}{2} \tag{13.7}$$

where α is the *chirp rate* in Hz/s, the pulse length is T, and the waveform resides on a carrier frequency f_c.

Typical values of these parameters for imaging radars might be $T \sim 10$ ms, and $\alpha \sim 10^{12}$. A chirp with these parameters has a bandwidth $B = \alpha T = 10$ MHz and a range resolution

$$\rho_{\text{range}} = \frac{c}{2B} \tag{13.8}$$

$$= \frac{3\times10^8}{2\times10^7} \tag{13.9}$$

$$= 15\ \text{m} \tag{13.10}$$

Now, if we can disperse the transmitted energy over an interval T· B longer than the equivalent-resolution short pulse, we reduce the peak power requirement by the same factor while maintaining the same transmitted energy, and hence SNR. This follows from examination of our imaging radar equation:

$$P_r = \frac{P_t G_t}{4\pi R^2} \times \frac{cT}{2} \times \frac{R\lambda}{D_{az}} \times \sigma^0 \times \frac{A_r}{4\pi R^2} \tag{13.11}$$

Note that P_r is proportional to the product $P_t \cdot T$. Hence if we can use a physically longer pulse we can reduce P_t and maintain the same received power level.

13.1.4 Spectrum of a Chirp Waveform

Since the chirp starts at some frequency $f_c - \frac{B}{2}$ and sweeps smoothly to $f_c + \frac{B}{2}$, ideally the spectrum is perfectly flat and contained in that interval, as shown in Figure 13.2. But infinitely sharp edges in the frequency domain require infinitely long pulses in the time domain, so in reality the spectrum of a chirp pulse with finite pulse length is more like that shown in Figure 13.3.

In fact, if you calculate the spectra of several chirp waveforms you will find that as the time-bandwidth product of the signal increases you approach the ideal spectrum. In practice, we can usually assume that a time-bandwidth product of 100 or greater is well-represented by the ideal spectrum.

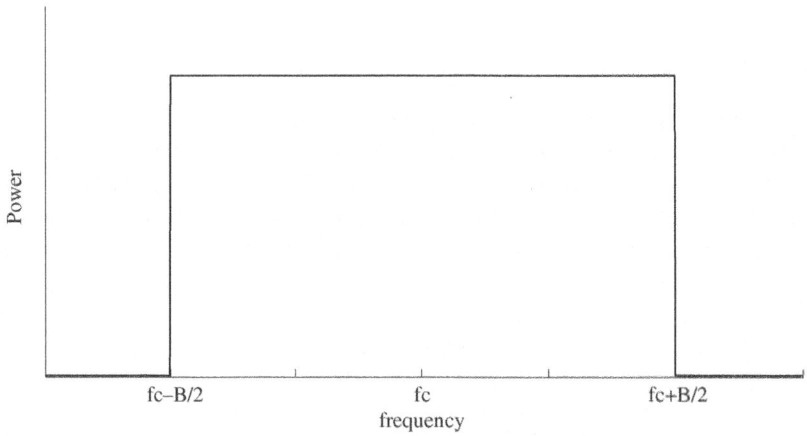

Figure 13.2 Ideal chirp spectrum. Spectral energy ranges from $f_c - \frac{B}{2}$ to $f_c + \frac{B}{2}$, at a constant level. The idealized chirp shown here requires an infinite time-bandwidth product.

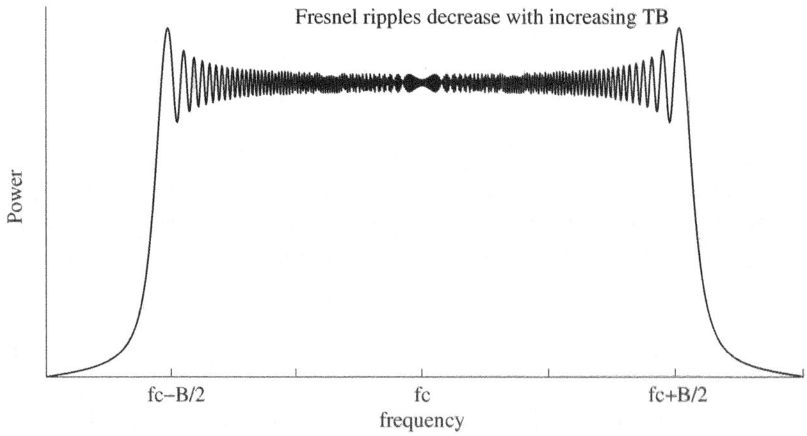

Figure 13.3 Finite length chirp spectrum. "Ideal" spectrum of Figure 13.2 typically appears as shown, with Fresnel ripples and non-instantaneous falloff at the edges of the spectrum. The rate of falloff increases and the amplitude of the ripples decreases as the time-bandwidth product of the chirp increases.

13.1.5 Impulse Response and Sidelobe Reduction

In the previous section, we briefly examined the sinc function impulse response for a signal of uniform spectral density and bandwidth B,

$$g(t) = \frac{\sin(\pi Bt)}{\pi Bt} \tag{13.12}$$

This function has infinite extent in time, and exhibits sidelobes that we may misinterpret as targets in a radar image. Even if we know that the false targets are artifacts of the data processing procedure, they always represent energy that should be placed at one location on the surface appearing at incorrect locations. We can analyze the effects of the sidelobes, and how to adjust our processing approach to minimize their impact, by first considering the magnitude of first sidelobe, $\frac{1}{\frac{3}{2}\pi}$, which is a level of −13.5 dB with respect to the main lobe. The magnitude of the remaining sidelobes falls off as $\frac{1}{t}$, so they will be observable for quite some distance away from the main object. We also note that the sign of the sidelobes reverses for each successive sidelobe, which will allow us to correct the impulse response, as we now see (Figure 13.4).

How can we lessen the magnitude of sidelobes in the impulse response? Suppose we convolve the impulse response with a function

$$w(t) = \frac{1}{2}\delta\left(t - \frac{1}{B}\right) + \delta(t) + \frac{1}{2}\delta\left(t + \frac{1}{B}\right) \tag{13.13}$$

which we plot in Figure 13.5. The result of this convolution replaces every value of the impulse response with the sum of the original value of the impulse response plus the average of the two points located ahead and behind of that point by the sidelobe spacing of the impulse response, $1/B$. Therefore each sidelobe is "cancelled out" by the sidelobes occurring just before and just after in the impulse response (see Figure 13.6). Therefore, this convolution will reduce sidelobe levels.

We note that the cancellation is only approximate, because the average of the magnitudes of the preceding and following sidelobes is not quite the same as the identified sidelobe. Because the

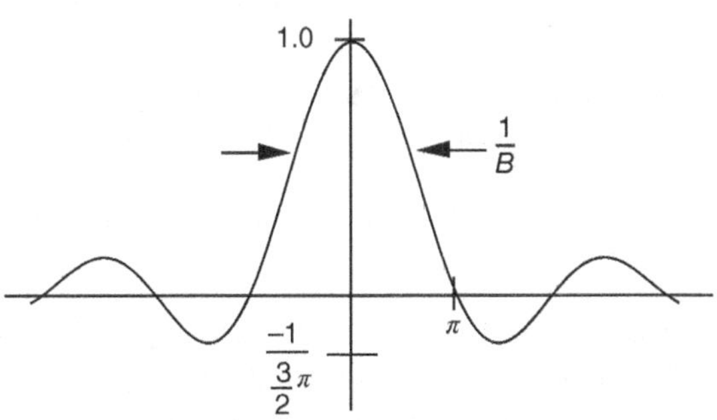

Figure 13.4 Sinc function response. An ideal chirp function produces an impulse response in the form of a sinc function, where the width of the main lobe of the sinc function is the reciprocal of the chirp bandwidth. The first zero crossing is at $1/B$, and the magnitude of the first sidelobe is $1/(3\pi/2)$. Non-ideal chirps produce impulse responses that are very nearly sinc-like in form.

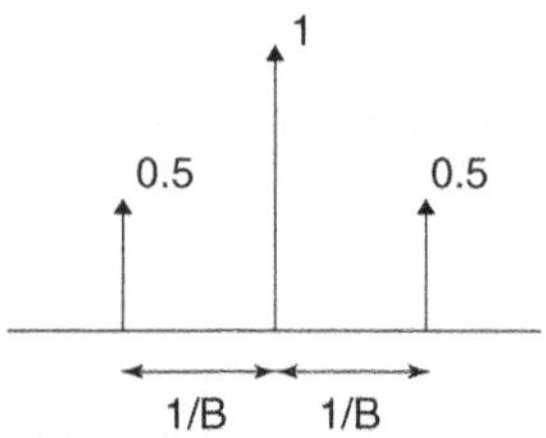

Figure 13.5 Weighting function $w(t)$, in time domain representation. This is convolved with the impulse response to lower sidelobes at the expense of broadening the main lobe width.

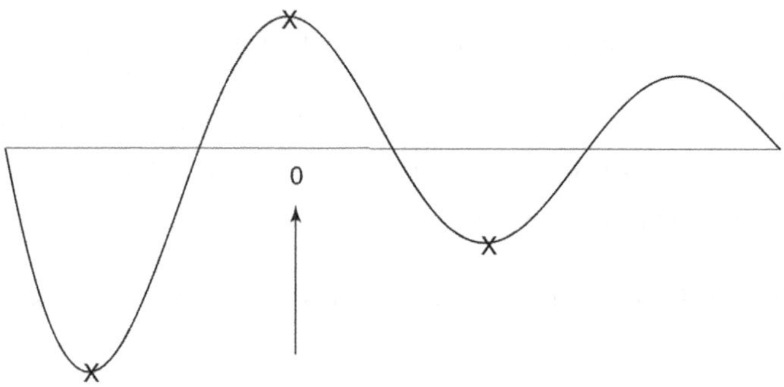

Figure 13.6 Sidelobe response before cancellation from convolution with weighting function. Since each sidelobe has amplitude nearly the average of its two neighbors, replacing each lobe with the weighted average of all three (sidelobe plus nearest neighbors) results in a very small residual sidelobe signal.

magnitudes fall off as 1/t, however, the cancellation becomes more accurate as time gets very large. The cancellation condition perfectly suppresses the sidelobes at ∞, and only somewhat suppresses those near the main lobe.

We could implement this sidelobe cancellation routine through a time-domain convolution as we have described above, but for reasons of efficiency we usually prefer to use frequency domain multiplication rather than time domain convolution. Hence we will compute the Fourier transform of the above convolution kernel to obtain the transfer function for the equivalent filter. The transform defines the weighting function we need to apply to the signal in the Fourier domain. So if we were to weigh the received spectrum in the processor so that $W(f)$ were (Figure 13.7),

$$W(f) = 1 + \cos \frac{\pi f}{\frac{B}{2}} \tag{13.14}$$

we would have low sidelobes in the impulse response.

This choice for $W(f)$ is called the *Hann* weighting function. Similar in form but different in detail is another common weighting function for sidelobe reduction, the *Hamming* function

$$W(f) = 0.54 + 0.46 \cos \left(\frac{\pi f}{\frac{B}{2}} \right) \tag{13.15}$$

which better suppresses the sidelobes near the main lobe but yields poorer sidelobe reduction at ∞.

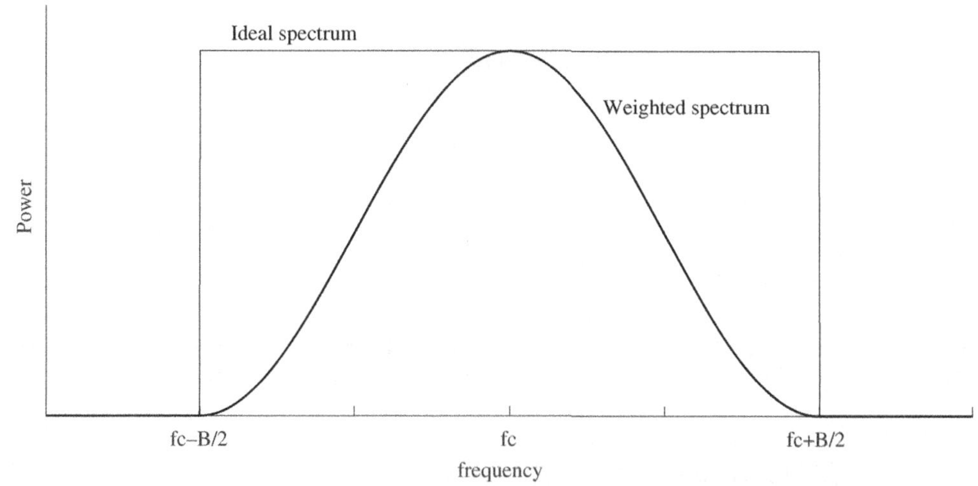

Figure 13.7 Weighted received spectrum. Application of the cosine function to the received signal spectrum "weights" each Fourier coefficient. This multiplication is equivalent to a convolution in the time domain, and reduces the sidelobe levels considerably.

13.1.6 Data Processor and Algorithm

Having defined the needed operations for range data processing for the imaging radar, we can now write down a recipe for a matched filter range processor. The radar system will receive an echo, convolve it with a function equal to the time-reversed transmitted signal, and optionally convolve the result with a sidelobe reduction filter. Noting that we are implementing two successive convolutions, we can combine both of these steps into a single convolution, which we will implement using Fourier domain techniques. The specific steps we will take to design a digital system are:

1. Digitize the echo at a rate sufficient for Nyquist
2. Calculate the received signal spectrum (using FFT algorithms)
3. Multiply by the complex conjugate of the transmitted chirp transform
4. Optionally multiply by weighting function W(f)
5. Inverse transform the product to obtain the time-domain reflectivity distribution

We can understand the feasibility of calculating the above on a computer by working through an example. Suppose a particular radar system has a slant range swath of 15 km, and a slant range resolution of 5 m. The bandwidth required is then 30 MHz ($B = c / 2\rho_{\text{slant range}}$), and we must sample at that rate if we obtain complex samples, or twice that if we obtain only real samples.

13.1.7 Offset Video or I/Q Detection

The received echo is a complex signal, and we must sample both the real and imaginary parts of this signal to fully characterize all the information on the signal. We have previously examined the analytic signal representation and found that we can either measure the complex envelope of the signal, or else use in-phase and quadrature detection to retain both the real and imaginary parts.

The radar system operates using a chirp modulation on a high-frequency carrier, with a spectrum as shown in Figure 13.8. Such a signal must be sampled at a very high rate ($\sim 2 \cdot f_c$) to preserve all information. But the information content is contained in the modulation, not the carrier, so usually

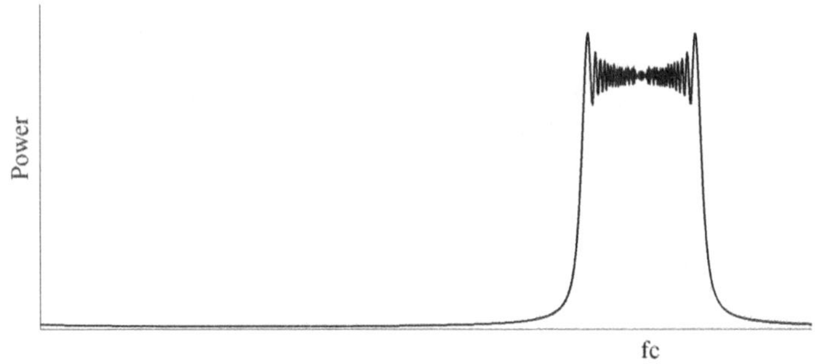

Figure 13.8 Spectrum of carrier signal with chirp modulation. Usually the carrier frequency is much higher than the modulation bandwidth, so that a narrowband approximation for the waveform is valid.

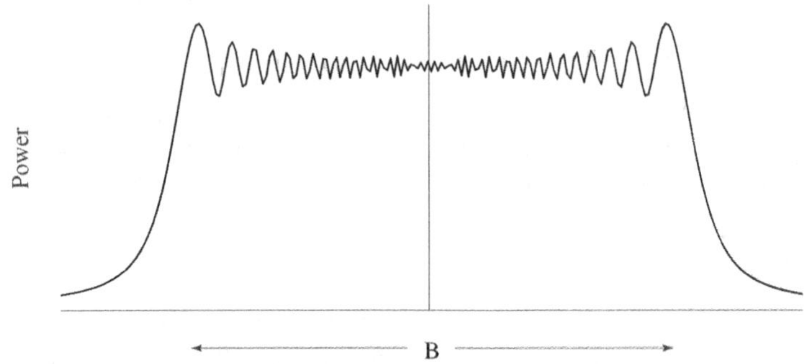

Figure 13.9 Baseband chirp spectrum. If the signal is mixed before sampling so that both positive and negative frequencies are needed to represent the spectrum, then the sampling system must be capable of complex (real and imaginary) digitization.

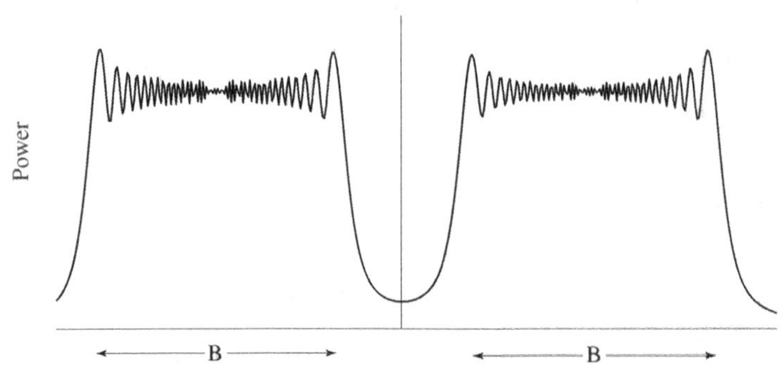

Figure 13.10 Offset video chirp spectrum. If the received signal is offset from the carrier before digitization, a real-valued sampling system suffices to capture all the information contained in the modulation. This is equivalent to stating that the received spectrum, before sampling, contains only positive or negative frequencies, but not both. The resulting spectrum of the real signal will then be two-sided.

the signal is heterodyned, or "beat down," to a lower value, as shown in Figure 13.9. From this figure it is quite clear that the bandwidth at which the sampler must operate is B, and independent of the carrier frequency (Figure 13.10).

The next question is how long the echo will last, information needed to size the memory and compute rate of the data processor. We next determine the time interval of the echo signal from the radar swath. This time is given by the conversion of the swath width in slant range into time delay rather than distance, where

$$t_{\text{echo}} = \frac{2\Delta R}{c} = \frac{2 \times 15,000}{3 \times 10^8} = 100\,\mu\text{s} \tag{13.16}$$

so the number of samples/echo = 100 μs × 60 MHz = 6000. Here we calculate the data volume assuming 60 MHz sampling of real signals, and note that we would get the same volume if we assumed 30 MHz sampling of complex signals. In fact, the echo would last one pulse length longer than 100 μs (left as an exercise for the reader), but we will ignore that extra time increment for now.

Step 2 above requires us to Fourier transform these 6000 samples. If we use an FFT algorithm, we need to calculate an 8192 length transform if we use an algorithm for data vectors of length 2^M. This type of FFT algorithm requires about $2N \log_2 N$ operations, N being the length of the data vector. In our case, we have a purely real signal so that a real FFT is all that is needed, which is the computational equivalent of a 4K, rather than 8k, transform. So step 2 needs $2 \times 4096 \times 12 \approx 100,000$ operations.

Step 3 needs 4096 operations.

Step 4 needs 4096 operations.

Step 5 again needs 100, 000 operations, so the total number of complex operations is of order 200, 000.

If the radar is operating at a pulse rate of 100 Hz, the range processing algorithm requires ~ 200 M flops rate. As you can see, high resolution radars are prodigious users of computer cycles.

13.1.8 Implementing Offset Video and I/Q Receivers

Whether a signal is digitized in the radar as an offset video signal or as an I/Q signal is largely a choice of the designer. As before, the information present in the radar echo is contained only in its complex envelope, not on how the signal is mixed and presented to the sampler. Sample backend block diagrams illustrate how each method may be built in system hardware: We refer to the signal as presented to the digitizer as the *video* signal. The video signal may be either a real signal on a carrier so that we may preserve the entire spectrum, or it may be a complex signal located anywhere in frequency space. Usually we choose the offsets of the signals from zero frequency to be as small as possible, so that our digitizing system may be run at the lowest speed possible, relaxing requirements of the sampling speed and data volume (Figure 13.11).

This dual approach follows from our representation of the analytic signal, where we found it equivalent to considering a real signal on a complex carrier or a complex signal with the real part on a cosine carrier and an imaginary part on a sine carrier. Either method works in practice, with the engineering challenges for each being:

Offset video sampling—higher A/D speeds for greater real samples, operating at twice the signal bandwidth

I/Q sampling—need to balance and time A/D converters exactly

13.1.9 I/Q and Offset Video Range Processing Algorithms

The implementation choice has a profound effect on the range processing algorithm. First consider the simpler I/Q case, where the signal is mixed so that its spectrum is approximately centered on zero frequency. This has the advantage of requiring the lowest possible data rate from the digitizer, relaxing any requirements on very high speed sampling. The spectrum of the video signal is as shown in Figure 13.12. We implement the matched filter receiver by calculating a reference signal

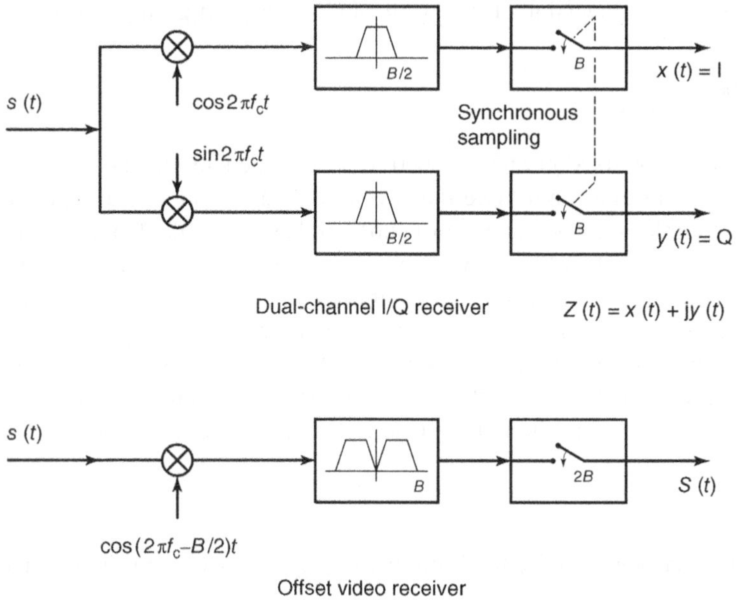

Figure 13.11 Receivers for in-phase/quadrature and offset video. Block diagrams illustrate back-ends for in phase/quadrature and offset video receivers. Offset video system requires only one analog-to-digital converter, but must be sampled at a rate twice the bandwidth of the received signal to avoid aliasing. The in-phase/quadrature system requires two samplers, offset by 90° in phase, but each digitizer runs at only the bandwidth of the signal.

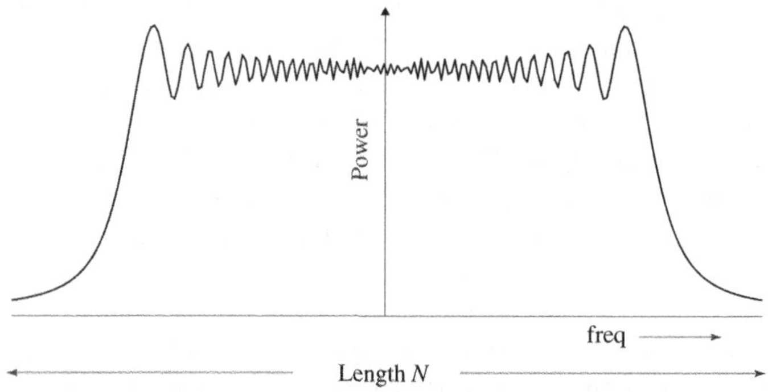

Figure 13.12 Spectrum of baseband signal for in-phase/quadrature receiver. The spectrum is usually centered on zero frequency to minimize rate requirements on the A/D converters.

based on the expected form of the echo, which means we must consider how the signal was mixed down to video in the hardware of the receiver. The spectrum of the reference signal, that is of the matched filter, is identical (Figure 13.13) to that of the echo. Our correlation receiver now derives the correlation between the reference signal and the receiver video, which we generally implement using frequency-domain techniques in the interest of efficiency. When we form the product of the reference and the video signal in the frequency domain in order to implement the matched filter receiver, the product will also have a spectrum covering the same range of frequencies, as shown in Figure 13.14. We complete the range processing by calculating the inverse transform of the product of the signal and reference spectra, resulting in an impulse response similar to that shown in Figure 13.15.

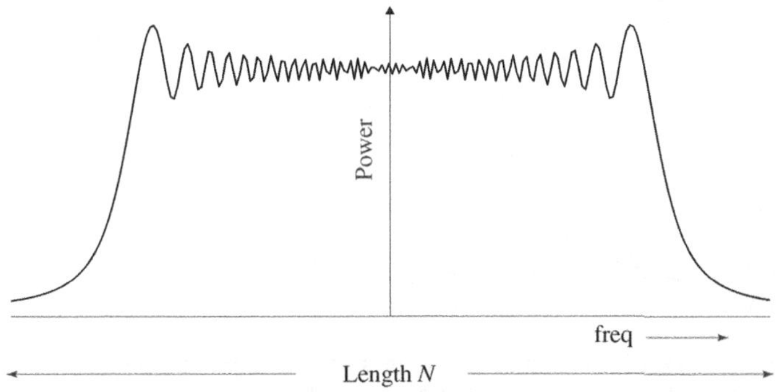

Figure 13.13 Spectrum, $R(f)$, of the reference signal. Note that the spectrum is the same as in the previous Figure 13.12. The spectrum here must agree exactly with the mixing approach used in the receiver in order to form the matched filter.

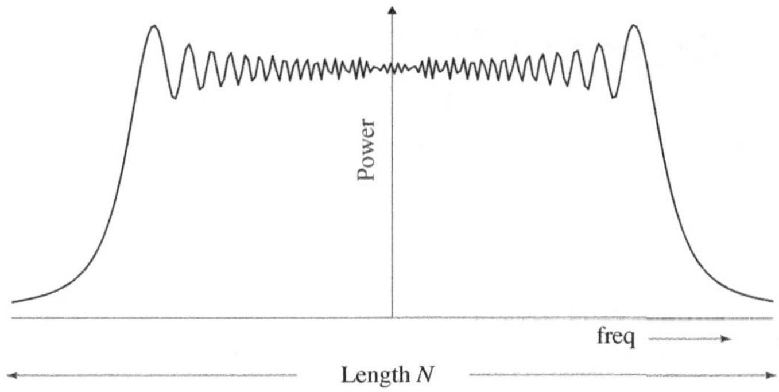

Figure 13.14 Spectrum of mixed signal and conjugate reference waveforms. Spectrum $S(f) \times R^*(f)$ is the product of the transforms of the video and the conjugate of the reference transform, and hence its amplitude distribution is the same as each of the received signal and reference functions, forming the matched filter. The resulting spectrum is the transform of the system impulse response.

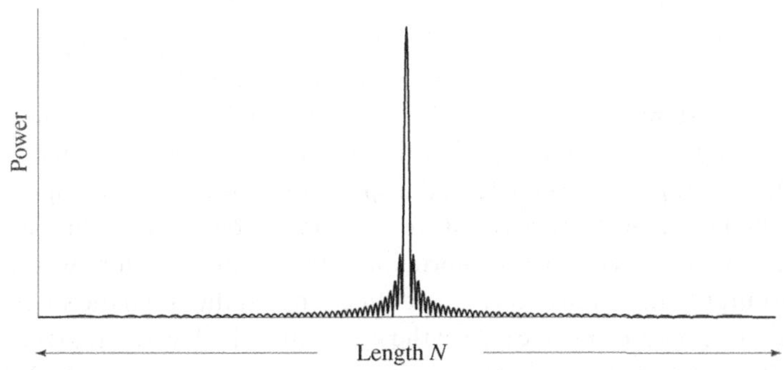

Figure 13.15 System impulse response is the inverse Fourier transform of $S(f) \times R^*(f)$. A rectangular spectrum produces a sinc response.

If the original signal consisted of a number of echoes at different time delays, by superposition the result of the above operations would yield something like we show in Figure 13.16.

The implementation of the offset video processor is a bit different, as the video signal contains two copies of the spectral information, one centered on a carrier at a positive frequency and one

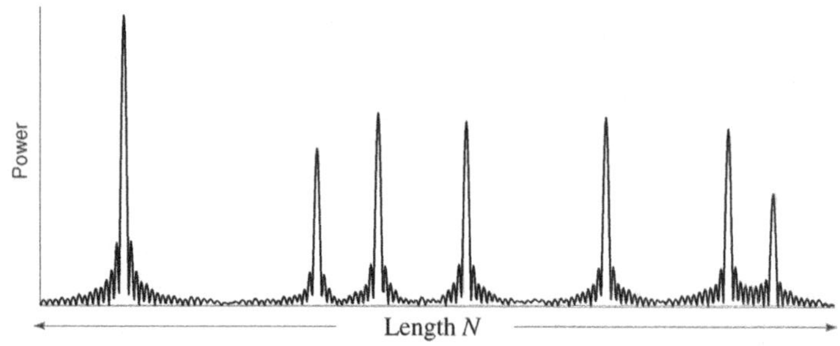

Figure 13.16 Output of matched filter for echoes with distinct time delays. Output signal from matched filter assuming that the input echo contained returns from several objects at different time delays. The total output signal is the superposition of the impulse responses from each scatterer in the antenna illumination, each weighted by its radar cross section amplitude.

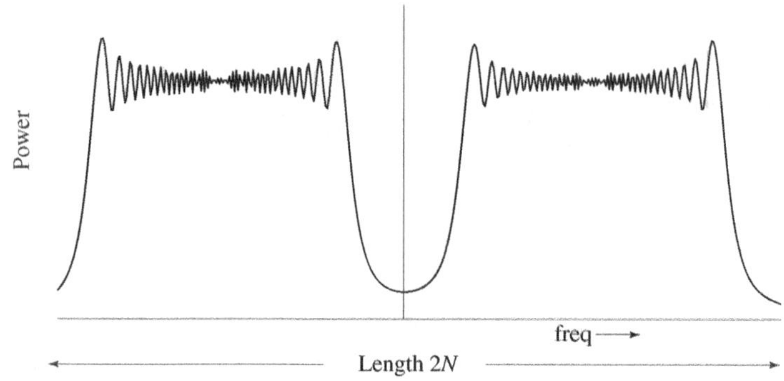

Figure 13.17 Offset video spectrum, $S(f)$. As this signal is sampled by real-only values, the spectrum contains both positive and negative frequency copies of signal. Hence its total bandwidth is twice the transmitted bandwidth, and it must be sampled at twice the bandwidth to avoid aliasing. In other words, the real-signal spectrum is Hermitian, and must be sampled at twice the rate of the I/Q system to preserve spectral information.

centered at the negative of that frequency. The spectrum $S(f)$ of the offset video signal thus has the form shown in Figure 13.17. Note that to ensure that there is no aliasing of the two spectra, the carrier must be at least half of the signal bandwidth. If the carrier is any lower in frequency, portions of the spectra start to overlap, causing aliasing with the result that the image signal is corrupted by artifacts that should not be present. As before, the spectrum of the reference signal is identical to that of the transmitted waveform and accounts for the offset video carrier. But half of this spectrum is unnecessary, and contains duplicate information. Processing as before would result in the superposition of two output images, each on a carrier frequency, rather than the single desired output image. We can therefore retain only one side of the spectrum, which we shift to zero frequency to remove the carrier effects, and then process the resulting spectrum in a manner similar to the I/Q case. So the frequency-domain product $S(f) \cdot R^*(f)$ we use is one of the two spectra shifted in frequency to remove the digitizing carrier (Figure 13.18). Note that either spectrum may be used, but one will be the conjugate of the other.

We can then inverse transform the result as before to obtain the time domain response (Figure 13.19). In both cases we end up with N valid complex measurements, even though in the offset video case we started with 2N real samples and for in-phase/quadrature we had N complex samples.

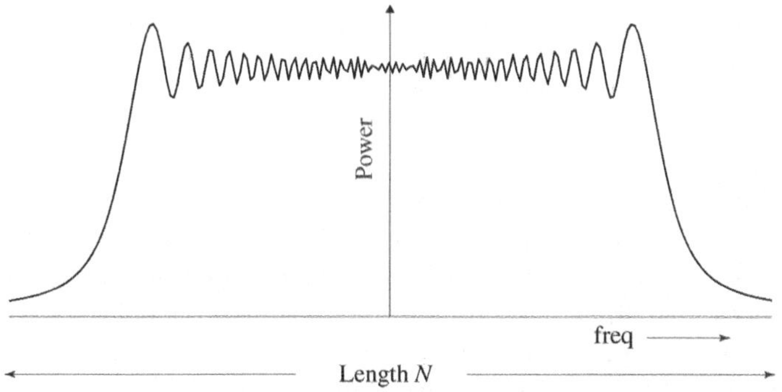

Figure 13.18 Shifted video spectrum $S(f)$. It is necessary to retain only one of the two halves of the spectrum, as one half is the conjugate of the other half. For simplicity of hardware design, usually zero frequency is redefined to be at the location in the spectrum previously occupied by the carrier. The shifted spectrum is now treated as a complex, rather than a real, signal.

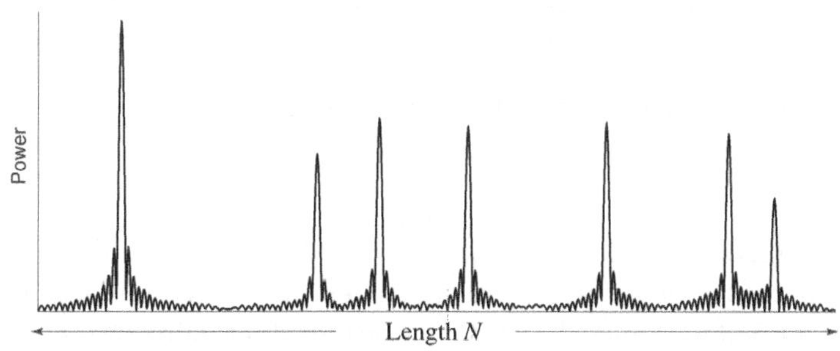

Figure 13.19 Offset video output. Output signal is the inverse transform of the selected half of the product spectrum, and is identical to that obtained in IQ processing.

13.2 Along-track, or Azimuth, Resolution

In our previous examples, we have used the physical aperture of the radar antenna to generate resolution in the along-track direction. This approach is fine for low-resolution, high frequency systems, but for most radars we want to achieve significantly finer resolution. Space systems in particular need sophisticated signal processing to be useful. Here we will first review the azimuth resolution afforded by real aperture systems, and then show how more sophisticated signal processing leads to very fine resolution systems.

13.2.1 Review: Real Aperture Radar (RAR)

We used the following expression to estimate our along-track resolution in the real aperture radar system:

$$\text{Resolution} \sim \frac{R\lambda}{D} \tag{13.17}$$

We can use this relation to calculate the resolution for a spaceborne system such as ERS-1, a radar operated by the European Space Agency (ESA), which acquired data from 1991 to 2000. For this radar system, the range R is about 850 km, the antenna is 10 m long, and the wavelength is 0.0567 m, therefore the real aperture resolution is

$$\rho_{az} = \frac{850,000 \times 0.0567}{10} = 4820 \text{ m} \tag{13.18}$$

For most applications, the coarse 5 km resolution will not suffice. We would only be able to distinguish city-sized objects in the images, whereas we usually want to be able to discern much smaller objects. We have shown previously that even if we illuminate a large, 5 km long area on the surface, we should be able to sort out the echoes from different spots in that area using the Doppler shift of the echoes. We will now see that using the Doppler information in the radar echo to improve our resolution is useful, albeit with some additional signal processing requirements.

Consider the geometry as shown in Figure 13.20, with a coordinate system centered on the radar flight track but offset along track. Let the radar system be at altitude z, and let the radar image a point at a distance y from the ground projection of the flight track. In this coordinate system, the ground point we wish to image is at the origin of the x-axis.

We first determine the range to the point, as before, and obtain

$$R^2 = z^2 + y^2 + x^2 \tag{13.19}$$

Note that the sensor position is linearly related to time, with $x = vt$. Rearranging the above

$$R(t) = (z^2 + y^2 + v^2t^2)^{\frac{1}{2}} \tag{13.20}$$

so that the range $R(t)$ is explicitly a function of time.

We are only interested in the time when the point is in the antenna beam– we do not need to know the range to the point when it is not illuminated. Usually the beamwidth is fairly narrow so

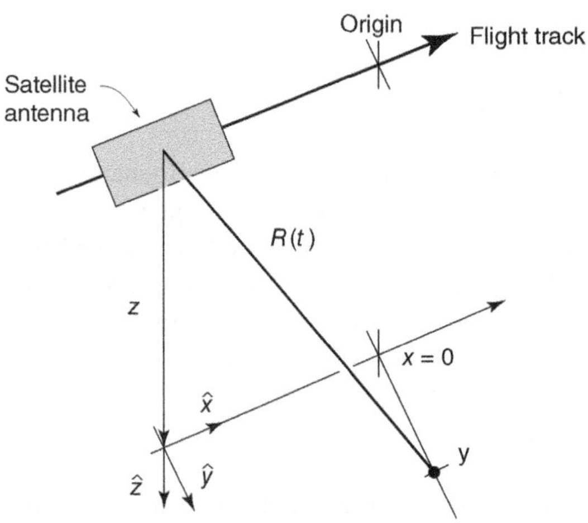

Figure 13.20 Satellite imaging geometry, defining a coordinate system with the origin centered on the radar flight track, but offset along track in the x-direction at the point to be imaged.

that the illuminated area on the ground is of a dimension small compared to the range R. Then $v^2t^2 << z^2 + y^2$, so we may approximate Eq. (13.20) by

$$\begin{aligned} R(t) &= \left(z^2 + y^2\right)^{\frac{1}{2}}\left(1 + \frac{v^2t^2}{z^2 + y^2}\right)^{\frac{1}{2}} \\ &\approx R_0\left(1 + \frac{1}{2}\frac{v^2t^2}{R_0^2}\right) \\ &= R_0 + \frac{1}{2}\frac{v^2t^2}{R_0} \end{aligned}$$

where we have used the definition $R_0^2 = z^2 + y^2$. Note that $R(t)$ is quadratic in time. We can write down the phase of the radar echo, assuming that 2π of phase is accumulated for each wavelength of propagation, and again we have two-way travel. Thus the phase $\phi(t)$ as a function of time is also quadratic:

$$\begin{aligned} \phi(t) &= \frac{4\pi}{\lambda}R(t) \\ &= \frac{4\pi}{\lambda}R_0 + \frac{4\pi}{\lambda}\frac{1}{2}\frac{v^2t^2}{R_0} \end{aligned}$$

Thus the phase of the radar echo as the platform flies over a scattering point has a very similar form to that we previously encountered in range processing when we used a chirp signal. The real signal $s(t)$ from the object is thus

$$s(t) = \cos\left(\phi_0 + \frac{2\pi}{\lambda R_0}v^2t^2\right) \tag{13.21}$$

If we plot the phase and $s(t)$ functions, we obtain signals as seen in Figure 13.21.

Again, the azimuth return $s(t)$ looks much like the chirp signal we studied earlier, which should not be surprising since both signals are quadratic in phase.

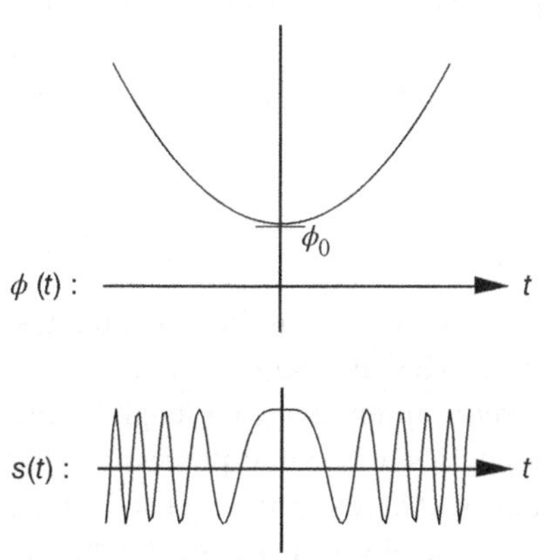

Figure 13.21 Phase and voltage of received echo. Phase, proportional to range, is parabolic. The received signal is the cosine (or exponential) of the phase, and is the chirp waveform we encountered in the discussion on range processing.

We wish to process this echo signal in order to obtain finer resolution than that afforded by the real aperture system. Suppose we simply integrate this signal $s(t)$, that is, we add up all the values for $s(t)$ for some time interval. We could consider the result of this operation as the convolution of the echo signal $s(t)$ with a rect() function whose width is approximately the width of the constant region near the origin of $s(t)$. Because the signal is nearly constant near the center of the return interval, while it fluctuates significantly for large time offsets, the echoes near the origin will add approximately in phase, whereas those from farther out on the axis will tend to cancel. The effect is that this integration is nonzero only when it operates on echoes from the central portion of the signal, while signals from farther out along the time axis tends to integrate to zero. This integration effectively selects only the portion of the return near the origin, improving the resolution by rejecting energy at higher Doppler frequencies.

This is called "unfocused SAR" processing, for unfocused synthetic aperture radar.

We can quantify the resolution we can achieve with unfocused SAR processing by determining what area on the ground corresponds to the central region of the azimuth return that contributes to the output of the integrator. If we assume that only signals with a phase deviation of less than $\frac{\pi}{4}$ over the integration interval contribute meaningfully to the point response, we can calculate which points on the surface will be contained in the resolution cell. This choice is a bit arbitrary, and other definitions are used occasionally. Hence, the resolutions quoted for unfocused SAR systems are dependent on the definition of which echoes are significant. Note that the resolution of the radar system itself does not change, only our interpretation of how large the azimuth resolution cell is changes as we change definitions. Under the first assumption above, the limits of integration are those for which

$$|\phi(t) - \phi(t = 0)| < \frac{\pi}{4} \tag{13.22}$$

from which we obtain

$$x = vt = \sqrt{\frac{\lambda R_0}{8}} \tag{13.23}$$

The length in space that the radar flies over which the echoes from a point on the ground will add in phase can be considered as an "antenna," by analogy to a physical antenna aperture that is designed to add up all the rays that intersect its aperture in phase. So the "synthetic aperture" length for the unfocused SAR is twice this distance, or

$$l_{\text{az}} = \sqrt{\frac{\lambda R_0}{2}} \tag{13.24}$$

This distance is also called the *Fresnel zone size*, which we first saw in our discussion on antennas.

The terminology synthetic aperture radar is thus derived from viewing the combination of a radar and a signal processor as producing an *effective*, or *synthetic*, antenna with a length determined not by physical construction but by a combination of physics and signal processing. For many approximate analyses, we will find that we can use the same sort of reasoning we used for the real aperture radar performance parameters, but rather than using the physical antenna length, we substitute the synthetic antenna length. This simplistic way of thinking about the synthetic aperture does, however, give results that are in error by a factor of two, when we derive the result more carefully.

It is straight-forward to calculate the resolution we can expect with a synthetic aperture. Our projected antenna beamwidth on the ground is as in the real aperture case, but with the new antenna length, so that

$$\rho_{\text{az}} = \frac{R\lambda}{2l_{\text{az}}} \tag{13.25}$$

$$= \frac{R\lambda}{2\sqrt{\frac{R\lambda}{2}}} \tag{13.26}$$

$$= \sqrt{\frac{R\lambda}{2}} \tag{13.27}$$

The resolution obtained with this approximation is the Fresnel zone size.

But the above resolution estimate depends on how much phase shift we can tolerate in forming the synthetic aperture. We can derive the resolution independent of this assumption from the Doppler shift directly. We start with the Doppler equation from before,

$$f_d = \frac{2v}{\lambda} \sin\theta \sin\phi = \frac{2v}{\lambda}\frac{x}{R} \tag{13.28}$$

Differentiating both sides of Eq. (13.28), we can relate the change in Doppler frequency to a change in position x along track,

$$\delta f_d = \frac{2v}{\lambda R}\delta x \tag{13.29}$$

which gives the resolution in frequency δf_d as a function of resolution in distance δx. The distance resolution δx is the distance over we have assumed that the echoes contribute in phase, which we see here is the region for which the Doppler frequency does not change much. Now, the change in time for a scatterer is just the reciprocal of its change in frequency,

$$\delta_t = \frac{\lambda R}{2v\delta x} \tag{13.30}$$

so we see that the resolution in time can be expressed in terms of the resolution in distance δx. Finally, note that the resolution in meters is simply the resolution in time multiplied by the velocity,

$$\rho_{az} = v \cdot \delta t = \frac{\lambda R}{2\delta x} \tag{13.31}$$

Note the similarity between this equation and the physical aperture relation $\rho_{az} = \frac{R\lambda}{D_{\text{az}}}$. Both equations have the same functional form but differ by a factor of 2 in resolution (see Eq. (13.17) and Table 13.1). The physical antenna beamwidth is derived by considering the region for which echoes add in phase, as we also considered above. For the synthetic aperture case, though, we consider two-way propagation which leads to a factor of two in total phase. Thus, summarizing,

Table 13.1 Limiting resolution of real and synthetic aperture radars. Resolution improves as we apply more sophisticated signal processing to digitized radar echoes. In this section we have compared the real aperture case to an unfocused SAR, in the next we will derive the result for the focused SAR case.

	Synthetic aperture	
Real aperture	**Unfocused**	**Focused**
$\rho_{\text{az}} = \frac{R\lambda}{D_{\text{az}}}$	$\rho_{\text{az}} = \sqrt{\frac{R\lambda}{w}}$	$\rho_{\text{az}} = \frac{R\lambda}{2l_{\text{az}}}$

We can now calculate the unfocused SAR system resolution for the ERS radar we considered earlier, and to compare this result to the real aperture case. Using Eq. (13.27), and using values appropriate for ERS-1,

$$\rho_{az} = \sqrt{\frac{R\lambda}{2}} \tag{13.32}$$

$$= \sqrt{\frac{850,000 \times 0.0567}{2}} \tag{13.33}$$

$$= 155 \text{ m} \tag{13.34}$$

Certainly a resolution of 155 m is much better than 5000 m, but it is possible to do better yet. We will apply a better matched filter to the azimuth history of the received signals, and produce much finer resolution images. In the next section we will examine focused synthetic aperture radar image formation.

13.2.2 Synthetic Aperture Processing

Because we have called the SAR processing just presented unfocused, you might expect that there exists a focused version of that processing approach that can achieve even better performance. This is indeed the case, and we will see that additional signal processing yields the increase in performance we seek. Recall the form of the echo we derived above:

$$s(t) = \cos\left(\phi_0 + 2\pi\frac{v^2t^2}{R_0\lambda}\right) \tag{13.35}$$

Previously, we obtained increased performance over the real aperture case by integrating the signal in azimuth, which resulted in the coherent addition of echoes over a longer, synthetic antenna aperture. The limits of integration were set here by selecting echoes whose phase remains within some fraction of a cycle of the value at the center of the observation period.

So, it makes sense to consider if we could improve performance by increasing those integration limits. This implies that we must now integrate echoes whose phase varies significantly from the small variation we allowed before. If we can use these echoes over a longer time period, the synthetic aperture will get longer, and hence the resolution will become finer.

Equation (13.35) shows that the variable phase term in the cosine depends only on known geometric values, so why not correct the received signals for a quadratic phase term before integration? Actually, we do not know the along-track time very well, so we must try a lot of different time offsets and pick the best one, that which gives a peak in the return. If echoes over a much longer interval may be added in phase because the imaging geometry is well-known, we can produce some very long synthetic apertures.

Unknown offsets are easy for us to handle, as we can implement the phase correction as a convolution. If we know the expected *phase history* of a point on the surface, we may convolve the signal with a reference signal of known offset, which forms the matched filter output. Given the imaging geometry, we can write down the output of the correlation receiver $g(t)$ directly as,

$$g(t') = \int_{-\infty}^{+\infty} s(t')e^{j\frac{2\pi}{\lambda R_0}v^2(t-t')^2}\,dt \tag{13.36}$$

or

$$g(t) = s(t)^* e^{j\frac{2\pi}{\lambda R_0}t^2} \tag{13.37}$$

The echo is convolved with a reference signal that has a quadratic phase in time, but now time corresponds to distance measured along the flight path rather than delay in the range direction. This is nonetheless the same equation we obtained in the range processing case where we also had a quadratic signal. The impulse response here, calculated by letting $s(t)$ also be this azimuth chirp signal, is once again a sinc() function.

So the impulse response is seen to be

$$h(t) \sim \frac{\sin\left[\frac{2\pi l_{az}}{R_0\lambda}(vt - x_0)\right]}{\frac{2\pi l_{az}}{R_0\lambda}(vt - x_0)} \tag{13.38}$$

Changing variables to $x = vt$ and centering at the origin gives a simpler form

$$h(x) = \frac{\sin\frac{2\pi l_{az}x}{R_0\lambda}}{\frac{2\pi l_{az}x}{R_0\lambda}} \tag{13.39}$$

We recognize this as a sinc with width $\frac{\lambda R_0}{2l_{az}}$, which is the same relation we obtained in the unfocused SAR case. But now l_{az} is much, much, longer, as it is not limited to the central part of the echo response. The limits here are set by how long we make our quadratic reference function, and how long the object is actually illuminated by our antenna.

We would like to choose the longest possible value for l_{az} as that choice will give us the finest resolution. Because we are correcting for all quadratic phase shifts, it may seem that we may make l_{az} arbitrarily long. However, we are limited by the physical antenna beamwidth. Each point remains in the antenna illumination pattern for only a finite time, and that time is determined by the antenna beamwidth in the azimuth direction and the velocity of the platform. In other words, we cannot get echoes outside the antenna beam. This sets our maximum integration time, and hence our limit in resolution.

Thus the largest value we can choose for the synthetic aperture will be the limits of the antenna pattern,

$$\rho_{az} = \frac{\lambda R_0}{2 \cdot \frac{\lambda R_0}{D_{az}}} \tag{13.40}$$

$$= \frac{D_{az}}{2} \tag{13.41}$$

Note that the achievable resolution is finer for smaller antennas, and independent of range. This remarkable result says that we can achieve arbitrarily fine resolution, down to half the physical antenna length, by applying signal processing to compensate for the known phase histories through the full antenna pattern.

Perhaps it is easiest to understand the independence on range by realizing that the projected antenna pattern on the ground is proportional to range, so that the farther away the radar is from the surface, the longer the permitted integration time is. This exactly cancels out the decrease in resolution with range we would have with a fixed antenna length.

To summarize, we obtain the finest resolution by convolving the received signal with the conjugate of the expected waveform. The processing can be completely analogous to our range chirp processing. Hence the processing strategy is to consider the echo data as forming a range/azimuth array, where each range line is first processed to a fine resolution range echo. Then for each range distance, convolve each azimuth line with a reference signal of quadratic phase that is determined by the imaging geometry (Figure 13.22).

Comparing resolutions now for all three azimuth approaches in the ERS example:

RAR	Unf. SAR	SAR
$\frac{R\lambda}{D_{az}}$	$\sqrt{\frac{R_0\lambda}{2}}$	$\frac{D_{az}}{z}$
5000 m	155 m	5 m

13.2.3 Number of Computations Needed to Form an Image

All of our improvements to resolution were obtained by assuming that signal processing could be applied to the data to compensate for shortcomings of the radar hardware. Thus we will need to determine whether a real data processor can be built to form images in a timely manner. We can evaluate the number of computations required in order to form a complete radar image by considering each processing step in turn. Assume we will process a set of data consisting of 4096 complex range samples/line and 4096 lines. Table 13.2 lists our processing recipe and number of real multiplies needed to implement the algorithm.

A spaceborne radar collects this quantity of data every 2 s! Thus the required real-time processing rate $\sim 10^{10}$ multiplies per second.

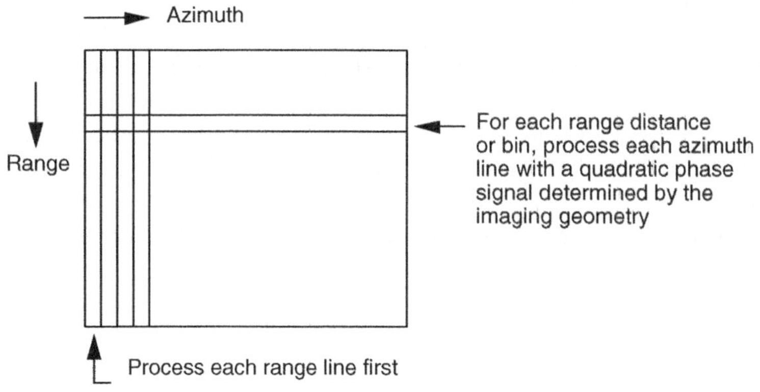

Figure 13.22 Range and azimuth lines.

Table 13.2 Computation load for SAR image processing.

Processing step	Real multiplies
1. Transform each line in range direction ($2N \log N \times 4096$ lines)	400×10^6
2. Multiply by the range reference function	64×10^6
3. Apply range weighting	6×10^6
4. Inverse transform	400×10^6
5. Transform in azimuth ($2N \log N \times 4096$ lines)	400×10^6
6. Multiply by azimuth reference function	64×10^6
7. Inverse transform	400×10^6
Total:	$\approx 18 \times 10^9$

13.3 Range Migration Processing

The algorithms described in the previous section will produce acceptable images in many situations. But for very high performance radar systems, especially those operating at wavelengths longer than about 10 cm, further refinements of the processing algorithm are needed to achieve the highest resolution possible. Chief among these is range migration processing, which we will see is a modification to the algorithm that accounts for the movement, or migration, of the radar return from a point target through several range bins as the distance from the radar to the target varies.

We found a quadratic dependence of phase with time for the radar echo which we could compensate in the processor, so that the reference filter duplicated the phase history of a point target. If we consider the source of this phase history, we will see that not only is the phase of the echo shifted with time, but also its delay. If the change in delay during the imaging integration period is greater than the size of a range resolution element, then the return has migrated from one range cell to the next.

The delay of the echo from a point scatterer changes because of the extra distance traveled by the echo compared to the distance when the target is exactly at right angles to the flight track. The phase is proportional to time delay, where the factor of proportionality is $4\pi/\lambda$. To determine if the shift in delay is of consequence for a particular radar system, we must calculate the distance involved.

From before, assuming that the target point is never observed very far away from the perpendicular to the flight direction,

$$R(t) = R_0 + \frac{1}{2}\frac{v^2t^2}{r_0} \tag{13.42}$$

The change in range with respect to time is

$$R(t) - R_0 = +\frac{1}{2}\frac{v^2t^2}{r_0} \tag{13.43}$$

This range change is denoted the "*range migration*." Range migration is not peculiar to radar systems, and is an issue in many imaging systems, including seismic imaging of the Earth's subsurface (see Figures 13.23 and 13.24) and acoustic sonar systems. These images are often cluttered with near-parabolic signatures from bright targets, and each depicts the motion of a point through the imaging aperture. Typically we may assume the trajectory to be parabolic to a good approximation.

For the subsurface seismic images, we generally express the migration in terms of distance along the surface x rather than time, as data are usually collected by geophones that are deployed to certain positions in space instead of along a flight track. Thus, and this holds for the radar case also, the migration in terms of position is

$$\text{migration} = \frac{1}{2}\frac{x^2}{r_0} \tag{13.44}$$

The intensity plot of a radar signal would look similar.

If the echo for a point scatterer moves through several range cells, then the processing algorithm we described earlier, where first we process in range, then convolve the returns with azimuth reference functions at right angles to the range processing, will not retain all of the energy reflected from the point. Some energy will be received at one range delay, and other parts of the return will be received at other range delays. If the resolution cell size is smaller than this spread, our processing will not be optimal. We can get an idea of how big an effect migration is by calculating a sample case. The imaging geometry is shown in Figure 13.25. Assume we are using a spaceborne radar,

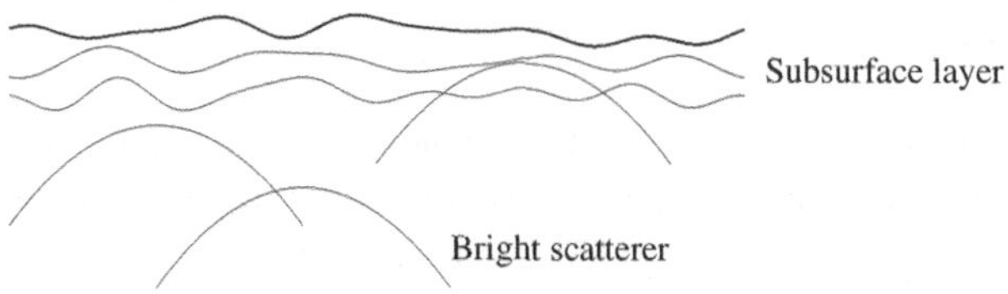

Figure 13.23 Seismic layers. Subsurface images usually delineate layers plus a set of near-parabolic returns from individual point scatterers. Layers consist of the superposition of many scatterers displaced along track but all at nearly the same range, so the almost-surface-parallel segments of the parabolas at closest approach dominate. Individual bright scatterers produce visible parabolic returns.

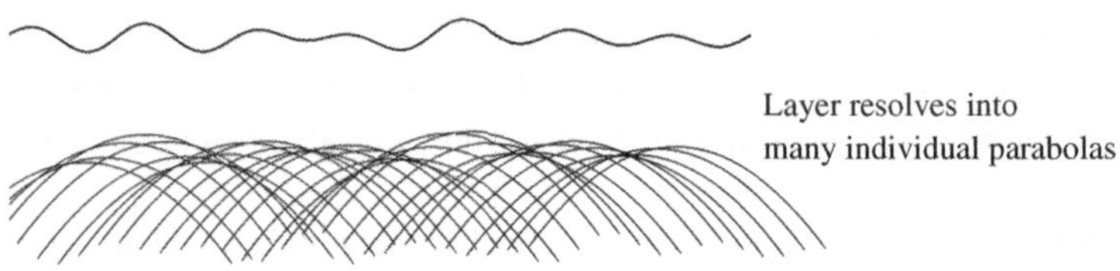

Figure 13.24 Single seismic layer. Close inspection of one layer shows it to be composed of the superposition of many point scatterers distributed in space at nearly fixed depth.

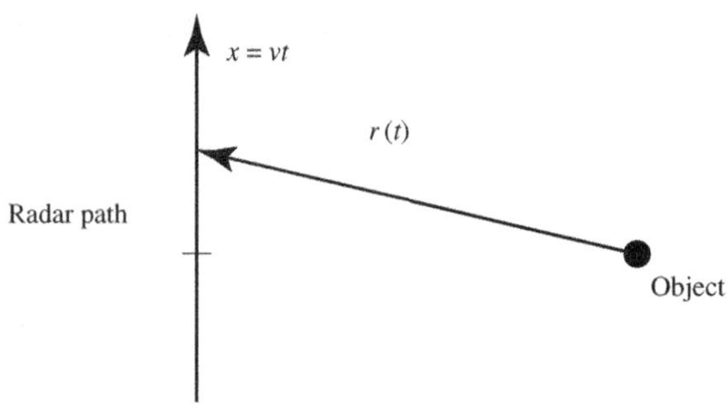

Figure 13.25 Range migration. As the radar flies by the object, the range from the radar to the point changes. Hence echoes from the object are received at different times. If the differences of these times are longer than the radar time resolution, the object is observed in multiple range bins. This apparent motion is called range migration, or range walk.

with a 24 cm-λ (L - band) wavelength, a 10 m antenna, operating at a range of 800 km: The limits of illumination for this radar are set by the antenna beamwidth, which is

$$vt_{\max} = \frac{R_0\lambda}{D} = 19,200\,\text{m} \tag{13.45}$$

and hence the maximum migration is

$$\text{migration} = \frac{1}{2}\frac{v^2t^2}{R_0} = 230\ \text{m} \tag{13.46}$$

Therefore, the signal moves 230 m from its location when the target is centered in the antenna beam to when it is located at the beam edge. If the resolution of the radar is finer than 230 m, the target will migrate from one cell to another. If the resolution is, say, 10 m, the target will migrate over 23 cells.

Let us plot the location of the echo in range-azimuth space by identifying the positions the echo will occupy. Recall that we are generating resolution in the across-track direction by the time delay, that is, we subdivide the echo into "range gates," each about one resolution element in size. We can plot the echo location in our gated and pulsed received data matrix as shown in Figure 13.26. Note that the signal is dispersed in range– we can't simply convolve each azimuth line with the theoretical azimuth chirp signal. The number of cells that the echo migrates over is determined by the resolution cell size in the range direction. If the migration exceeds one resolution element, we must compensate for this effect before applying the azimuth matched filter.

One way to correct the data matrix is to predistort the measurement matrix for one migration parabola, as shown in Figure 13.27. The row which is shifted now contains the entire echo history for the point target, and we can obtain full azimuth resolution. Unfortunately, the shape of the parabola is range dependent (migration $= \frac{1}{2}v^2t^2/R_0$), therefore we cannot simply shift all the data as shown for the entire data set, we must vary the shift with range. For each line processed, it is necessary to derive a new parabolic trajectory for the interpolation routine. In other words, the migration path and the azimuth reference function of the system are space variants. This makes the azimuth processing step of the data processor very computationally intensive.

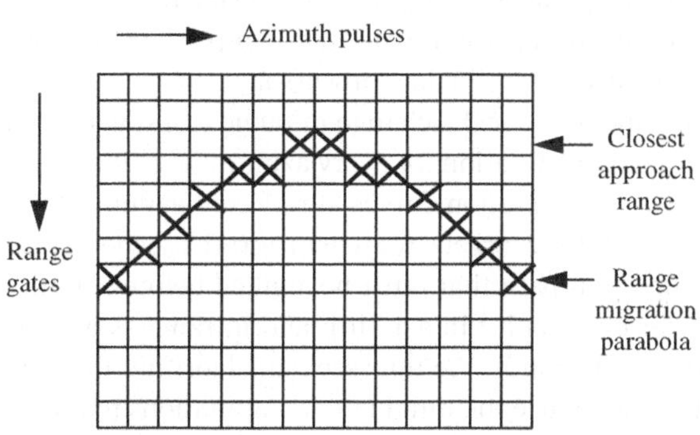

Figure 13.26 Curve of range migration. The echo from a point scatterer will appear in different range gates as a function of azimuth position. As the variation in range is second-order with respect to the azimuth displacement of radar platform, the curve is locally parabolic. Actual curve for a flat Earth is hyperbolic (see text).

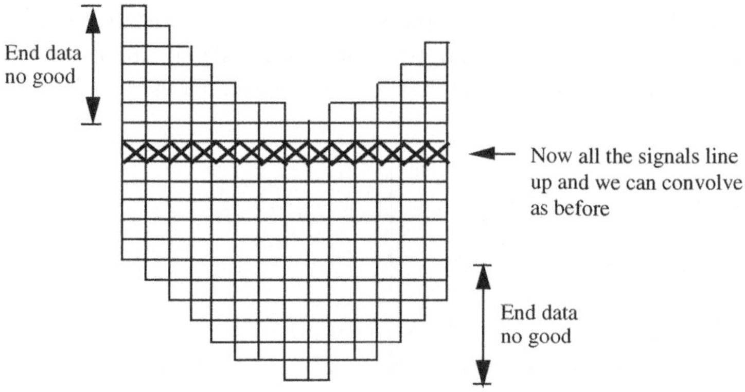

Figure 13.27 Range migration adjustment. Matrix of echo data shifted to compensate for migration parabola, forcing return from the desired object to lie within a single range bin. Convolution in the azimuth direction can then properly account for entire phase history, obtaining theoretical SAR resolution. Note that the array is incomplete near top and bottom.

Of course, on a computer we do not really have to shift the data at all. We simply keep track of how much shift we need as we select data from each azimuth line, and sample appropriately. Thus, some kind of interpolation algorithm is needed to select the proper data for each azimuth line.

Many interpolation algorithms exist for processing radar data with the range migration correction. It is generally required to trade off interpolation accuracy for processing speed to optimize any design, and this turns out to be one of the fundamental decisions in processor design.

Perhaps the simplest method of interpolating for SAR processing, and in fact the first approach used in digital SAR systems, is nearest neighbor interpolation. Here, rather than interpolating between data points, we select the value in the array closest to the computed location. The range migration path will be arbitrarily located in the array, whereas the signal itself is known only at integral grid points. Nonetheless, the nearest neighbor approximation works well enough for many applications.

Nearest neighbor interpolation for SAR range migration processing is sometimes called *cut-and-paste*, because it is reminiscent of cutting up the data matrix and pasting the pieces back together after applying the proper shifts, as suggested by Figure 13.28. Cut-and-paste algorithms are fast, but inaccurate. Some more complex interpolation approaches are spline fits, or sinc interpolators, which are much more accurate by also very time consuming. Interpolation to compensate for range migration is often the most computationally intensive parts of a data processor.

The above range migration algorithms can compensate for the change in range to a single point target. But the image contains objects at the same range location for every azimuth location in the array. Consider Figure 13.29, which depicts only two migration curves at different azimuth locations. If we apply the migration correction for one of these targets, we will distort the second curve even more. Clearly, each target has its own migration path that must be applied before azimuth matched filtering. Applying a different curve for every point in azimuth and in range is quickly going to break the bank in terms of computational demands. We thus seek an algorithm that will allow us to process many points together and hence reduce the number of computations required. Let us seek a convolutional process to address all points at the same range together and thus save computational time.

Consider the data acquired from many point targets, all at the same range but dispersed in azimuth. Each such azimuth line consists of a superposition of echoes chirped in accordance to imaging geometry, from scatterers at different azimuth locations, as suggested by Figure 13.30. Each of the azimuth "chirps" is identical except for its amplitude and position, which we wish to recover. Let us now describe the properties of the azimuth transform of this signal, and from that derive an algorithm that will allow us to process many of these echoes simultaneously.

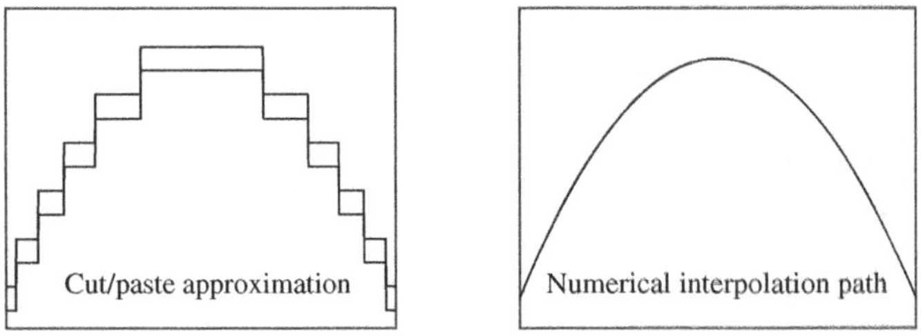

Figure 13.28 Interpolation methods for correcting range migration. Schematic representations of the paths approximating the range migration curve for cut-and-paste and fine-interpolation algorithms. The interpolation approach much more closely mimics the actual migration curve, but it consumes many more computational cycles to implement.

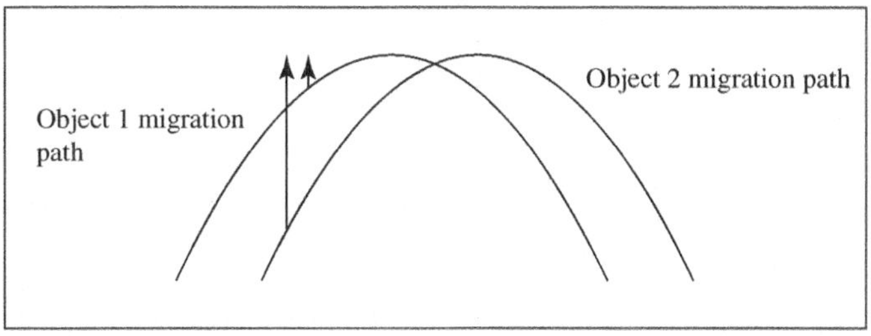

Figure 13.29 Offset range migration curves. Migration curves for two scatterers at the same range but different azimuth locations have the same shape, but are displaced in the azimuth direction. Arrows show required displacement, for example, points on each curve, but we can only apply one shift to the data at each location. Hence we will distort one curve as we correct for the other.

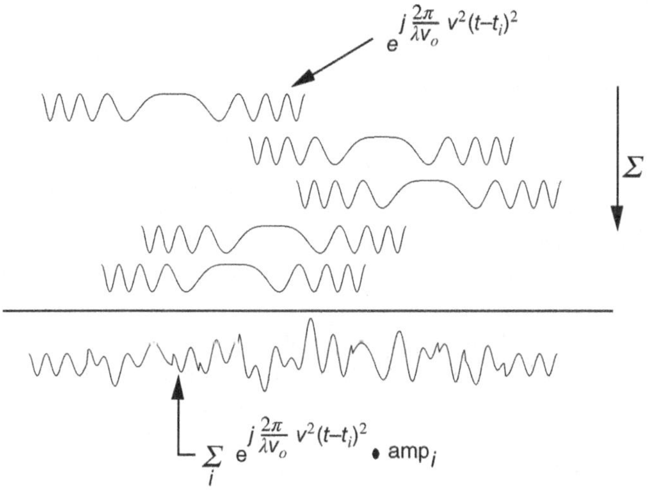

Figure 13.30 Superposition of azimuth chirps. An azimuth line consists of the superposition of azimuth chirp echoes of various magnitudes and at various offsets.

Note that by the shift theorem, the transforms of the echo from each point, as long as the range at closest approach R_0 is held constant, will be identical to the others save for a linear phase term dependent on azimuth position, and an amplitude associated with the reflectivity at the scattering points. Therefore the spectra will all lie on top of each other when viewed as a frequency domain plot. In other words, the range migration history as a function of frequency will be the same for each azimuth location, and the single range migration curve will contain the superposition of all the azimuth scatterers. The only difference between the echoes will be that linear phase term. Thus if we can redesign the interpolation algorithm in terms of frequency, rather than azimuth position, we can predistort the *azimuth* spectrum, rather than the signal itself, once, and interpolate a single curve in transform space to correct migration at all azimuth locations.

We can calculate the migration history as a function of frequency starting with the migration as a function of time as before:

$$R(t) - R_0 = \frac{1}{2}\frac{v^2 t^2}{R_0} \tag{13.47}$$

which gave us the phase history

$$\phi(t) = \phi_0 + \frac{2\pi}{\lambda R_0} v^2 t^2 \tag{13.48}$$

Since frequency is $\frac{1}{2\pi}$ times the phase derivative with respect to time,

$$f_d = \frac{1}{2\pi}\frac{d\phi}{dt} = \frac{v^2 t^2}{\lambda R_0} \tag{13.49}$$

or

$$t = \frac{f_d \lambda R_0}{2v^2} \tag{13.50}$$

Combining Eqs. (13.47) and (13.50), we obtain

$$\begin{aligned} R(t) - R_0 &= \frac{1}{2}\frac{v^2}{R_0}\left[\frac{f_d \lambda R_0}{2v^2}\right]^2 \\ &= \frac{f_d^2 \lambda^2 R_0}{8v^2} \end{aligned}$$

Note that the range migration as a function of frequency is quadratic in form, so in frequency space we still have the parabolic interpolation issue. But one interpolation is sufficient for all azimuth points. The parabola is still range dependent, however, and must be recomputed for each range gate.

The ability to correct for range migration in the frequency domain fits in well with our azimuth convolutional processing, where we are already calculating the azimuth transform in the process of creating the matched filter for the azimuth chirp. We can now state a complete azimuth processing algorithm, including the migration correction:

1. Transform data in azimuth direction
2. Resample to correct for range migration
3. Multiply line by conjugate of reference function
4. Inverse transform result to get time domain output image line

While the recipe is simple, in practice the resampling step is very time-consuming. Usually much more time is required here than for the transform operations.

We present an end-to-end processor block diagram in Figure 13.31.

With a processing approach now well-defined, we can examine processing rates and other design considerations in the next section.

13.4 Design and Performance of a SAR Processor

We have already examined processing speed requirements for range processing of SAR data. The azimuth processing algorithms require that we operate on the transform of many azimuth lines at once in order to implement range migration, so that large computer memories are needed. For a 4K × 4K image, for example, to represent a 4096 × 4096 complex floating point image or spectrum requires 4096 × 4096 × 8 bytes/sample = 128 Mbytes memory. So, not only are the computational rates high, but large memories are also required. Thus data rate and volume both become important design considerations.

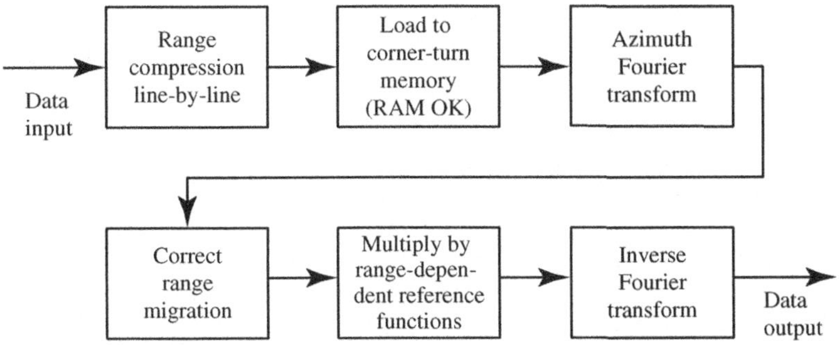

Figure 13.31 Processor block diagram. Data flow through a SAR processor begins with the range compression algorithm. The data are then transposed to be in azimuth-line order rather than range-line order; this operation is often referred to as a "corner-turn" operation. Then the data are transformed to the azimuth frequency domain, and corrected for range migration. At this point the azimuth reference functions are applied, and the result is transformed back into the time domain to form the final image.

We can calculate the throughput of a processor for the above-used 4096 × 4096 image acquired from a satellite radar. If the antenna length is 10 m, then the maximum expected Doppler frequency would be

$$f_{\mathrm{d}} = \frac{2v}{\lambda} \cdot \frac{\lambda}{D_{\mathrm{az}}} = \frac{2v}{D_{\mathrm{az}}} \approx \frac{2 \times 7500}{10} = 1500 \text{ Hz} \tag{13.51}$$

Thus we can sample at 2048 Hz and still allow for modest oversampling, and we will acquire the 4K × 4K data set in 2 s.

We found before that about 2×10^{10} operations were required to process the image in the simpler, non-range-migrated case. Suppose that range migration interpolation triples the number of operations, then we require 6×10^{10} multiplies to create the image. We just found that 128 Mbytes of RAM were required to hold one buffer with the signal from the scene. Usually another factor of two in memory is needed for other operations.

Assume that we implement the data processor on a workstation, with more than 256 MB RAM and a processor speed of 10^8 complex multiplies per second. Our expected execution time to create the image is

$$\frac{6 \times 10^{10} \text{ operations}}{10^8 \text{ operations/s}} = 600 \text{ s} \tag{13.52}$$

So for this processor to reduce the 2 s worth of data, 600 s are needed, a slowdown by a factor of 300. But the cost of the machine to do this is only about $5000.

It is worthwhile to examine how much better we can do using a supercomputer. For a supercomputer, typical computation rates may approach a few times 10^9 operations per second. Then the execution time becomes

$$\frac{6 \times 10^{10}}{3 \times 10^9} = 20 \text{ s}, \tag{13.53}$$

or only 10 times real time throughput. But the cost of this computer is more like $5,000,000.

So you can see why real-time radar operation requires special purpose processing hardware.

14

Advanced Processing Topics

In the previous chapter, we have seen that the application of signal processing methods to the radar signals can significantly improve image quality. Range coding allowed us to decouple range resolution from signal to noise ratio, so that we can generate high-SNR images and still attain the fine resolution associated with very short transmit pulses. In the azimuth direction, using the unfocused or focused synthetic aperture algorithms permitted reduction of the azimuth resolution cell size below the real antenna aperture limits by several orders of magnitude.

Now we present several extensions to the previous discussion that allow us to characterize the response of the radar to special imaging situations. First, we examine the output from an imaging radar when the objects under study are moving with respect to the background. Next, we quantify image degradation due to sampling effects related to the pulse repetition rate, which lead to ambiguities in both the range and azimuth dimensions. Finally, we examine the case when the radar is not pointed exactly broadside to the flight direction, so that our azimuth phase history is not quite what we had assumed in the previous chapter. We present a means to correct the processor algorithm for this change in geometry, and yet retain the high image quality standards enabled by SAR processing.

14.1 Moving Objects

When we formed high resolution images using synthetic aperture techniques, we made use of the phase history of the scatterers on the surface. Although we used the phase formulation, this is equivalent to using Doppler frequency to sort echoes along track. And since we are using Doppler frequency to sort the various targets by azimuth location, one might ask if there is any effect on system performance if the object is moving and producing additional Doppler shifts on the echoes. The answer is yes, and we consider those effects now.

We have already seen in our discussion of the ambiguity function that for a chirped waveform, there exists an uncertainty between object velocity and distance, with the slope of the line presenting error in time as a function of error in velocity proportional to the chirp slope. Recall the form of our convolutional matched filter, where we image an object by the operation

$$g(t) = s(t) \ * \ \exp\left(j\frac{2\pi v^2}{R_0\lambda}t^2\right) \tag{14.1}$$

and hence $g(t)$ is the system response in azimuth to the signal $s(t)$.

Suppose our object is a single reflector at location $t = t_0$, that is, the viewing direction from the radar to the point is at a right angle to the flight track at $t = t_0$. Then,

$$g(t) = \exp\left(- j\frac{2\pi}{\lambda R_0}(t - t_0)^2\right) * \exp\left(j\frac{2\pi v^2}{R_0\lambda}t^2\right) \tag{14.2}$$

Writing out the complete integral,

$$g(t) = \int_{-\infty}^{\infty} \exp\left(- j\frac{2\pi}{\lambda R_0}v^2(t' - t_0)^2\right) \exp\left(j\frac{2\pi}{\lambda R_0}v^2(t' - t)^2\right)dt' \tag{14.3}$$

where we have again ignored amplitude effects in calculating the received signal, which would simply scale the result.

Expanding terms in the integral, we obtain:

$$\begin{aligned} g(t) &= \int_{-\infty}^{\infty} \exp\left(-j\frac{2\pi v^2}{R_0\lambda}[t'^2 - 2t_0t' + t_0^2 + 2tt' - t^2]\right) dt' \\ &= \exp\left(-j\frac{2\pi v^2}{\lambda R_0}({t_0}^2 - t^2)\right) \int_{-\infty}^{\infty} \exp\left(-j\frac{2\pi v^2}{R_0\lambda}(2t')(t - t_0)\right) dt' \end{aligned}$$

The magnitude of this function will be a sinc function centered at $t = t_0$, as expected, with a width dependent on range, wavelength, and velocity. The leading phase term has magnitude unity and may be ignored if we are displaying only the intensity of the signal.

Now, compare this return to that of an object, also at location t_0, that is moving toward the radar with a velocity such that it experiences a Doppler shift in frequency equal to f_d. Here the received signal $s(t)$ will be,

$$s(t) = \exp\left(-j\frac{2\pi v^2}{\lambda R_0}(t - t_0)^2\right) \exp(-j2\pi f_d t) \tag{14.4}$$

Rewriting Eq. (14.4) as follows

$$\begin{aligned} s(t) &= \exp\left(-j\frac{2\pi v^2}{\lambda R_0}(t^2 - 2tt_0 + t_0^2)\right) \exp\left(-j\frac{2\pi v^2}{\lambda R_0}\left(f_d \cdot \frac{\lambda R_0}{v^2}\right)t\right) \\ &= \exp\left(-j\frac{2\pi v^2}{\lambda R_0}\left(t^2 - 2tt_0 - f_d\frac{\lambda R_0}{v^2}t + t_0^2\right)\right) \\ &= \exp\left(-j\frac{2\pi v^2}{\lambda R_0}\left(t^2 - 2t\left[t_0 - \frac{f_d\lambda R_0}{2v^2}\right] + t_0^2\right)\right) \end{aligned}$$

we obtain an expression for the echo similar to the form of the echo from a stationary target, but with a different time of closest approach. The form could be made simpler by completing the square to add an additional phase term which would not affect the magnitude of the convolution integral. However, the important part is that the true t_0 has been replaced by an effective t_0 equal to

$$t_{0,\text{eff}} = t_0 - \frac{f_d\lambda R_0}{2v^2} \tag{14.5}$$

so that the impulse response will now peak at a value offset by $\frac{f_d\lambda R_0}{2v^2}$. In other words, the image of the moving target will appear at the wrong azimuth location, with the offset dependent on its Doppler shift, and hence its velocity.

We may envision the displacement of a moving object in another way by examining the Doppler equation directly. The Doppler frequency associated with a point on the ground is:

$$f_d = \frac{2v}{\lambda}\frac{x}{R_0} = \frac{2v^2 t}{\lambda R_0} \tag{14.6}$$

So an error in Doppler measurement will simply result in a position error of magnitude,

$$t_{\text{error}} = \frac{\lambda R_0}{2v^2} f_{d\ \text{error}} \tag{14.7}$$

How significant is this effect? Suppose we are imaging a railroad locomotive moving toward the radar at 50 miles per hour. Our radar measures a Doppler frequency of,

$$f_d = \frac{2v}{\lambda} \cdot 50\ \text{mi/h} = \frac{2}{\lambda} \cdot 22.4\ \text{m/s} \tag{14.8}$$

Thus the time error will be:

$$t_{\text{error}} = \frac{\lambda R_0}{2v^2} \cdot \frac{2}{\lambda} \cdot 22.4\ \text{m/s} \tag{14.9}$$

The λ's cancel, and if $R_0 = 10^4$ m and $v = 250\ \text{m s}^{-1}$, all typical aircraft parameters,

$$t_{\text{error}} = 3.58\ \text{s} \tag{14.10}$$

and the offset is 896 m.

For a spaceborne radar, typical imaging geometries yield a value for R_0 more like 8×10^5 and v of approximately $7500\ \text{m s}^{-1}$, hence we get,

$$t_{\text{error}} = 0.32\ \text{s} \tag{14.11}$$

which is a much smaller offset in time but it corresponds to a larger offset in distance, 2.4 km, because of the much greater velocity of the sensor.

The effect on the image of this displacement is illustrated in Figure 14.1. This is sometimes called the "train off the track" phenomenon. There are many other times when this effect is observed in radar images. Two common situations are shown in Figure 14.2, where cars driving over a bridge appear out in the middle of the water, and for a moving ship which is observed being displaced from its wake. Sometimes this effect is exploited as a moving target detector or velocity measurement tool.

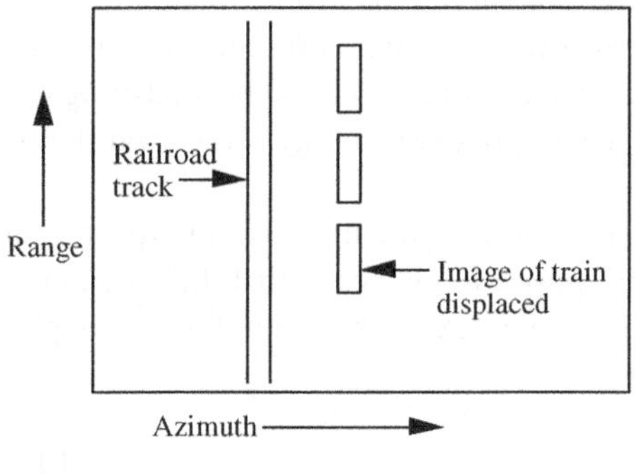

Figure 14.1 Displacement of moving train in SAR image. A moving object imaged by a SAR system is displaced in proportion to its Doppler shift. The displacement is entirely in the azimuthal direction. A train moving train toward the radar has a positive Doppler shift, and is displaced in the direction of radar motion.

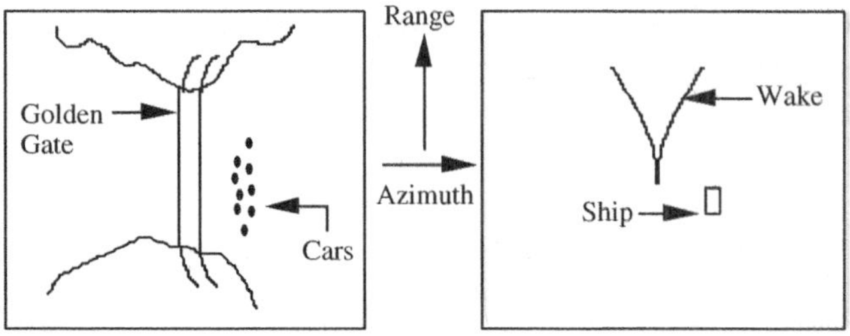

Figure 14.2 Other example displacements. Two other examples in which an object is obviously displaced in a radar image. Left, cars are shifted off the bridge they are driving on and over the water; right, a ship is displaced from its wave.

Thus, object motion produces a shift in position from where its image should appear in the result. If we are not aware that the object under study is moving, we will produce a distorted map that does accurately represent the radar reflectivity of the surface. On the other hand, in many cases, we have outside knowledge of the true position of the object, such as we know that the train must be located in the railroad tracks or that the wake must be "attached" to the ship. Here we can use the measured displacement and the imaging geometry to determine the velocity of the object.

14.2 Azimuth Ambiguities

The imaging radars we have discussed are pulsed systems. Therefore, issues of sampling adequacy are important in their design. We obtain azimuth resolution by analyzing the return from multiple pulses, in order to measure the Doppler or phase history of points on the ground. It is necessary to increase the sampling rate of the radar in order to measure the complete azimuth spectrum of the radar echoes, or else aliasing will lead to the incorrect determination of the Doppler shifts, and distorted images. Echoes with a true Doppler shift greater than the sample rate (pulse rate, or PRF) of the system will be aliased and imaged at the wrong location on the ground.

Since the maximum positive Doppler shift from a point on the surface is $2v/\lambda$, and the minimum negative Doppler shift possible is $2v/\lambda$, we could simply pulse the radar at $4v/\lambda$ to unambiguously resolve the shift from every point on the surface. However, this would often lead to very large data rates. Moreover, such rapid pulsing would also make the achievable range swath width quite small as it is proportional to the interpulse period. The radar antenna illumination pattern generally selects only a fraction of the entire surface for viewing, so that it makes more sense to choose the pulse rate based on the illuminated area rather than the range of Doppler frequencies over the entire Earth surface. Hence, we usually operate the radar at a lower rate, derived by analyzing our antenna pattern. But antenna patterns are not ideal, so inevitably some energy is received that is aliased by the system.

Consider an example of a sinc-shaped antenna pattern in azimuth, which we would expect from a uniformly weighted aperture along track, as shown in Figure 14.3. We can label the abscissa of the plot in Figure 14.3 in several ways, including position along track on the ground or in Doppler frequency, which are coupled through the linear relation

$$fd = \frac{2v}{\lambda}\frac{x}{R_0} \tag{14.12}$$

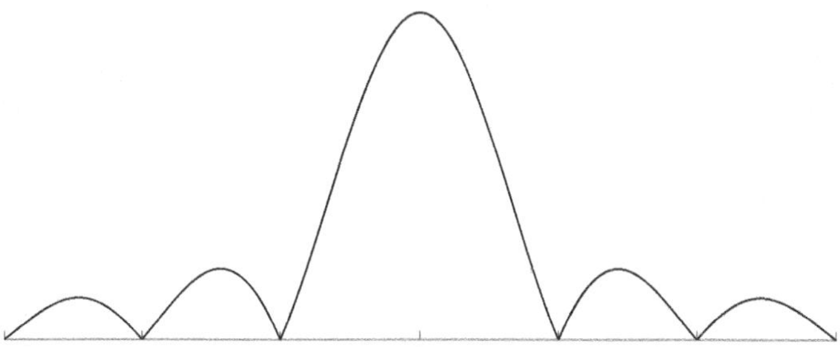

Figure 14.3 Sinc-shaped azimuth antenna pattern. A uniformly weighted aperture results in a sinc(θ) voltage response in azimuth.

The location of the first null in the antenna pattern occurs at $x = \frac{R\lambda}{D_{az}}$ referenced against position, and in Doppler frequency at $\frac{2\upsilon}{D_{az}}$. Because the pulsed system samples the Doppler spectrum at some finite value, and the reflected energy extends to very high Doppler frequencies in the far sidelobes of the antenna, there will be some aliasing of highly shifted echoes into the return from the main lobe of the antenna.

For example, suppose we choose a pulse rate such that the pulse repetition frequency (PRF) is $\frac{4\upsilon}{D_{az}}$, and label the axis accordingly as in Figure 14.4.

With aliasing at the PRF, the bandwidth of the sampled spectrum is also $\frac{4\upsilon}{D_{az}}$, and all signals greater than that frequency fold back into the base spectrum, as illustrated in Figure 14.5. Hence, if we have an object at position $x + 1.5\frac{R\lambda}{D_{az}}$, with a true Doppler frequency of

$$fd = \frac{3\upsilon}{D_{az}} \tag{14.13}$$

it will be aliased in frequency to location $\frac{-\upsilon}{D_{az}}$. In other words, energy from the first antenna sidelobe appears in the main lobe spectrum after the signal is sampled. Remember that if the sample rate is $\frac{4\upsilon}{D_{az}}$ centered at zero, anything outside of $\pm\frac{2\upsilon}{D_{az}}$ will be aliased. Thus an object located out of the main beam, in an antenna sidelobe, will nonetheless be passed by the matched filter that is designed to allow only energy in the fundamental region of the spectrum to be retained.

This example object is exactly centered on the first antenna sidelobe, so its amplitude is −13.5 dB relative to the antenna pattern peak. This will result in a "ghost" image 13.5 dB smaller than the peak, located at a position offset from its true position by an amount equal to the distance corresponding to a Doppler shift of the PRF. Nevertheless, the ghost image appears in our output array. We call the shift in position from where the object should be seen the *ambiguity distance*. We can calculate the ambiguity distance by setting our Doppler shift to the PRF in Eq. (14.12).

$$x = \text{PRF} \cdot \frac{\lambda R}{2\upsilon} \tag{14.14}$$

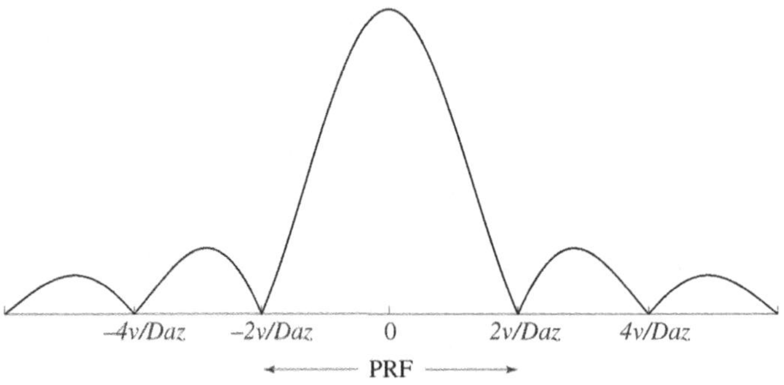

Figure 14.4 Relationship of PRF to antenna aperture pattern. Aliasing in the azimuth direction from undersampling the Doppler spectrum results in folding of the antenna pattern. Often the PRF is chosen so that the folding points of a uniformly weighted array fall at the locations of nulls in the antenna pattern.

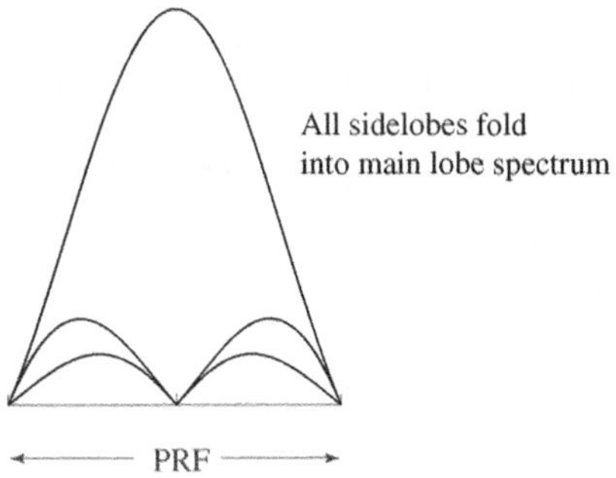

Figure 14.5 Antenna pattern with aliasing. For uniformly weighted antenna aperture the side lobes fold perfectly into the main beam if the PRF is chosen as in Figure 14.4.

Clearly, this aliased energy is going to degrade our image at the least, or even destroy our interpretation of what we see. There are several methods to reduce the effect of azimuth ambiguities, including

1. Sample at a higher pulse rate, so that ambiguous echoes come from sidelobes located further from the main lobe and hence lower in magnitude. The two main drawbacks to this approach are that it increases the data rate and that it reduces the available range swath width.
2. Reduce the azimuth processing bandwidth to lower the ratio of ambiguous to main lobe energy (see Figure 14.6). This approach selects the region of the azimuth spectrum containing the least amount of aliased signal. Here we pay the price of reduced bandwidth, which implies poorer azimuth resolution.

If the PRF is chosen properly so that the minima of the antenna pattern fold back to zero frequency, as in Figure 14.6, the relative energy in the ambiguities can be made acceptably low. But because the bandwidth of the processed azimuth signal is reduced, this second approach trades off ambiguities for resolution.

It is a useful engineering parameter to determine the amount of energy in the image that has been aliased in this way. We can calculate the total amount of energy in the azimuth ambiguities with respect to the energy in the base spectrum by comparing the total energy in the integrated

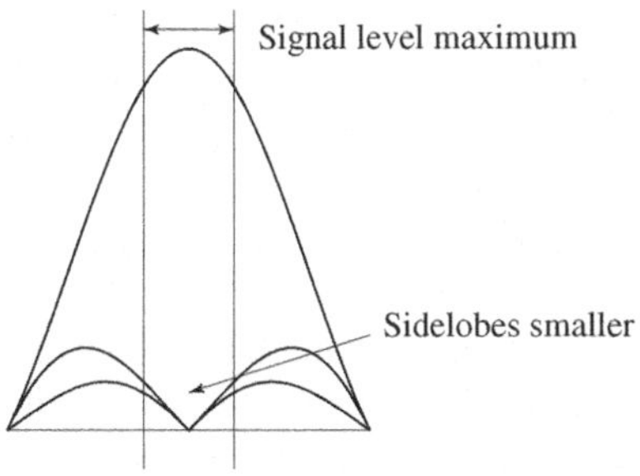

Figure 14.6 Effect of reducing processed bandwidth. Reduction in processed bandwidth increases the ratio of main beam to sidelobe antenna response.

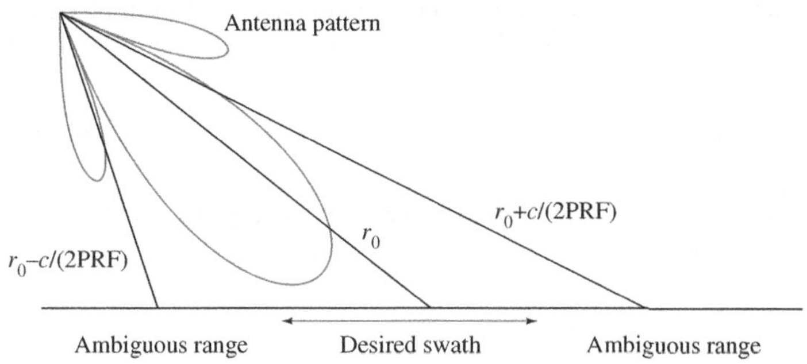

Figure 14.7 Antenna range pattern and PRF ambiguities. Aliasing in range occurs in the usual way for side-looking systems. The strength of aliased features to the those in the intended image swath is determined by the antenna pattern in the "range" dimension.

antenna pattern for each ambiguity to the integrated value of energy in the main antenna lobe over the processed bandwidth, as indicated in Figure 14.6. We can define an *azimuth ambiguity ratio* as

$$\frac{\sum_{i\neq 0}\int_{-PBW/2}^{PBW/2} G^2(f+i\cdot PRF)df}{\int_{-PBW/2}^{PBW/2} G^2(f)df} \tag{14.15}$$

where PBW is the processed bandwidth in azimuth and $G(f)$ is the antenna intensity pattern as a function of Doppler frequency (through Eq. 14.12). Usually an acceptable level of ambiguous energy is about −20 dB compared to the main lobe energy, so that 1% of the return is in the ambiguities.

14.3 Range Ambiguities

There is another problem with increasing the PRF to combat azimuth ambiguities, especially for spaceborne geometries. A second type of ambiguous return is associated with running the radar at too high a PRF, causing energy from the range sidelobes of the antenna pattern to fold into the imaged area. Consider the viewing geometry as shown in Figure 14.7, which is our usual picture of the swath geometry but with the range antenna pattern sketched in.

Recall that we achieve resolution in the across-track direction by discriminating time delays, but as before, there is an ambiguous range for any pulsed system corresponding to the time delay equal to the inverse of the PRF. In other words, targets at a distance greater than or less than the true range by an integral multiple of this ambiguous range will fall within the desired swath. As with the azimuth ambiguities, these will generally be received by the antenna sidelobes, so that they appear with reduced energy, but they may be objectionable nonetheless.

For range ambiguities, increasing the PRF decreases the distance of the ambiguous ranges to the desired set of swath distances, and usually increases the total ambiguity level.

So, we always have a trade-off between range and azimuth ambiguities. Raising the PRF lowers azimuth ambiguities, but increases range ambiguities. Design optimization requires us to quantify the range of ambiguity effects, similar to how we analyzed ambiguities in the azimuth sense, so that we can reasonably make the correct design choices. In the range case, because ambiguities are received from points at different incidence angles, we must not only consider the antenna pattern but also the changes in the scattering behavior as represented by σ^0.

We begin by integrating the signal from the imaged swath alone. The total power we might expect from the desired swath in range can be expressed as proportional to

$$P_{\text{range, desired}} = \int_{R_0=R_1}^{R_2} G^2(\theta)\sigma^0(\theta)dR_0 \tag{14.16}$$

where implicitly θ depends on R_0 (recall $\cos\theta = z/R_0$). Thus we can rewrite Eq. (14.16) as

$$\int_{R_0=R_1}^{R_2} G^2(R_0)\sigma^0(R_0)dR_0 \tag{14.17}$$

We know that the ambiguous range is $c/(2 \cdot PRF)$, hence we can express the total power of ambiguities in range as

$$\int_{R_0=R_1}^{R_2} \sum_{i\neq 0} G^2\left(R_0 + i\frac{c}{2\cdot PRF}\right)\sigma^0\left(R_0 + i\frac{c}{2\cdot PRF}\right)dR_0 \tag{14.18}$$

We can then define a *range ambiguity ratio* much the same as we did for the azimuth case:

$$\frac{\sum_{i\neq 0}\int_{R_1}^{R_2} G^2\left(R_0 + \frac{ic}{2\cdot PRF}\right)\sigma^0\left(R_0 + i\frac{c}{2\cdot PRF}\right)dR_0}{\int_{R_1}^{R_2} G^2(R_0)\sigma^0(R_0)dR_0} \tag{14.19}$$

Typically, design practice again sets this limit at −20 dB or less. Note that it is determined by the scattering law, antenna pattern and imaging geometry, as well as PRF.

14.4 Squinted Geometries and Autofocus

If the radar look direction is not at right angles to the flight track, the radar is said to operate in a squinted geometry, or that the radar is a squinted system. Sometimes this is intentional, such as when we wish to obtain information about the surface ahead of the radar. Other times, squinted geometries follow from uncontrollable or unanticipated changes in the antenna orientation. Squinting the antenna ahead or behind the usual direction perpendicular to the flight track, implies that the range of Doppler frequencies within the main antenna beam will no longer be centered about zero, so we must compensate for this in the data processor.

The degree of squint can be specified either by the angle of the antenna boresight off the right angle to flight track direction, or by the Doppler frequency at the center of the antenna beam. The latter is particularly useful for spaceborne geometries, because it can include other contributions to observed offsets in Doppler frequency. For example, the Earth's rotation under a satellite is a significant contributor to Doppler frequency shift for satellite sensors.

We can visualize the important quantities for analyzing squinted systems if we use the imaging geometry as shown in Figure 14.8, where the plane of the page contains the radar look direction and flight track, and we indicate the hyperbolic path followed by a point target as the radar flies by. The range at closest approach is R_0, and the radar itself is pointed somewhat ahead of the closest approach point. The location of the intersection between the radar look vector and the range history curve can be specified either as an angular offset angle θ_{sq} or by the Doppler frequency associated with that point. Here the Doppler frequency range of the illuminated points in the antenna beam is not centered on zero, but around some other Doppler centroid frequency f_{DC}. We still have our relation between along-track position and Doppler frequency

$$f_d = \frac{2v}{\lambda}\frac{x}{R} \tag{14.20}$$

where we can formally define a *squint angle* by $\sin\theta_{sq} = \frac{x}{R}$ that includes both of the previously-used look angle and yaw angle. The yaw angle is the projection of the composite squint angle into the plane of the surface (Figure 14.9). The squint angle θ_{sq} lies in the plane defined by the radar, the object to be imaged, and the point of closest approach for that object.

In a squinted geometry, we still illuminate a beamwidth determined by the wavelength and the length of the antenna $\sim \frac{\lambda}{D_{az}}$, but now it is centered on the Doppler centroid, or at angle θ_{sq} ahead

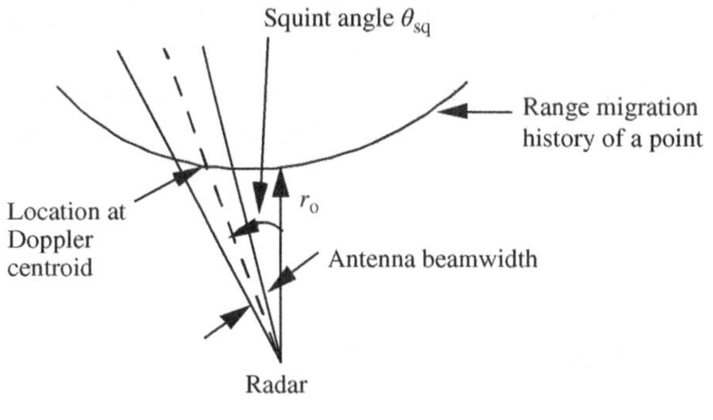

Figure 14.8 Squint angle geometry. View is in the +*z* direction, looking down on the image swath. Note that the echo will be received from an aperture that may or may not contain the point of closest approach, and so the spectrum may not contain energy from the zero Doppler point.

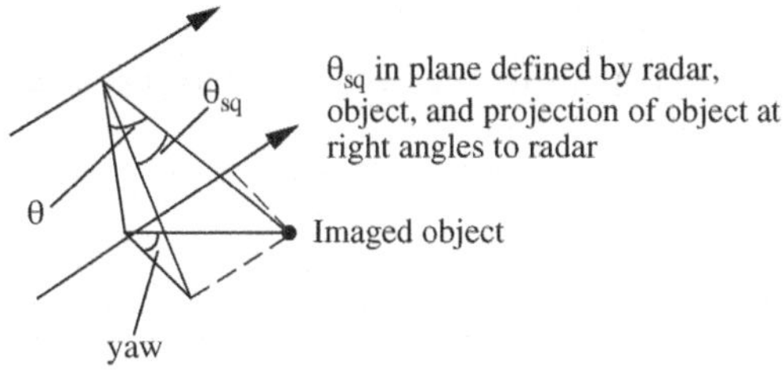

Figure 14.9 Yaw angle. Coordinate is defined as angle between two planes containing the *z*-axis. Reference plane is perpendicular to the flight path. Yaw plane contains the center of the antenna beam.

of the usual point of closest approach. Thus the range of Doppler frequencies from points in the main lobe illumination pattern is from

$$f_d \min \approx \frac{2v}{\lambda} \sin\left[\theta_{sq} - \frac{\lambda}{2D_{az}}\right] \tag{14.21}$$

to

$$f_d \max \approx \frac{2v}{\lambda} \sin\left(\theta_{sq} + \frac{\lambda}{2D_{az}}\right) \tag{14.22}$$

The range of frequencies thus reduces to the previously-discussed case for $\theta_{sq} = 0$.

Note that the azimuth chirp no longer need contain the zero frequency point, that is, its frequency spectrum may not contain the origin if the offset in frequency exceeds the illuminated bandwidth.

We can still plot the observed Doppler spectrum (Figure 14.10), assuming a very high sample rate in azimuth, and we will see the same amplitude pattern driven by the antenna response as before. The only exception is that the pattern is no longer centered on zero. We get the same pattern as before but centered at f_{DC}. The antenna still illuminates a band of frequencies, which now ranges from $f_{DC} - \frac{BW}{2}$ to $f_{DC} + \frac{BW}{2}$.

So a plot of the real part of the chirp signal might look like the waveform in Figure 14.11, which ranges from a high frequency to a lower frequency, without passing through zero frequency.

The complete azimuth chirp signal still depends on the imaging geometry and time:

$$s(t) = e^{-j\frac{2\pi v^2}{\lambda R_0}t^2} \tag{14.23}$$

However, since the selected portion of the azimuth history may not be centered around zero, the range of times for which this equation applies is now arbitrary. It is often helpful to define the time limits as fixed at $(-t_{az}/2, t_{az}/2)$, where t_{az} is the azimuth integration time, especially when we are allocating computer memory, for the calculation of the matched filter. We can thus expand

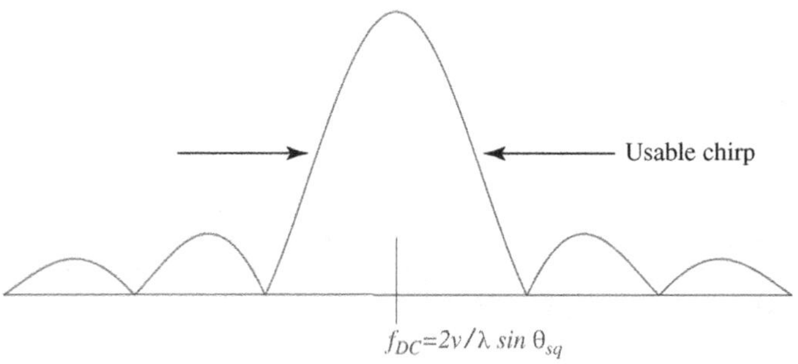

Figure 14.10 Usable chirp bandwidth. The extent of the usable Doppler related frequency/phase excursion is limited by the azimuthal width of the antenna beam.

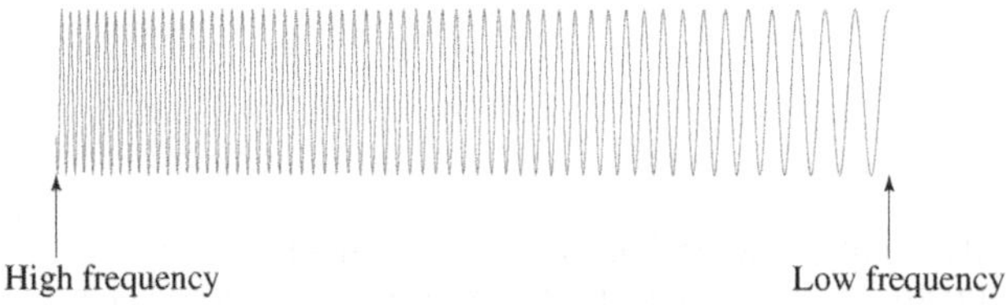

Figure 14.11 Example chirp for yawed antenna. An antenna yawed in the direction of platform motion will bias the observed Doppler chirp spectrum toward positive frequencies.

Eq. (14.23) about the time associated with the Doppler centroid t_{DC} by a change of variables $t = t_{DC} + \tau$, to obtain,

$$s(\tau) = e^{-j\frac{2\pi v^2}{\lambda r_0}(t_{DC}+\tau)^2}$$

$$= e^{-j\frac{2\pi v^2}{\lambda r_0}\left[t_{DC}^2+2\tau t_{DC}+\tau^2\right]}$$

Ignoring the phase factor which is independent of τ we can rewrite this as having two terms:

$$s(\tau) = e^{-j2\pi\left[\frac{2v^2 t_{DC}}{\lambda R_0}\tau+\frac{v^2}{\lambda R_0}\tau^2\right]} \tag{14.24}$$

The first term is an additional linear term due to squint geometry, while the second term is the same quadratic term as before. We can thus derive our matched filter as equal to the earlier filter but with an additional frequency shift to account for the squint. Thus other than accounting for the frequency shift, then, the processing proceeds as before and the same algorithms can be used.

This parameterization has, in fact, another advantage for processor implementation—it can yield more accurately focused images than the single term quadratic form with variable limits. Recall that the quadratic phase history was an approximation assuming that along-track offsets are small, that is, the squint angles are low. The two-term factorization described earlier can be viewed as an expansion about an origin centered at the Doppler centroid, so that the approximation now is dependent on the magnitude of the x displacement from the centroid, rather than from the point of closest approach. Thus a processor defined in terms of both the centroid and the quadratic phase term will better match the true phase history of each point.

14.5 Errors in the Parameters f_{dc} and f_{rate}

In the previous section we parameterized the azimuth chirp function in terms of only two parameters, f_{DC}, the Doppler centroid, and a quadratic phase term. It is often simpler to think of the quadratic term as a single *chirp rate* rather than as a combination of wavelength, range, and velocity. Hence, we may define the chirp rate f_{rate}, or Doppler rate, as

$$f_{rate} = \frac{-2v^2}{\lambda R} \tag{14.25}$$

From before, $f_{dc} = \dfrac{2v^2 t_{DC}}{\lambda R}$. Then we can rewrite a simplified representation of the reference azimuth signal

$$s(t) = e^{-j2\pi f_{dc} t} e^{-j\pi f_{rate} t^2} \tag{14.26}$$

We can then design the matched filters directly from our knowledge of f_{dc} and f_{rate}.

Hence, it is important to quantify the performance of the data processing algorithms if we use values for these parameters that are slightly in error, which could happen if we did not know the squint angle exactly, for instance, or if the geometry were uncertain. The latter may occur, for example, if an aircraft platform experiences significant turbulence, or if the orbit parameters of a satellite system are not known precisely.

Clearly, we must know the correct values for f_{dc} and f_{rate} if we are deriving the proper matched filter and obtain proper focusing of the image. We will see that in first order, errors in f_{dc} result in loss of energy or ghosting in the image, while errors in f_{rate} lead to lack of focus, that is, the image is blurry.

The following approach may help you visualize how errors in the Doppler centroid or rate affect the point response of the SAR system. We can refer to this approach as time–frequency plot analysis, but note that no formal nomenclature for the method exists in the literature. We have already shown that a linear relationship between Doppler frequency and time exists, at least to a reasonable approximation. If we plot Doppler frequency as a function of time and label the important points on the graph with quantities used in the data processing, we can determine the time–frequency history of a given point in the radar illumination pattern, as shown in Figure 14.12.

Our matched filtering operation in the processor requires matching the time–frequency plot of the point to an ideal radar echo. Consider convolving the signal in Figure 14.12 with a reference signal of equal slope. Thus, we simply "slide" the reference along the axis until the lines overlay one another. Because the output from the correlation receiver is the integral of the product of the signal as a function of their offset, the output from the matched filter corresponds to the area of overlap of the two curves, and in this case peaks at $t = t_0$, as shown in Figure 14.13.

Using these plots and the sliding approach just described, we can determine what happens if f_{dc} is not correct. Consider the previously described signal $s(t)$ acquired at a squint angle that is slightly in error from the expected angle. Then the waveform will appear to be shifted, and thus be on a carrier frequency preventing exact alignment in frequency of the signal and the reference. We can represent the carrier here with the f_{dc} term, and we process the shifted signal with a reference defined on a carrier of zero frequency. The situation is illustrated in Figure 14.14.

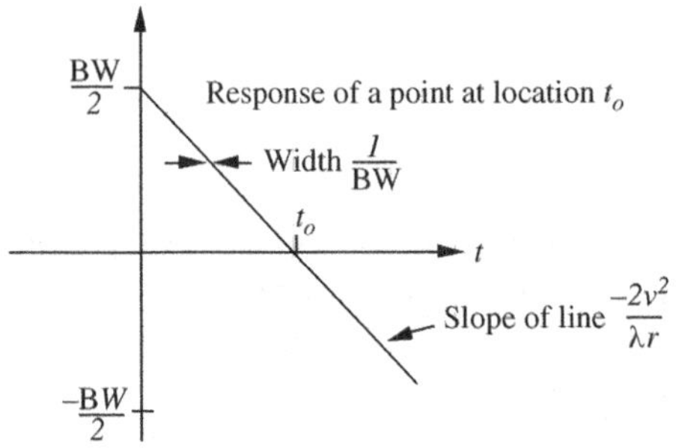

Figure 14.12 Doppler time–frequency history. There is a linear relationship between Doppler frequency and time. This is just the definition of a chirp, as discussed in the text. The slope of the line in the time–frequency plot is the chirp rate f_{rate}.

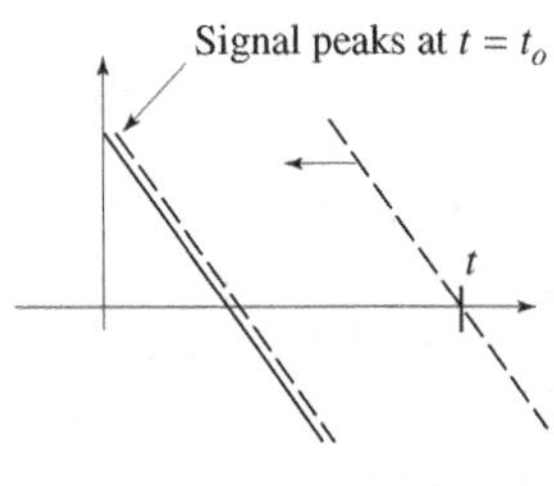

Figure 14.13 Sliding match to Doppler time–frequency history. Matching of the echo and reference signals can be accomplished by sliding the reference signal with respect to time. This is equivalent to adjusting the effective location of the reference target point.

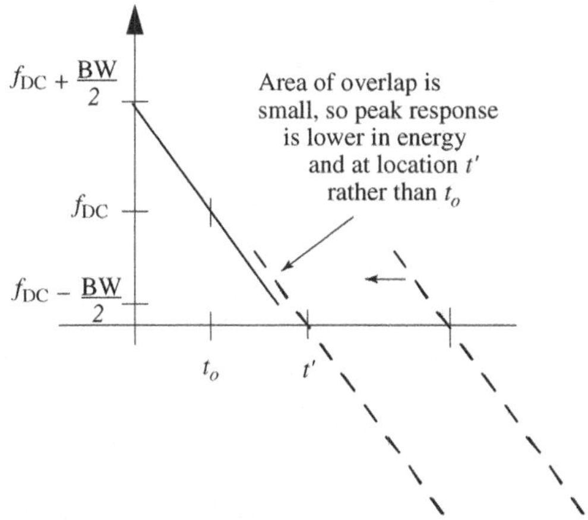

Figure 14.14 Effect of error in f_{dc}. Effect of incorrect value for f_{dc} can be easily visualized by comparison of time-frequency histories for echo signal with erroneous reference signal.

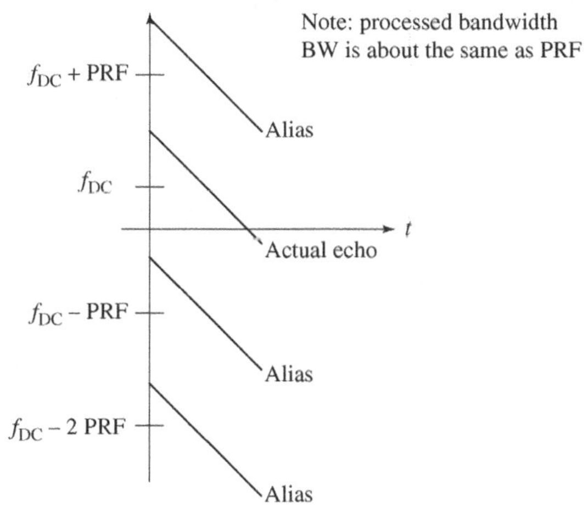

Figure 14.15 Spectrum with aliases. Multiple copies of chirp line exist at different Doppler offsets due to sampling. Note that the bandwidth of the chirp is approximately the same as the PRF.

The area of overlap for the two lines is small due to the shift, so the peak response is lower in energy. Note that it also appears at the wrong location, peaking at t' rather than t_0. The return is thus smaller and shifted in position because the wrong processing parameters were chosen.

But, remember that we have a sampled system. The pulses of the radar can then lead to aliasing of the azimuth return, further degrading the impulse response of the processor. If the prf is chosen to approximate the bandwidth of the illuminated portion of the surface, we cannot tolerate very much of a shift in Doppler centroid. The pulsed signal will thus have a time–frequency plot that is replicated in frequency, as is the case with any sampled signal spectrum. We show the replicated spectra in Figure 14.15, where we have assumed that the processed bandwidth is comparable in magnitude to the prf.

This is the signal that must be convolved with our reference, not the simpler spectrum of Figure 14.14. Now when we convolve with the sampled signal, the result will be more like that illustrated in Figure 14.16. As you can see from the time–frequency plot, the result of an f_{dc} error

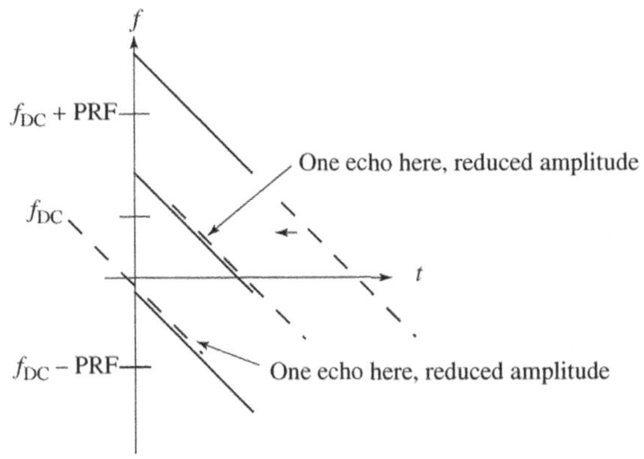

Figure 14.16 Convolution of aliased spectrum with reference. Presence of aliased components in the echo spectrum results in multiple filter outputs. Location and intensity can be visualized by matching time–frequency histories.

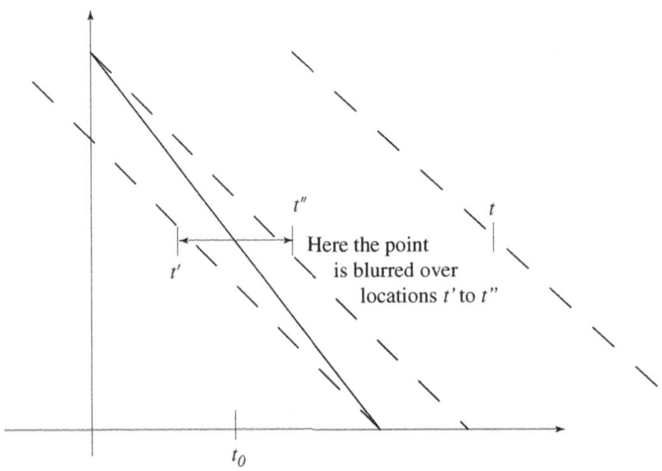

Figure 14.17 f_{rate} error. Effect of Doppler frequency rate error can be visualized by comparing time–frequency histories of actual and erroneous reference signals.

is that two images of each point appear in the output, both in the wrong place, and each of lower amplitude. In contrast, if there is no f_{dc} error, there is only one image, and it is located correctly.

We can next discuss the result of an f_{rate} error using the same method. Recall that this means the quadratic phase term is in error. Here the time–frequency history is in the right place, but the slope is wrong. The impulse response will now be blurred a bit in time, as the two lines overlap for quite a wide range of times, ranging from t' to t'', as shown in Figure 14.17. Thus an f_{rate} error is often called a focusing error, because the image appears out of focus.

14.6 Autofocus Algorithms, or How to Determine the Correct f_{dc} and f_{rate}

Since f_{dc} and f_{rate} depend on the orbital (or flight -track) knowledge and the exact orientation of the radar platform, if we monitor these imperfectly we will induce processing errors. We can usually recover the proper parameters by examining properties of the final images. This bootstrapping procedure is called autofocus, as it can automatically determine the correct processing parameters from the data themselves.

First, consider errors in the centroid f_{dc}. These might result from errors in the squint angle knowledge, for example. There are many algorithms to obtain estimates of f_{dc} from the data. One common algorithm is simply to plot the spectrum and calculate its centroid, as illustrated in Figure 14.18.

We can evaluate the centroid using the weighted integral

$$\text{centroid} = \frac{\int_0^\infty f_{\mathrm{dc}} \cdot S(f_{\mathrm{dc}}) df_{\mathrm{dc}}}{\int_0^\infty S(f_{\mathrm{dc}}) df_{\mathrm{dc}}} \tag{14.27}$$

But since this system was sampled, one must first "unwrap" the spectrum with respect to PRF and choose the correct "ambiguity." Usually we know the squint well enough to do this unwrapping step.

Another algorithm follows from simply comparing the phases of successive pulses, and estimating the average phase lift per pulse. The phase shift for each known time increment is then easily translated to a Doppler frequency, and the frequency component from the center of the antenna beam is assumed dominant. Consider a sequence of signal samples s_i along an azimuth line. Since the data are on a "carrier" of frequency f_{dc}, forming the sum and arctangent

$$\arctan\left[\sum_{\mathrm{i}} s_{\mathrm{i}} s_{\mathrm{i}-1}^*\right] \tag{14.28}$$

yields the average phase change per line. A shift of 2π phase per line corresponds to a Doppler carrier at the same frequency as the sample rate, or PRF. This is also a measurement modulo the prf, so once again the ambiguity must be resolved.

It is also possible to deduce f_{rate} errors. Here we turn to the time–frequency plots once again. Suppose rather than processing with the complete azimuth reference function, instead we break the reference into two halves and process a separate image with each. The resulting images are described in the plots in Figure 14.19, where first we set the slope of the reference correctly. Two images will be formed here, and, if compared, they will overlay each other exactly.

But suppose instead that we selected the chirp slope incorrectly. We already know that this will cause some defocusing, but now in addition we will observe a shift in position for each image. One image will be shifted ahead in time, while the other will appear late (Figure 14.20).

Notice that from the offset we can calculate the correct value f_{rate}. If we process the two subimages this way, and measure the shift between them, we can automatically compensate for errors in f_{rate}, and perfectly focus the final image.

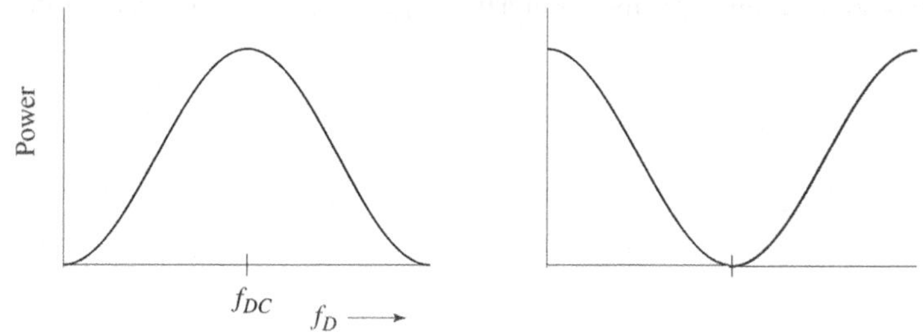

Figure 14.18 Echo spectra for shifted and unshifted centroids. Left, a spectrum centered on a Doppler carrier f_{DC}. Right, spectrum shifted to be centered on zero frequency. Zero in this plot is at the beginning of the frequency axis, which is the convention for the FFT algorithm.

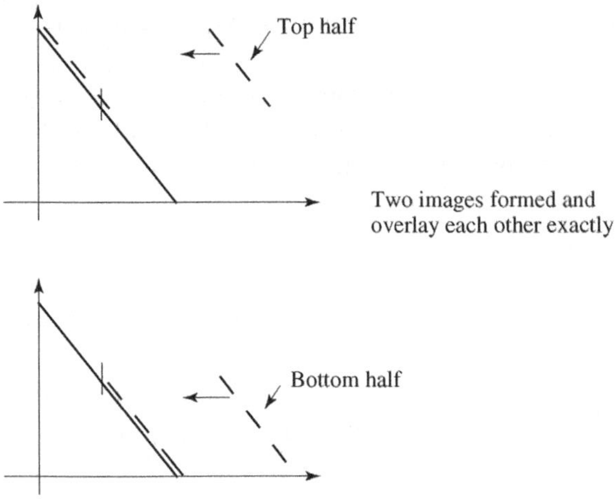

Figure 14.19 Split reference test. Processing echoes twice using the two halves of the reference signal, and then the comparing the resulting images, permits an evaluation of the f_{rate} error. The two images will appear displaced in azimuth with respect to one another if the f_{rate} is incorrect. See Figure 14.20.

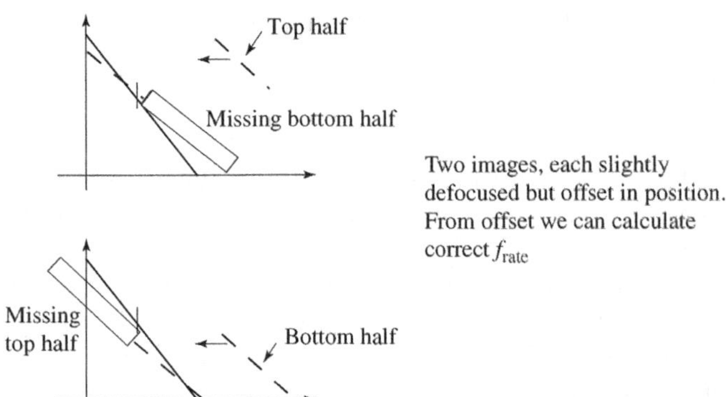

Figure 14.20 Detail of split reference test. Two slightly defocused and shifted images result from an incorrect chirp slope.

Most practical radar processors include algorithms to refine initial estimates of f_{dc} and f_{rate} along the lines we have described here. This lessens the need for expensive monitoring systems to record all the details of platform motion, attitude, and position. For your information, f_{dc} estimation routines are often called “clutter lock” algorithms, because they “lock” onto signals from the ground, or clutter. As we stated previously, similar algorithms to determine f_{rate} are referred to as autofocus, for obvious reasons.

15

Wave–Surface Interactions

Imaging radars produce high-resolution maps of the radar reflectivity of surfaces. Although these maps resemble pictures of the ground, they are not pictures in the conventional, optical sense at wavelengths $\lambda \sim 0.1$–$10\ \mu$m. Rather, the patterns of brightness and dark in these maps are expressions of the radar scattering properties of the ground cover for $\lambda \sim 0.001$–10 m. Additionally, the illumination in conventional images is almost always either from a range of directions or diffuse emission from the surface itself, whereas since most radars carry the source illuminator, the illumination is antiparallel to the observed scattering. We would therefore like to consider just what properties of the surface are reflected in the radar echoes. In other words, what do we see when we examine radar images?

15.1 Specific RCS: σ^0

The observed brightness in a radar image is typically displayed as proportional to some power of the backscattered energy, and is related to the physical composition of the surface through the strength of the backscatter or the polarization. It is also dependent on the illumination geometry, for usually a surface is brighter when viewed at an angle which is closer to the normal to the mean surface than to grazing incidence, as illustrated in Figure 15.1.

In addition to questions of viewing geometry, it is also necessary to consider how to characterize the scattering properties of an extended surface. For surface mapping, the interesting part includes the variations of RCS with location on the surface. So we characterize not only the scattering from the surface, but how this changes from one point on the surface to another. In these circumstances, our calculation of the RCS of a very large plane is not of much use as (i) it applies only to norm incidence and (ii) does not address the question of variations. To get at this problem, the key concept is that the radar must be able to isolate or distinguish between signals from different parts of the surface. There are several practical ways to do this—as we shall see in later chapters. For now, it is sufficient to consider the side looking radar method introduced in Section 2.4, in which a narrow beam along track in combination with pulsed range resolution is used to isolate signals from various points within an extended swath along the ground. By use of such methods, we can attribute the echo power received at various ranges and points along the SAR flight path to specific areas on the surface in the construction of a radar map or image of the surface. As different radars have different resolution areas on the surface, and even a given radar usually has different surface resolutions at different look angles, we need a uniform measure of the surface scattering that is independent of variations in radar performance. For this purpose, we use the *specific, or normalized, radar cross section*, σ^0 defined as RCS per unit area of the mean scattering surface. This normalized unit will

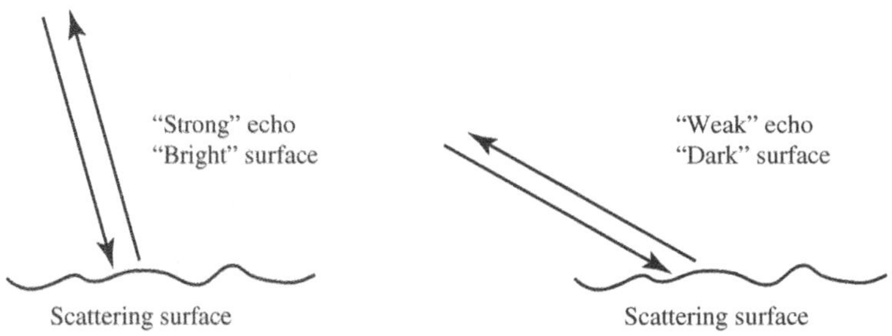

Figure 15.1 Dependence of surface RCS on viewing angle. Same surface may appear bright when viewed at near normal incidence (left) and dark when viewed closer to grazing incidence (right).

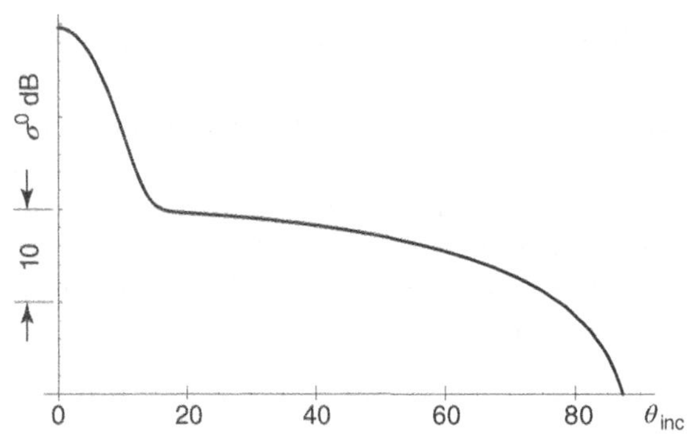

Figure 15.2 Typical backscatter law. Highest RCS occurs at zero incidence and decreases with increasing incidence angle. Change in shape of curve at $\theta_{inc} \approx 15°$ is result of change in controlling mechanism of scattering.

be the same for a given surface viewed with the same geometry independently of the areal resolution of the radar. With this definition, the actual RCS for a given area is $\sigma = \sigma^0 dA$, where dA is the resolved area. In calculations of echo power, we need only replace σ in Eq. (3.1) with $\sigma^0 dA$. Thus, the received echo power is proportional to σ^0 and embodies both the properties of the scattering surface itself and properties of the imaging geometry. Decoupling of the effects of surface material from those of surface shape and viewing geometry can be difficult and lies at the core of radar image analysis.

The angular variation of backscatter, i.e., $\sigma^0(\theta_{inc})$, where θ_{inc} is the angle of incidence, usually follows an angular scattering law with features similar to those shown in Figure 15.2. The strong enhancement of σ^0 near normal incidence, a transition to a broad slowly varying falloff as θ_{inc} increases, and the marked decrease in scattering strength as $\theta_{inc} \rightarrow 90°$ are typical of backscatter from natural terrains.

This dependence of the radar backscatter on angle may be considered as the composite effect of several different scattering processes, some dominant for near-nadir scattering geometries and others dominant at larger incidence angles. The balance between these various processes gives the characteristics of the surface.

Bistatic surface scattering in general has a very different behavior with the strongest echoes occurring in the near forward direction.

In remote sensing applications, radar echo intensity depends primarily on the surface shape at the wavelength and longer scale, and only secondarily on the surface electrical properties. That is, surface roughness and topographic shape play a much more important role than variations in dielectric constants, although the latter is a result of a rather limited range of material properties found in dry geological materials. Scattering from wet vs. dry surfaces, liquid water, and water ice are principal exceptions where the extreme differences between wet and dry, and liquid and solid states are important factors controlling the RCS. There are, of course, other special cases when ϵ or μ changes dominate, but these occur very infrequently.[1]

15.2 Near-normal Incidence Scattering: Facet Models

One of the simpler scattering models is quite important for small angles of incidence. That model is the *facet* surface model which explains the primary properties of near vertical incidence backscatter. In this model, the surface is decomposed into small regions, or "facets," which lie along the surface. If a facet faces the radar directly, its echo is relatively strong, while if it lies at an oblique angle to the radar look direction its echo will be weaker. When excited by an incident wave, each facet may be thought of as a small antenna illuminating the radar, where the radar cross-section is proportional to the radiation pattern generated by the facet, as suggested in Figure 15.3.

The return from a facet is greatest when it is illuminated normally, and the total power in the echo can be expressed as the sum of the power from each of the individual facets. Since the power is also related to the Fresnel reflection coefficient of each facet, the dielectric constant plays a small role, but the major effect will be from the slope distribution of the surface. Because scattering from the facets is a composite of mirror-like reflections, this type of scattering is referred to as *quasi-specular*, in order to distinguish it from "specular" reflection of a large smooth plane. Quasi-specular scattering is common in near-vertical incidence observations of lightly vegetated regions of the Earth, as well as the moon and terrestrial planets.[2]

For most surfaces, the probability that a facet has a slope greater than a few degrees is quite small, so facet models would predict little returned power for $\theta_i \geq 15$–$20°$. For some surfaces, such as water, incidence angles θ_i greater than 1–2° are rare. Hence a smooth surface, such as most water surfaces, will exhibit high backscatter within only a narrow region around zero incidence, while a

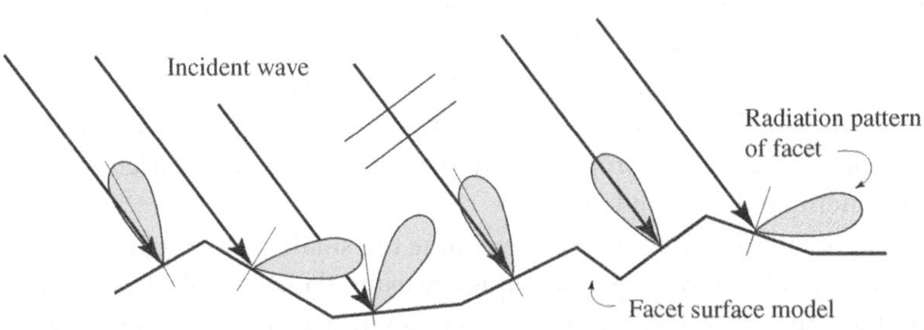

Figure 15.3 Facet scattering model. Surface is approximated by system of flat facets that independently backscatter energy to the radar. Strength of backscattered energy is greatest when facet is normal to radar line-of-sight, and considered negligible when angle between surface normal and line-of-sight exceeds diffraction lobe width of facet.

---

1 An example is the effect of tellurium, which apparently accounts for observations of very strong radar echoes from regions near the tops of mountains on Venus. As a semiconductor, tellurium is highly conductive under Venusian conditions, it is the only known material with a phase transition to the solid under the conditions where these radar bright areas are observed.

2 Mercury, Venus, Earth, and Mars. Scattering from the icy polar regions of Earth and Mars is distinctly not quasi-specular, however.

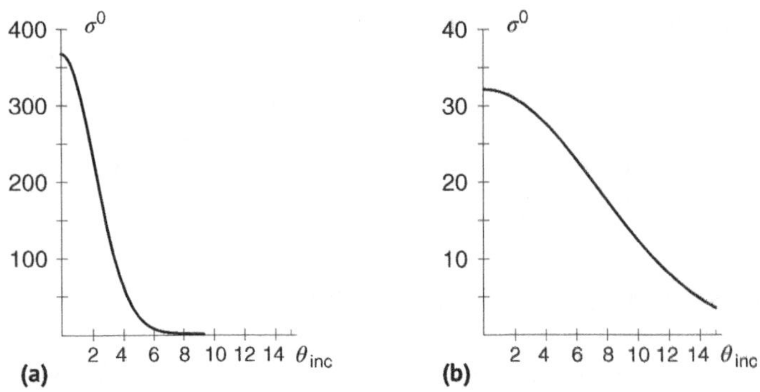

Figure 15.4 Scattering from smooth and rough surfaces. Facet scattering model predicts that for a smooth surface most of the backscattered energy is observed primarily at small incidence angles (a), while rougher surfaces produce significant returns at higher angles (b). Curves are drawn for Gaussian distribution of surface slopes with 3° and 10° in panels (a) and (b), respectively. Note 10:1 difference in vertical scales.

rougher surface with much greater average slopes, such as dune fields, will evidence backscattered energy over a much wider range of θ_{inc}. These phenomena allow us to distinguish between smooth and rough surfaces by the width of the quasi-specular part of the σ^0 curve. Two curves of σ^0 shown in Figure 15.4 illustrate the variation in form that the facet model predicts for varying roughness.

15.3 Moderate Incidence Scattering: Perturbation Models

For larger incidence angles, the number of facets facing the radar with near normal incidence is very small, with the result that other effects scattering processes dominate the radar return. Several models are commonly invoked to model σ^0 for an incidence angle θ_i greater than about 20°, the most common of these models being based on *Bragg scattering*, where certain spectral components of the surface height $\xi(x, y)$ couple strongly to the incident wave.

For the Bragg scattering model, we consider a surface broken up into a large-scale, faceted component as before plus a residual small-scale roughness component (see Figure 15.5). This decomposition keeps the amplitudes of the roughness component small, so that perturbation methods may be used to evaluate scattering behavior. This is not always required, though, if specialized numerical methods are used to calculate the scattering properties. We find the small-size restriction helpful in our brief analytical treatment, and adopt it here.

We consider the faceted component in a moment. First examine the small-scale roughness component and break it up into surfaces of differing spatial frequency using Fourier analysis, as shown in Figure 15.6, where each of the Fourier components is relatively small-scale in amplitude. Usually the analysis requires that the rms height h of the roughness component be less than $\frac{\lambda}{8}$ in amplitude, which is why we removed the large-scale facets before the decomposition. We will put the effects of the large-scale surface shape back in after considering the small-scale surface scattering.

In Bragg scattering models, only one Fourier component of the surface shape, the component for which the distance Λ as shown in Figure 15.7 is equal to $\frac{\lambda}{2}\sin\theta_i$, will result in significant backscatter. As suggested by the figure, the return echoes from each cycle of this component of the surface height distribution add up in phase. In other words, we assume that the incident wave, when projected onto the surface, is resonant with this single component of the surface spectrum. Therefore, surface variations at this spatial frequency are dominant in affecting the amount of reflected

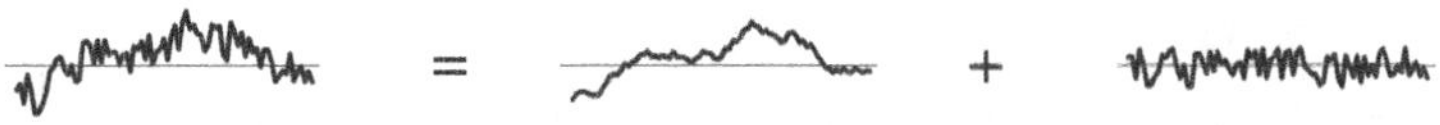

Figure 15.5 Two-scale model with facets. Rough surface is decomposed into two components: (i) a large-scale facet model that describes the overall shape of the surface plus; (ii) a small-scale roughness component with low values of residual amplitude.

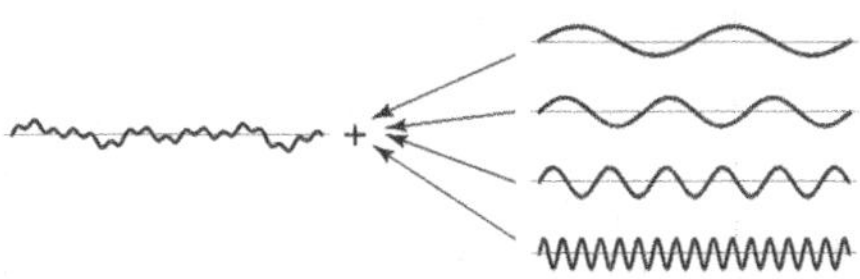

Figure 15.6 Fourier decomposition of surface roughness. Small-scale roughness surface may be decomposed into many components of differing spatial frequency. Relative spatial frequencies in figure are 2, 3, 11/2, and 33/2 cycles/unit length; corresponding amplitudes of composite surface are 1,1,1, and 0.75, respectively, where the component with the lowest spatial frequency (2 c/l) substitutes for facets in the previous model.

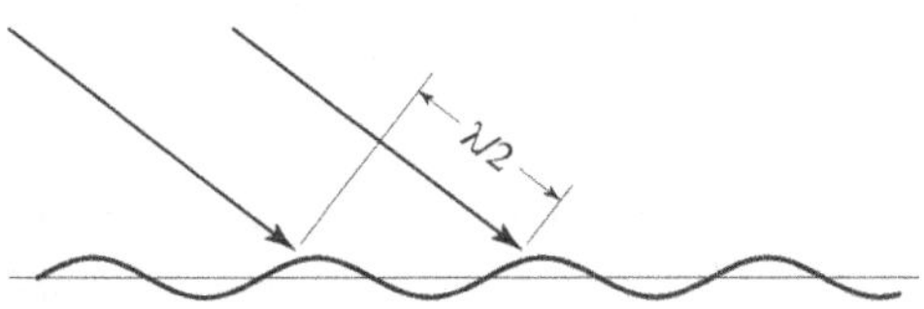

Figure 15.7 Bragg scatter model. This model is based on strong coherent reinforcement of backscatter by the spatial component of the surface roughness for which the projected incident wave resonates.

energy. All other terms of the surface height distribution will not resonate, and provide very little return signal.

Scattering properties of the Bragg model can be evaluated formally by the use of perturbation methods, leading to an expression of the normalized cross-section as a function of surface electrical and physical parameters as σ^0 as,

$$\sigma^0_{\text{pol}} = 8k^4h^2\cos^4\theta|\alpha_{\text{pol}}|W(2k\sin\theta) \tag{15.1}$$

where α_{pol} depends on the wave polarization and $W(\cdot)$ is the power spectrum of the surface. The $\cos^4\theta$ term expresses the incidence angle dependence of the cross-section for fixed $W(\cdot)$. It is clear from Eq. (15.1) that only the component of the surface spectrum meeting the Bragg condition affects the reflectivity.

15.4 Discrete Scatterer Models

Another model valid over a wide range of incidence angles, but most useful at larger angles, simply considers each resolved area to consist of a finite set of discrete scatterers, which are considered as point scatterers (Figure 15.8). Echoes received from each scattering point on the surface within the resolved area add coherently, but the randomness of the locations introduces a statistically random phase to the composite signal from the resolved area, as we saw before when we evaluated the cross section of complex objects and aggregates of multiple scatterers. From Eq. (5.55), taking each scattering point to have the same scattering strength σ_{sp}, and letting N represent the number of scattering points per unit area on the ground, we can write down directly that,

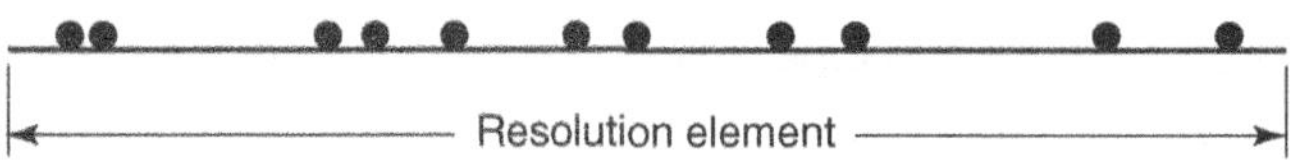

Figure 15.8 Discrete surface scattering model. Model represents surface as a collection of independently scattering elements, with the total echo simply the coherent sum of all of the individual echoes.

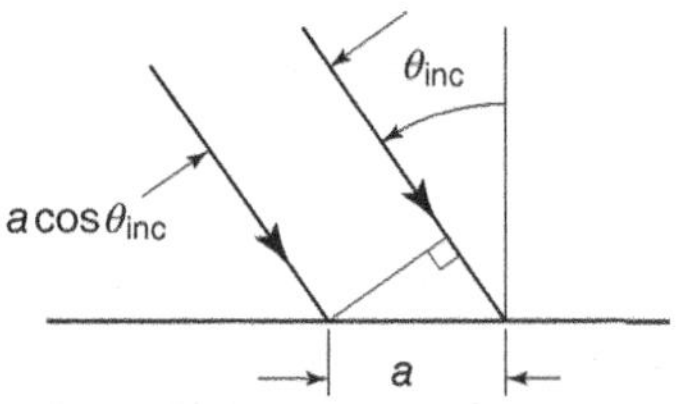

Figure 15.9 Wavefront projection on mean scattering surface. Two rays with angle of incidence θ_{inc} separated by a distance a along the surface surface are separated by only $a \cos \theta_{inc}$ when measure perpendicular to the wavefront. Thus, the power per unit area of surface decreases decreases with increasing θ_{inc} as a decreasing total power, $\mathcal{S}_{inc} a \cos \theta_{inc}$, is distributed over a fixed surface area.

$$\sigma^0(\theta) = N \cdot \sigma_{sp} \cdot \cos \theta_{inc} \tag{15.2}$$

where the obliquity factor, $\cos \theta_{inc}$ is needed to account for the reduction in incident power per unit area of the illuminating wavefront, Figure 15.9.[3] Eq. (15.2) holds for isotropic scattering elements. If, on the other hand, the reradiation patterns of the scatterers are not isotropic, the proper radiation pattern for each scatterer should be substituted. For horizontal dipole scatterers lying in the plane of scattering, the radiation pattern is $\cos^2\theta_{inc}$, in which instance the composite scattering function becomes,

$$\sigma^0(\theta) = N \cdot \sigma_{dipole} \cdot \cos^3\theta_{inc} \tag{15.3}$$

Both of these models, Eqs. (15.2) and (15.3), fall off with an incidence angle even more slowly than the Bragg model, and are used for very rough surfaces where the small-height restrictions of the Bragg model are exceeded.

15.5 Slope Modulation: Brightness Effects from Topography

We now consider the effects on the backscatter function $\sigma^0(\theta_{inc})$ of large scale variations of the surface, such as those due to topographic variations on the surface. We begin by considering a typical composite backscatter model, such as that shown in Figure 15.2 earlier. Suppose that we are imaging a topographic hill, which we show schematically in Figure 15.10a. We presume that the surface shown in Figure 15.10 represents the large scale average of a scattering surface with a known scattering law. In order to evaluate the brightness of different regions of this surface, we first determine the local incidence angle at each point on the large scale surface. These angles are identified in Figure 15.10b, where the smallest incidence angle is associated with the front side of the hill, the largest incidence angle with the back side of the hill, and intermediate angles correspond to

3 There is a subtle point here. The obliquity correction strictly applies only when the scattering surface is treated as a continuum. Scattering from a very loose collection of discrete objects on the surface would not be expected to experience the wave projection effect until θ_{inc} is greater than some threshold value. To understand this, consider an instance with only a few scattering points in a particular surface versus a continuously scattering surface.

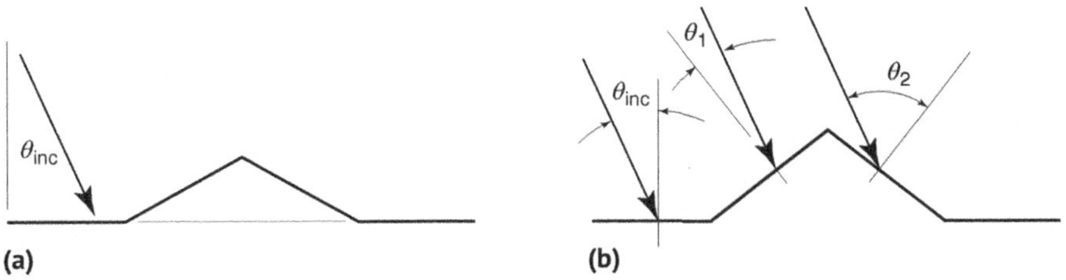

Figure 15.10 Surface slope modulation of RCS. Topographic hill has smallest angle of incidence on leading edge, which appears as the brightest area in the image. The back side of the hill is the darkest region, with flat slopes intermediate in brightness.

the horizontal surfaces. Recall that the backscatter function typically is greatest for small angles, and decreases monotonically, Figure 15.2. Consequently, the topographic hill feature appears the brightest on the side toward the radar and darkest on the side away from the radar. This slope-dependent shading effect makes even relatively subtle changes in topography quite pronounced in a radar intensity image.

15.6 Surface Roughness Variations

Now examine the effect on the scattering of radar signals arising from changes in the roughness of the scattering surface. Increasing the surface roughness has two effects on the radar echoes that can be inferred from the discussions on facet models and on Bragg scattering in the previous sections. First, increasing roughness spreads the reflected energy from facets over a larger range of scattering angles because an increase in roughness is usually accompanied by an increase in the width of the surface slope distribution. Second, Bragg scattering intensity increases as it is directly proportional to the square of the rms height of the surface, Eq. (15.1), thereby increasing the contribution to the RCS from small-scale surface roughness.

These two physical effects tend to result in backscattered energy that is more concentrated in a single narrower scattering lobe for the smoother surfaces, as suggested by Figure 15.11. The image for a rougher surface will be darker for small incidence angles and brighter for large incidence angles than for a smoother surface. Sample comparative plots of the backscatter intensity function are shown in Figure 15.12. Note the cross over of the $\sigma^0(\theta_{inc})$ curves a $\theta_{inc} \simeq 6°$. Rougher surfaces appear brighter for θ_{inc} larger than the cross over angle, while smoother surfaces are brighter for θ_{inc} less than the cross over angle.

Observe that because topographic shading depends on the slope of the cross section as a function of incidence angle $\dfrac{\partial \sigma^0(\theta)}{\partial \theta}$, the shading effect is strongest in smoother areas viewed at near-normal incidence than in rougher areas viewed at moderate and higher angles of incidence.

Scattering from vegetation can be thought of either as scattering from an extremely rough surface, or as resulting from the previously-described discrete element scattering (see Figure 15.13). Either of these models predict only a slight dependence on the normalized cross-section σ^0 on angle of incidence. Thus, scattering from a distributed canopy that is very dense. Backscattering from the Amazon rain forest, for example, has little dependence on θ_{inc}. In fact, the near independence of σ^0 on θ_{inc} for scattering from dense, uniform vegetation makes such areas useful for antenna pattern measurement and calibration (see Figure 15.14 for sample σ^0 plots for dense vegetation).

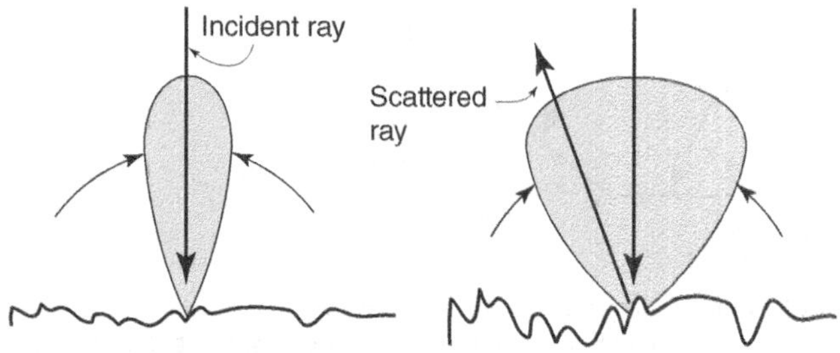

Figure 15.11 Effect on RCS of varying surface roughness. Smooth surfaces result in backscattered energy concentrated in a relatively narrow scattering lobe, while rougher surfaces spread the same incident energy over a wider range of angles. Strength of scattering lobes here are the same for clarity. Generally, for the same surface materials, maximum strength of scattering in wider lobes will be less than that of narrow lobes.

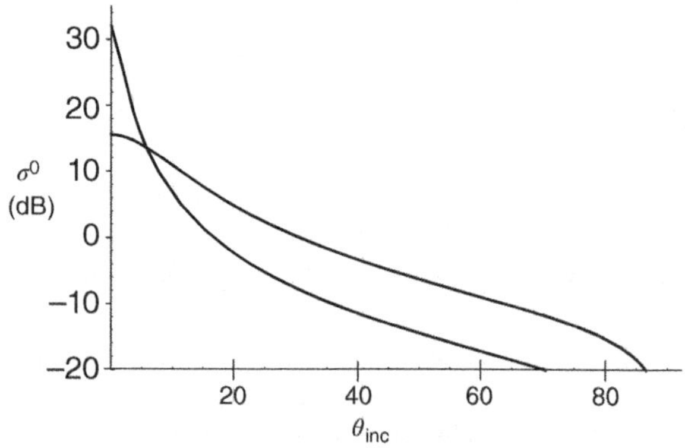

Figure 15.12 Variation of backscatter function with surface roughness. Backscatter function of smooth surface is more peaked near zero and lower at higher incidence angles than the backscatter function of a rougher surface. Figure depicts surfaces with nominal RMS slopes of 1/6 and 1/40 for the relatively flat and steep curves, respectively. Computation is based on the Hagfors' Law.

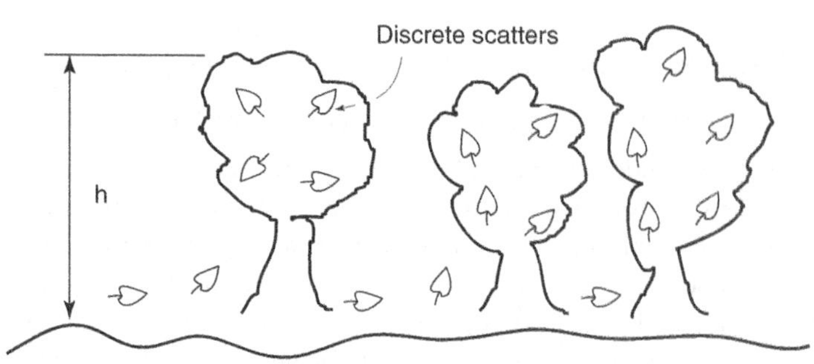

Figure 15.13 Volume scattering by vegetation. Vegetation layer can be viewed as a collection of discrete scatterers in volume of thickness h. Such layers exhibit little angular dependence.

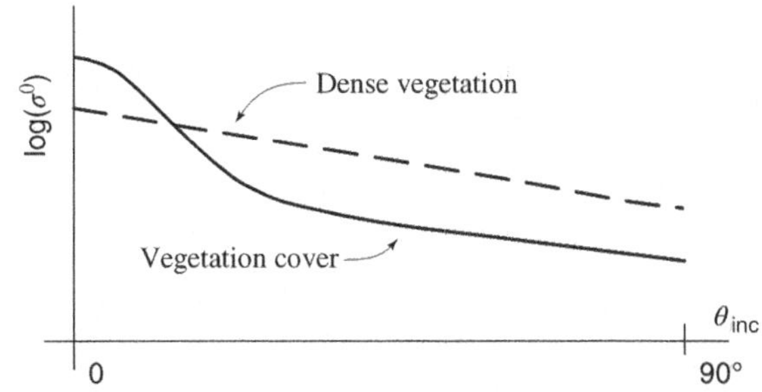

Figure 15.14 Use of vegetation for radar calibration. Regions of thick vegetation are often used for calibration of radar systems as their cross section is less dependent on imaging geometry than other regions. Hence small errors in such factors as incidence angle or antenna beamwidth are less than they would be if RCS variations were more rapid, as is typical of most scattering surfaces.

15.7 A Distribution of Discrete Scatterers

It is said that radar scattering from a surface or volume is dominated by the echoes from objects similar in size to the radar wavelength. In practice, this rule of thumb is often correct; it follows as a consequence of the interplay between the variation in scattering efficiency of objects with size and their natural size distribution. We can understand why this is so if we first examine the relative cross section of particles of various sizes.

We have seen that the G.O. RCS of large spherical objects equals the object's projected area, Section 4.2.1.3, p. 54. Figure 15.15 shows the variation in RCS of conducting spheres normalized by the projected area for a wide range of particle size, for spheres with radius of about $\lambda/30$ and larger. This ratio is referred to as the *scattering efficiency*. For scatterers of radius significantly less than $a/\frac{\lambda}{2\pi}$, i.e., less than about 1 wavelength in circumference, the scattering efficiency is proportional to a^4, while it is essentially unity for much larger objects.

We call the region of sizes where scattering efficiency decreases quickly with decreasing size the Rayleigh region, because Rayleigh (small-particle) scattering dominates here.[4] The region for which the scattering efficiency is essentially unity is the geometrical optics region, as the geometrical optics result holds. In between is the resonance region, where behavior is more complicated and forms a transition between the two other regions.

For most classes of natural objects, the number and density of scatterers of various sizes follows a power law distribution,

$$N(a) \propto a^{-n} \tag{15.4}$$

where a is the characteristic size and the exponent n is the power law index. When presented on log-log coordinates n is the slope of the resulting straight line, as shown in Figure 15.16. The value of the index has a large possible range, but for most problems of interest here, $2 < n < 4$.

When we have a collection of independent scatterers of different sizes, we estimate the total cross section by first evaluating the individual cross section of each particle and then adding up the results. The curve in Figure 15.15 tells us how to weight each of the geometrical cross sections for the various scatterers when we sum the total cross section.

4 Rayleigh scattering is an important regime for many disciplines. Scattering cross section for *any* particle with a characteristic dimension $\ll \lambda$ is proportional to a^6/λ^4. That is, the scattering of small objects varies as V^2/λ^4, where V is the volume.

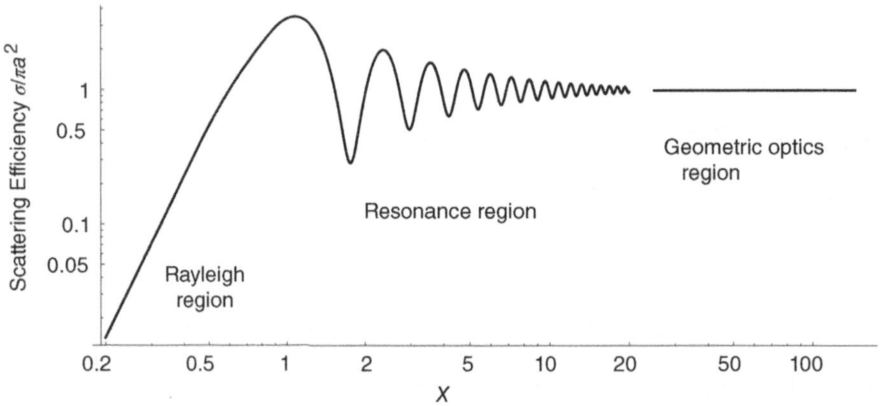

Figure 15.15 Scattering efficiency of a sphere. RCS normalized by projected area ($\sigma/4\pi^2$). Size of sphere is expressed in terms of the size parameter, $X = 2\pi a/\lambda$, where a is the radius, λ the wavelength. Three distinct scattering mechanisms are important in this plot: (i) Rayleigh scattering for small particles of any shape, $x \leq 1$; (ii) resonance scattering for sizes comparable to a wavelength, $1 < x \leq 10$; and (iii) geometric optics scattering for large objects, $x < 10$. Boundaries of resonance and geometric optics regions are indistinct and depend on details of object material and shape. Scattering behavior illustrated by a sphere is representative of scattering from many objects.

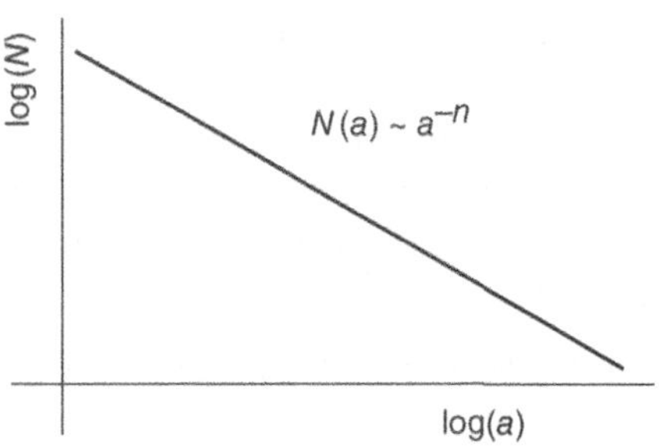

Figure 15.16 Power-law distribution function. Natural surfaces and volumes usually possess a distribution of size scales that is well-represented by a power law.

As an illustration, consider the distribution of scatterers forming a tree. The tree is composed of many objects of differing size, from small needles and leaves up to the large trunk. On most trees there are many more leaves than trunks and branches. We can create a typical scattering model "tree" by tabulating the size and number density of each component part. Table 15.1 gives results for an r^{-3} tree, where r is the characteristic dimension of the part, as that is the approximate power law expressing the change in number density for a given change in scatterer size on a reasonable tree. Other "trees" or other ensembles may be modeled on the basis of different power laws.

In order to calculate the radar cross section, we must first determine the physical and radar cross sections of each tree element. We first assume that we are studying the r^{-3} tree with a radar operating at a wavelength of 12.5 cm, with the results given in Table 15.2, which gives results for "total area" and "RCS" corresponding to the part types, sizes, and numbers in Table 15.1. In calculating the RCS we have assumed that the RCS of each tree part is equal to its area if the size parameter for that part is greater than unity, i.e., $X > 1$, and that the RCS follows the Rayleigh law for $x \leq 1$. Study of Figure 15.15 shows that this approximation is accurate except in the limited range of size parameter $0.5 < X < 5$, and that the maximum errors are not important for purposes of illustration.

Table 15.1 Equivalent size and number of scattering objects for an r^{-3} tree. Table gives numbers of various parts for power law tree with index $n = 3$. As most actual trees have only a single trunk, the table begins with this value. Sizes of other parts are arbitrarily set by powers of two, for convenience. This table created for illustration only.

Tree part	Equivalent size	Number
Trunk	100 cm	1
Large branches	50 cm	8
Small branches	25 cm	64
Twigs	$12\frac{1}{2}$ cm	512
Leaves	$6\frac{1}{4}$ cm	4096

Table 15.2 Area and total RCS of tree parts ($\lambda = 12.5, 25$ cm). Total area is obtained by taking the projected area as πa^2 for each tree part and multiplying the result by the number of parts. Estimates of RCS are based on a two part approximation to Figure 15.15 with transition at $X = 1$ (see text). Note that "leaves" are in the Rayleigh scattering regime $X \leq 1$ while larger parts have $X > 1$.

			$\lambda = 12.5$ cm	$\lambda = 25$ cm
Tree part	**Number**	**Total area**	**Total RCS**	**Total RCS**
Trunk	1	π	π	π
Large branches	8	2π	2π	2π
Small branches	64	4π	4π	4π
Twigs	512	8π	8π	$\frac{\pi}{8}$
Leaves	4096	16π	$\frac{\pi}{4}$	$\frac{\pi}{256}$

The combined effect of a size distribution that favors smaller objects and a scattering efficiency that discounts the return from objects smaller than a wavelength in size leads to the greatest contribution from the wavelength-size objects. For $\lambda = 12.5$ cm the peak contribution to RCS comes from twigs, whose physical sizes are comparable to the radar wavelength. If, instead, we use a radar with a longer wavelength of 25 cm–λ, we obtain different values for the cross section of each set of component parts—at 25 cm–λ the greatest contribution to the RCS is from the small branches, which are twice the size of the twigs. Once again, the radar echo is dominated by objects comparable to a wavelength in size. Use of several different radar wavelengths allows us to probe objects at different size scales.

The assumption of an r^{-3} is convenient for computation but is not necessary for the result that the strongest contributors are roughly λ-size. There is a range of indices for which this result obtains, but it is limited. We summarize the effects of changing n in Table 15.3.

Of course, a real tree does not exhibit an infinitely extended r^{-3} power spectrum. There are no scatterers larger than the trunk, leading to a large-scale cutoff, and if the small objects are placed very close together they exhibit bulk properties which can change the results of the simple scattering calculations here. But this model is illustrative and is accurate enough for some purposes. Therefore, we would expect most vegetated surfaces to look different at different wavelength radars.

Table 15.3 Summary behavior of aggregate scattering for power law distributions. Contributions of various size objects to total RCS depends on the relative numbers in each size range, weighted by the scattering strength for each size. $n = 2$ results in a uniform contribution for large objects because product of numbers per size unit and area of each size is constant.

Power law	Results for total RCS
r^{-1}	Large objects dominant
r^{-2}	All objects larger than $X \approx 1$ contribute equally
r^{-3}	λ-size objects dominant
r^{-4}	Strong selection of $r \approx \lambda$ size objects
r^{-7}	Smallest objects in distribution dominant

An implicit assumption for this type of model is that each object in the distribution scatters independently of all the rest. This same assumption applies in the analysis of scattering from multiple objects in Section 5.5. This is valid because the volume density of the objects is low and/or the scattering from individual objects is weak. Certainly, when objects are more closely spaced than about 10 λ or 10 radii, they begin to interact electromagnetically regardless of other considerations, and the assumption of independent scattering fails.

As it happens, scattering from rough surfaces also varies with wavelength, with the apparent or "radar" roughness generally decreasing as the wavelength is increased. Although the power law example provides an important insight for collections of discrete scatterers, the problem of wavelength dependence of scattering from rough surfaces is much more complex, and remains an area of active research.

15.8 Some Scattering "Laws"

Characterization of the relationship between surface roughness and electromagnetic scattering is difficult for all but the simplest of surfaces. No exact solutions to Maxwell's equations valid for all wavelengths exist even for easily described deterministic surfaces such as harmonic undulations or uniform square waves, although limited solutions applicable for a restricted range of wavelengths are available. Restricted solutions also exist for stochastic surface models. In some instances, a statistical solution is more readily obtained than is a deterministic result, as a result of simplifications associated with the averaging process.

With the understanding that exact solutions are unusual, it is useful to be aware of a few scattering "laws" that provide representative results for various surface types and which are also in wide use. Each of these is a simple analytical formula describing how the normalized radar cross section σ^0 depends on incidence angle θ. Four such laws are the exponential, the Gaussian, the Hagfors law, and the "cosine-to-the-n." The first three of these are often referred to quasispecular scattering laws, in which the response is computed using physical optics and various assumptions on the probability distributions of surface height and slope. These approximations are most accurate in the near-specular direction, that is, near normal incidence ($\theta = 0$) for backscatter and near the forward specular direction for bistatic geometries. The cosine-to-the-n approximation is used most often when the volume scatter of one sort or another is significant. Many surfaces may be approximated by one of these laws for a subset of incidence angles and another for the remaining incidence angle range. All of these have been in analytical use as models for specifying engineering performance in radar designs.

15.8.1 Quasispecular Scattering

Quasispecular, or near-specular, direction scattering laws derive from a physical optics approach called the Kirchhoff approximation. Under this approximation, each point on the scattering surface is treated as a point on an infinite tangent plane made up of same material as the surface under study. Surfaces of sufficiently small roughness, with limited surface slope and height distributions, satisfy boundary conditions mathematically. However, it is common practice to use the approximations beyond their formal limits as scattered power measurements have often been observed empirically to follow the predicted angular dependence as closely as other means for modeling the echo. Using several well-defined distributions for height and slope variations yields many different scattering functions, some of which will match measurements better than others.

A criterion for quasispecular scattering is that the surface is smooth, which under the Rayleigh criterion means that the root-mean-square height h_{rms} is such that

$$h_{rms} < \frac{\lambda}{8 \cos \theta} \tag{15.5}$$

The rms height h_{rms} is the first descriptor of the surface under observation. As noted earlier, even if the rms height exceeds this value, measurements still tend to follow the derived angular dependence, so that these laws are more robust than the mathematical assumptions would imply.

The second descriptor needed is the roughness in the horizontal direction, rather than in the vertical direction of h_{rms}, and is usually quantified as the surface autocorrelation function $C_f(R)$. This function will possess an autocorrelation length, R_0, the distance at which the autocorrelation function falls by a factor of e^{-1}.

The rms height and autocorrelation functions are related to the distribution of slopes $p(\theta)$. The slope distribution is also characterized by a rms value, s, which would be zero for a perfectly flat surface and yields a finite value for surfaces with realistic roughnesses. The rms surface slope s depends on the autocorrelation function and the rms height,

$$s = h_0 \sqrt{-2C_f(0)''} \tag{15.6}$$

where $-2C_f(0)''$ is the second derivative of the autocorrelation function, evaluated at $R = 0$.

The scattered power depends not only on the distribution of slopes and heights, but also on the dielectric properties of the surface. Because the quasispecular approximation takes scattering as resulting from a plane tangent to the surface, and only planes normal to the wavefront contribute, we can compute the scattered power from a perfectly conducting surface of the same shape as the surface under study, scaled by the Fresnel reflection coefficient ρ of the surface material. For a surface of bulk dielectric constant ϵ we can compute ρ from

$$\rho = \left[\frac{\sqrt{\epsilon} - 1}{\sqrt{\epsilon} + 1} \right]^2 \tag{15.7}$$

We will also find it convenient to parameterize the angular extent of the backscattered lobe, which we will denote C. Not to be confused with the autocorrelation function of similar notation, traditionally C has been defined as

$$C = \frac{\lambda^2 R_0^2}{16\pi^2 h_{rms}^4} \tag{15.8}$$

The roots of this definition are historical, but the inverse square root of C has often been taken to be the rms surface slope:

$$s = \frac{1}{\sqrt{C}} \tag{15.9}$$

Typical assumptions for both the surface heights and autocorrelation functions are either exponential or Gaussian forms, with different combinations resulting in different scattering functions. Tabulating exponential and Gaussian forms for each,

$$p_e(h) = \frac{1}{h_0} \exp\left[\frac{-h}{h_0}\right] \tag{15.10}$$

$$p_G(h) = \frac{1}{h_0\sqrt{2\pi}} \exp\left[\frac{-(h-\bar{h})^2}{2h_0^2}\right] \tag{15.11}$$

$$C_{fe}(R) = \exp\left[\frac{-R}{R_0}\right] \tag{15.12}$$

$$C_{fG} = \exp\left[\frac{-R^2}{R_0^2}\right] \tag{15.13}$$

whereas before $p(h)$ is the pdf of heights and $C_f(R)$ is the surface autocorrelation.

Using these distributions, we can compute the rms slope for the exponential correlation function as

$$s = h_0\sqrt{\frac{1}{R_0 R_{min}}} \tag{15.14}$$

where R_{min} is the minimum horizontal size scale of the surface that meaningfully contributes to scattering. This may seem confusing, but it is just stating that the echo power is not dependent on scatterers much smaller than a wavelength, as we saw earlier in the discussion on scattering from trees. R_{min} is customarily taken such that the rms slope s is equal to the width parameter C.

The result for the Gaussian correlation function is simpler,

$$s = h_0\frac{\sqrt{2}}{R_0} \tag{15.15}$$

At this point we are ready to examine the exponential, the Gaussian, the Hagfors laws. Each follows from a particular combination of height, slope, or correlation distribution.

15.8.1.1 Exponential Law

The exponential law follows if we use an exponential function for both the height and slope distributions:

$$\sigma_e^0(\theta) = \frac{3\rho C}{\cos^4\theta} \exp(-\sqrt{6C}\tan\theta) \tag{15.16}$$

15.8.1.2 Gaussian Law

For the Gaussian law, we use a Gaussian distribution for both the heights and the autocorrelation function:

$$\sigma_G^0(\theta) = \frac{\rho C}{\cos^4\theta} \exp(-C\tan^2\theta) \tag{15.17}$$

15.8.1.3 Hagfors Law

If we instead use a Gaussian distribution for the heights and the exponential autocorrelation function, we obtain Hagfors law:

$$\sigma_H^0(\theta) = \frac{\rho C}{2}(\cos^4\theta + C\sin^2\theta)^{3/2} \tag{15.18}$$

We can compare these different forms for normalized radar section σ^0 by plotting them for similar sets of surface parameters. We let the Fresnel reflection coefficient be unity, wavelength 10 cm, h_{rms} 1 cm, and s 8° (Figure 15.17). (These selections imply that $C = 51.3$, and $R_0 \approx 9$ cm, implying that the surface decorrelates over a distance comparable to the wavelength.)

15.8.2 Diffuse Scattering: Cos^n Law

The quasispecular laws all derive from considering backscatter by facets normal to the incident wavefront, and hence are only valid for angles corresponding to physically realizable angles of repose. In other words, because local slopes rarely exceed 10°–20°, these functions predict very little return from larger angles of incidence. But we in fact measure quite large signals at these larger angles, so other mechanisms may be in play. Scattering from large angles is often referred to as diffuse scattering, as it can represent, for example, multiple scattering events that still return energy to the illuminating radar. Analytical solutions for diffuse scatter can be very complex and specialized, so it is common practice to characterize these with a combination of analytical modeling and empirical observation.

Perhaps the most common diffuse scattering "law" is the $\cos^n\theta$ form, whose shape is such that it falls off slowly over the full range of $0 < \theta < \frac{\pi}{2}$ (see Figure 15.18). Observed values for the exponent n range from unity, corresponding to an isotropic scatter from the surface, to 2, which represents a surface scattering according to the optical remote sensing derived Lambert's cosine law, where energy scattered from a surface is concentrated with cosinusoidal dependence. Coupled with the cosine inherent in the definition of σ^0 whose units are m^2/m^2-surface, the exponent assumes a value of 2. Natural diffusely scattering surfaces rarely exhibit exponents greater than 2. The simplest form, then, for a diffuse scattering component, is

$$\sigma^0 = A\cos^n \tag{15.19}$$

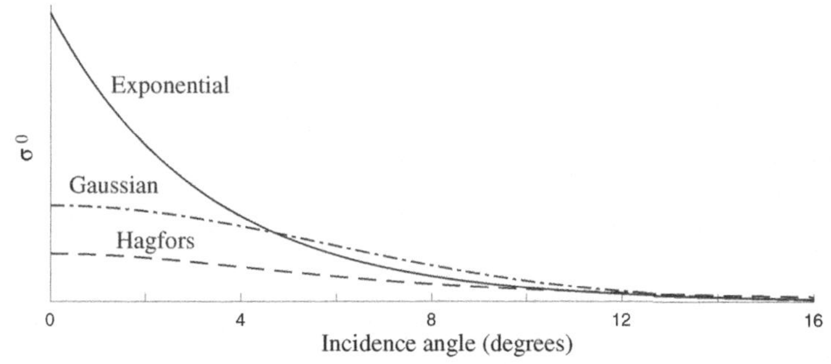

Figure 15.17 Quasispecular models. Normalized radar cross section σ^0 as a function of incidence angle for exponential, Gaussian, and Hagfors Laws. These models are valid for near-normal incidence, note that the range of angles plotted is for $\theta < 16°$. These curves are computed assuming $\lambda = 10$ cm, rms surface height $h_{rms} = 1$ cm, and rms slope $s = 8°$.

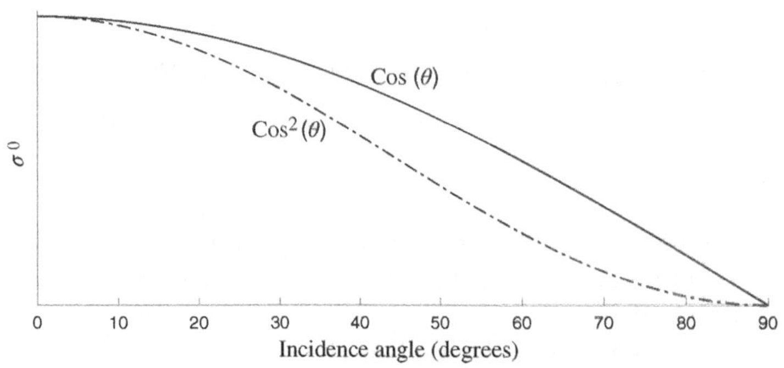

Figure 15.18 Cos^n models. Normalized radar cross section σ^0 as a function of incidence angle for $\cos^n\theta$ models. Depicted here are results for isotropic scatter ($\cos\theta$) and Lambertian scatter ($\cos^2\theta$).

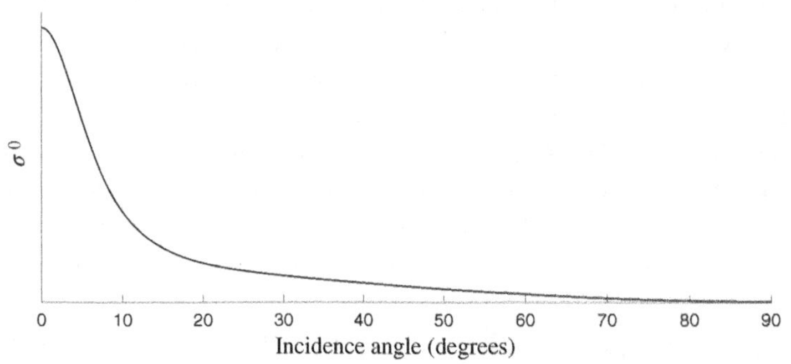

Figure 15.19 Composite model. Normalized radar cross section σ^0 as a function of incidence angle for composite surface with both quasispecular and volume scattering components. Plotted here is sum of Hagfors and $\cos^2\theta$ models.

Real surfaces are generally more complex than these simple models might indicate. It is useful to combine one or more models to represent scatter accurately over a large angular range. For example, we might use Hagfors Law for small incidence angles and a diffuse component for the larger angles, with the total scattering function being a weighted sum of the two models (see Figure 15.19).

16

Polarimetric Radar Systems

Up to this point, we have seen how we can use the basic principle of transmitting a burst of energy and receiving power reflected back to the instrument in order to determine characteristics of an object or surface under study, including its location, velocity, size or composition. In all the examples we have seen, the radar configuration consists of a single transmitting antenna and a single receiving antenna, which may or may not be co-located. We now move to systems with multiple antennas, where we can exploit different antenna properties of imaging geometries in addition to the basic observables we have considered to date.

If we use more than one antenna where each has a different polarization sensitivity, we can characterize the polarization properties of a radar echo. If we change the geometry by using two or more antennas, we will be sensitive to details of the target shape or change in shape. We will denote systems satisfying the former as polarimetric radars, or polarimeters, and the latter as interferometers. In both cases, we generate additional observables by combining the signals received at two or more radar antennas.

16.1 Polarimetric Radars

We have seen that frequency diversity allows us to obtain different views of the surface through sensitivity to different size scale scatterers that are present on the surface. Now let's consider how the radar echoes are modified in polarization by interaction with the surface.

We'll call a radar with polarization discrimination capability a polarimeter, and if it produces images, an imaging radar polarimeter. We will find that different surface types cause the polarimetric signature of the echoes to have very different characteristics. First, we'll define some basic polarization terms.

16.2 Polarization Definitions

Consider this plane wave solution to Maxwell's equation:

$$\underline{E} = Re\left[(E_h\hat{\mathrm{h}} + E_v\hat{\mathrm{v}})e^{-j(wt-kz)}\right] \tag{16.1}$$

where $\hat{\mathrm{h}}$ and $\hat{\mathrm{v}}$ are linear, orthogonal basis polarization directions. We could have used any basis, though. The linear basis convention we select here has the advantage of being readily relatable to horizontal and vertical structures on the surface, such as tree trunks (vertical) vs. randomly oriented branches (which will be at some combination of horizontal and vertical.) Here E_h and E_v

are complex field amplitudes given by

$$E_h = a_h e^{-j\phi h}$$
$$E_v = a_v e^{-j\phi v}$$

where a_h and a_v are positive amplitudes in the $\hat{h}$ and $\hat{v}$ directions are the ϕ's are phases. Any wave can be represented by vector

$$\underline{E} = \begin{pmatrix} E_h \\ E_v \end{pmatrix} \tag{16.2}$$

Our plane wave solution is the real part of the phasor of Eq. (16.1), which we can rewrite as

$$a\cos(\phi + wt - kz) = Re\left[(a_h e^{-j\phi h}\hat{h} + a_v e^{-j\phi v}\ \hat{v})e^{-j(wt-kz)}\right] \tag{16.3}$$

To recognize this as a parametric definition of an ellipse, we express the real solution in terms of its components in the $\hat{h}$ and $\hat{v}$ directions:

$$\begin{aligned} A_h &= a_h \cos(\phi_h + wt - kz) \\ A_v &= a_v \cos(\phi_v + wt - kz) \\ A_z &= 0 \end{aligned} \tag{16.4}$$

Eliminating the variable part of the phase follows if we rewrite Eq. (16.4) using the trigonometric sum formula:

$$\begin{aligned} \frac{A_h}{a_h} &= \cos(\phi_h)\cos(wt - kz) - \sin(\phi_h)\sin(wt - kz) \\ \frac{A_v}{a_v} &= \cos(\phi_v)\cos(wt - kz) - \sin(\phi_v)\sin(wt - kz) \end{aligned} \tag{16.5}$$

which leads to

$$\begin{aligned} \frac{A_h}{a_h}\sin\phi_h - \frac{A_v}{a_v}\sin\phi_v &= \cos(wt - kz)\sin(\phi_v - \phi_h) \\ \frac{A_h}{a_h}\cos\phi_h - \frac{A_v}{a_v}\cos\phi_v &= \sin(wt - kz)\sin(\phi_v - \phi_h) \end{aligned} \tag{16.6}$$

If we square and add the two equations 16.6 we now have

$$\left(\frac{A_h}{a_h}\right)^2 + \left(\frac{A_v}{a_v}\right)^2 - 2\frac{A_h}{a_h}\frac{A_v}{a_v}\cos(\phi_v - \phi_h) = \sin^2(\phi_v - \phi_h) \tag{16.7}$$

We recognize Eq. (16.7) as the parametric description of an ellipse, leading to a geometric interpretation of polarization. Most authors make use of a set of auxiliary angles defined by

$$\tan\alpha = \frac{a_v}{a_h}$$

$$\tan 2\psi = (\tan 2\alpha)\cos(\phi_v - \phi_h)$$
$$\sin 2\chi = (\sin 2\alpha)\sin(\phi_v - \phi_h)$$

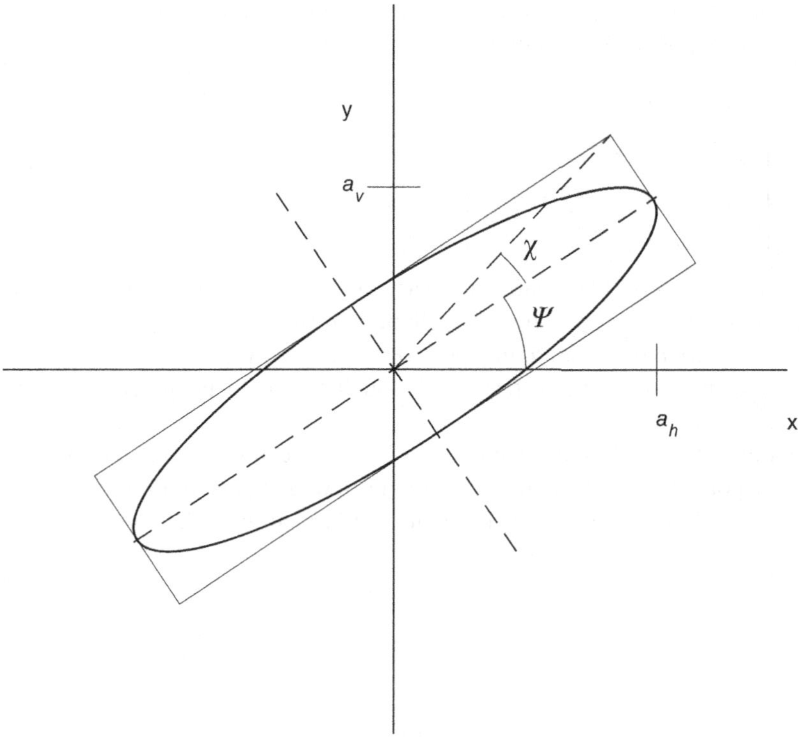

Figure 16.1 Polarization ellipse. Ellipse describes the path of a single frequency electric field vector in plane perpendicular to the wave vector. Three parameters, orientation, ψ, the ratio of semimajor to semiminor axis, or axial ratio, tan χ, and size usually of semimajor axis, a describe the ellipse. In addition a fourth parameter is required to specify the sense of rotation, either right- or left-hand (RCP or LCP); the direction of rotation is coded in the sign of χ. Altogether three numbers or parameters are required.

leading to a picture of the elliptical path followed by the end point of the vector ($x = A_h, y = A_v$) of the polarized wave as it propagates in the z-direction as in Figure 16.1. If the ellipse collapses to a line, that is $\chi \to 0$, then the polarization is called linear with orientation ψ. $\psi = 0$ is horizontal polarization, $\psi = 90°$ is vertical. If $a_h = a_v$ or $\chi = 45°$, the polarization is called circular. Depending on the direction of rotation, the circular polarization is right-handed or left-handed.

Not only waves, but antennas exhibit polarization states that can be defined either in terms of complex amplitudes or geometrical parameters χ, ψ. Thus, we can represent the polarization effects of scattering by a surface S as:

$$\begin{pmatrix} E'_n \\ E'_v \end{pmatrix}^{\text{sc}} = \frac{e^{ikr}}{kr} \begin{pmatrix} Sh'h & Sh'v \\ Sv'h & Sv'v \end{pmatrix} \begin{pmatrix} E_h \\ E_v \end{pmatrix}^{\text{inc}} \tag{16.8}$$

Here sc and inc denote the scattered and incident waves, with h, v and h', v' the bases of the two waves, which may be the same. Thus, the element $Sh'h$ describes the transformation of the E_h component to $E_{h'}$ during the scattering event. Others are defined similarly. The $Sx'x$ are typically complex numbers but are often real.

The relationship of χ and ϕ to polarization is expressed as the Stokes vector, which we'll define as follows (many definitions are possible depending on bases):

$$F = \begin{bmatrix} |E_h|^2 + |E_v|^2 \\ |E_h|^2 - |E_v|^2 \\ 2Re(E_h E_v^*) \\ 2Im(E_h E_v^*) \end{bmatrix} = \begin{bmatrix} I \\ Q \\ U \\ V \end{bmatrix} = \begin{bmatrix} I \\ I\cos 2\psi \cos 2\chi \\ I \sin 2\psi \cos 2\chi \\ I \sin 2\chi \end{bmatrix} \quad (16.9)$$

The Stokes vector neatly summarizes the polarization properties of the wave. Not only does it define the h and v components of the wave, as does the wave vector (Eq. 16.2), but it also tells us what fraction of the signal power is unpolarized. For a fully polarized wave, $I = \sqrt{Q^2 + U^2 + V^2}$. But if it contains energy without a defined polarization state, we denote it a partially polarized wave, and find that $I > \sqrt{Q^2 + U^2 + V^2}$.

Very commonly, radar antennas are selected to record *cross-polarized* echoes. This means that the receive polarization is orthogonal to the transmit polarization. Viewed at polarization ellipses, an orthogonal polarization is an ellipse rotated 90° in the *xy*-plane and rotating in the opposite sense. If antennas are represented by Stokes vectors, then

$$\begin{aligned} I_{cross} &= I \\ Q_{cross} &= -Q \\ U_{cross} &= -U \\ V_{cross} &= -V \end{aligned}$$

It is instructive to look at an example. A horizontally polarized wave has $E_h = 1, E_v = 0$. What is its stokes vector F?

$$\begin{aligned} I &= |Eh|^2 = 1 \\ Q &= |Eh|^2 = 1 \\ U &= 0 \\ V &= 0 \end{aligned}$$

Here are a few other commonly seen cases.
Vertically polarized wave:

$$E = \begin{pmatrix} 0 \\ 1 \end{pmatrix}; \text{ yielding } F = \begin{bmatrix} 1 \\ -1 \\ 0 \\ 0 \end{bmatrix} \quad (16.10)$$

Right hand circular polarization:

$$RCP : E = \frac{1}{\sqrt{2}} \begin{pmatrix} 1 \\ j \end{pmatrix}; \text{ yielding } F = \begin{bmatrix} 1 \\ 0 \\ 0 \\ -1 \end{bmatrix} \quad (16.11)$$

Left hand circular polarization:

$$LCP : E = \frac{1}{\sqrt{2}}\begin{pmatrix} j \\ 1 \end{pmatrix}; \text{ yielding } F = \begin{bmatrix} 1 \\ 0 \\ 0 \\ 1 \end{bmatrix} \tag{16.12}$$

Thus any incident wave can be expressed as one of these two forms, as can any scattered wave. Our next issue is how these waves are related to various objects, that is, what is S for different scattering types? As with the surface scattering laws we discussed in the previous chapter, calculation of these can be very challenging. But for very simple forms, we can create nominal scattering matrices, and these ideal cases are surprisingly effective in reproducing backscatter observed from a real surface.

Now let's consider what our received wave power looks like as a function of the polarizations of the transmit and receive antennas, respectively. Thus, if we know S for a point on the ground, we can determine the received power for any combination of transmit and receive polarizations.

This variation of P with polarization is called the *polarization signature.*

Consider a plot (Figure 16.2) of the polarization signature of an object where E_r and E_t are the same, plotted as a function of χ and ϕ from the polarization ellipse. This describes a sphere with S modeled using the matrix of Table 16.1. The signal is highest for $\chi = 0$ or linear polarizations and 0 for $\chi = \pm 45°$, circular polarizations.

We can easily distinguish scattering mechanisms by looking at the polarization signature. For example, in Figure 16.3, we plot the co-polarized signatures for a rough surface and for a dihedral corner reflector. Each of these is distinct, and we can recognize the scattering mechanism directly from the shape of the signature.

For targets that contain a significant unpolarized part in the echo signal, we cannot use the scattering matrix approach easily because all waves are modeled as polarized, by a *scattering matrix* (Eq. 16.8) of the surface. But as noted earlier, the Stokes vector incorporates the randomly

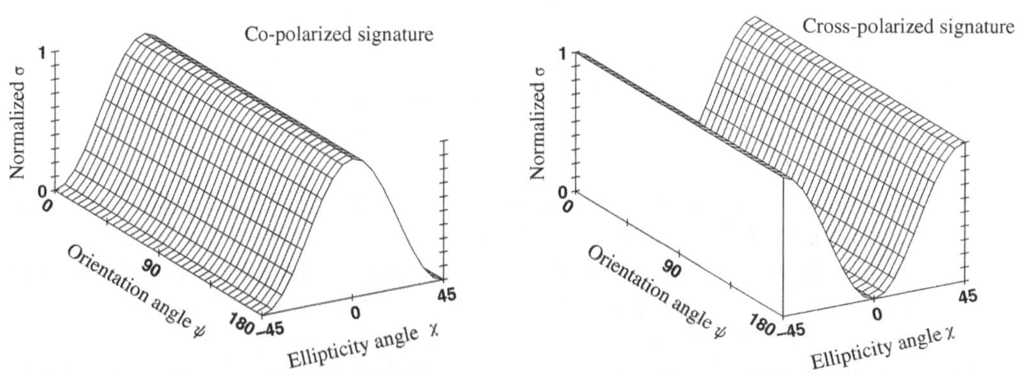

Figure 16.2 Polarization signature of sphere. At left, the co-polarized signature (transmit and receive polarizations identical). At right, cross-polarized signature (receive polarization orthogonal to transmit polarization). Surface represents received backscatter echo power from a sphere vs. orientation, ϕ, and ellipticity, χ.

Table 16.1 Scattering matrices for some simple mechanisms.

Scattering mechanism	Scattering matrix	Notes
Sphere	$\begin{bmatrix} a & 0 \\ 0 & a \end{bmatrix}$	a real, no cross-polarized return
Bragg model	$\begin{bmatrix} a & 0 \\ 0 & b \end{bmatrix}$	$b > a$, slightly rough surface
Real dielectric dihedral corner reflector	$\begin{bmatrix} -a & 0 \\ 0 & b \end{bmatrix}$	$a > b$, both real

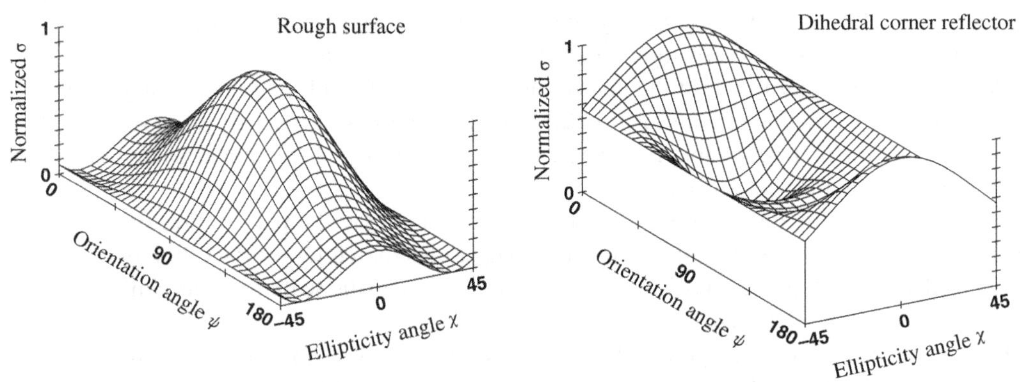

Figure 16.3 Polarization signatures of rough surface and dihedral reflector. Co-polarized signature for a rough surface scattering according to Bragg model (left), same for a dihedral corner reflector (right). Each of these is distinct from the co-polarized sphere signature of Figure 16.2.

polarized part of wave, so we can extend the idea of a polarization signature by describing the surface in terms of a matrix relating power at distinct polarization states to a *Stokes matrix*,

$$P = F_r^T M F_t \tag{16.13}$$

where the elements of M relate to the components of F_t and F_r. We can write the elements of M in terms of spatial averages of the cross products of elements of S, as in Table 16.2. Figure 16.4 shows the co- and cross-polarization signatures from a combination of a dihedral corner reflector and a volume scattering mechanism. Here you can see the dihedral signature from above, which sits on a pedestal of polarization-independent echo power. This type of signature is observed from some forest terrain, where the trunk/ground interaction creates a dihedral reflection, and the canopy provides multiple scatter targets. The unpolarized component often arises from such multiple scatter. The resulting polarization signature then sits on a "pedestal" of unpolarized energy.

Table 16.2 Stokes matrix definition.

Stokes matrix element	Definition
M(1,1)	$\frac{1}{4}\left[S_{hh}S_{hh}^{*}+S_{vv}S_{vv}^{*}+2S_{hv}S_{hv}^{*}\right]$
M(1,2)	$\frac{1}{4}\left[S_{hh}S_{hh}^{*}-S_{vv}S_{vv}^{*}\right]$
M(1,3)	$\frac{1}{2}\left[Real(S_{hh}^{*}S_{hv})+Real(S_{hv}^{*}S_{vv})\right]$
M(1,4)	$\frac{1}{2}\left[Imag(S_{hh}^{*}S_{hv})+Imag(S_{hv}^{*}S_{vv})\right]$
M(2,1)	M(1,2)
M(2,2)	$\frac{1}{4}\left[S_{hh}S_{hh}^{*}+S_{vv}S_{vv}^{*}-2S_{hv}S_{hv}^{*}\right]$
M(2,3)	$\frac{1}{2}\left[Real(S_{hh}^{*}S_{hv})-Real(S_{hv}^{*}S_{vv})\right]$
M(2,4)	$\frac{1}{2}\left[Imag(S_{hh}^{*}S_{hv})-Imag(S_{hv}^{*}S_{vv})\right]$
M(3,1)	M(1,3)
M(3,2)	M(2,3)
M(3,3)	$\frac{1}{2}\left[(S_{hh}S_{hv}^{*})+Real(S_{hv}^{*}S_{vv})\right]$
M(3,4)	$\frac{1}{2}\left[Imag(S_{hh}^{*}S_{vv})\right]$
M(4,1)	M(1,4)
M(4,2)	M(2,4)
M(4,3)	M(3,4)
M(4,4)	$\frac{1}{2}\left[S_{hv}S_{hv}^{*}-Real(S_{hh}^{*}S_{vv})\right]$

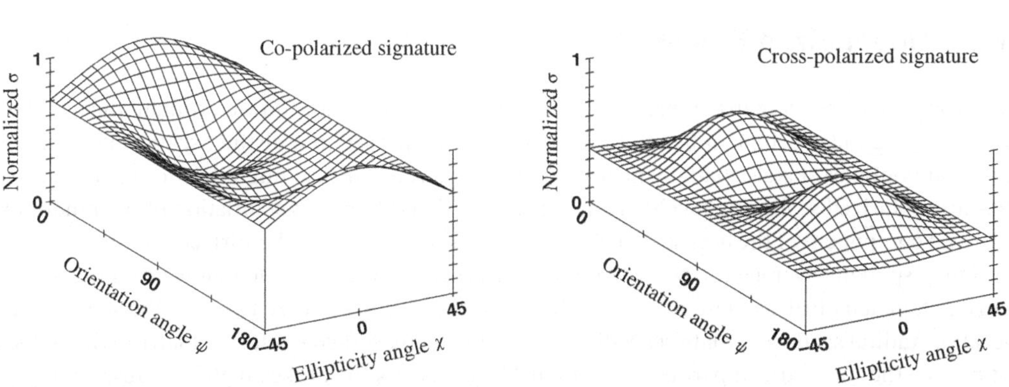

Figure 16.4 Polarization signature of a dihedral reflector plus volume scatter. For many natural surfaces, echo contains an unpolarized component in addition to any well-defined mechanism. Return from a forest, for example, might be a dihedral component from trunk/ground scattering, plus a volume term from multiple scatter in the canopy. In the presence of an unpolarized, i.e., randomly polarized, component of the echo, the polarization signature appears raised by a constant power. The fraction of polarized to unpolarized echo power becomes part of the polarization signature, although the characteristics of the polarized part still largely retain their original interpretation.

16.3 Contrast Enhancement

Let's say we want to distinguish two different types of terrain in an image, say to make a geologic map. If we measure the polarization signature $P(\chi, \phi)$ for each, we can form the ratio:

$$\frac{P_1(\chi, \phi)}{P_2(\chi, \phi)} \tag{16.14}$$

and plot this quantity as a function of χ, ϕ. We could then find the maximum of this ratio, yielding the values of χ, ϕ that separates them best in the image. Generating an image with this polarization then permits the easiest mapping of those units.

16.4 Generating Polarimetric Data

The aforementioned applications require knowledge of S (or M) for each and every point in an image. Conversely, if we know S for each point, we can implement many polarization-based analyses. So, how do we determine S observationally?

Remember S consists of the four combinations of transmit and receive polarizations (Eq. 16.8.) Suppose we have two antennas on our radar, one h-pol and one v-pol. Transmit h and receive h, that measure Shh. The others follow similarly. In fact, we can transmit h and receive both h and v simultaneously in separate channels, then repeat transmitting from the v antenna. Two pulses are needed to construct the four elements of the scattering matrix.

Hence, we measure the elements of S by using two orthogonal - basis antennas, typically $\hat{\mathrm{h}}$ and $\hat{\mathrm{v}}$ but any could be used. We generate four complex radar images of each scene, one for each S element, which we can then combine to produce polarization signatures for each point in the scene.

16.5 Unpolarized Echoes

Since each measured S matrix represents a fully polarized wave, how do we estimate the depolarized part? The radar echo can be modeled as a polarized part with a distinct scattering matrix S plus a random part. Pixels spatially nearby each point on the surface might be viewed as sharing the same polarized scattering mechanism, but each with a different instantiation of the randomly polarized component of the backscatter. So if we assume the region on the surface is homogeneous, we simply spatially average the powers in the polarization signature. Note that we sum the Stokes matrices for each pixel, as these retain both the polarized and unpolarized parts of the signature in the sum. Adding scattering matrices S directly results in an average scattering matrix, which does not convey the unpolarized part of the echo. If M_i represents the Stokes matrix corresponding to scattering matrix S_i, the average Stokes matrix for all of the pixels added together results in a matrix M_{acc} containing both the polarized and unpolarized parts:

$$M_{\mathrm{acc}} = \sum M_i \tag{16.15}$$

Then the polarization signature for co-polarized antennas is $P(\chi, \phi) = F_{\mathrm{ant}} M_{\mathrm{acc}} F_{\mathrm{ant}}^T$.

We note that we can achieve the same result by computing the polarization signature for each scattering matrix pixel and then adding the signatures. The signature contains the power represented by the scattering matrix at each pixel. Summing these causes the shared scattering mechanism at each point to add constructively, whereas the unpolarized, random part will vary as a function of polarization and average to a mean, nonzero value. For the copolarized signature, then, if E_{ant} is the antenna polarization (synthesized),

$$P(\chi, \phi) = \sum |E_{ant} S E_{ant}^T|^2 \tag{16.16}$$

This procedure is equivalent to taking looks at a conventional radar. If we store the data as a Stokes matrix in the first place, things are more straight-forward.

16.6 Example Observations

As an example of what actual polarization signatures from real instruments look like, we plot in Figure 16.5 the co-polarized signatures of slightly rough surface without much vegetation and of an area in an urban setting. These data were acquired by the NASA/JPL UAVSAR airborne instrument, which operates at L-band in a small jet aircraft. By comparison with Figure 16.3, we can see that the dominant mechanisms in each case are i) the Bragg resonance model and ii) a real dielectric corner reflector formed by a building and the street, respectively. In addition to the pure mechanism suggested by the signatures, an unpolarized component is also present. This could result from any inhomogeneity in the pixels averaged to create the signature, or if there are any multiple scattered components in the return echo. The single signature shown here for each case does not allow us to distinguish these or other causes; further experiments and analyses are needed.

We can also use the Stokes matrices measured at every point in the image to synthesize any desired polarization transmit/receive combination. Using the Stokes matrix, we pre- and post-multiply it with the Stokes vectors for the receive and transmit polarizations, respectively,

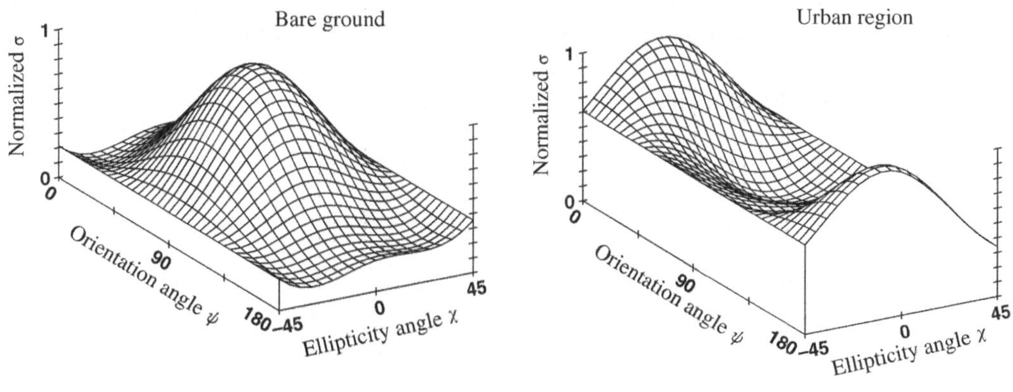

Figure 16.5 Polarization signatures of actual terrain. Measured co-polarized signatures of a patch of bare ground and of an urban area north of Los Angeles, California, as observed with the NASA/JPL UAVSAR polarimetric radar on 23 April 2012. Note the similarity to Figure 16.3. We interpret these as identifying the Bragg mechanism as dominating for rough surfaces, and dihedral scatter as dominant in urban regions with many buildings. Note the presence of a large volume scatter component in each.

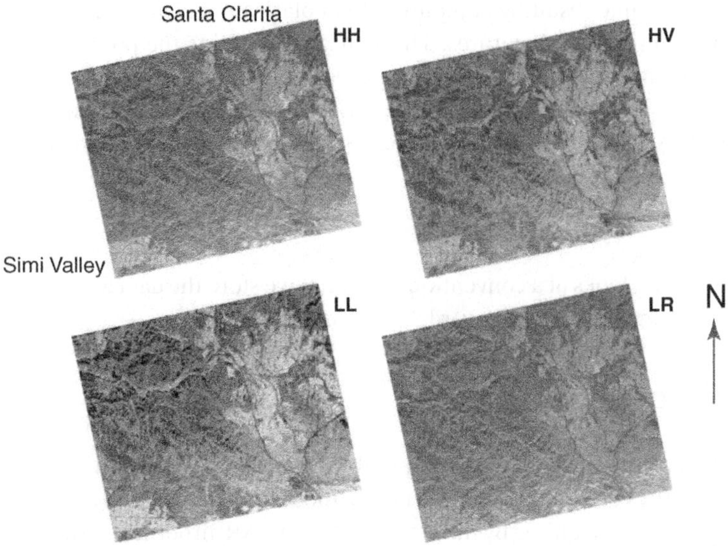

Figure 16.6 Synthesis of various polarization combinations imaging terrain. The figure displays four examples derived from the UAVSAR run used for the polarization signatures in Figure 16.5. The images, from top left in clockwise order, represent polarization pairs HH, HV, LR, LL, where H, V, L, and R are horizontal, vertical, left hand circular, and right hand circular polarizations. The first letter is the transmit polarization, the second the receive polarization. It is evident that various terrains behave differently as the polarization is varied.

and record the measured power at each pixel to form the image. Figure 16.6 shows four examples derived from the UAVSAR run described earlier. The images, from top left in clockwise order, represent polarization pairs HH, HV, LR, LL, where H, V, L, and R are horizontal, vertical, left hand circular, and right hand circular polarizations. These images show how differing terrains behave differently as the polarization is varied. Depending on what ground feature is to be modeled or analyzed, descriptive polarization states may be chosen. It would also be possible to estimate the amount of the radar echo that would be considered depolarized, which may be related to how much volume scattering may be occurring in different parts of the image.

17

Radar Interferometers

The next set of radar applications we are going to introduce is interferometric radar. For the environmental applications we have been emphasizing here, interferometric analyses have found their most widespread use in imaging radars. And because imaging radars generally require fine resolution and synthetic aperture processing, the systems we are most interested in are interferometric synthetic aperture radars. Thus, you will often see the nomenclature InSAR as an acronym (ISAR was already in common usage as an abbreviation for inverse synthetic aperture radar.)

We will see that a geometric interpretation of radar phase is a very useful way to visualize and model interferometric echoes. While the development of classical interferometers at light or radio wavelengths precedes InSAR by several or more decades, much of this work was aimed at making interferometry work using technologies that did not permit ready measurement of signal phase. In today's digital world, we can sample and record complex signals easily, bypassing the need for intermediate conversion to power quantities that characterizes classical interferometry. Nonetheless, it is worthwhile to briefly examine classical development for its precursory role in radar interferometry.

An interferometer is a device that analyzes EM waves coherently at two observation points, that is, the phase difference of two waves is studied. Combining the signals from both antennas to determine the phase difference forms the interferometer. As it was difficult to measure the two signal phases in the pre-digital era, there evolved two main implementations that exploited amplitude quantities dependent on phase adding interferometers and multiplying or correlation interferometers. Both have been used at radar wavelengths, although the latter is more common today. Because the adding interferometer was developed first, and is perhaps conceptually simpler, it is worthwhile to take a look at the basic principles of interferometry from this viewpoint.

17.1 Adding Interferometer

We have seen that we can measure the amplitude and frequency of an arriving EM wave, which in the case of a radar is an echo reflected from an object of interest. It is also important in many applications to know the direction of arrival as well, such as in mapping the location of a radioastronomical source or of an incoming nuclear missile. Examine the geometry of a wave field on a pair of antennas as shown in Figure 17.1. The wave travels a distance $\delta = B \cdot u_{inc} = B \sin \alpha$ longer to reach the second antenna, accumulating additional phase in the amount $2(\pi/\lambda)\delta$. Note that the projection of B in the incident direction is written using a sin() rather than cos() because we have defined the angle of arrival α in the off-zenith direction rather than with respect to the horizon. The factor of 2 becomes a factor of 4 if we have two independent radar systems, one at the

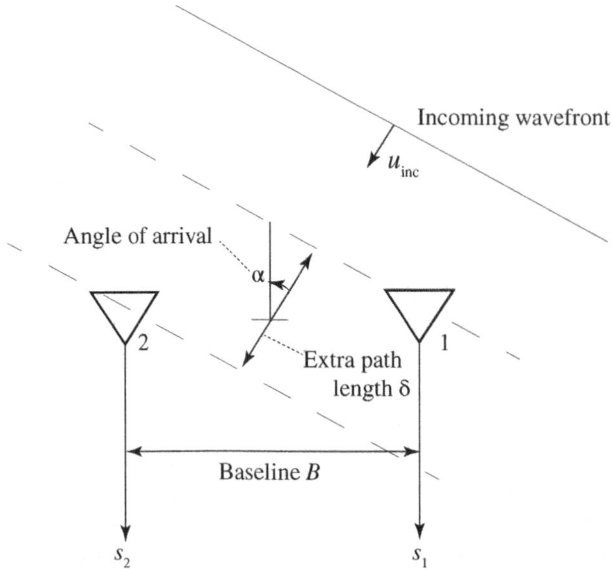

Figure 17.1 Wavefront on two antennas. A plane wave in the direction of unit vector u_{inc} impinges on two antennas separated by a baseline B. The extra path length traveled by the wave from the first antenna encountered δ is equal to $B \cdot u_{inc} = B \sin \alpha$, and accumulates extra phase in the amount $2(\pi/\lambda)\delta$. If each antenna both transmits and receives, then the path difference is traversed twice, and the phase shift is $4(\pi/\lambda)\delta$. An interferometer sensitive to the phase difference can thus determine the location of the source.

location of antenna 1 and one at the location of antenna 2, as the extra path length is traversed twice. For a cosinusoidal incoming wave, we will find it convenient to consider the signal from the first antenna as advanced by half of this phase and the signal from the second antenna as delayed by half of this amount:

$$s_1(t) = a \exp\left[-j\left(2\pi f t + \frac{2\pi}{\lambda}\frac{\delta}{2}\right)\right] \tag{17.1}$$

and

$$s_2(t) = a \exp\left[-j\left(2\pi f t - \frac{2\pi}{\lambda}\frac{\delta}{2}\right)\right] \tag{17.2}$$

where a is the amplitude of the incoming wave. The adding interferometer obtains its output signal by summing the two waves coherently,

$$\begin{aligned} c_{sum}(t) &= s_1(t) + s_2(t) \\ &= a \exp\left[-j\left(2\pi f t + \frac{2\pi}{\lambda}\frac{\delta}{2}\right)\right] + a \exp\left[-j\left(2\pi f t - \frac{2\pi}{\lambda}\frac{\delta}{2}\right)\right] \\ &= a \exp\left[-j\left(2\pi f t\right)\right] 2 \cos\left(\frac{2\pi}{\lambda}\frac{\delta}{2}\right) \\ &= a \exp\left[-j\left(2\pi f t\right)\right] 2 \cos\left(\frac{2\pi}{\lambda}\frac{B \sin \alpha}{2}\right) \end{aligned} \tag{17.3}$$

The received signal amplitude is multiplied by a cosine factor that depends on angle of arrival, so that measuring the power of the signal tells us something about its location. In fact, Eq. (17.3) shows that the amplitude of the sum signal is proportional to the component of the source distribution with a spatial frequency determined by the baseline length and the angle of arrival.

17.2 Monopulse Radar

The adding interferometer converts phase information to amplitude. This is especially valuable for radioastronomical instruments where sensitivity to a single or set of Fourier coefficients of the sky distribution is desired. For a radar system, where there may be only a single point target of interest, however, this observation of signal power depends not only on arrival angle but also on intrinsic brightness (denoted a earlier) of the object, which is usually unknown. A variation of the adding interferometer, denoted monopulse because it can be implemented on a single radar echo, removes the dependence on knowledge of signal amplitude by considering the ratio of two quantities, so that the absolute amplitude of the echo cancels out.

When we formed the adding interferometer signal (Eq. 17.3), we computed the sum of the signals from the two antennas. For monopulse, we also form the difference of the two signals, as

$$\begin{aligned} c_{diff}(t) &= s_1(t) - s_2(t) \\ &= a \exp\left[-j\left(2\pi f t + \frac{2\pi}{\lambda}\frac{\delta}{2}\right)\right] - a \exp\left[-j\left(2\pi f t - \frac{2\pi}{\lambda}\frac{\delta}{2}\right)\right] \\ &= a \exp\left[-j\left(2\pi f t\right)\right] 2j \sin\left(\frac{2\pi}{\lambda}\frac{\delta}{2}\right) \\ &= a \exp\left[-j\left(2\pi f t\right)\right] 2j \sin\left(\frac{2\pi}{\lambda}\frac{B\sin\alpha}{2}\right) \end{aligned} \tag{17.4}$$

Now we can form the ratio of the magnitude of each of the difference and sum measurements, and obtain

$$\frac{\sin\left(\frac{2\pi}{\lambda}\frac{B\sin\alpha}{2}\right)}{\cos\left(\frac{2\pi}{\lambda}\frac{B\sin\alpha}{2}\right)} = \tan\frac{2\pi}{\lambda}\frac{B\sin\alpha}{2} \tag{17.5}$$

Thus the amplitude a of the incoming wave cancels out, so that the ratio of the difference and sum amplitudes depends solely on the angle of arrival. Such a differential measurement permits a much less ambiguous determination of angle of arrival.

17.3 A Simple Multiplying Interferometer

Accurate measurement of the phase of an incoming wave on two antennas allows us to determine the angle of arrival. With today's technology, we can measure this phase directly, without requiring the intermediate step of conversion to magnitude, and derive the angle of arrival more accurately.

Let's examine a radar interferometer design for measuring surface topography. We will see that this is equivalent to measuring the angle of arrival from a point of interest on the surface, as long as we know the platform altitude and the baseline between the two radar antennas. Consider a radar geometry similar to the adding interferometer figure but altered so as to put the antennas on an observing platform looking down at the surface, as in Figure 17.2. Similar to the adding interferometer, the extra path length to the second antenna is the projection of the baseline in the radar line of sight:

$$\delta = B\sin(\theta - \alpha) \tag{17.6}$$

For this and the examples in the rest of this chapter, let us assume a system configuration where we treat each antenna as a complete radar system, that is, we transmit and receive signals from both positions and record the echo from each at each. This will usually be the appropriate model for spaceborne platforms, where synchronization of multiple satellites is challenging, though not

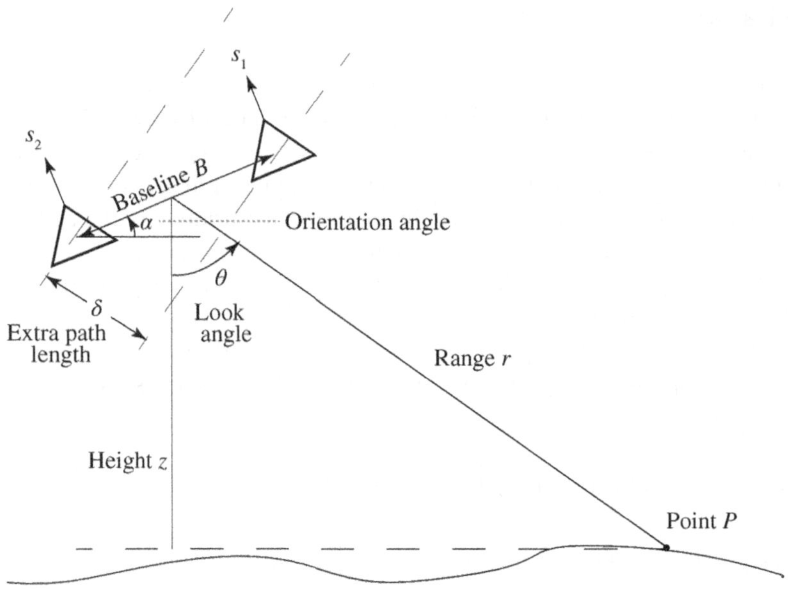

Figure 17.2 Radar interferometer geometry. Two radar antennas on an airborne or spaceborne platform illuminate a point on the surface below. The range from the antenna array center to the point is r, and the range difference from the first to the second antenna is δ, where δ as before is the projection of the baseline B in the line of sight from the radar to the point P.

impossible. For our purposes, the only difference between this approach and one in which we transmit at a single position and receive at two locations is the change from a factor of 2 to a factor of 4 in the phase accumulated for each wavelength of extra path length δ. Under the two transmitter assumptions, the factor is 4. Hence, the measured signals at antennas 1 and 2 respectively, are then

$$s_1(t) = a \exp\left[-j\frac{4\pi}{\lambda}\left(r(t) - \frac{\delta(t)}{2}\right)\right] \tag{17.7}$$

and

$$s_2(t) = a \exp\left[-j\frac{4\pi}{\lambda}\left(r(t) + \frac{\delta(t)}{2}\right)\right] \tag{17.8}$$

where both r and δ vary with echo time as we are imaging a surface. Instead of adding these two signals, we form the product of s_1 with the conjugate of s_2, to form the interferometric image i:

$$\begin{aligned} i(t) &= s_1(t) \cdot s_2(t)^* \\ &= a \exp\left[-j\frac{4\pi}{\lambda}\left(r(t) - \frac{\delta(t)}{2}\right)\right] \cdot a \exp\left[+j\frac{4\pi}{\lambda}\left(r(t) + \frac{\delta(t)}{2}\right)\right] \\ &= a^2 \exp\left[j\frac{4\pi}{\lambda}B\cos(\theta(t) - \alpha)\right] \end{aligned} \tag{17.9}$$

The measured interferogram phase $\phi(t)$ for a point on the surface then yields θ for that point, if the other parameters are known:

$$\phi(t) = \arg\ [i(t)] \tag{17.10}$$

$$\theta(t) = \arcsin\left[\frac{\lambda\phi(t)}{4\pi B}\right] + \alpha \tag{17.11}$$

Recall that our goal here is to measure the topography at point P. Completing the triangle shown in Figure 17.2, we see that we can obtain the height from P to the altitude z of the platform as

$$z = r\cos\theta(t) \tag{17.12}$$

In other words, the InSAR phase at each point on the surface can be interpreted as the elevation of that point, or more precisely, the distance below the platform of every point.

A more subtle point here is that we have assumed that the two echoes received at a time t correspond to exactly the same point on the ground, which will not be true if the ranges from the point to the two antennas differ. The signals must be aligned in time for the phase to have a relation of Eq. (17.9), so that rather than a simple multiplication, we compute the correlation of the two signals and select the phase at the correlation peak. Hence, we are in fact implementing a *correlation interferometer* instead of the multiplying interferometer. Once the signals are properly aligned so that echoes from each point on the surface match in both channels, the multiplying equation results hold.

17.4 Geometrical Interpretation of Phase Difference

The discussion of the multiplying interferometer in Section 17.3 uses a simplification in which we modeled the incoming wavefront as a plane wave, an approximation that holds for a target at infinite distance. We can be more accurate and accommodate objects at any distance if we use a more general geometry and interpret the measurements geometrically as path length differences rather than as angle of arrival. The more general configuration is shown in Figure 17.3. We note that we quantify the distances to the two antennas as r_1 and r_2, without requiring plane wave incidence, so that the signals we receive and measure as interferometric phase may be given in terms of the difference $r_1 - r_2$:

$$s_1(t) = e^{-j\left(\frac{4\pi}{\lambda}r_1 - wt\right)}$$

$$s_2(t) = e^{-j\left(\frac{4\pi}{\lambda}r_1 - wt\right)}$$

$$s_1(t)s_2^*(t) = e^{-j\frac{4\pi}{\lambda}(r_1 - r_2)}$$

so the interferometer phase $\phi = \dfrac{4\pi}{\lambda}(r_1 - r_2)$. Note that we have omitted the signal amplitude a in these equations as we have already established that it does not affect the phase retrieval.

17.5 Measuring Topography Without the Parallel Ray Approximation

Accurate determination of topographic elevation requires solving for the two triangles illustrated in Figure 17.3. Consider first the triangle formed by the two antennas and the point to be imaged P. We relate the ranges r_1 and r_2 to the baseline length $|B|$ using the law of cosines:

$$\begin{aligned} r_1^2 &= r_2^2 + |B|^2 - 2r_2|B|\cos\left(\frac{\pi}{2} - \theta + \alpha\right) \\ &= r_2^2 + |B|^2 - 2r_2|B|\sin(\theta - \alpha) \end{aligned}$$

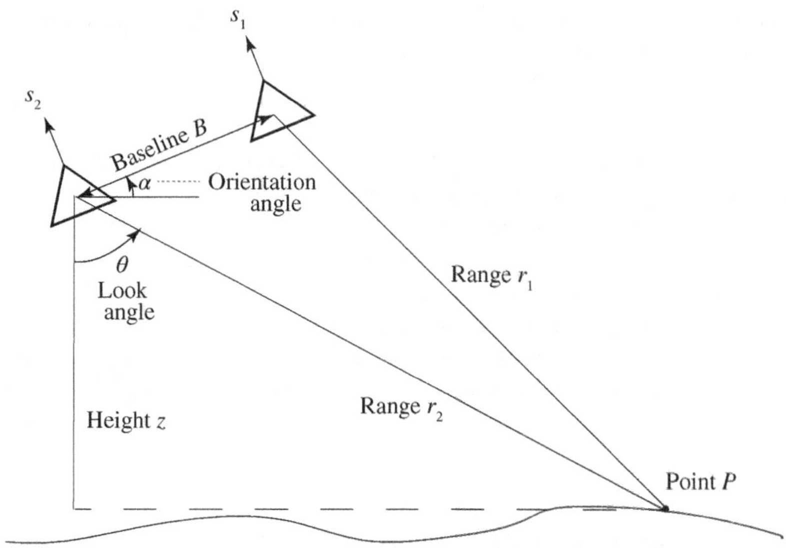

Figure 17.3 Path length differences. The distances from the point of interest to the two antennas are r_1 and r_2, and the InSAR phase is the difference of the ranges multiplied by $\frac{4\pi}{\lambda}$.

Rearranging to find the enclosed angle,

$$2r_2|B|\sin(\theta-\alpha)=r_2^2-r_1^2+|B|^2 \tag{17.13}$$

$$\begin{aligned}|B|\sin(\theta-\alpha) &= \frac{r_2^2-r_1^2+|B|^2}{2r_2}\\ &= \frac{(r_2+r_1)(r_2-r_1)+|B|^2}{2r_2}\\ &\doteq \frac{2r_2(r_2-r_1)+|B|^2}{2r_2}\\ &= \frac{\delta 2r_2+|B|^2}{2r_2}\\ &= \frac{\phi\lambda}{4\pi}+\frac{|B|^2}{2r_2}\end{aligned} \tag{17.14}$$

Once we have calculated $\sin(\theta-\alpha)$, we use that quantity to solve for the height z as before. It is also worth noting that Eq. (17.14) reduces to our previous result (Eq. 17.11) when $r_2 >> |B|$.

17.6 Measurement Accuracy

We have seen that we can relate the phase observed in a radar interferometer to the topographic elevation of a surface. Let us now see how accurately we can make that determination, and what factors in a radar system contribute most significantly to the error in measurement. Note that we can make several kinds of errors, two of which are (i) phase errors (ϕ) and (ii) tilt errors (α). The effect of each error class is different: phase errors are uncorrelated from point to point, while tilt errors tilt the entire output image.

We can quantify the uncertainty of our height (z) determination by computing the partial derivatives of the height with respect to our system parameters.

Measuring phase. The interferometer resolves height through measurement of the phases of radar echoes. We need to be able to relate the accuracy of these phases to system parameters, such as SNR. To quantify this, refer to Figure 17.4. The phase we observe ϕ is the angle of the vector sum of the signal and a noise term. The ratio of the lengths of the signal and noise is the square root of the signal to noise ratio, because the lengths of the vectors are amplitude and not power quantities:

$$\frac{|s|}{|n|} = \sqrt{(SNR)} \tag{17.15}$$

The measured phase error $\delta\phi$ will be

$$\begin{aligned}\delta\phi &= \arctan \frac{|n| \sin \theta}{|s| + |n| \cos \theta} \\ &= \arctan \frac{\sin \theta}{\sqrt{(SNR)} + \cos \theta}\end{aligned}$$

The noise term $|n|$ and its direction θ will not be constant for every measurement, thus the phase error $\delta\phi$ can only be described by its statistics. In a model where the noise is assumed to be drawn from a bivariate Gaussian distribution, we can model the pdfs of the angle as a uniform random variable and of the amplitude as a Rayleigh random variable. For unit variance of the bivariate Gaussian in both axes, we obtain

$$p_{\theta}(\theta) = \frac{1}{2\pi}, \quad \theta = [0, 2\pi) \tag{17.16}$$

and

$$p_{|n|}(|n|) = 2|n| \exp(|n|^2), \quad |n| = [0, \infty) \tag{17.17}$$

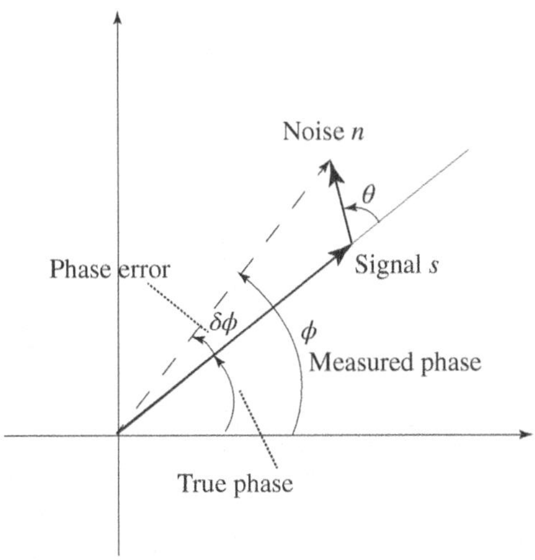

Figure 17.4 Phase noise. We measure a phase ϕ that is the angle of the sum of a complex (vector) signal plus a noise term. The higher the signal to noise ratio, that is the ratio of the vector lengths, the smaller the phase error will be.

Then the first and second moments of $\delta\phi$ would be

$$\mu_1 = \int_0^\infty \int_0^{2\pi} \arctan \frac{|n| \sin \theta}{\sqrt{(SNR)} + |n| \cos \theta} p_\theta(\theta) p_{|n|}(|n|) d\theta d|n| \tag{17.18}$$

and

$$\mu_2 = \int_0^\infty \int_0^{2\pi} \left(\arctan \frac{|n| \sin \theta}{\sqrt{(SNR)} + |n| \cos \theta} \right)^2 p_\theta(\theta) p_{|n|}(|n|) d\theta d|n| \tag{17.19}$$

from which we can calculate the standard deviation σ_ϕ. These integrals are complicated to derive analytically, but can be readily computed numerically. Figure 17.5 shows numerical evaluations (black dots) for several values of SNR, along with a good approximation that is often used for InSAR phase noise:

$$\sigma_\phi = \frac{\sqrt{1 + 2SNR}}{SNR} \frac{1}{2\sqrt{N}} \tag{17.20}$$

where N is the number of complex pixels averaged, or "looks." Increasing the number of looks has a similar effect as increasing SNR as both represent a coherent sum that adds signal in phase and noise by power only.

Phase errors. Now let us return to the accuracy of our elevation estimates. We derived a relation between the height z of a surface and InSAR phase in two coupled equations, which we repeat:

$$|B| \sin(\theta - \alpha) = \frac{\phi\lambda}{4\pi} + \frac{|B|^2}{2r_2} \tag{17.21}$$

and then solve for height:

$$z = r \cos \theta \tag{17.22}$$

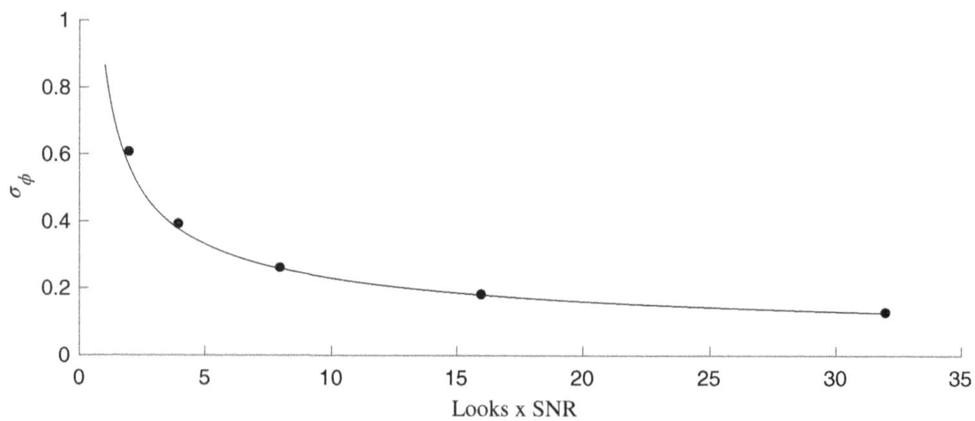

Figure 17.5 Phase noise as a function of SNR. InSAR phase noise decreases as the SNR increases. Black dots show numerical evaluation of Eq. (17.19), and the solid line shows the useful approximation Eq. (17.20). The approximate expression includes the number of looks (here 1), as averaging over pixels to form looks leads to the signal adding coherently with the noise adding incoherently. Thus increasing looks has a similar effect as increasing SNR. Both improve the accuracy as the square root of the increase.

Our needed sensitivity is how z varies with ϕ, or the derivative $\frac{\partial z}{\partial \phi}$. But our two equations do not yield this result straight-forwardly, so instead we compute intermediate derivatives $\frac{\partial \theta}{\partial \phi}$ and $\frac{\partial z}{\partial \theta}$, whose product is the needed relation. So our approach is to calculate first

$$\frac{\partial}{\partial \theta}|B| \sin(\theta - \alpha) = \frac{\partial}{\partial \phi}\left[\frac{\phi\lambda}{4\pi} + \frac{|B|^2}{2r_2}\right]$$

$$|B| \cos(\theta - \alpha)\, \partial\theta = \frac{\lambda}{4\pi}\partial\phi$$

and then infer

$$\frac{\partial \theta}{\partial \phi} = \frac{\lambda}{4\pi}\frac{1}{|B| \cos(\theta - \alpha)} \tag{17.23}$$

Noting that uncertainties in α and θ are equivalent save for change in sign, repeating a simpler argument for $\frac{\partial z}{\partial \theta}$,

$$\frac{\partial z}{\partial \theta} = r \sin\theta$$

Now we compute the product to get the desired derivative:

$$\frac{\partial z}{\partial \phi} = \frac{\partial z}{\partial \theta}\frac{\partial \theta}{\partial \phi} = \frac{\lambda}{4\pi}\frac{r \sin\theta}{|B| \cos(\theta - \alpha)} \tag{17.24}$$

Stating this in terms of the uncertainties in z and ϕ, σ_z, and σ_ϕ, respectively,

$$\sigma_z = \frac{\lambda}{4\pi}\frac{r \sin\theta}{|B| \cos(\theta - \alpha)}\sigma_\phi \tag{17.25}$$

We can get a feel for how accurate this might be in practice by looking at some nominal system parameters. Suppose we can measure ϕ to an accuracy of about 0.1 rad, or about 6°. In an aircraft system, assuming an orientation angle $\alpha = 0$, with $r = 10$ km, a wavelength of 0.06 m, look angle 30°, and baseline distance 1 m, our height accuracy would be

$$\sigma_z = \frac{0.06}{4\pi}\frac{10000 \cdot 0.5}{1 \cdot 0.866} \cdot 0.1$$

$$= 2.8\text{m}$$

So with such an aircraft system, the contribution to measurement error from phase noise is 2.8 m. A similar set of computations for the error due to tilt, and noting that uncertainty in α has the same effect but with a negative sign as uncertainty in θ, follows from the derivative of z with respect to θ we calculated earlier. In terms of uncertainties then,

$$\sigma_z = r \sin\theta\, \sigma_\theta$$

which for the aircraft system example here, and assuming we can measure the tilt to 0.001 rad,

$$\sigma_z = 10000 \cdot 0.5 \cdot 0.001$$

$$= 5\text{m}$$

Thus, you can see that for this system, if the uncertainties in the system are as used here, the overall accuracy will be dominated by tilt rather than phase noise errors.

In a spaceborne implementation, we will very often use two separate satellites, or a single satellite in a nearly repeating orbit, to form the interferometer baseline. The use of multiple platforms allows us to choose the baseline length, since we are not restricted to a single structure supporting two antennas, which is constrained by the size of the orbiting platform. Why is this helpful? Suppose we can measure the relative orbit position accuracy across-track to 10 cm, which is easily achievable for many orbital systems today. Then the accuracy of the virtual tilt angle α is 10 cm/$|B|$ rad. For 100 m satellite separation, $\sigma_\alpha = 0.1/100 = 10^{-3}$, while for a 1 km baseline $\sigma_\alpha = 10^{-4}$.

For this configuration, both the uncertainty due to phase noise (Eq. 17.25) and tilt decrease with baseline length. So it might seem that the best accuracy would result from allowing the baseline length to grow as large as possible. For an aircraft system there is a limit to the baseline length set by the dimensions of the aircraft itself, but for two satellites, very large separations are possible. But, as we are about to see, longer baselines lead to more noise in the phase measurement, and hence an engineering trade-off exists in the baseline length design.

17.7 Decorrelation

We see that the error in our measurements is minimized for as $|B|$ becomes larger and larger, so it might seem that we should always seek to maximize $|B|$. But there is another effect that increases phase noise if we let $|B|$ grow without limit. This additional noise source is called baseline decorrelation noise, and it arises because we have viewed the surface from two view points that are dissimilar.

Physically, what is happening is at each resolution element we measure the coherent sum of reflected waves from each individual scattering center in that element. If we change the viewing geometry, the waves will add in a different fashion and the average phase will not be preserved (Figure 17.6). We want to determine the degree to which the phases from the sums of all the points ϕ are related to each other, or, more formally, correlated.

For this calculation, we consider the radar return from a resolution element defined using ground range coordinates where the along-track (azimuth) coordinate is x, and the across-track y. Referring to Figure 17.7, the left-right dimension represents a slice in the y direction, with the x direction

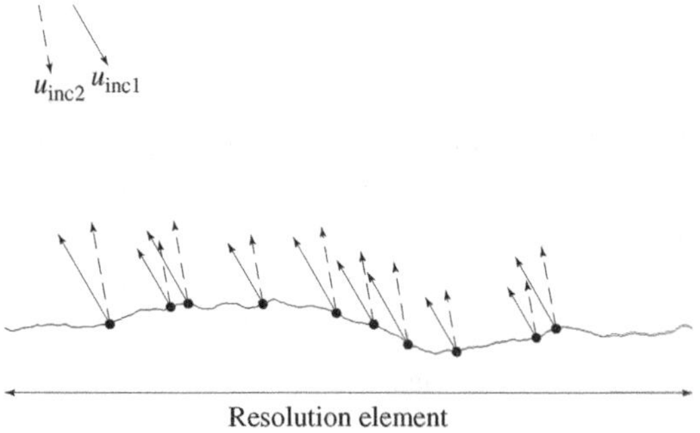

Figure 17.6 Spatial decorrelation noise. For each resolution element, we observe the coherent sum of the reflected waves from each scattering center in the element. If we change the incidence angle, here from u_{incl} to u_{inc2}, the waves will add to a different value as their line of sight paths differ. The mean phase of the total return wave (ϕ) will vary as the wave directions change.

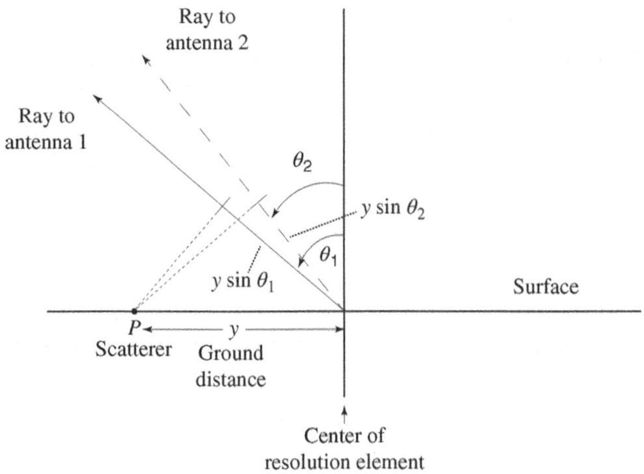

Figure 17.7 Spatial decorrelation geometry. Quantities needed to calculate the correlation between two radar echoes as we vary the separation of the incidence angles for each signal. The phase of a reflected wave from a single scatterer P changes as the line of sight distance $y \sin \theta$ from the center of the resolution element changes.

into the page. We model the total signal s_1 measured at antenna 1, at incidence angle θ_1 corresponding to u_{incl}, as the sum of returns from all points on the surface:

$$s_1 = \int \int f(x - x_0, y - y_0) \exp\left[-j\frac{4\pi}{\lambda}(r + y \sin \theta_1)\right] w(x, y)dxdy \tag{17.26}$$

where r is the range to the center of the resolution element (x_0, y_0), $w(x, y)$ is the impulse response of the system, and $f(x - x_0, y - y_0)$ is the complex reflectivity of surface position.

Similarly, for s_2:

$$s_2 = \int \int f(x - x_0, y - y_0) \exp\left[-j\frac{4\pi}{\lambda}(r + y \sin \theta_2)\right] w(x, y)dxdy \tag{17.27}$$

The complex product $s_1 s_2^*$ is thus the following complicated-looking integral:

$$\begin{aligned} s_1 s_2^* = \int \int \int \int & f(x - x_0, y - y_0) f^*(x' - x_0, y' - y_0) \cdot \\ & \exp\left[-j\frac{4\pi}{\lambda}y(\sin \theta_1 - \sin \theta_2)\right] w(x, y)w^*(x', y')dxdydx'dy' \end{aligned} \tag{17.28}$$

We can greatly simplify this expression if we assume that the scattering at each point on the surface is uncorrelated from every other point, so that the autocorrelation function of the surface reflectivity is a δ-function:

$$< f(x, y)f^*(x', y') >= \sigma^0 \delta(x - x', y - y') \tag{17.29}$$

reducing the integral to

$$< s_1 s_2^* >= \sigma^0 \int \int \exp\left[-j\frac{4\pi}{\lambda}y \cos \theta \delta\theta\right] |w(x, y)|^2 dxdy \tag{17.30}$$

where $\theta = (\theta_1 + \theta_2)/2$ and $\delta\theta = \theta_1 - \theta_2$. Normalizing the average $< s_1 s_2^* >$ to unity at zero separation defines the *spatial correlation* of the interferogram, which tells us how well related two echoes are as we vary the separation of the look angles. We define the correlation of the two signals as the magnitude of the normalized cross product $| < s_1 s_2^* > |$. We recognize Eq. (17.30) as the Fourier transform with respect to $\delta\theta$ of the square of the impulse response. If the impulse response is $\text{sinc}(x) = \frac{\sin \pi x}{\pi x}$, then the transform of its square is a triangle function, which is normalized to express correlation yields,

$$\rho_{spatial} = 1 - \frac{2 \cos \theta \delta_y d\theta}{\lambda} \tag{17.31}$$

δ_y being range resolution projected onto the ground. Changing the dependence on angle to dependence on baseline length, we can define $B_\perp$ as the component of the baseline perpendicular to the line of sight, and thus the difference angle.

$$d\theta = \frac{B_\perp}{r} \tag{17.32}$$

Combining this with Eq. (17.31) we obtain a similar triangle function that goes to zero when the baseline reaches a critical value $B_c = \frac{\lambda r}{2 \cos \theta \delta_y}$ (Figure 17.8). The critical baseline represents the maximum distance allowed between antennas if the interferogram phase is to be of nonzero average. In a practical system, we must limit the baseline to a value somewhat smaller than the critical value to keep the correlation at an acceptable level. Thus, again we have a trade-off to make: the baseline needs to be long enough to be sensitive to topographic variation but not so long as to be poorly correlated.

It is useful to keep a few typical values as a starting point for the design of an InSAR system. Table 17.1 lists several values of B_c for airborne and spaceborne platforms at common wavelengths and moderate (~5–10 m) resolution. Keeping the noise from decorrelation at an acceptable level, we would prefer to operate at about 10–20% of B_c. But that value tends to be too long to implement

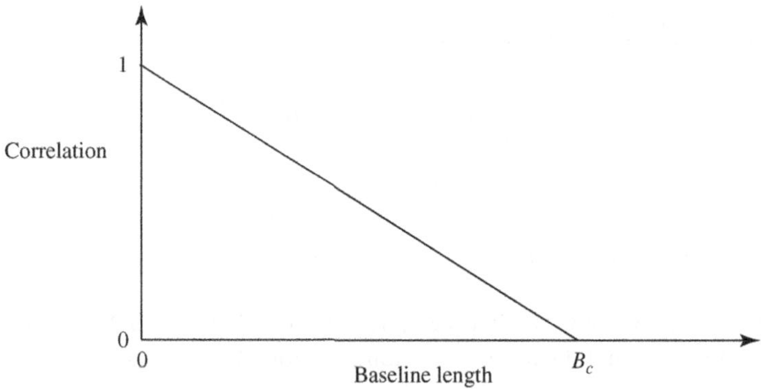

Figure 17.8 Correlation from idealized impulse response. If the radar impulse is well-characterized by a sinc() function, the degree of correlation as we increase the baseline length falls off as a triangle function, reaching zero at a maximum baseline length B_c, denoted the *critical baseline*.

Table 17.1 Typical InSAR critical baselines.

Aircraft C – band system:	~	100 m
Satellite L – band system:	~	5000 m
Satellite C – band system:	~	1250 m

on an aircraft or a single satellite. Thus, for aircraft we usually have a relatively short baseline and make up for the ϕ-sensitivity with high SNR, while for spacecraft we often prefer a single satellite in a repeat-pass mode, our next topic.

17.8 Decorrelation and SNR

It may seem that decorrelation and signal to noise ratio are related, and in fact they are different ways of describing the same statistics of radar measurements. We can readily derive one from the other.[1]

Let's begin by defining two different measurements of a repeating signal plus a variable noise component, as we illustrated in Figure 17.4. Denote these two measurements as

$$m_1 = s + n_1$$
$$m_2 = s + n_2$$

where s is the common signal vector and n_1, n_2 are the noise components of each. As before, the ratio of signal power to noise power is the SNR:

$$\text{SNR} = \frac{|s|^2}{|n_i|^2} \tag{17.33}$$

We calculate the correlation ρ of the two measurements through the usual definition:

$$\begin{aligned}\rho &= \frac{\langle m_1\, m_2^* \rangle}{\sqrt{\langle m_1\, m_1^* \rangle}\sqrt{\langle m_2\, m_2^* \rangle}} \\ &= \frac{\langle (s+n_1)\,(s+n_2)^* \rangle}{\sqrt{\langle (s+n_1)\,(s+n_1)^* \rangle}\sqrt{\langle (s+n_2)\,(s+n_2)^* \rangle}}\end{aligned}$$

We can assume that the signal and noise components are independent and thus uncorrelated, as are the two noise instances. The expectations in Eq. (17.34) were then reduced to,

$$\begin{aligned}\rho &= \frac{|s|^2}{|s|^2 + |n|^2} \\ &= \frac{1}{1 + \text{SNR}^{-1}}\end{aligned}$$

As we might expect, if the SNR grows very large, ρ approaches unity. And if the SNR goes to zero, so does the correlation. We can use either SNR or ρ to describe the statistics of our phase measurements. Sometimes it will be more useful to employ the SNR measure, and sometimes the decorrelation will prove more valuable.

17.9 Repeat-track Orbital Interferometry

We have just seen that we would like to use baselines of several hundred meters in length for spaceborne interferometric observations, and such distances are too large for single spacecraft construction. The way around this is either to use two spacecraft flying in formation (often very expensive!) or to use one spacecraft flying in a repeat orbit.

1 We often mix the terms correlation and decorrelation, which is fine as long as we remember that correlation (ρ) = 1 - decorrelation.

Recall that we have modeled the return from a resolution element as the sum of echoes from a randomly distributed set of scatterers on the ground. What happens if we view the same region again at a later time? If none of the scatterers have moved, the return will be exactly the same except for the system noise component. Thus repeat-orbit geometries can form interferometers over unchanging terrain.

What if the scatterers have indeed moved around? If the motions are random and on the order of a wavelength or more in size, the sum of the echoes will have a different propagational phase and there will no longer be correlation between the two radar echoes. It will be the same as if we had imaged the area with a spatial baseline exceeding the critical value.

But if the random motion is much smaller, we have a tool for characterizing surface changes at the wavelength scale. Using an argument following Eqs. (17.26–17.30) and allowing the position of each scatterer to move according to a Gaussian-distributed distance in each dimension, we can show that the amount of decorrelation from surface movement is related to amount of motion and wavelength by

$$\rho_{\text{temporal}} = \exp\left[-\frac{1}{2}\left(\frac{4\pi}{\lambda}\right)^2 [\sigma_y^2 \sin^2\theta + \sigma_z^2 \cos^2\theta]\right] \tag{17.34}$$

where we define ρ_{temporal} as the correlation for temporally-spaced observations, and σ_y and σ_z are the rms y and z motions. As was the case for spatial decorrelation, the temporal decorrelation will be at an acceptably low level as long as the motions are smaller than a large fraction of a wavelength.

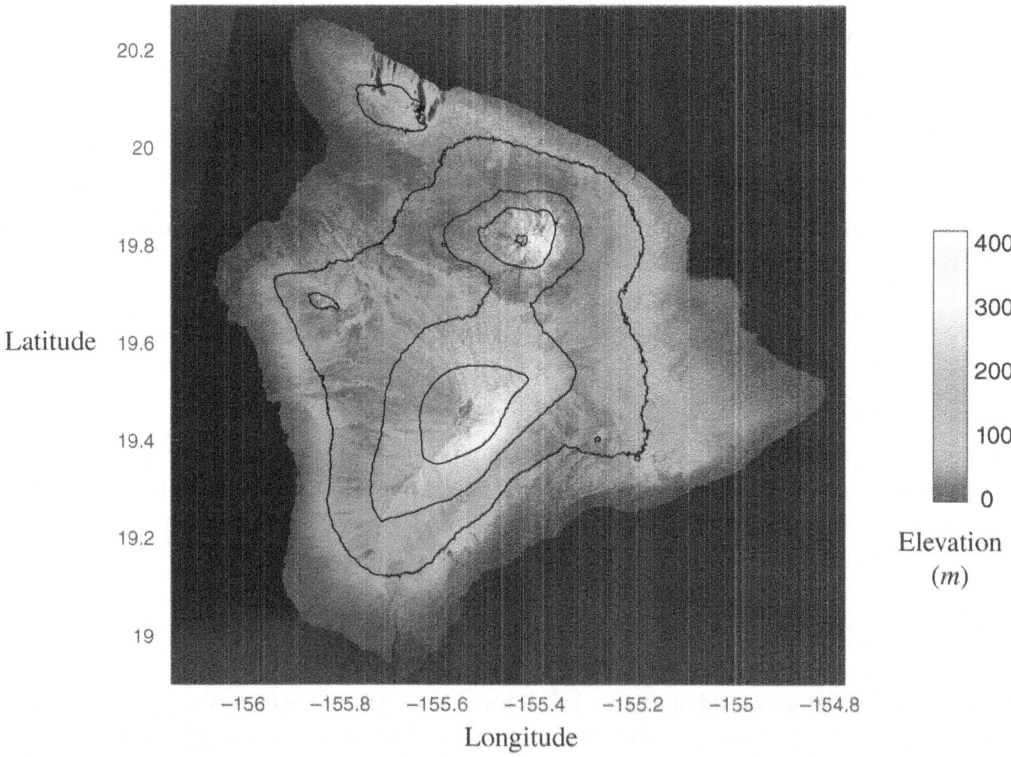

Figure 17.9 Elevation map of island Hawaii derived from repeat pass Sentinel-1A and -1B radar satellites. Estimates from several interferograms are averaged to reduce system noise. Brightness is radar backscatter magnitude, height coded as color. Contour lines depict heights at 1000, 2000, 3000, and 4000 m.

As long as the temporal correlation is high, which means that the random motion of scatterers in each resolution element is small, we can proceed as we did in the fixed array case earlier and form interferograms from the spaceborne images. We need to preserve the phase of each image to be able to calculate the phase differences, so we will require complex valued images, as we needed for the multiplying interferometer described earlier.

An example of an elevation (topography) map derived from a set of repeated pass spaceborne radar images is shown in Figure 17.9. These data were acquired over the island of Hawaii by the Sentinel-1A and -1B spacecraft between Nov. 2019 and March 2020. The complex interferogram resulting from the multiplying interferometer formed by two satellite passes yields the height, and in this figure several interferograms were averaged to reduce system noises. Elevation is shown as color, with contour lines added every 1000 m of altitude.

17.10 Motion Measurements

The main difference between the repeat orbit implementation and the fixed two-element array is that the two images are acquired at different times. This leads to sensitivity to mean motion of each resolution element, as the phase difference can change systematically, shifting the mean value of the phase difference but without appreciably affecting its variance.

To see this, suppose that instead of random motion, the surface scatterers move together toward or away from the radar. In this case, as a good approximation the phase of each scatterer's echo changes by the same amount. Thus, there will be no decorrelation, but instead an average phase change will be observed in the interferogram.

We can illustrate an InSAR system measuring the surface at two different times by an example of an aircraft measuring ocean currents (Figure 17.10). In this implementation, a patch of ocean is first observed at time t_1 and range r_1 (solid line airframe) by an antenna mounted forward on the aircraft body, and at a second time t_2 with range r_2 (dotted line airframe) by an antenna mounted aft on the aircraft. In the case when there is no motion of the surface, the two measurements will be identical and the interferogram phase will be zero. If there is a current u on the surface, the range will change by $\Delta x \sin\theta = u \cdot (t_2 - t_1) \cdot \sin\theta$ in the time it takes the aircraft to fly the distance between its two antennas. For an aircraft traveling at velocity v, with antennas separated by baseline B, $t_2 - t_1 = B/v$. The observed phase difference will be proportional to the line-of-sight component of the current velocity, and scaled by the sine of the incidence angle:

$$\begin{aligned}\Delta\phi &= \frac{4\pi}{\lambda}\,\Delta x \sin\theta \\ &= \frac{4\pi}{\lambda}\,u\,(t_2 - t_1)\,\sin\theta \\ &= \frac{4\pi}{\lambda}\,\frac{u}{v}\,B\sin\theta\end{aligned}$$

So for a swath-imaging radar interferogram, a map of $\Delta\phi$ gives the surface current distribution toward the radar. Computing the accuracy of a velocity measurement using the same logic we employed for the elevation interferometer, we derive:

$$\sigma_u = \frac{\lambda}{4\pi}\,\frac{v}{B\sin\theta}\,\sigma_\phi \tag{17.35}$$

For typical aircraft system parameters,

$$0.1\ \text{m/s} \approx \frac{0.06}{4\pi}\,\frac{200}{2\sin 30}\,0.1\ \text{rad} \tag{17.36}$$

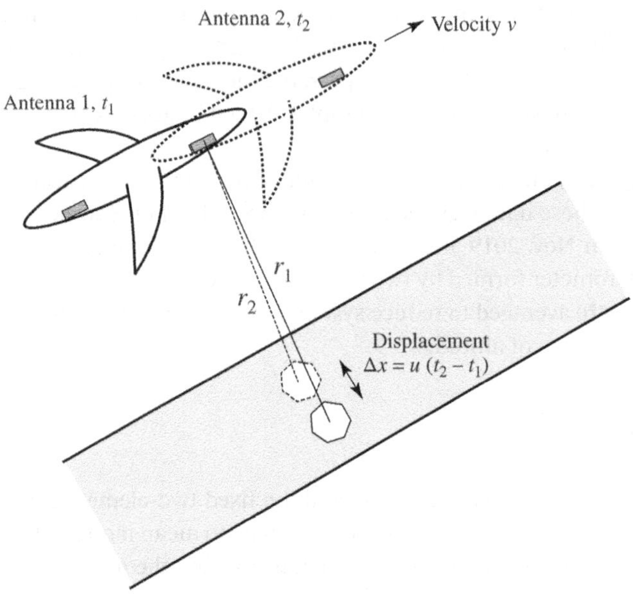

Figure 17.10 Ocean current geometry. An aircraft system for observing ocean currents in which a patch of ocean is first observed at time t_1 and range r_1 (solid line airframe) by an antenna mounted forward on the aircraft body, and at a second time t_2 with range r_2 (dotted line airframe) by an antenna mounted aft on the aircraft. If there is a current u on the surface, the range will change by $\Delta x \sin\theta = u \cdot (t_2 - t_1) \cdot \sin\theta$ in the time it takes the aircraft to fly the distance between its two antennas.

The velocity uncertainty of about 0.1 m/s is just what is needed for ocean current observations. But if we use a much longer temporal baseline than the short times available on the single aircraft implementation, we can greatly improve the sensitivity to subtler motions. Let us then look at a satellite implementation, where the time between observations can be weeks to years. In this case, we will find it more useful to use a phase sensitivity equation of form:

$$\Delta\phi = \frac{4\pi}{\lambda}\, u_{LOS}\, (t_2 - t_1) \tag{17.37}$$

where u_{LOS} is the radar line of sight component of velocity. We can also recognize $\Delta r = u_{LOS}\,(t_2 - t_1)$ as the total line of sight displacement over the time interval $\Delta t = (t_2 - t_1)$, which leads to a slightly different form of the sensitivity equation:

$$\sigma_r = \frac{\lambda}{4\pi}\, \sigma_\phi \tag{17.38}$$

The deceptively simple equation shows the power of the repeat pass satellite implementation. Let us look at a spaceborne C-band radar satellite interferogram, for which

$$0.005\ \text{m} \approx \frac{0.06}{4\pi}\, 0.1\ \text{rad} \tag{17.39}$$

For a modest phase uncertainty of 0.1 rad, we can measure a displacement of 5 mm. Depending on how long a time span we can use to form an interferogram, we can measure very slow rates indeed. For example, a one year interferogram yields 5 mm/yr sensitivity. This delicate sensitivity to motion permits observations of very detailed motions.

An example deformation image is shown in Figure 17.11. This image depicts the line of sight displacement coincident with an eruptive event on the island of Hawaii during the first half of 2018.

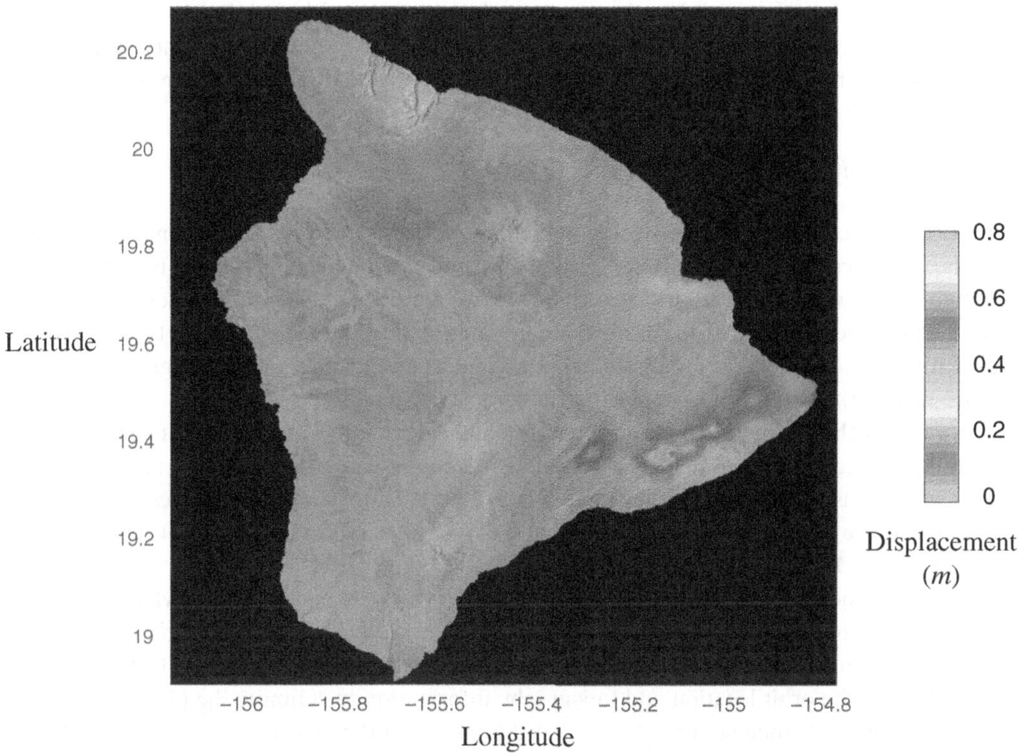

Figure 17.11 Surface deformation at Kilauea volcano in Hawaii. Image presents line of sight displacement coincident with an eruptive event on the island of Hawaii during the first half of 2018. Data were acquired by the Sentinel-1A/1B spacecraft between 6 March 2018 and 28 June 2018. The phase difference in each interferogram scaled by $\lambda/4\pi$ yields the change in range. Motion is seen mainly along the east rift zone of Kilauea volcano. The rest of the image is fairly stationary, but changes in atmospheric water vapor content lead to apparent motions. For this image, multiple interferograms were averaged to reduce the errors from the variable atmosphere.

Data were acquired by the Sentinel-1A and -1B spacecraft between 6 March 2018 and 28 June 2018; the phase difference of each interferogram was scaled by $\frac{\lambda}{4\pi}$ to yield the change in range. Motion is seen mainly along the east rift zone of Kilauea volcano. The rest of the image is fairly stationary, but changes in atmospheric water vapor content lead to apparent motions. For this reason, multiple interferograms are averaged for many geophysical investigations, as was done here.

17.11 Nonzero Baselines

In the two previous examples, we derived elevation and deformation estimates from radar interferograms. We found that separating antenna locations spatially permits us to infer heights, while separating antennas temporally yields sensitivity to motion. It is rather straightforward to create a spatial baseline with zero temporal offset by using a physical array of antennas, but when we try to implement a temporal baseline it is very challenging to have zero spatial offset. It is generally hard to have a spacecraft, for example, return to exactly the same spot in space at both acquisition times. So we must ask how this will affect our motion measurements.

We are asking what happens when the across-track component of the baseline is not zero. In that case, we get contributions from both motion and topography. We need to determine the relative sizes of these contributions, and how they can be separated in order to achieve the greatest accuracy. We begin with an equation for total phase that includes both effects.

$$\phi = -\frac{4\pi}{\lambda}|B|\sin(\theta - \alpha) - \frac{4\pi}{\lambda}\Delta r \tag{17.40}$$

The total phase measured has two components, one from antenna separation in space that depends on surface topography (through $z = r\cos\theta$), and one that depends on any line of sight displacement Δr between observation times. There are two main approaches in common use to isolate each other, one denoted the *two-pass method* and one the *three-pass method*. For this discussion, let us assume that we are most interested in surface deformation and that the topographic signal is what we would like to remove from our interferogram.

Two-pass method. The first term on the right hand side of Eq. (17.40) is the contribution from the parallax between the two incidence directions. This phase goes to zero if the baseline length is zero, which was the assumption in the derivation we used for a motion interferometer. But it is hard to achieve this in practice, as even what we refer to as repeat orbits do not repeat exactly. Separations from 100's to 1000's of meters are common.

Even if we cannot control satellites to near zero baselines, cross-track orbit knowledge, and thus the InSAR baseline, is often available at the cm level. Further, for many applications, and many parts of the Earth, the topography itself may be known to a few meters or less. In the two-pass method, we use the orbit location and topography information to compute the phase from variations in elevation. We then subtract the modeled phase from the measured total phase and what remains is the contribution from deformation.

We found in Eq. (17.25) how σ_z depends on σ_ϕ, which inverted yields

$$\sigma_\phi = \frac{4\pi}{\lambda}\frac{|B|\cos(\theta - \alpha)}{r\sin\theta}\sigma_z \tag{17.41}$$

and thus we can compute how much phase error we might expect from uncertainty in topography or orbital position. Substituting nominal spacecraft C-band repeat pass parameters, we let $|B| = 1000$ m, $\alpha = 0$, $\theta = 45°$, $r = 800$ km, and $\sigma_z = 1$ m, and obtain $\sigma_\phi \approx 0.26$ rad. This corresponds to a line of sight deformation uncertainty of $\lambda/4\pi \cdot 0.26 = 1.3$ mm. In many cases, the remaining phase error is small compared to the desired deformation phase.

Three-pass method. We will not always be able to remove the topography-related artifact from our interferograms. This will most often happen when we do not know the topography sufficiently well, or at all, when we analyze an interferogram for crustal deformation. This can be the case for parts of the Earth when topography is either unknown or changeable, and is almost always the case when we are studying other planets. All is not lost, however, as we can use the measurements themselves to derive and remove the topography, as long as we have three satellite passes rather than two.

Refer to Figure 17.12. Using the parallel ray approximation to keep things simple, the phase from the unprimed coordinates will be

$$\phi = -\frac{4\pi}{\lambda}|B|\sin(\theta - \alpha) \tag{17.42}$$

and the phase from the primed coordinates is

$$\phi' = -\frac{4\pi}{\lambda}|B'|\sin(\theta - \alpha') \tag{17.43}$$

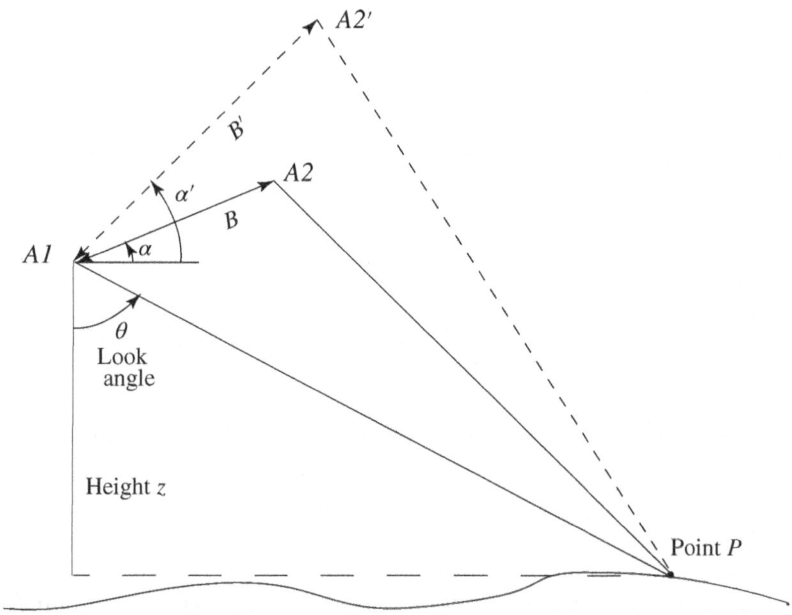

Figure 17.12 Three-pass method. This geometry pertains if we have three passes over our area of interest, rather than two. We form two interferometers, one from $A1$, $A2$, and P (solid lines), and one from $A1$, $A2'$, and P (dashed lines). Comparing the phases from each allows us to remove the topography phase term (see text).

Note that the two measurements each yield the same estimate of θ, or implied elevation. Our goal is to compute one from the other in order to eliminate the topographic dependence. Their ratio still depends on topography through θ

$$\frac{\phi'}{\phi} = \frac{|B'| \sin(\theta - \alpha')}{|B| \sin(\theta - \alpha)} \tag{17.44}$$

but we can eliminate that dependence by correcting the measured phase for the "flat-Earth" phase, which is the phase that would have been measured in the absence of topography[2] (Figure 17.13). Defining as ϕ_{flat} the corrected phase after subtracting the flat-Earth phase at look angle θ_{flat}

$$\phi_{flat} = -\frac{4\pi}{\lambda}[|B| \sin(\theta - \alpha) - |B| \sin(\theta_{flat} - \alpha)] \tag{17.45}$$

Since $\theta \approx \theta_{flat}$, we can simplify Eq. (17.45) using

$$\theta = \theta_{flat} + \delta\theta \tag{17.46}$$

$$\begin{aligned} \sin(\theta - \alpha) &= \sin(\theta_{flat} - \alpha + \delta\theta) \\ &\approx \sin(\theta_{flat} - \alpha) + \cos(\theta_{flat} - \alpha)\delta\theta \end{aligned}$$

so that

$$\phi_{flat} = -\frac{4\pi}{\lambda}|B| \cos(\theta_{flat} - \alpha)\delta\theta \tag{17.47}$$

2 Note that if we are using a spherical Earth rather than a truly flat Earth, the reference would be along the curved surface.

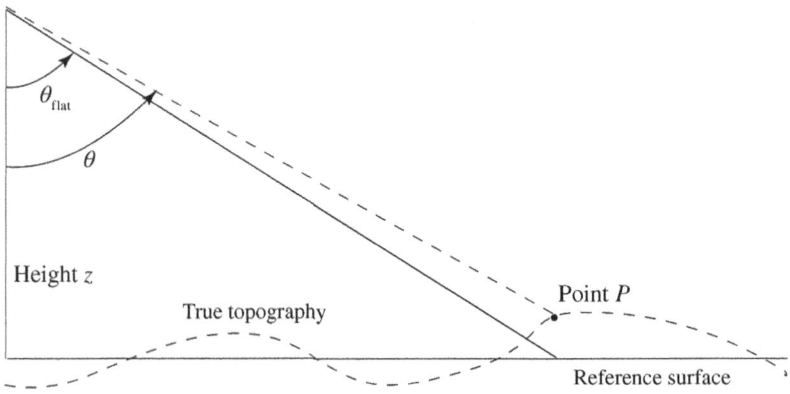

Figure 17.13 Flat-Earth phase: the phase pattern that would be observed using an imaging geometry identical to the acquisition under question but assuming that there is no topography present. The look angle for the flat-Earth phase is denoted θ_{flat}.

Under these approximations, the flattened phase is now proportional to the perpendicular component of the baseline rather than the parallel component, and to the topographic distortion $\delta\theta$ directly. So now when we form the ratio

$$\frac{\phi'_{flat}}{\phi_{flat}} = \frac{|B'|\cos(\theta_{flat} - \alpha')}{|B|\cos(\theta_{flat} - \alpha)} = \frac{B'_{\perp}}{B_{\perp}} \tag{17.48}$$

it no longer depends on topography. Consider again the total phase equation resulting from both topography and deformation, but this time in terms of flattened phase:

$$\phi_{flat} = -\frac{4\pi}{\lambda}|B|\cos(\theta_{flat} - \alpha) - \frac{4\pi}{\lambda}\Delta r \tag{17.49}$$

Now, further suppose that the time interval corresponding to the primed coordinates is when the point P moves closer to the radar by a distance δr, with no motion in the time covered by the unprimed interferogram. Then we can restate the deformation phase equation using the two flattened phase measurements:

$$\phi'_{flat} - \frac{B'_{\perp}}{B_{\perp}}\phi_{flat} = -\frac{4\pi}{\lambda}\Delta r \tag{17.50}$$

From this we see that we can acquire two interferograms over the same area, and use one to correct for the topographic phase term to yield the deformation phase. This does require knowledge of when deformation occurs, or at least a common rate of displacement, to be able to isolate it from the topography. But it does allow us to be able to form images of displacement when the actual surface shape is unknown.

A

The Ambiguity Function

We now return to the questions, first raised in Chapter 10, concerning the interplay of simultaneous radar determinations of range and velocity of a scattering object as inferred from a series of time-delay and Doppler-frequency observations. Earlier we saw that extending a particular waveform repetitively in time permits accurate determination of the Doppler shift of radar echoes from a coherent series of simple short pulses that are otherwise broadly spread in frequency. Likewise, use of a repetitive FM-CW swept-frequency waveform permits determination of a linear combination of time delay and Doppler frequency from a continuously transmitted signal. In the first example, employing pulses echo signal time delay and frequency possess a discrete system of ambiguities time and frequency. In the second example, the FM-CW waveform leads to a fundamental ambiguity in the coupling of time-delay and frequency-shift for even a non-repetitive waveform, which is replicated at the repetition frequency when the waveform is extended.

A.1 Recasting the Matched Filter

Previously, we saw that two extreme examples of waveforms, the FM-CW swept frequency signal, and pulsed wavetrains, each provide a simultaneous measure of target range and velocity. While a simple CW waveform is perfectly satisfactory for observing target velocity only, adding range capability with modulation introduces ambiguities. Similarly, the use of simple impulsive waveforms works well for observation of range only, but increasing the total signal length by coherent repetition of the waveform in order to refine the measure velocity introduces an ambiguity. Also, repetitious waveforms of either type introduce ambiguities in observational range-velocity space where there were none before. We now wish to generalize these specific results regarding the mensuration properties of simple signals to arbitrary waveforms.

A.1.1 Waveform Representation

For consistency with the literature, we introduce some new notation. Here we use $u(t)$ to represent the complex envelope of a narrow band radar signal, where $\omega_0 = 2\pi f_0$ represents the corresponding radian center or reference frequency for the analytic signal. The real and associated analytic forms are listed below:

	Real	Analytic	
Signal	$s(t)$	$u(t)e^{j\omega_0 t}$	(A.1)
Impulse Response	$h(t) = s(t_\mathrm{d} - t)$	$u^*(t_\mathrm{d} - t)e^{-j\omega_0 t_\mathrm{d}}e^{j\omega t}$	

With this notation, when $u(t)$ is the complex envelope of $s(t)$, $u^*(t_d - t)$ is the complex envelope of a filter matched to $s(t)$.

A.1.2 Output of a Matched Filter

With the representations above, the continuous time output of a matched filter can be written,

$$s_0(t) = \Re\left\{\underbrace{\left[e^{-j\omega_0 t_d}\int_{+\infty}^{-\infty} u(t')u^*(t'-t+t_d)\,dt'\right]}_{\text{Complex envelope of } s_0(t)} e^{\omega_0 t}\right\} \tag{A.2}$$

In the integrand, the product,

$$u(t)u^*(t'-t+t_d)\bigg|_{t=t_d} = |u(t)|^2 \tag{A.3}$$

displays complete cancellations of time-domain phase. This is the time domain equivalent of the behavior we saw earlier in the frequency domain, Eq. (7.41).

Ambiguities in range and velocity measurements occur when the filter output displays a significant response to range or velocity other than that which is intended in the design. Consider a tapped delay line model of the matched filter discussed in Section 7.2, as reproduced in Figure A.1. We recognize that the signal duration, t_d, could be greater than the tap spacing, Δ. In the model, we expect an output signal $s^{(n)}(t) = s(t - T_0 - n\Delta)$ from the tap that aligns the signal at time $t - T$ with $t - T_0$, i.e., $T = T_0 - n\Delta$, so that $s^{(n)}(0)$ is the start of the delayed echo aligned with the nth tap. This certainly occurs when there is an echo signal at the range corresponding to the prescribed delay. Now that we allow the signal duration to exceed the tap delay, we must also consider the effect of other signals on the output of the tap of interest. Let $t_d = 3\Delta$, for example. In this instance, the input signals from three different delays overlap, so that the output of the nth tap becomes the sum of the desired signal plus misaligned signals from two other delays,

$$\begin{aligned} s_{\text{tot}}(t) &= s(t - T_0) + s(t - (T_0 + \Delta)) + s(t - (T_0 + 2\Delta)) \\ &= s^{(n)}(t) + s^{(n+1)}(t) + s^{(n+2)}(t) \end{aligned} \tag{A.4}$$

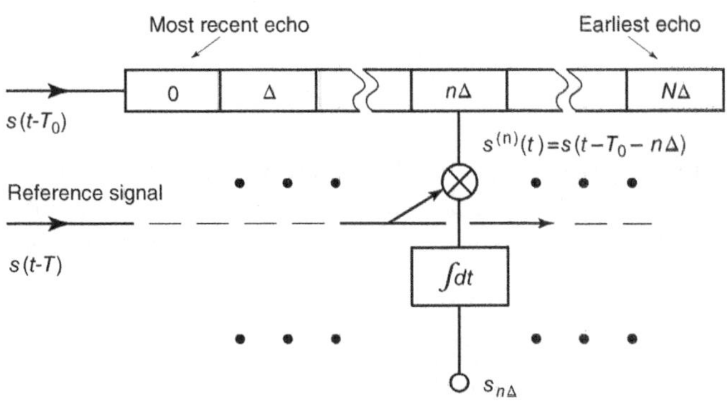

Figure A.1 Segment of a tapped delay line. At any delay the tapped delay line is a simple correlator driven by the reference signal.

where $s^{(n)} = s(t - (T_0 + n\Delta))$ represents an echo at a delay $n\Delta$ earlier than T_0, and for this example the only residual echoes occur for $0 \leq n \leq 3$. Note that at the nth the only signal $s^{(n)}(t)$ is aligned with the reference signal $s(t - T)$, while signals $s^{(n+1)}(t)$ and $s^{(n+2)}(t)$ are misaligned by times Δ and 2Δ. Whatever discrimination in delay that occurs among these three temporally overlapping signals is the result of the correlation applied at the output of the tap to produce $s_{n\Delta}$. The alternative is to change the spacing of the delay line taps to, say, $\Delta' = 3\Delta$; such a change completely isolates the signals at each tap, but at the cost of increasing the minimum delay resolution by a factor of 3:1, defeating the goal of maximizing the temporal resolution. Keep in mind, as this example illustrates, that in using matched filter signal selection there is no requirement that the waveform employed be short or have a simple structure, although there is the obvious limit for a repetitively pulsed waveform. Also remember that we only need to look at the filter outputs at the anticipated times of maximum response, which occurs after additional delay for the correlator integration.

Echoes from ambiguous ranges can also produce an output at the nth tap. This is very easy to visualize if one is thinking in terms of, say, short pulses with one echo from a target within the unambiguous range and a second echo from a target at an ambiguous range offset from the desired range by a multiple of R_{app}, and works the same way as in Section 10.3.

Now suppose that we wish to measure range and Doppler shift simultaneously. How would we go about this? Clearly, we need to extend the functioning of our processor to the frequency domain. We already know that the optimal filter in terms of maximum SNR for any waveform is the matched filter. So our best choice for the detection of a Doppler shifted signal is a filter matched to our transmitted waveform but adjusted for reception at the Doppler shifted frequency.

We accommodate the Doppler shift by changing the reference waveform employed in our tapped delay line model, where the new reference signal s_{f_D} is obtained by shifting the transmitted waveform in frequency by f_D. For this we need to add a new output tap matched to a Doppler shifted signal at some particular time delay, as before. With these considerations, the actual signal used for the correlation is the delayed echo waveform from a point reflector shifted in frequency by a particular frequency, or $s_{f_D}(t - T)$. There is no change to the basic tapped delay line structure other than adding a signal path and reference signal for correlation with Doppler shifted waveform. The form of the required additional taps is shown in Figure A.2.

We extend this approach to cover all possible Doppler shifts within the aliased bandwidth of a set of repetitive radar pulses. This is accomplished with a bank of "Doppler-matched" filters created

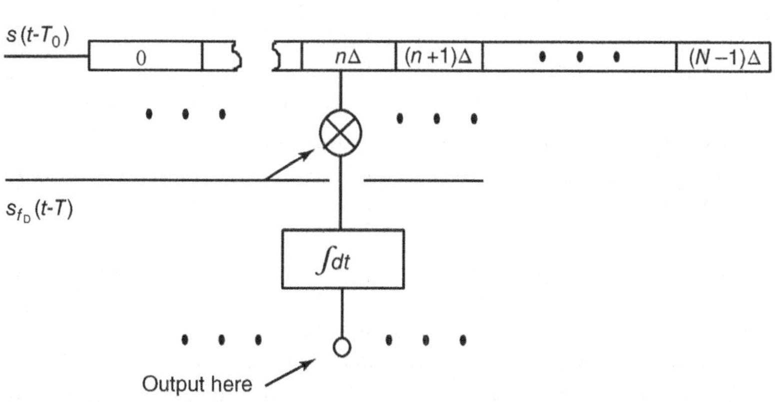

Figure A.2 Tapped delay line with Doppler shifted reference signal. Tapped delay line matched filter can be "tuned" to Doppler shifted echo signal by changing the carrier frequency of the multiplying, or correlator, waveform. Here the frequency shift is f_D. This is the only change from system shown in Figure A.2.

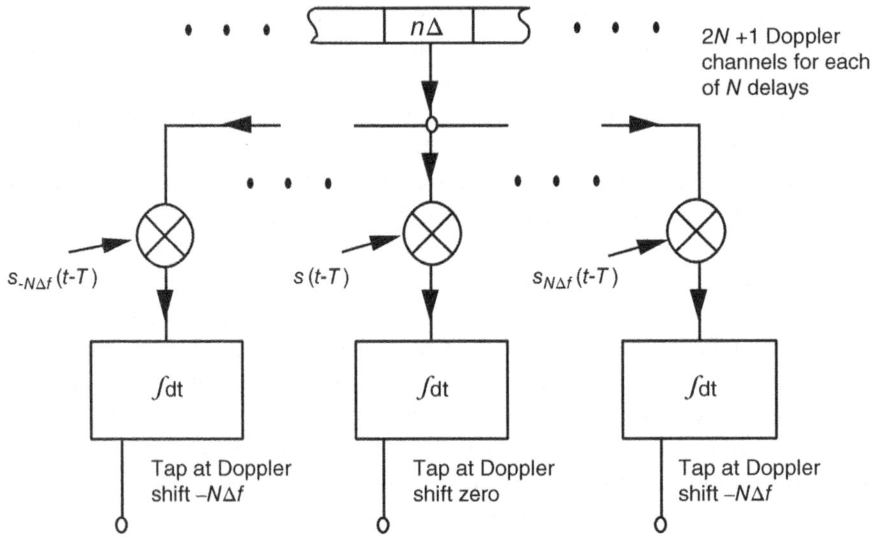

Figure A.3 Fanout of Doppler filters. A bank of matched filters can be synthesized at each delay by distributing the tap outputs to a bank of multipliers and integrators. Each multiplier in the bank must be driven by an appropriately "Doppler-shifted" waveform. The use of $2N + 1$ filters provides a reasonably uniform response in the frequency domain.

by replicating the form of Figure A.2, fanning out the taps from each delay to a corresponding set of filters, as indicated in Figure A.3.

How much complexity is involved in implementing such a scheme? For a radar transmitting simple pulses the spectrum of a single echo is determined essentially by the inverse of the pulse duration $\approx 1/\Delta$, which determines the finest frequency resolution that can be achieved, as discussed in Section 10.5. On the other hand, if the signal duration is 3Δ, as in our example above, then our discussion in Sections 10.5 and 10.6 suggests that the frequency resolution can be improved by a factor of three, in keeping with the increase in observation time. We will see that in general that the coherent observation time is the determinant of the frequency resolution that can be obtained, and ultimately the number of Doppler filters that are required.

For a radar emitting a repetitive waveform, the maximum duration of observation from each individual transmission is the interpulse interval T, implying that the finest frequency resolution that can be achieved of $1/T$, which is the same as the PRF. Recall that for coherent waveforms the PRF is also the spectral line spacing. This implies that the number of frequencies that can be estimated independently between two aliased lines in the frequency spectrum is one. More formally, for an observation time T, dividing the frequency span to be observed by the minimum channel width δf gives,

$$\mathrm{PRF} \div \delta f = \left(\mathrm{PRF} = \frac{1}{T}\right) \div \left(\mathrm{Min\_channel} = \frac{1}{T}\right) = 1 \tag{A.5}$$

Interpreting this, it is not possible in a single observational interval to improve the frequency resolution. In order to do better, we must increase the effective observation time. Extending the observation time by the inclusion of multiple echoes from a sequence of transmission into a single frequency observation allows us to improve the frequency resolution by the number of pulses employed, M,

$$\mathrm{PRF} \div \delta f = \left(\mathrm{PRF} = \frac{1}{T}\right) \div \left(\mathrm{Min\_channel} = M\frac{1}{T}\right) = M \tag{A.6}$$

Such a processing system can be achieved by generalizing the tapped delay line so that the correlator integration extends over multiple transmissions—perhaps many such intervals. This requires treatment of a sequence of echo signals as sample data from the scattering object. Implementation is achieved by adding switches between the multipliers and integrators in the vertical paths in Figure A.3 in order to isolate the echoes from each delay for purposes of data processing. Processing of signals from multiple transmissions requires the use of a system that is "coherent" in the sense of the discussion of Section 10.5.

A.1.3 A Bank of Doppler-matched Filters

When we extend the fan out of Doppler filters in Figure A.3 to every delay tap, then we create a filter bank spanning some portion of possible range delays and Doppler shifts. In this instance, the output of the radar processor array can be represented by a two-dimensional matrix corresponding to pairs of time-delayed and Doppler-shifted values of the received signal waveforms. The output of such a processor is referred to as a "delay-Doppler map" of the relevant target space. Ambiguous values of Doppler shift appear at the frequency aliased Doppler taps in this matrix, in a manner similar to echo signals at ambiguous ranges in our example of range ambiguity with no, i.e., zero, Doppler shift. In order to understand this behavior, consider the power spectrum of a repetitive, coherent waveform, such as that depicted in Figure 10.25, and the Doppler shifted version of such a spectrum as in Figure 10.26. In order to understand the Doppler ambiguity here, consider that the lines of a coherently repeated wavetrain are separated by the PRF, as in the referenced figures. Additionally, the input to the tapped delay lines is shifted through the various delay bins and read once per transmitted waveform, so that the output of the taps is coherently sampled at the PRF, and the spectrum computed from a sequence of echoes at each tapped range delay. As a result, the spectrum of the output at each range is aliased in frequency by multiples of the PRF. Consequently, aliased spectra all fall directly on top of each other, so that a signal with $f_D = 0$, for example, appears only at the DC position in the computed sampled spectrum, and not as a series of repetitive lines. The same is true for a signal at any other Doppler shift. We examine the implications of this structure for the choice of waveform and fundamental limitations on radar observations below.

We emphasize that the tapped delay line is only one of several ways in which the required processing can be implemented. All such schemes are equivalent to versions of the tapped delay line, however. Remember, that in an actual system we only have access to signals at the output taps, regardless of how the internal processing is arranged. So, for studies of the principles involved we need not be too concerned with details of the particular implementation, other than for illustration. While we initially drew the tapped delay line in connection with short pulse waveforms, this restriction on the waveform is not necessary. The structure represented by Figures A.2 and A.3 above is perfectly general with regard to the necessary functions.

A.2 Woodward's Ambiguity Function $|X(\tau, \nu)|$

Returning to our original concern, what is the degree of confusion that might result from range-velocity (time-delay-Doppler) ambiguities inherent in a general waveform? In terms of the specific receiver processor model above, the central question is *the degree to which a filter tuned to one delay and Doppler shift responds to signals at other delays and Doppler shifts*. With this understanding we ask, "What is the response of a matched filter to a signal that is mismatched in time-delay and frequency?"

To answer this question, consider a filter that is designed to match a signal

$$u(t)e^{j\omega_0 t} \tag{A.7}$$

but the signal that arrives at the receiver is,

$$u(t-\tau)e^{j(\omega_0+2\pi\nu)(t-\tau)} \tag{A.8}$$

That is, the received signal is “too late” in time by τ, and “too high” in frequency by ν with respect to the filter design. The complex envelope of the waveform $u(t)$ is fundamentally unchanged, however.

The complex envelope of the output signal under such circumstances is

$$X(\tau,\nu) = \int_{-\infty}^{+\infty} u(t')e^{j2\pi\nu t'}u^*(t'-\tau)\,dt' \tag{A.9}$$

where ν is frequency shift and τ is the delay beyond that planned, both with respect to the reference signal $u(t)$. In this formulation we have suppressed the total signal delay and the carrier or reference frequency as these quantities do not affect the form, i.e., the shape, of the filter output envelope. Obviously, for $\tau = 0, \nu = 0$, we regain our original matched filter and maximally retrieved SNR. For all other values of $\tau,\ \nu$ we calculate the matched filter response to a *mismatched* input signal.

The quantity $|X(\tau,\nu)|$ is called the *ambiguity function* for a given waveform; its plot is called the *ambiguity diagram*. This is discussed in many places; one is Skolnik.[1] There is some variation in the definition of the ambiguity function among authors. Sometimes the sense (sign) of the time-delay and frequency shifts are changed. Sometimes the squared modulus, $|X(\tau,\nu)|^2$ is called the ambiguity function, rather than the modulus $|X|$ itself. These minor variations are of no real significance, but obviously it is important to keep straight which is in use.

The ambiguity function can be expressed both in terms of the impulse response and in terms of the frequency transfer function,

$$|X(\tau,\nu)| = \left|\int_{-\infty}^{+\infty} u(t')u^*(t'-\tau)e^{j2\pi\nu t'}\,dt'\right| \tag{A.10}$$

$$= \left|\int_{-\infty}^{+\infty} U^*(f)u(f-\nu)e^{+j2\pi f\tau}\,df\right| \tag{A.11}$$

Note that,

$$|X(0,0)|^2 = |s_0(t)|^2\,\Big|_{t=\text{time of maximum output}} \tag{A.12}$$

A.3 Examples of Ambiguity Functions

As has been the case many times before in these pages, it is worthwhile to look at some simple cases of ambiguity functions in order to get a feel for how they vary and what information they present.

A Pulse

The ambiguity function for a “cw” pulse is a triangle in time delay and a sinc function in frequency (Figure A.4).

$$s(t) = \sqcap\left(\frac{t}{t_p}\right)\cos\omega_0 t = \sqcap\left(\frac{t}{t_p}\right)\left[\frac{e^{jw_0 t}+e^{-jw_0 t}}{2}\right] \tag{A.13}$$

---

1 Skolnik, Section 11.4, pp. 411-ff. Note that Skolnik uses tr for our τ , and f_d for our ν .

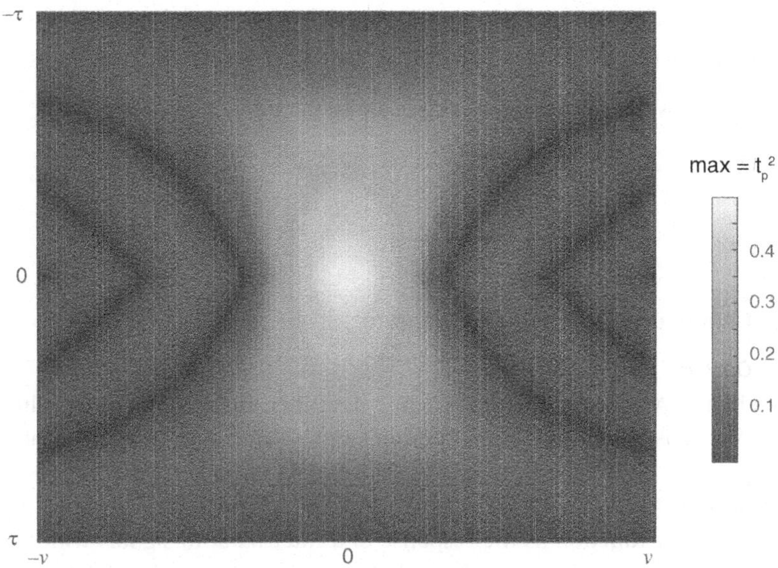

Figure A.4 Ambiguity function of a single pulse. On the principle axes the function is a triangle in delay and a sinc in Doppler. Nulls form arcs pointing away from the τ, ν origin.

$$U(t) = \sqcap\left(\frac{t}{t_p}\right) \tag{A.14}$$

$$x(\tau,\nu) = \int_{-\infty}^{+\infty} \sqcap\left(\frac{t}{t_p}\right)\sqcap\left(\frac{t-\tau}{t_p}\right) e^{j2\pi\nu t}\,dt \tag{A.15}$$

$$X(\tau,0) = \Lambda\left(\frac{\tau}{t_p}\right); \qquad X(0,\nu) = t_p \mathrm{sinc}(\nu t_p) \tag{A.16}$$

Simple Chirp

$$s(t) = \sqcap\left(\frac{t}{t_p}\right)\cos[2\pi f_0 t + 2\pi\frac{\alpha}{2}t^2] \tag{A.17}$$

$$u(t) = \sqcap\left(\frac{t}{t_p}\right) e^{j2\pi\frac{\alpha}{2}(t-\tau)^2} \tag{A.18}$$

$$\begin{aligned} X(\tau,\nu) &= \int_{-\infty}^{+\infty} \sqcap\left(\frac{t}{t_p}\right)\sqcap\left(\frac{t-\tau}{t_p}\right) e^{j2\frac{\pi\alpha}{2}t^2} e^{-j2\pi\frac{\alpha}{2}(t-\tau)^2} e^{j2\pi\nu t}\,dt \\ &= e^{j2\frac{\pi\alpha}{2}\tau^2}\int_{-\infty}^{+\infty} \sqcap\left(\frac{t}{t_p}\right)\sqcap\left(\frac{t-\tau}{t_p}\right) e^{-j2\pi(\nu+\alpha\tau)t}\,dt \end{aligned} \tag{A.19}$$

Evaluation of Eq. (A.19) is not easy, and no closed form exists. For zero delay, however, we can see the behavior with frequency,

$$X(0,\nu) = e^{-j2\pi\frac{\alpha}{2}\tau^2}\int_{-\infty}^{+\infty} \pi^2\left(\frac{t}{t_p}\right) e^{+j2\pi\nu t}\,dt \tag{A.20}$$

$$= t_p e^{-j2\pi\frac{\alpha}{2}\tau^2}\mathrm{sinc}(\nu t_p) \tag{A.21}$$

which is easy to sketch (Figure A.5).

The general expression for $|X(\tau,\nu)|\Big|_{\text{chirp}}$ is very messy, and we will not write it down here. It should be clear from the above, however, that the magnitude along the line $\nu = -\alpha\tau$ falls as

$$|X(\tau, -\alpha\tau)| = \Lambda\left(\frac{\tau}{t_{\text{p}}}\right) \tag{A.22}$$

Plots of the ambiguity diagram are computed readily, however.

A.3.1 Points of View and Ambiguity Function and Diagram

A.3.1.1 Mismatched Filter Output

With respect to the filter in Figure A.6, the ambiguity function describes the behavior of the filter output, s_0^2, in the τ, ν plane, which is also known as the "delay-Doppler" or "range-Doppler" space.

A.3.1.2 Response of Filter Bank to s(t)

We can also consider the ambiguity function as the response of a filter bank to signals with the complex reference envelope over a range of frequencies and delays, as indicated in Figure A.7. Note that in this instance the "τ" of the convention $h(\tau)$ is not the offset delay of the temporal mismatch.

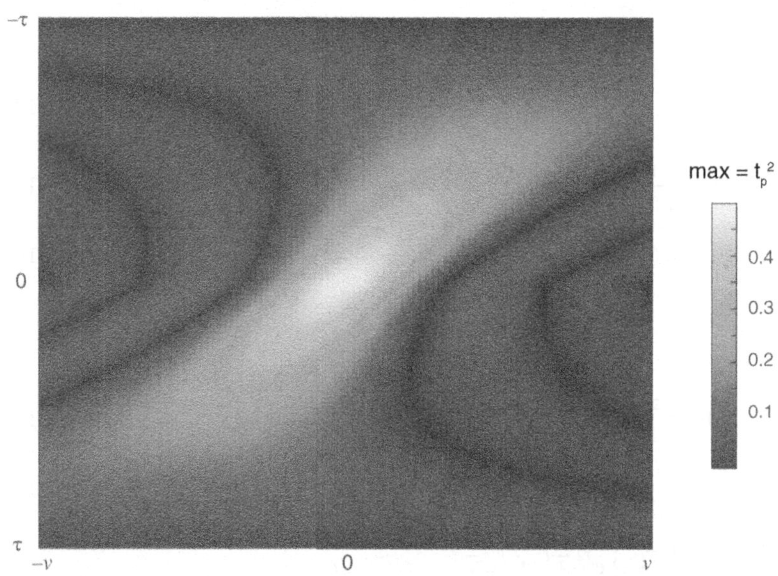

Figure A.5 Ambiguity function of a chirp. While this resembles the ambiguity function of the cw pulse, it shows explicitly the ambiguity (confusion) between changes in time delay and chirp slope errors.

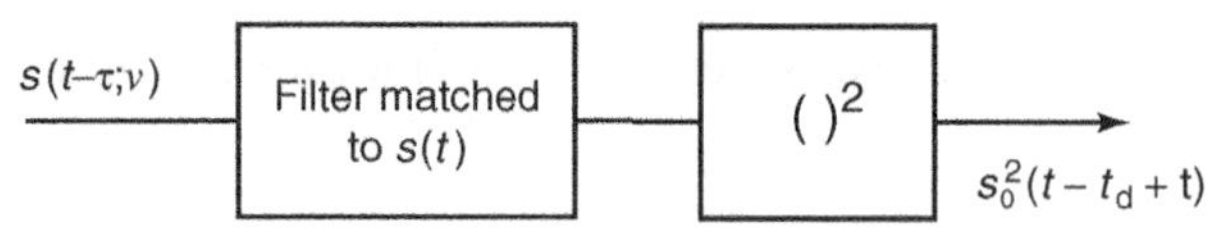

Figure A.6 Mismatched filter. The ambiguity function describes the behavior of the filter in the τ, ν plane. Here the output is $|X|^2$.

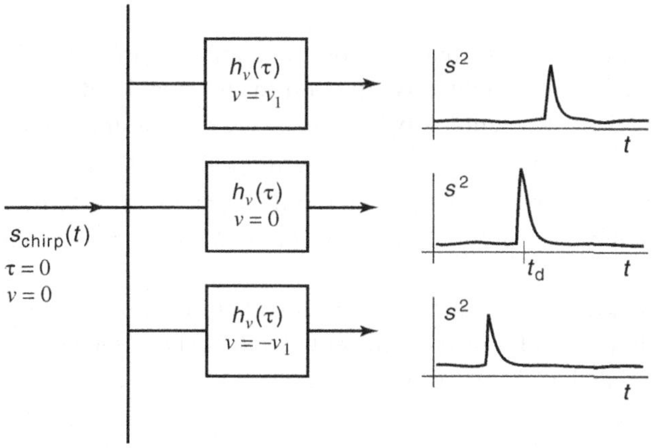

Figure A.7 Filter bank representation of the ambiguity function. The sketched response corresponds to a chirp waveform input. This filter bank gives the continuous time response for sample values of the offset frequency.

A.4 Properties of the Ambiguity Function

A.4.1 Four Theorems

We now state without proof four basic theorems describing fundamental properties of the ambiguity function,

$$|X(\tau, \nu)| = \left| \int_{-\infty}^{+\infty} u(t)u^*(t-\tau)e^{+j2\pi\nu t}\,dt \right| \tag{A.23}$$

The proofs are all straightforward, from direct manipulation the definition above.

1. Localization:

$$|X(\tau, \nu)| \leq |X(0,0)| = 1 \qquad \text{(normalized case)} \tag{A.24}$$

2. Volume Invariance:

$$\iint_{-\infty}^{+\infty} |X(\tau, \nu)|^2 d\tau dr = |X(0,0)|^2 \tag{A.25}$$

3. Symmetry:

$$|X(-\tau, -\nu)| = |X(\tau, \nu)| \tag{A.26}$$

4. Shearing:

$$u(t) \leftrightarrow |X(\tau, \nu)| \tag{A.27}$$

$$u(t)e^{j2\pi kt^2} \leftrightarrow |X(\tau, \nu + k\tau)| \tag{A.28}$$

A.4.2 Shearing Effect

As stated above under Theorem 4, "Shearing," the introduction of a quadratic phase factor in the time domain linearly expands or contracts the ambiguity function in the Doppler domain. Figures A.8 and A.9 illustrate this characteristic. The ambiguity function $|X(\tau,\ \nu)|$ of a simple long pulse is shown in Figure A.8. Multiplying the complex envelope of the pulse by the factor $e^{j\pi kt^2}$ has the effect shown in Figure A.9, in which the ambiguity diagram is pulled progressively farther from the abscissa with increasing values of τ; for constant τ, every value of $|X|$ at $\tau,\ \nu$ is shifted vertically by an amount $k\tau$, as shown.

If $f(u(t)) = u(f)$, then, $u(f)e^{+j\pi kf^2}$ also introduces a shear, but in the other direction, along τ. This is illustrated in Figure A.10 You should work out the constants for this shear. Obviously, sheared ambiguity functions are closely related to chirp waveforms.

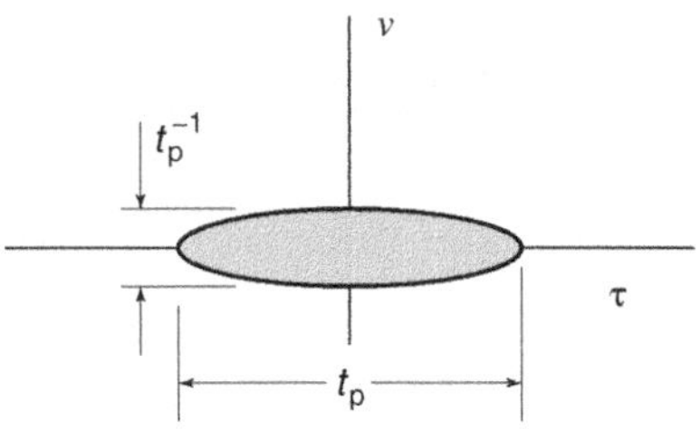

Figure A.8 Ambiguity function of a long pulse. A long pulse is well-defined in frequency but less so in time. The ambiguity function reflects this characteristic.

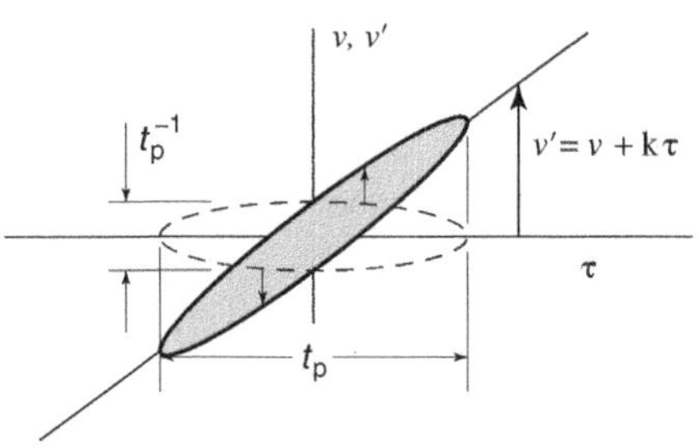

Figure A.9 Frequency-sheared ambiguity function. Frequency, or ν, shearing is associated with a swept frequency, or chirp waveform. It is the result of introducing a quadratic phase characteristic in the time domain.

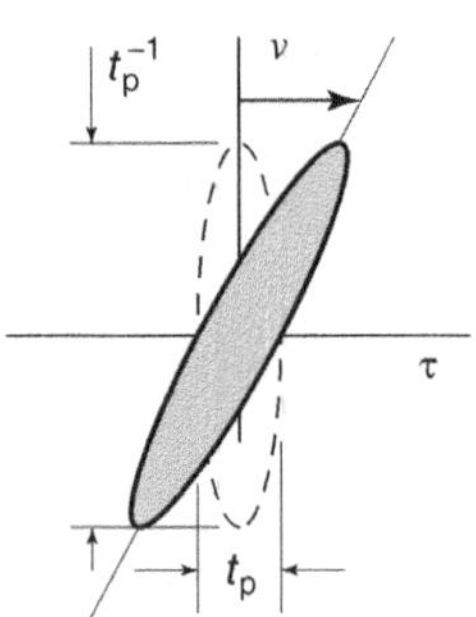

Figure A.10 Time-sheared ambiguity function. This is the result introducing a quadratic phase characteristic in the frequency domain.

A.5 Some Basic Ambiguity Functions

We now compare the ambiguity diagrams of three waveforms: the simple pulse, the simple chirp, and the coherent pulse train.

A.5.1 Simple Pulse

For a simple pulse,

$$u(t) \sim \sqcap\left(\frac{t}{t_p}\right)\cos(2\pi f_c t) \tag{A.29}$$

we might expect an ambiguity diagram some like that of Figure A.11, where the elliptical contour represents a level cut through the three-dimensional shape, and the diagram suggests a "short pulse" with relatively wide bandwidth. For this waveform the product of the waveform duration in time, $T = t_p$ and its spread in frequency, B, is BT $\simeq 1$. this is representative of the uncertainly relation for the Fourier transform.

A.5.2 Chirp Waveform

Not all waveforms are so simple, however. The single chirp waveform,

$$u(t) \sim \sqcap\left(\frac{t}{t_p}\cos\left(2\pi f_c t + \frac{1}{2}2\pi k t^2\right)\right) \tag{A.30}$$

possesses the sheared waveform discussed above, and shown below in Figure A.12. Because the shearing process widens the waveform in one dimension without affecting the other, the time-bandwidth product is necessarily increased, typically BT >> 1 for these waveforms.

A.5.3 Coherent Pulse Train

Similarly, the coherent pulse train,

$$u(t) \sim \cos(2\pi f_c t)\sum_{n=1}^{N}\sqcap\left[\frac{t - nT_R}{t_p}\right] \tag{A.31}$$

also must have BT ≥ 1 because the repetition of the simple pulse lengthens the duration of the signal in time. A typical ambiguity diagram is shown in Figure A.13

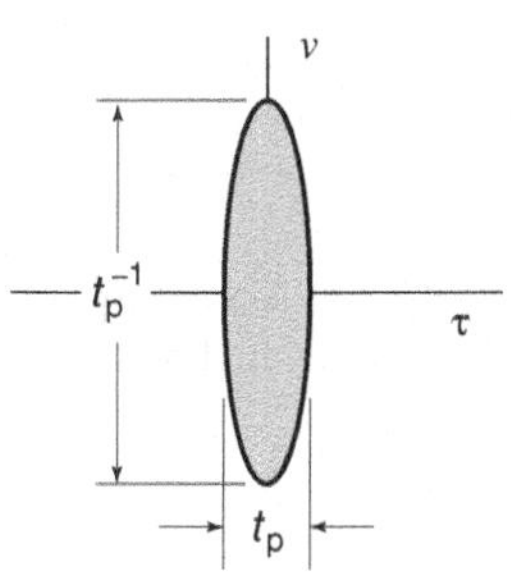

Figure A.11 Ambiguity diagram of a simple pulse. Note that the product of the time duration and frequency spreading is near unity ($BT \simeq 1$).

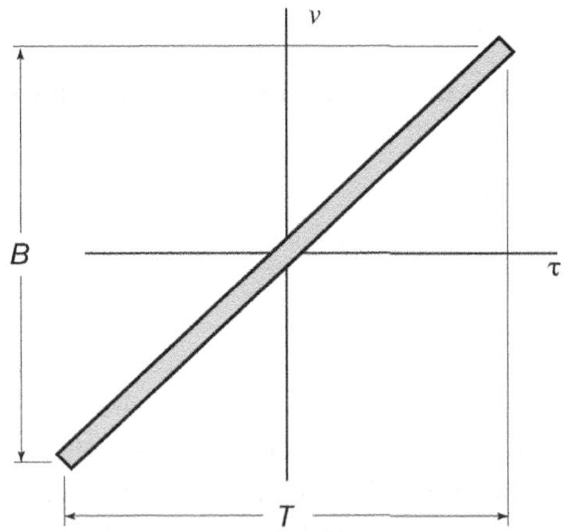

Figure A.12 Ambiguity diagram of a single chirp. Chirp waveforms have instantaneous frequency-time spread that is much less that total frequency range or time duration of overall waveform; in such instances the product of the time duration and frequency spreading is greater than unity ($BT > 1$).

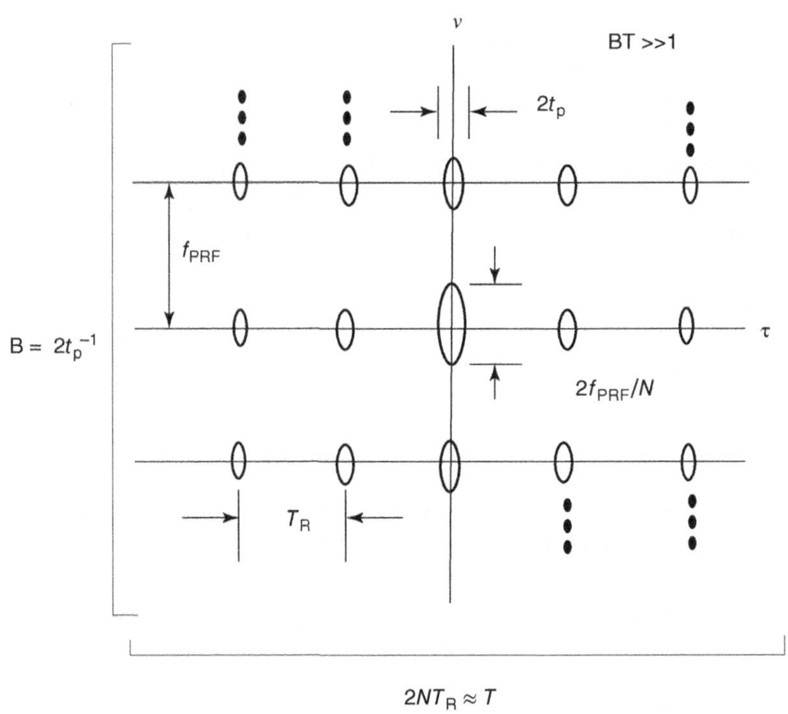

Figure A.13 Ambiguity diagram of coherent pulse train. The array of peaks is set by the PRF, while the dimensions of the peaks are determined by the combination the burst duration and the individual pulse length.

A.6 Implications of Volume Invariance

What are the implications of Theorem 2, "volume invariance" of $|X(\tau,\ \nu)|^2$? Consider the normalized ambiguity function

$$|X(0,0)|^2 = 1 \tag{A.32}$$

Given that the peak value of $|X|$ is unity, then by the volume invariance cited above, the integrated squared-magnitude is unity also,

$$\iint_{-\infty}^{+\infty} |X(\tau,\nu)|^2\, d\tau d\nu = 1 \tag{A.33}$$

Surprisingly, perhaps, this has profound implications for our ability to separate radar targets and to observe targets in $\tau,\ \nu$ space without ambiguity. To see this, consider idealized ambiguity function designed to produce maximal range-Doppler resolution, such as that shown in Figure A.14. By choosing the area of the central peak to be small we obtain the highest possible definition of the location of a point target within the observation space. At the same time, we are forced by the volume constraint to place the remaining response in the surrounding pedestal. Such a function is referred to as a "thumbtack" ambiguity function since it approximates those small pins in shape. Here we have chosen a peak value and a volume of unity, so our examples represent $|X|^2$, and are consistent with the theoretical constraints. Within the constraints of practical signals and volume invariance, we are free to design ambiguity functions of arbitrary shape. Our ability to do so depends only on the selection of the complex envelope $u(t)$. So while the example is extreme in that it has sharp edges, it approximates forms within the realm of possibility.

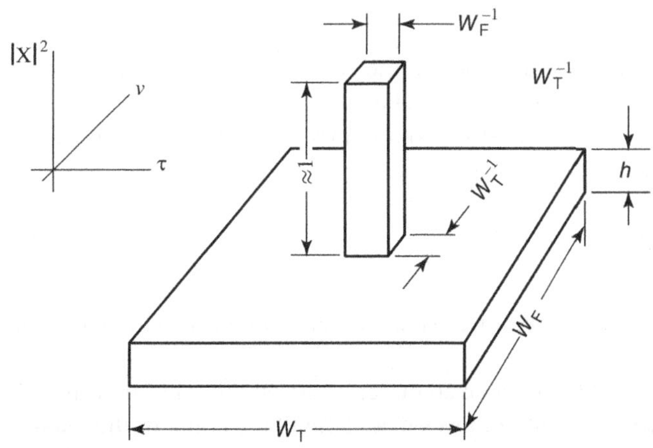

Figure A.14 Idealized "thumbtack" ambiguity function. This idealized ambiguity function is intended to provide the greatest possible resolution in τ, ν space, so the area of the peak is set at the minimum expected for a waveform of duration W_T and bandwidth W_F. Volume of this peak is much less than unity, when the height of the peak is normalized. Consequently, the required volume must be distributed elsewhere. In order to minimize the response from the extra volume, it is spread in a broad pedestal. The total volume of the function shown is unity. Creation of a sharp peak to provide maximum range-Doppler resolution results in a large pedestal containing most of the ambiguity function volume. While thumbtack-type function provides excellent resolution for a single target, the self-noise associated with the pedestal is limiting in a dense target environment.

Such an ambiguity function and its underlying waveform can be highly effective when used to observe[2] a single target. But it is limited by self noise when the number of targets is large! How do we understand this?

We approach this question by calculating the total number of resolution elements within the function pedestal, and then see how many of these can be occupied before the net response from other targets within the range of our system overwhelms the signal from the desired target. For a conventionally well-designed waveform, the area of the peak is determined by the time-bandwidth product of the reference signal, while the extent of the signal in the range-Doppler space is determined by the factors of the time-bandwidth product, i.e., by the signal duration and the highest frequency present in the complex signal envelope. Given this,

$$\text{Area of Base:} \quad W_{\text{T}}W_{\text{F}} = A_{\text{B}} \quad \text{(typically} \gg 1) \tag{A.34}$$

$$\text{Area of Peak:} \quad \frac{1}{W_{\text{T}}}\frac{1}{W_{\text{F}}} = \frac{1}{A_{\text{B}}} \tag{A.35}$$

$$\text{Pedestal Height:} \quad h = \frac{A_{\text{B}} - 1}{A_{\text{B}}^2 - 1} \simeq \frac{1}{A_{\text{B}}} = \frac{1}{W_{\text{T}}}\frac{1}{W_{\text{F}}} \tag{A.36}$$

We also need the number N_{R} of resolution elements, or "cells," or size $(W_{\text{T}}W_{\text{F}})^{-1}$ within the pedestal of $|X|^2$.

$$N_{\text{R}} = \frac{W_{\text{T}}W_{\text{F}}}{D\frac{1}{W_{\text{T}}W_{\text{F}}}} = (W_{\text{T}}W_{\text{F}})^2 \tag{A.37}$$

On the other hand, the signal-to-self-noise, calculated when looking at a single target is,

$$\frac{\text{Signal Power}}{\text{Self-noise}} = \frac{1}{D\frac{1}{W_{\text{T}}W_{\text{F}}}} = W_{\text{T}}W_{\text{F}} \tag{A.38}$$

This is the ratio of the signal power when the target is within the point, or spike, of the thumbtack to that when the same target is within the surrounding pedestal.

What if we have two discrete targets of equal echo strength occupying distinct resolution cells in τ, ν space? Then, the ratio of the signal-to-self-noise becomes,

$$\frac{\text{Signal Power}}{\text{Self-noise}} = \frac{1}{D\frac{2}{W_{\text{T}}W_{\text{F}}}} = \frac{W_{\text{T}}W_{\text{F}}}{2} \tag{A.39}$$

which compares the signal power when one target is on the point of the thumbtack, with that when neither is on the point.

What number, N_{max}, of cells can be filled before a given target is lost, in the sense that there is very little change signal output as a particular target moves in and out of the point, or the radar is "tuned" to a particular target? We find that for N_{max} cells the signal-to-self-noise is about unity. To calculate the maximum number we use the results, Eqs. (A.38) and (A.39), for one and two targets and generalize to N targets, assuming, as above, that all targets are the same strength in the receiver,

$$\frac{W_{\text{T}}W_{\text{F}}}{N_{\text{max}}} \simeq 1 \Rightarrow N_{\text{max}} \simeq W_T W_F \tag{A.40}$$

or

$$N_{\text{max}} \simeq \sqrt{N_{\text{R}}} \tag{A.41}$$

2 Dare we say "look at"?

Thus, the number of targets that can be accommodated in range-Doppler space before the self-noise overwhelms the signal is approximately of the time-bandwidth product of the signal. This number is much less than the number of resolution elements available! The clash between the increase in self noise and improvement of τ, ν resolution represents a fundamental constraint on the performance of radars operating in range-Doppler space. A similar result is obtained if a distribution of target strengths is assumed.

In considering the implications of volume invariance, keep in mind that the theorem does not constrain the shape of the ambiguity function. It is possible to design waveforms that suppress the pedestal in local regions of τ, ν space, and designers are free to do when when there is sufficient *a priori* information regarding the target characteristics to do so. Similarly, in mapping radars the design of the antenna can be used to suppress interference in τ, ν space. Finally, in some instances, one can use detailed knowledge of the shape of $|X(\tau, \nu)|^2$ to remove pedestal effects in post detection processing of the signals. Some illustrations of these points will be given in class.

B

Targets in τ, ν Space

B.1 Target Categories

Targets, also, can be described in τ, ν space. A small, stationary target is described as a *point target*, or a *point non-fluctuating target*. Such targets occupy a single point at $\tau, \nu = 0, 0$. They do not modify the signal structure in any way, although they affect signal strength since they can have varying strength and occur at various ranges. Point targets are represented as Dirac deltas in τ, ν space, Figure B.1.

Some targets are spread in range, but introduce neither a frequency shift nor frequency spreading. These targets are referred to as *point-spread* targets because they stretch the echo in the range dimension, as illustrated in Figure B.2. Physically, these targets are usually larger than the range resolution of the radar by a significant amount, but are stationary with respect to the radar.

On the other hand, it is also possible to have point targets with no significant range depth but which nevertheless modulate the radar echo, thereby creating a spread frequency signal. These targets are referred to as *point-fluctuating* targets, since the frequency spread is associated with a concomitant temporal variation in the RCS (Figure B.3).

Finally, some targets are spread in both delay (range) and Doppler (frequency). These are referred to as *spread-fluctuating* targets; they occupy some area in τ, ν space, as illustrated for one example in Figure B.4. Such targets are characterized by a time-bandwidth product $B \cdot T$.

B.2 Target-signal Interactions

What is the response of a radar to a spread-fluctuating target characterized by $\sigma(\tau, \nu)$ in delay τ and frequency shift ν? In order to address this, assume that each element of the target space $\sigma(\tau, \nu)\, d\tau d\nu$ scatters independently of all others. Let the transmitted or illuminating waveform be given by,

$$s_T(t) = \Re[u(t)e^{j2\pi f_c t}] \tag{B.1}$$

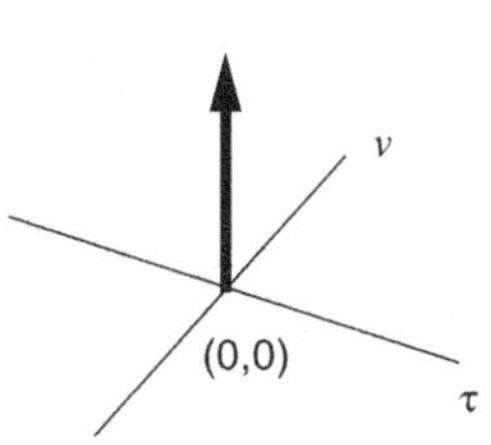

Figure B.1 Point non-fluctuating target. Sometimes referred to as a "point target." Such a target does not change the signal structure.

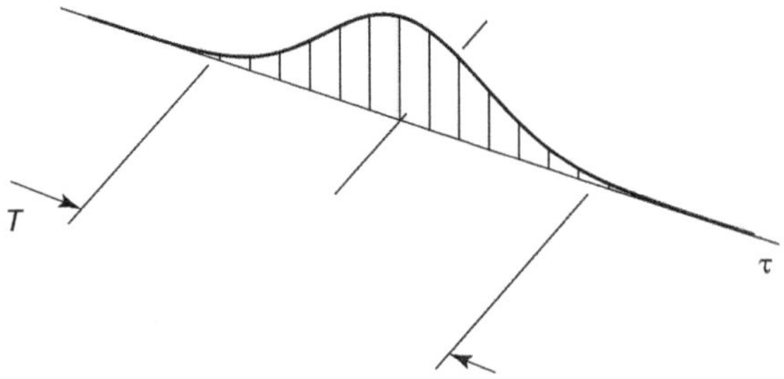

Figure B.2 Point spread target. This target spreads the signal in time, but does not increase the bandwidth. Obviously this is an idealization.

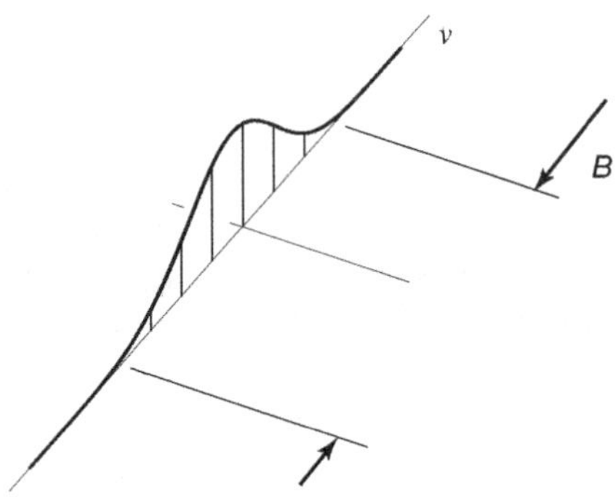

Figure B.3 Point fluctuating target. Target fluctuation represents an increase in scattered signal bandwidth. This target does not smear the signal in time, however.

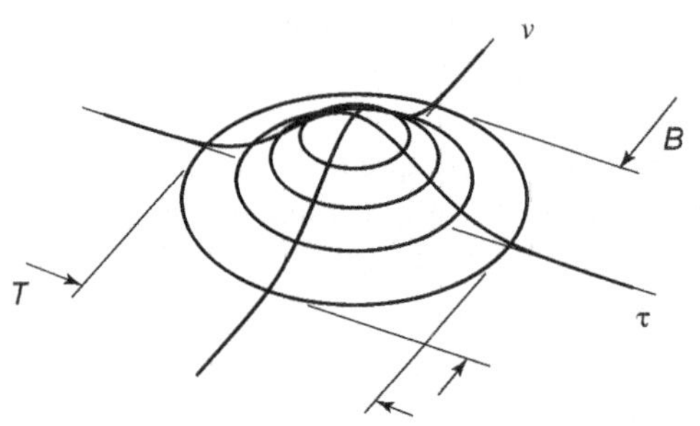

Figure B.4 Spread fluctuating target. This is the most general case. Interaction of a signal with this target increases both the signal-delay spread and an increase in spectral width.

For target element at $\tau_\mathrm{k}, \nu_\mathrm{k}$ we have,

$$s_\mathrm{R}(t) = Re[a(\tau_\mathrm{k}, \nu_\mathrm{k})e^{j\phi(\tau_\mathrm{k}, \nu_\mathrm{k} u(t-\tau_\mathrm{k})}e^{j2\pi(f_\mathrm{c}+\nu_\mathrm{k})(t-\tau_\mathrm{k})}] \tag{B.2}$$

where

$$\sigma(\tau_\mathrm{k}, \nu_\mathrm{k}) = < |a(\tau_\mathrm{k}, \nu_\mathrm{k})|^2 > \tag{B.3}$$

Obviously, we have suppressed the multiplicative factors of $P_\mathrm{T}, G_\mathrm{T}, A_\mathrm{R}, R^{-1}$, etc., so $a\, e^{j\phi}$ is assumed to be corrected for variations associated with these.

The receiver is matched to the same waveform, albeit perhaps for a different τ, ν pair.

$$s_{\text{match}}(t) = \Re[u(t-\tau_0)e^{j2\pi(f_c+\nu_0)(t-\tau_0)}] \tag{B.4}$$

The output of the matched filter is

$$u_0(t)e^{j2\pi f_0 t} = a(\tau_k,\nu_k)e^{-j2\pi(f_c-\nu_0)(t+\tau_k-\tau_0)}\frac{1}{2}\int_{-\infty}^{+\infty} u(t')u^*(t'+t \\ +\tau_k-\tau_0)e^{j2\pi(\nu_k-\nu_0)t'}\,dt' \tag{B.5}$$

Equation (B.5) is obtained from

$$Z_0(t) = \frac{1}{2}\int_{-\infty}^{+\infty} W(t')Z_i(t-t'\,dt') \tag{B.6}$$

which leads to

$$u_0(t) = \frac{1}{2}\int_{-\infty}^{+\infty} a(\tau_k,\nu_k)e^{j\phi(\tau_k,\nu_k)}u(t'-t+\tau_k)e^{j2\pi(f_c+\nu_k)(t'-t+\tau_k)}u^* \\ (t'-\tau_0)e^{-j2\pi(f_c+\nu_0)(t'-\tau_0)}\,dt' \tag{B.7}$$

Simplification leads to the form given above in Eq. (9.19)

Continuing with the calculation, the integral in the expression for $u_0(t)$ is just $X[(\tau_0-t+\tau_k),(\nu_k-\nu_0)]$ so,

$$u_0(t) - \frac{a(\tau_k,\nu_k)}{2}e^{j\phi(\tau_k,\nu_k)}e^{-j2\pi(f_c+\nu_0)(t-\tau_k-\tau_0)}X[(\tau_0-t+\tau_k),(\nu_k-\nu_0)] \tag{B.8}$$

For K scatterers this is readily extended to

$$u_{0_K}(t) = \sum_{k=1}^{K} a_k e^{j\phi k}e^{-j2\pi(f_c+\nu_0)(t-\tau_k-\tau_0)}X[(\tau_0-t+\tau_k),(\nu_k-\nu_0)] \tag{B.9}$$

which is a general result.

If the K scatterers are densely packed, approaching a continuum in τ, ν space, the meaning of the individual $a_k e^{j\phi_k}$ is lost. In this instance, only the overall envelope of the output can be detected. This is given by,

$$|u_{0_K}(t)|^2 = \sum_{k=1}^{K}|a_k|^2|X[(\tau_0-t+\tau_k),(\nu_k-\nu_0)]|^2 + \sum_{k\neq l} a_k a_l e^{j(\phi_k-\phi_l)}e^{-j2\pi(f_c+\nu_0)(\tau_k-\tau_l)} \\ \times X_k \cdot X_l^* \tag{B.10}$$

The average output of square law detection becomes

$$\lim_{\Delta\tau,\Delta\nu\to 0} < |U_{0_K}(t)|^2 >= \int\int_{-\infty}^{+\infty} \sigma(\tau,\nu)X[(\tau_0-t+\tau),(\nu-\nu_0)]|^2\,d\tau d\nu \tag{B.11}$$

provided that

$$< a_k a_l e^{j(\phi_k-\phi_l)} >=0 \tag{B.12}$$

that is, if the returns from individual elements of $\sigma(\tau,\nu)$ are uncorrelated. Thus, we see that the power response of a matched filter to a spread-fluctuating target *is the convolution of the target function with the ambiguity function of the radar signal.* This provides us with a convenient means for visualizing the effect of the target-signal interaction in delay-Doppler space.

For an "ideal" ambiguity function, $X(\tau, \nu) = \delta(\tau, \nu)$ we have the result,

$$\begin{aligned} < |u_0(t)|^2 > &= \int \int \sigma(\tau, \nu)\delta[(\tau_0 - (t - \tau), (\nu - \nu_0)]\, d\tau d\nu \\ &= \sigma(t - \tau_0, \nu_0) \end{aligned} \tag{B.13}$$

So, a matched-filter response can represent a distributed target. The target might be mapped by a bank of filters, each set to a slightly different frequency offset, ν_0.

B.3 Realistic Limits of Performance

B.3.1 Performance Dependence on Target Properties

How well can we do, realistically, in range-Doppler mapping? Figure B.5 compares the ideal, delta function response which recovers the target function exactly (Eq. B.13), with a more realistic thumbtack-type response consistent with possible ambiguity function realizations.

Consider a spread-fluctuating target such as that in Figure B.6. How many *distinct* targets might there be in such a representation? Stated another way, "How many resolution elements can we achieve in such a target?"

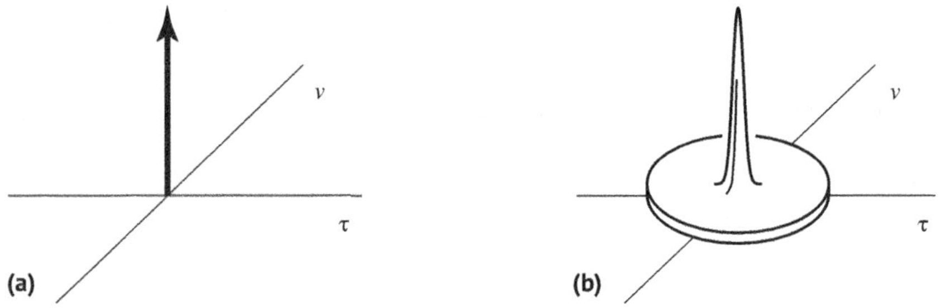

Figure B.5 Ideal and realistic ambiguity functions. (a) The Dirac delta represents the ideal ambiguity function. Such an ideal form cannot be achieved, however, as a result of the volume constraint on the characteristics of ambiguity functions. (b) Thumbtack function represents a realistic form of ambiguity function.

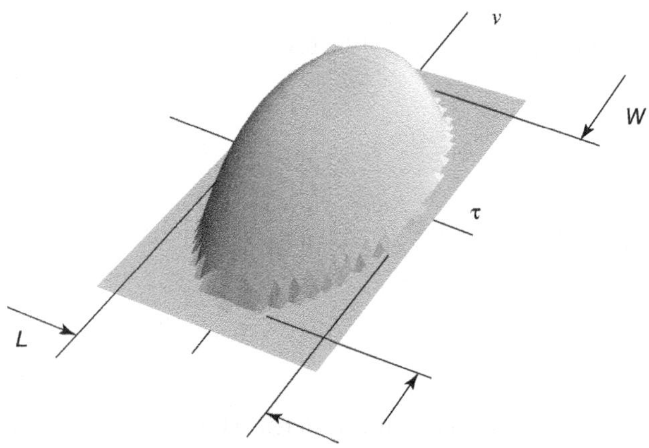

Figure B.6 Spread fluctuating target. Target is characterized by a time-bandwidth product $W \times L$.

Recall that the resolution of a general waveform is $(W_F W_T)$, so the number of possible resolution elements in WL is,

$$N_{\text{res elements}} = N_{\text{targets}} = \frac{WL}{D\frac{1}{W_F W_T}} = (WL)(W_F W_T) \tag{B.14}$$

We have shown, however, that the maximum number of cells that can be filled before self noise overwhelms the signal from any one target is roughly

$$N_{\text{Max}} = W_F W_T \tag{B.15}$$

Applying the requirement that $N_{\text{targets}} \leq N_{\text{Max}}$ implies then that

$$WL \leq 1 \tag{B.16}$$

That is, we have the surprising result that the resolution achievable in τ, ν space depends on the target, and that a widely spread target is *more difficult to resolve than a less spread target!*

This leads to the following subdivision of target space properties:

- $WL < 1$: The target is *underspread*; under this condition, the target can be resolved with arbitrary accuracy.
- $WL \simeq 1$: The target is *critically spread*; self-noise approximates the signal from an average target.
- $WL > 1$: The target is *overspread*; self-noise overwhelms the system.

B.3.2 Considerations in Waveform Design

As to choice of PRF, pulse length, etc. a consideration of the target properties plus signals from unwanted targets, called *clutter*, often either dictates or strongly constrains waveform design. For example, if signal Doppler shift is to be measured without ambiguity, then there is a minimum PRF—set by $v_{\text{radial}}\lambda/2$—that can be used without aliasing of the received signal in frequency. This requirement might or might not be compatible with the needed R_u. Conversely, if the unambiguous range dictates the PRF, then the Doppler might have ambiguities.

Other considerations include interference, limits on achievable transmitter power, cost, operating range, and the estimation theory constraints on observation time for a given measurement precision.

B.3.2.1 Example of Mars as an Overspread/Underspread Target

A simple example of the interaction of the target spread parameter WL, the PRF, and range-Doppler ambiguity is provided by the problem of radar range-Doppler imaging of Mars from the Earth. Mars rotates on its axis slightly more slowly than Earth, with a period,

$$T_{\text{Mars}} = 24^{\text{h}}37^{\text{m}} = 88,650 \text{ sec} \tag{B.17}$$

The corresponding rate of rotation is

$$\Omega = \frac{2\pi}{88,650} \simeq 7.1 \times 10^{-5} \qquad \text{rad/s} \tag{B.18}$$

where the direction of rotation is the same as for Earth (CCW when viewed from above the planes of the two orbits). We also need the radius of Mars, which is approximately 3.4×10^6 m. Referring to Figure B.7, an observer "looking" at Mars from Earth sees the western limb approach, and

Figure B.7 Radar looking at Mars. Radar, at left, illuminates Mars, here assumed to be rotating perpendicular to the radar line-of-sight. The bandwidth of the echo signal is increased by the range of Doppler shifts from different locations on the planet's surface. Extremes of the Doppler broadening occur at the limbs of the planet; the approaching limb (top) provides the extreme increase in frequency, while the receding limb, (bottom), provides the extreme decrease. (See Section B.3.2.1.)

the eastern limb recede. We can calculate the velocity of the limbs of Mars and hence the band width W, assuming for simplicity that the Earth is located at a position normal to Mars' axis of rotation,

$$v_{\text{limb}} = R_{\text{Mars}}\Omega = 3.4 \times 10^6 \cdot 7.1 \times 10^{-5} = 241 \text{ m/s} \tag{B.19}$$

$$W = \frac{2v_{\text{limb}}}{D\frac{\lambda}{2}} = \frac{4v}{\lambda} \simeq \frac{10^3 f}{c_0} = \frac{f}{3 \times 10^5} \text{ Hz} \tag{B.20}$$

where B is the total bandwidth, as measured by difference in Doppler frequency, limb-to-limb. We also need the range depth, or delay spread, L,

$$L = \frac{2R_0}{c_0} = \frac{23 \times 3.4 \times 10^6}{3 \times 10^8} = 23 \text{ m/s} \tag{B.21}$$

From this we have a formula for the target spreading of Mars,

$$WL \simeq \frac{10^3}{\lambda} \cdot 23 \times 10^{-3} \simeq \frac{23}{\lambda} \tag{B.22}$$

which tells us that the spreading is proportional to transmitter carrier frequency. Mars is *overspread* for a frequency such that $WL > 1$.

$$f_c > 13\text{MHz} = f_{\text{critical}} \tag{B.23}$$

The corresponding critical wavelength is 23 m.

Putting in numbers makes the meaning of this clear. Take, for example, $\lambda = 0.1$ m, $f_c = 3 \times 10^9$ Hz, which is a practical operating range for a large radar capable of detecting other planets in our solar system. Then calculate W,

$$W \simeq \frac{10^3}{\lambda} \simeq 10^4\text{Hz}, \quad WL = 230 \tag{B.24}$$

The Nyquist criterion requires a sample rate of twice this value to characterize a signal of this bandwidth. In a pulse radar one sample of the echo signal is obtained per pulse transmission. But at,

$$f_{\text{PRF}} = 2 \times 10^4, \qquad T_{\text{R}} \frac{1}{f_{\text{PRF}}} = \frac{1}{2} \times 10^{-4} \tag{B.25}$$

so the interpulse interval and unambiguous range are,

$$T_R = 50\,\mu\text{s}, \qquad \Rightarrow R_u = 7.5\text{ km} \tag{B.26}$$

We find that an unavoidable consequence of adequately sampling the echo fluctuation bandwidth is that the unambiguous range is much less than the radius of Mars! In fact,

$$\frac{R_{\text{Mars}}}{R_{\text{u}}} \simeq \frac{3.4 \times 10^6}{7.5 \times 10^3} \simeq 453 \tag{B.27}$$

This is the number of pulses overlapping the visible hemisphere of Mars at only one time. So there are 453 ambiguous ranges associated with observing Mars at 10 cm wavelength if we choose to avoid aliasing, i.e., ambiguity, in frequency.

If, however, we choose to reduce the ambiguity in range, then we unavoidably introduce ambiguity in frequency. This occurs for any operating frequency greater than $f_{\text{critical}} \simeq 13$ MHz, as determined above. Operating below the critical frequency the ambiguity problem associated with incompatible sampling requirements disappears. This type of problem is not limited to large, solid rotating bodies such as planets. It can occur in trying to sort out, in range and velocity, the internal motions of large storms, for example. Unfortunately for planetary work, the other system requirements for observations at such a low frequency, antenna size for example, preclude construction of practical systems at this time.[1]

1 Clever and determined researchers have found a couple of unconventional ways around the Mars overspread problem. How would you attack this?

C

Atmospheric Propagation Effects on Radar Systems

Some radars operate under very nearly *free space* conditions, i.e., conditions in which the propagation is close to that of the vacuum. While there are no true examples of free space, there are many instances for which the "free space approximation" is a good one. The question is one of the precision of measurement required and the effect of the propagation medium on the radiation, which is often a function of frequency. Space-based radars "looking at" other objects also in space or on the surface of an airless planet qualifies. Even then, the tenuous quantity of ionized gas streaming from the sun can be important.[1] The free space approximation is a good one for Earth-based planetary mapping radars for which the propagation path is predominantly in space. Even in this instance, the small fraction of the path within the atmosphere can be extremely important. At the other extreme, a few radars, called *ground penetrating radars* are designed to operate within or at least looking into the upper layers of the Earth, although only to depths of a few tens of meters.

The overwhelming majority of radars, however, operate entirely within or looking through Earth's atmosphere. Although the effect of the atmosphere is locally always a weak one, when the radar range is large and the propagation is predominantly horizontal with respect to Earth's surface the overall impact on the atmospheric gas becomes significant. Consider, for example, that atmospheric refraction extends the time of sunset by about two minutes, and that the sun is actually below the geometric horizon before we see it disappear. Similarly, the distortion of the sun's shape when it is near the horizon results from variations in the total refraction along paths with slightly different angles to parts of the sun at different heights above the line of the horizon. Radar signals suffer from similar distortions of the propagation path affecting both the direction and apparent length of the radar path. These effects can vary from those producing minor errors in angle of arrival of echoes to those that dramatically affect the range and visibility of radar system. For example, radar signals can become trapped in waveguide-like structures in the lower atmosphere, called "ducts," in which they can propagate for thousands of kilometers with relatively little loss of signal strength. Similarly, radar signals at wavelengths longer than about 1 m can be returned to Earth by the ionized gas in the upper atmosphere, referred to as the "ionosphere," an effect that is critical for a class of radars designed to "look" over the horizon (OTH) for remote studies of ocean waves and long distance tracking of ships a planes.

---

1 The space between the planets contains about 5 free electrons and the corresponding ions per cm^3. Statistical variations in this small quantity of material are the limiting error source in the tracking of interplanetary spacecraft, which depend on radar-like systems using spacecraft radio transponders to measure range and range-rate.

As the effects of the atmosphere on radar performance and design can be highly significant, it is important that we understand the effects of atmospheric paths on the radar signals. We discuss this subject in the context of monostatic systems; the principles applied are readily extensible to the bistatic case.

C.1 Considerations for Modeling

We think of air as transparent, that is, as having no effect on electromagnetic propagation other than a slight slowing of the waves—and to a very great extent this is an accurate description. In the optical region of the spectrum where our vision operates ($\lambda \approx 0.5$ μm), electromagnetic waves can travel for tens to a hundred miles in "clear air" with very little attenuation. Such losses as occur are primarily the result of scattering by small particles, called aerosols, suspended in the gas, or of water droplets in clouds. Additional attenuation is the result of very weak lossy interactions with molecules of the atmospheric gas. Some loss is the result of absorption when the moisture content is high, as on hazy days of high humidity, when the air might seem transparent but objects only several miles away are not seen clearly. Refraction by the atmosphere in the optical—as contrasted with scattering and absorption—is primarily a function of atmospheric structure and is not significantly affected by the factors just mentioned.

Things are quite different in the radio region of the spectrum where the wavelength is greater than roughly 1 mm, or $\lambda > 10^{-3}$ m. At these longer wavelengths aerosols have almost no effect, and wavelengths as short as a few centimeters can readily penetrate clouds, typically comprising droplets of up to about 10 μm in size. At frequencies below about 20 GHz the primary loss mechanism is scattering. The refractivity of the atmosphere is, on the other hand, strongly influenced by the presence of water vapor, and since the concentration of H_2O vapor within the atmosphere is quite variable with space and time, propagation of radar signals also varies markedly from time-to-time and place-to-place. Thus, our visual experience is not a good indicator of the characteristics of radio and radar propagation—a ray path that is connected by a "direct-line-of-sight" visually may not exist at radio and microwave wavelengths, and vice versa. Again, this is a result of the inhomogeneity of atmospheric properties at long wavelengths, as compared with the optical region.

The longest wavelength molecular resonance with radiation in Earth's atmosphere occurs for the H_2O vapor at 22.3 GHz. Well-below this frequency, the refractivity of water vapor is easily modeled. At frequencies approaching that of the H_2O resonance and above the absorptive losses become significant, although there are relatively clear frequency intervals, or "microwave windows." Here, our discussion is restricted primarily to frequencies below about 20 GHz ($\lambda > 0.15$ m) so that absorptive losses can be neglected. In this longer wavelength region of the spectrum radio signals can propagate much greater distances than usually occurs in the optical region. Over long paths, the role of refraction becomes important as even very small refractive path deviations can build up to significant levels.

These characteristics of the atmosphere can be summarized as follows:

1. Losses are negligible for $f < 20$ GHz;
2. Refraction by the dry gas is significant;
3. H_2O vapor is a significant source of variation in refractive properties;
4. Losses can be significant due to absorption by H_2O vapor and other gases above ≈ 20 GHz.

C.2 Refractivity of Air

Let $v = c_0/n$ represent the speed of light in air.[2] Because the value of the refractive index in air is very close to unity, $n \simeq 1.0003$ at sea level, it is convenient to work with a scaled value of the difference of n from unity, called the *refractivity*,

$$N = (n-1) \times 10^6 \tag{C.1}$$

that is, the refractivity is one million times the difference between the refractive index and its free-space value. In air,

$$N = \frac{77.6}{T}\left(p_{\text{dry}} + \frac{4810}{T} p_{\text{vapor}}\right) \tag{C.2}$$

where T is the temperature in kelvins, p_{dry} is the dry air pressure in hectopascals (hPa),[3] and p_{vapor} is the partial pressure of H_2O vapor, also in hPa. We can rewrite Eq. (C.2) to make the temperature effects more clear,

$$N = \underbrace{77.6\,\frac{p_{\text{dry}}}{T}}_{\text{dry value}} + \underbrace{77.6 \times 4810\,\frac{p_{\text{vapor}}}{T^2}}_{\text{H}_2\text{O vapor contribution}} \tag{C.3}$$

We should note that in most places near Earth's surface $T \approx 290$ K. With this, comparing the first and second terms of Eq. (C.3),

$$\frac{77.6}{T} << \frac{77.6}{T} \cdot \frac{4810}{T} \tag{C.4}$$

Evaluating this we see that H_2O vapor typically contributes about 17 times more to the refractivity per unit of partial pressure than does dry air.

Because the effect of refractivity of H_2O vapor is much greater than that of air (per molecule present) even the relatively small amount of H_2O present on a typical day can have a large effect. Today, near sea level in the room where this is being written,[4] $p_{\text{vapor}} \sim 10^{-2} p_{\text{dry}}$. This leads to,

$$\begin{aligned} D\frac{77.6\ p_{\text{dry}}}{290} &\simeq 0.27\ p_{\text{dry}} &\rightarrow\quad & N_{\text{dry}} \sim 270 \\ D\frac{77.6 \cdot 4810}{290 \cdot 290} \times 10^{-2}\ p_{\text{dry}} &\simeq 0.04\ p_{\text{dry}} &\rightarrow\quad & N_{\text{vapor}} \sim 40 \\ && & N_{\text{tot}} \sim 310 \end{aligned} \tag{C.5}$$

This value is near $N \simeq 300$, mentioned above for a typical day with $n \simeq 1.0003$ at mean sea level. As the example shows, at typical radar wavelengths about 12% of the total refractivity is due to a very small quantity of H_2O vapor mixed with the major constituents nitrogen and oxygen. It is also worth noting that the spatial variations in water vapor are typically much greater than those in dry atmosphere gases, and this is an important effect in interpreting imaging radar measurements.

---

2 $c_0 = (\mu_0 \epsilon_0)^{-\frac{1}{2}} \simeq 3 \times 10^8$ m sec^{-1}; n = refractive index $\sqrt{\frac{\mu_a \epsilon_a}{\mu_0 \epsilon_0}}$, where subscript "a" indicates air, is the ratio of the speed of light in vacuo to that in the gas.

3 1 hectopascal = 100 Pa = 1 millibar pressure

4 Nice, Fr

C.3 Refractivity of the Average Atmosphere

The distribution of gas in the atmosphere is approximately exponential with a slowly varying decay rate called the *scale height*, which is about 8 km in the lowest approximately 20 km of Earth's atmosphere called the troposphere. The exponential variation with altitude is the result of the increased weight of gas compressing the lower portions of the air column relative to those portions higher up. Since the refractivity varies almost directly with density, in the lower atmosphere the refractivity is also exponential with altitude as well, provided that the atmospheric gases, including H_2O vapor, are well-mixed. A typical refractivity profile is shown in Figure C.1, which is drawn for $N(z) = N_0\ \exp(-z/H)$, where $H = 8$ km is the scale height, and $N_0 = 300$ is taken as a characteristic value at Earth's surface. To emphasize, the ideal exponential form and the values for surface conditions (N_0, H) are only "typical," or "characteristic." No location in the atmosphere has this particular profile as its fixed value, and when such profiles occur, they are not likely to persist—the atmosphere is a constantly changing entity.

C.4 Propagation in a Stratified Atmosphere

Although the values of refractivity are small, the change with height is quite significant for the propagation of radio-wavelength signals. To appreciate this significance, consider the implications of Snell's Law for the effect of the change in N with height. At a sharp boundary between regions of differing refractive indices such as that illustrated in Figure C.2 we have,

$$n_1 \sin \phi_1 = n_2 \sin \phi_2 \tag{C.6}$$

$$n = \frac{\lambda_0}{\lambda} = \frac{c_0}{v} \tag{C.7}$$

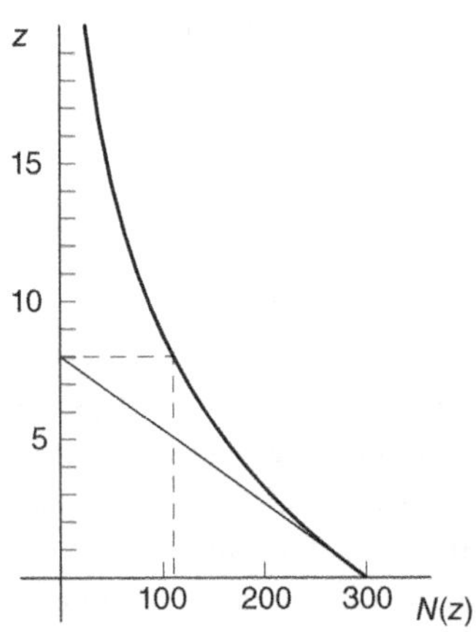

Figure C.1 Variation of refractivity N_{air} with height. Curve is exponential in z with $H = 8000$ m. Values on abscissa are in N units. Initial slope at surface is $dN/dz = -300/8000 = 0.0375$ m^{-1}. Curve shown is characteristic of lowest ≈ 20 km of Earth's atmosphere.

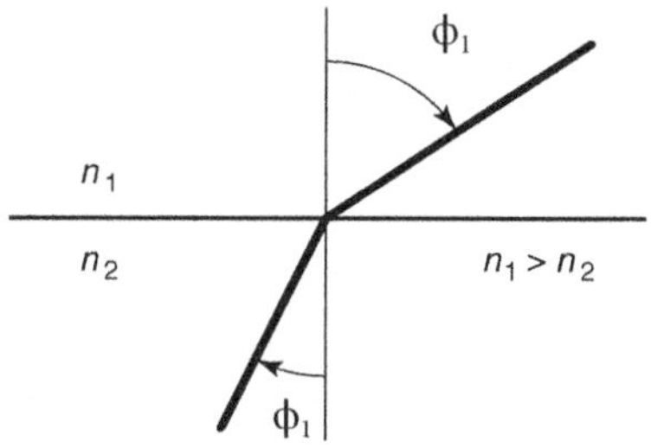

Figure C.2 Snell's law. Refraction at a plane dielectric interface is governed by the boundary conditions for electromagnetic waves. At a plane interface between two dielectric media the change in ray direction is described by Snell's law (see Eqs. (C.6) and (C.7)).

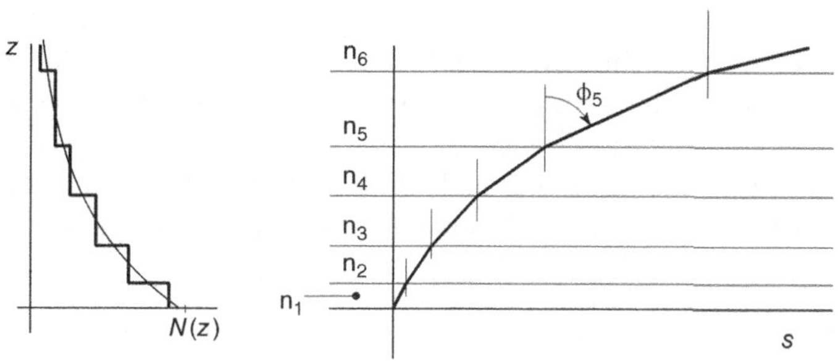

Figure C.3 Step approximation to Snell's law. Snell's law can be applied to a continuously varying media by use of a stepwise approximation, shown at the left. Propagation through such a medium would progress by piecewise linear approximation to the actual ray path. Difficulties can occur at horizontal turning points. Angle ϕ_5 is typical (see Eq. (C.8)).

In order to apply Snell's Law to the propagation problem, the atmosphere might be modeled as a series of steps, as in Figure C.3. Applying Eq. (C.6) at each level we see that,

$$n_1 \sin \phi_1 = n_2 \sin \phi_2 = n_3 \sin \phi_3 = \cdots \tag{C.8}$$

Since all the products $n_i \sin \phi_i$ in Eq. (C.8) are equal independently of the value of refractive index, these products must all be the same. That is, $n(z) \sin \phi = c$, a constant! Further, this conclusion does not depend on the thickness of the layers, so it holds as the layer thicknesses go to zero, leading to a result for a continuous function that $n(z) \sin \phi = c$, also. Therefore, since we determined Eq. (C.8) on the basis of a series of plane interfaces, this is the law governing propagation by ray optics over a *flat* Earth.

C.4.1 Ray Trajectories Over a Flat Earth

What is the behavior of ray trajectories following this law? Still considering a flat Earth. What does Snell's law tell us concerning a ray trajectory? First, in any transition across a refractivity interface, the angle of incidence, ϕ, is always larger on the side with the lower refractive index. Hence, an upward traveling ray launched with $\phi > 0$ in an atmosphere in which the refractive index decreases with height always bends toward the horizontal. At some point the ray actually turns to a level, i.e., horizontal orientation. Beyond this point the ray curves downward with decreasing $|\phi|$.

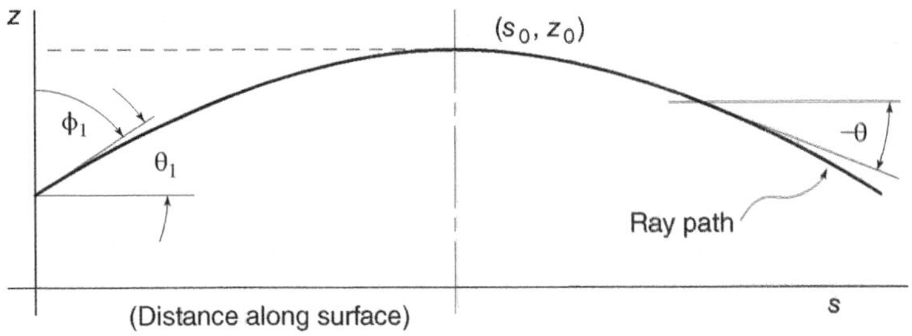

Figure C.4 Ray trajectory over a flat Earth. Ray trajectories in a stratified medium over a flat Earth must be symmetrical about the apogee, located at s_0, z_0. Note definitions of the zenith angle, ϕ, and elevation angle, θ.

By symmetry, the path over a flat Earth must be symmetrical about the horizontal turning point of the ray.[5] Figure C.4 illustrates this symmetry where the location of its highest point on the ray is (s_0, z_0).

Assuming that we know the refractive index $n(z)$ and angle with respect to the horizontal $\theta = \pi/2 - \phi$, Figure C.4, at any one point, we can find a solution for all other points along the ray path. To see this write,

$$n(z_1)\cos\theta_1 = n(z_0) = n(z)\cos\theta = C = \text{constant} \tag{C.9}$$

Manipulate this by substituting for $\cos\theta$ to obtain,

$$n(z_1)\left(\pm\sqrt{1-\sin^2\theta_1}\right) = \mathrm{C} = n(z)\left(\pm\sqrt{1-\sin^2\theta}\right) \tag{C.10}$$

where we must be careful about choosing the sign for the radical, which depends on whether or not we are on the ascending or descending branch of the arc. Now use the definition of the derivative to relate Eqs. (C.9) and (C.10),

$$\frac{dz}{ds} = \tan\theta = \frac{n(z)\sin\theta}{n(z)\cos\theta} = \pm\frac{\sqrt{n(z)^2 - C^2}}{C} \tag{C.11}$$

where the independent variable s is distance along the ground under the curve. We have found a differential equation for the trajectory $z(s)$, subject to the boundary condition that we know a stationary point slope on the curve,

$$\frac{dz}{ds} = \pm\frac{\sqrt{n(z)^2 - C^2}}{\mathrm{C}} \tag{C.12}$$

This result applies to the path followed by a ray—one could also say a photon—through a stratified atmosphere over a flat Earth.

Despite the protestations of The Flat Earth Society, Earth is not flat! How does this change things?

5 To see this, consider that a ray launched horizontally, but reversed in direction, at top of the trajectory must return to the original source. That is, Snell's law applied at every interface on the reversed ray would produce the same angles as for the rising ray, but in the reversed sequence, so that the path is reversed. Hence the shapes of the ascending and descending trajectories are the same. Thus, flipping the descending ray to the source to continue the forward-propagating ray leads to a symmetric curve.

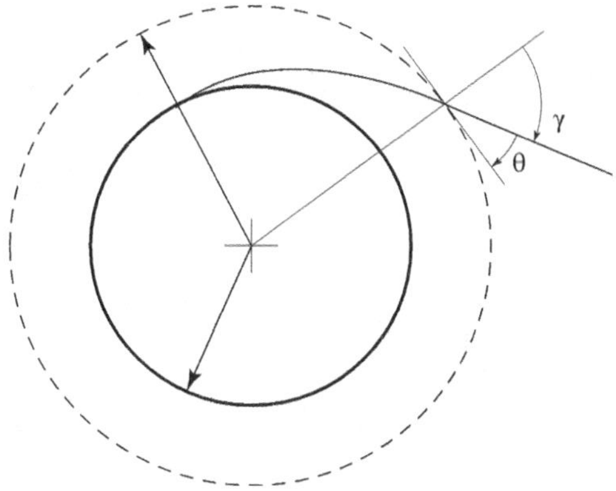

Figure C.5 Geometry of Snell's law under spherical symmetry. Zenith angle, ϕ, and elevation angle, θ, are defined with respect to the local vertical and horizontal, respectively. Refractivity is assumed to be spherically stratified and described by $n(r)$.

C.4.2 In the Case of a Spherical Earth

In this instance, we assume that the Earth is round and that the atmosphere is spherically symmetric, at least in the region of interest. Under these conditions, the refractive index becomes a function only of the radius, r,

$$n = n(r) \tag{C.13}$$

The geometry is defined in Figure C.5. In spherical coordinates it is more convenient to use the angle of the ray from the *local* horizontal, as indicated above. Using θ Snell's Law for spherical symmetry takes the form,

$$\frac{r}{a}n(r)\sin\phi = \mathrm{C} = \tilde{n}(r)\cos\theta \tag{C.14}$$

where $\tilde{n}(r) = r \cdot n(r)/a$. This is the same form as that for the flat earth, except that the refractive index has been scaled by r/a.

C.4.2.1 Equation of Ray Path for Spherical Symmetry

Figure C.6 provides a detailed diagram of the ray trajectory, and defines the angle γ, which is the angle subtended by the rays between two points of interest at Earth's center. The distance along the surface beneath the ray remains s, but the meaning is slightly different from that of the flat earth case, as the ray path is longer with respect to the sub-ray distance than previously. We continue to refer to "height" or "altitude" z, which we now measure from reference radius a, *i.e.* $z = r - a$. Normally a represents the locations of Earth's surface. The derivation of the differential equation of the ray is essentially the same as before.

We express the derivative of the radius with respect to the angle γ by inspection of Figure C.6,

$$\tan\theta = \frac{dr}{r \cdot d\gamma} \tag{C.15}$$

where the variable γ is related to the radius and the distance along the surface by,

$$\gamma = \frac{s}{a} \tag{C.16}$$

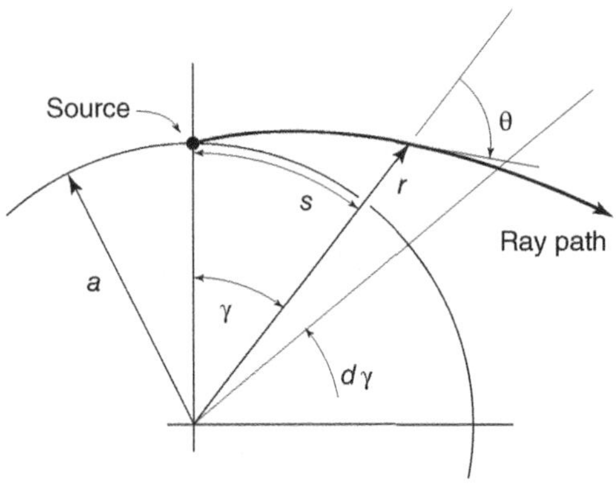

Figure C.6 Geometry of a ray trajectory in a spherically symmetric refractivity field. Progress of the ray is measured in terms of its distance along the spherical surface, s, or by the corresponding angle, γ. Ray height is $z = r - a$. (See text.)

while, typically,[6] $z/a << 1$. Continuing, we can rewrite $\tan\theta$ as,

$$\tan\theta = \pm\frac{\sqrt{\tilde{n}^2 - C^2}}{C} = \frac{dr}{r \cdot d\gamma} = \frac{dz}{(a+z)}ds/a \tag{C.17}$$

Finally, then, from the figure,

$$\tan\theta = \frac{1}{(1+z/a)}\frac{dz}{ds} \tag{C.18}$$

C.4.3 Flat-earth Equivalent Spherical Path

The similarity between the forms of Snell's Law for the linearly and spherically stratified cases, Eqs. (C.9) and (C.14) allow us to connect the two. The result permits us to transform the spherical case into rectangular coordinates (z, s). In order to accomplish this, note that the only difference between the formulas connecting the ray angle to refractive index, Eqs. (C.11) and (C.17), is the modification of $n(r)$ by the factor r/a to give $\tilde{n}(r)$. Change n in the flat-earth equation to $\tilde{n}$ from the spherical-earth equation and make the substitution $r = z + a$. Then, treating $n(r)$ as a function of z we have,

$$n \to \tilde{n}(r)\Big|_{\text{flat earth}} = \frac{r}{a}n(r) = \left(\frac{a+z}{a}\right)n(z+a) = n \cdot (1 + z/a) \tag{C.19}$$

That is, we can use a modified $n(z)$—the rightmost expression in Eq. (C.19)—to obtain the result for the effective flat-Earth $n(r)$ corresponding to a spherical earth. Alternatively, viewing this a simple change of coordinates in which $n(r) \to \tilde{n}(z)$ applied to the spherical case, Eq. (C.17), leads to the same result.

Consider a comparison of two distinct ray paths in each of the two systems as a means of gaining a physical understanding of the relationship of the flat-Earth and spherical-Earth solutions. Figure C.7 gives two cases for a spherical Earth. We plot the same two rays in an equivalent flat-Earth form, Figure C.8. Ray A, which is a straight line when shown in the spherical system, curves

6 Always, in Earth's neutral atmosphere.

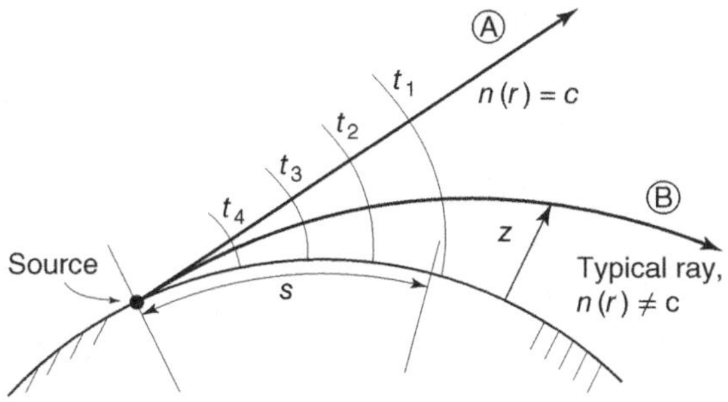

Figure C.7 Two example rays over a spherical Earth. Ray A is a straight line corresponding to propagation through a homogeneous refractivity field. Ray B represents a typical ray in earth's atmosphere. Coordinate s is measured along surface. Wavefronts shown indicate equal intervals of time since transmission. Note that in terms of its height coordinate, $z(s)$, the height of the Ray A increases with distance as a result of the curvature of Earth's surface.

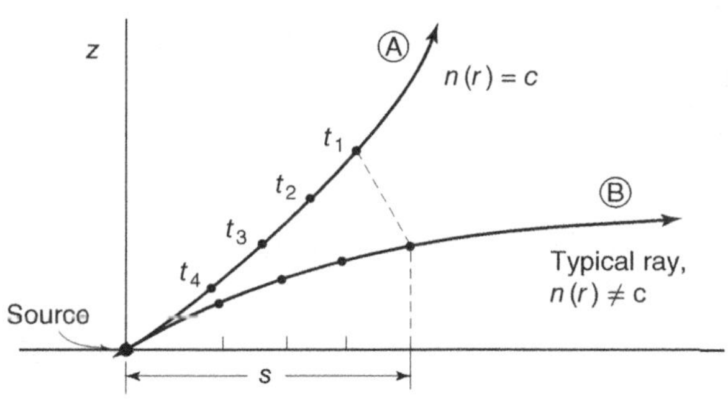

Figure C.8 Two example rays over an equivalent flat Earth. Rays A and B correspond to rays with same labels in Figure C.7; surface coordinate s is same in two figures. Ray A is a straight line in space, but here appears to curve away, upward, from the surface; actually, the surface is "falling away" from the ray. Ray B retains its downward curvature here over most of the arc, but degree of curvature is less as compared with Figure C.7. Indicated pairs of points labeled t_1, etc., correspond to points on same wavefront in spherical coordinates. (Not to scale.)

upward in the flat-earth representation. This upward curvature follows from the choice of flat-earth coordinates in which z is measured from the curved surface of the spherical planet. In actuality, the planet is curving away from the ray and not the other way around. Ray B, however, appears to follow a less curved trajectory in the flat-earth representation because its downward curvature causes it to roughly parallel the curved, spherical surface.

We can solve the differential equation for the ray in equivalent flat-earth coordinates exactly, for some specified $n(r)$. The solution would behave as indicated in Figures C.7 and C.8. This is the proper procedure for evaluating the ray behavior in instances for which the most accurate results are required.

C.4.4 Useful Approximations

C.4.4.1 A Simplifying Coordinate Choice

Suppose we assume that the variation of $n(r)$ is linear. This is an approximation to the exponential variation of our model atmosphere, valid in the lower portion of the first scale height of altitude.

Using a Taylor series expansion and retaining only the first term,

$$n = n(r) = n_0 + \frac{dn}{dz} \cdot z, \frac{dn}{dz} \simeq \text{constant} \tag{C.20}$$

Expanding the equivalent spherical refractivity, Eq. (C.19), above,

$$\tilde{n}(r) = n \cdot \left(1 + \frac{z}{a}\right) \doteq n_0 + \frac{dn}{dz} \cdot z + n_0\frac{z}{a} + \frac{dn}{dz} \cdot \frac{z^2}{a} \tag{C.21}$$

$$\left.\tilde{n}(r)\right|_{z<H} = n \cdot \left(1 + \frac{z}{a}\right) \doteq n_0 + \frac{dn}{dz} \cdot z + n_0\frac{z}{a} \tag{C.22}$$

We only need to keep the first three terms in Eq. (C.21), however, as we are interested only in the lower atmosphere and our expression is accurate in the derivative, which is the parameter controlling ray behavior, Eq. (C.22). Combining the terms below, observe that the *form* is exactly the same as that for $\tilde{n}$, when $n(r) = n_0$ is constant,[7]

$$\tilde{n} \doteq n_0\left(1 + \underbrace{\left[\frac{1}{n_0}\frac{dn}{dz} + \frac{1}{a}\right]}_{Da_{\text{eq}}^{-1}} z\right) = n_0\left(1 + \frac{1}{a_{\text{eq}}}z\right) \tag{C.24}$$

Comparing Eq. (C.24) with Eq. (C.19), radius a of the actual earth has been replaced by a_{eq}. Obviously,

$$a_{\text{eq}} = D\frac{1}{D\frac{1}{a} + D\frac{1}{n_0}\frac{dn}{dz}} = a\frac{n_0}{Dn_0 + aD\frac{dn}{dz}} \tag{C.25}$$

The adopted value for refractive lapse rate is,

$$\left.\frac{dn}{dz}\right|_{z=0} = -3.94 \times 10^{-8}\ \text{m}^{-1} = -1.2 \times 10^{-8}\ \text{ft}^{-1} \tag{C.26}$$

For this value,

$$\left.a\frac{dn}{dz}\right|_{z=0} \simeq -0.25, (a = 6370\ \text{km}) \tag{C.27}$$

Therefore,

$$a_{\text{eq}} = \frac{4}{3}a = 8493\ \text{km (5280 mi)} \tag{C.28}$$

when the effects of sphericity and refraction are combined!

Evaluation of Eq. (C.24) tells us that on a typical day the path of a curved ray propagation resulting from a variable $n(r)$ over a curved earth can be approximated by a straight ray propagating in a constant $n = n_0$ refraction field over a curved earth of 4/3 the radius of the actual Earth. This result is called the "$\frac{4}{3}$-Earth" rule.

---

7 A useful way to view this is to consider the approximation is,

$$n\left(1 + \frac{z}{a}\right) \doteq n + \frac{z}{a} \text{ since } n \doteq 1.00 \tag{C.23}$$

This approximation is valid to about three significant figures in $\tilde{n}$, and more accurate in the derivative.

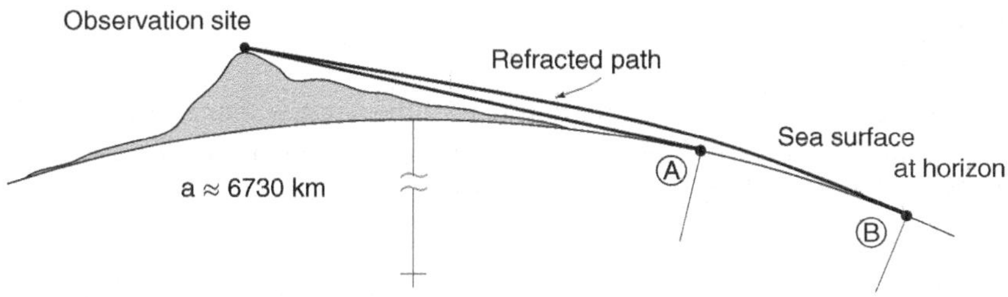

Figure C.9 Apropos the horizon seen through Earth's atmosphere. Sketched is cross section of the view from a ridge overlooking the sea. The horizon is the point from which a ray tangent to the spherical surface of the Earth reaches the observer. Looking west, the geometric horizon is located at position "A," while the apparent horizon is farther away, at "B." The greater distance to point B results from the curvature of the refracted ray, which shifts the tangent to a more remote point than that of the straight line.

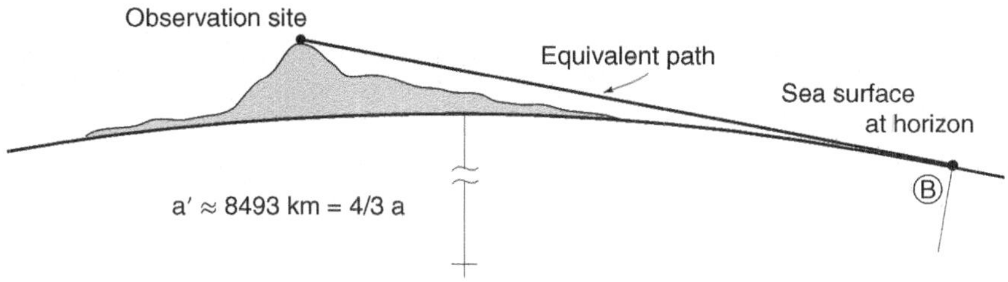

Figure C.10 Horizon for the equivalent Earth path. Typical effects of the refractive index lapse rate can be accounted for by use of an equivalent radius for the earth, in combination with the surface value of the index of refraction. When this is done, the propagation path is treated as a straight line. The distance from the observer along the mean surface to the horizon at point B here is, characteristically, the same as in Figure C.9 (see text).

C.4.5 Interpretation of $\frac{4}{3}$-Earth Rule

Consider the view from a hilltop overlooking the sea. Figure C.9 provides a schematic section of the view to the horizon. A straight line path to the geometric horizon leads to A. The radio horizon, owing to refraction, is farther away, at B. The location of the radio horizon, on a reference typical day,[8] can be accurately approximated by use of a_{eq} from Eq. (C.24). We accomplish this by using a straight line path through an atmosphere of constant refractive index n_0 to the geometric horizon on an "Earth" of radius $4/3a = 8493$ km. This works for the geometry of all points, not just the horizon; we retain the definition of z as height above the reference surface, but distort the geometry by changing the radius of curvature of the sphere, and also holding the refractive index constant, Figure C.10.

Obviously, on any given day an equivalent radius specific to that day can be found by changing dn/dz in Eq. (C.25). Straight line calculations then yield accurate geometric results for the conditions modeled. Use of the equivalent radius should be limited to instances in which the ray paths are confined to the lower portions of the first scale height in the atmosphere. Critical calculations should be implemented by accurate numerical or ray tracing methods. Nevertheless, the equivalent radius is an important tool for quick estimates, while the underlying concept is useful for visualizing the effect of an exponential refractivity on the ray path.

---

8 That is, on a day with conditions corresponding to Eqs. (C.26) and (C.27). These conditions represent an "adopted average" atmosphere.

C.5 Modified Refractivity

As before, it is awkward to work with n, as the important part of the value is four to six places to the right of the decimal. To mitigate this with the spherically symmetric refractive field, we employ a modified refractivity, M,

$$M = \left(n - 1 + \frac{z}{a}\right) \times 10^6 \tag{C.29}$$

which is the spherical equivalent to N. This works because the derivative of the refractive index, dM/dz, is the important part. Note that for z large,

$$n(r) \to 1 \tag{C.30}$$

while

$$\frac{10^6}{a} \times \left.\frac{dM}{dz}\right|_{z\text{ large}} = 0.157\ \text{m}^{-1} \tag{C.31}$$

or, expressed in English units,

$$0.048\ \text{ft}^{-1} \tag{C.32}$$

Near the surface, at $z \sim 0$

$$\left.\frac{dM}{dz}\right|_{z=0} < 0.157\ \text{m}^{-1} \tag{C.33}$$

$$\simeq 0.12 \quad \text{m}^{-1}\ (0.036\ \text{ft}^{-1}) \tag{C.34}$$

Note that $dM/dz > 0$, while $dN/dz < 0$.

An alternative to use of the modified index M is to use the spherical Snell's Law with a linear lapse rate. From this point of view, we can quickly rederive the 4/3-Earth rule. Recall

$$a\frac{dn}{dz} \simeq -0.25 = -\frac{1}{4} \tag{C.35}$$

Let

$$n(z) = n_s + \frac{dn}{dz} \cdot z \simeq n_{\text{surf}} - z/4a \tag{C.36}$$

Then,

$$n(z)\frac{r}{a}\cos\theta = n(z)\left(1 + \frac{z}{a}\right)\cos\theta \tag{C.37}$$

Therefore,

$$n(z)\left(1 + \frac{z}{a}\right) \simeq (n_{\text{surf}} - z/4a)(1 + z/a) \doteq n_{\text{surf}}\left(1 + \frac{3z}{4a}\right) \tag{C.38}$$

Therefore,

$$n_{\text{surf}}\frac{4}{3}a\cos\theta_{\text{surf}} \simeq n_{\text{surf}}\left(\frac{4a}{3} + z\right)(r + z)\cos\theta \tag{C.39}$$

That is, we can use the spherical solution for a constant refractive index $n = n_{\text{surf}}$ and $a_{\text{eq}} = 4a/3$.

C.6 Atmospheric Structure and Modified Refractivity

Figure C.11 presents the modified refractivity, M, for two example atmospheres, the approximate standard atmosphere with refractive lapse rate of 0.12 m^{-1}, and an atmosphere with *critical* refraction, for which the radius of curvature of the ray is equal to its radius from the center of the planet. The standard atmosphere, when described in terms of M, is described by a straight line representing *increasing* refractivity with height. Although the physical refractivity must decrease (see Figure C.1), the modified refractivity increases in order to account for the curvature of the earth. That is, a ray launched into the standard atmosphere will curve upward relative to the surface; M increases upward in order to reflect this behavior because rays always bend toward a region of increasing refractive index. The atmosphere with critical refraction has a segment of $M(z)$ that is vertical. A ray launched horizontally in this region will travel horizontally in this coordinate system; in actuality, such rays bend toward the surface at the same rate, per unit surface distance covered, as the surface. As a consequence, the altitude of these rays never changes. The name "critical refraction" refers to this property of the medium as it is a transitional case in that curves of $M(z)$ with a lower, i.e., negative, slope than critical will result in rays that curve into the surface.

Figures C.12 and C.13 present the behavior of rays in a standard and critically refracting atmosphere. A straight line ray, corresponding to a constant value of n with height, is shown for comparison. The two figures demonstrate the differences by comparison of the ray curvature with the spherical surface curvature. If the rays are launched perpendicular to the direction of the center (i.e., horizontally), then they will diverge as shown in Figure C.13.

C.6.1 Atmospheric Ducting

The real atmosphere is remarkably varied. Over warm oceans it is common for *ducts* to form. Ducts are a type of dielectric waveguides usually created by the variable distribution with height of water vapor. Figure C.14 illustrates $M(z)$ for a typical instance, and indicates the behavior of the rays in the duct. The case shown corresponds to a *surface duct* since the rays are trapped between the surface and a critical layer at an altitude indicated by the dotted line.

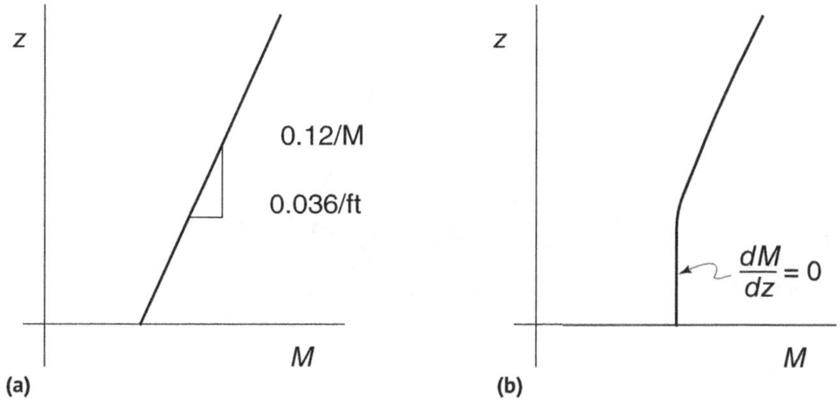

Figure C.11 Standard atmosphere and critical refraction. Modified index of refraction $M(r)$ at left, (a), is characteristic of the lower atmosphere under standard conditions. On right, (b), is example atmosphere with "critical" refraction near surface; the lower, vertical segment of this curve implies no ray curvature over an equivalent flat earth. (See text.)

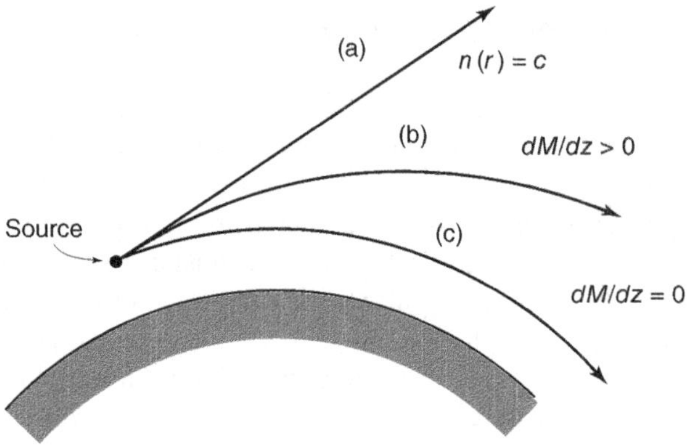

Figure C.12 Three rays in a spherical atmosphere. The three rays are indicative of the behavior of (a) an atmosphere with no variation in refractive index, $dM/dz = a \times 10^6$, (b) the standard atmosphere for which $dM/dz > 0$, and (c) critical refraction in which the curvature of the ray is the same as its radius, as $dM/dz = 0$. (See text.)

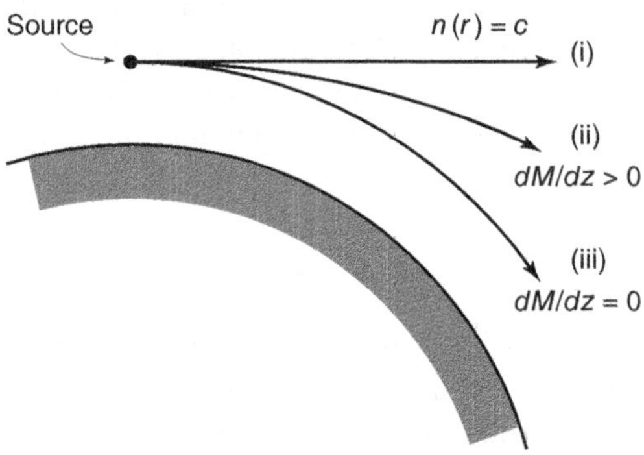

Figure C.13 Three rays launched horizontally. Three rays shown correspond to the three conditions of Figure C.12, except that the rays are launched in the direction of the local horizontal at their point of origin.

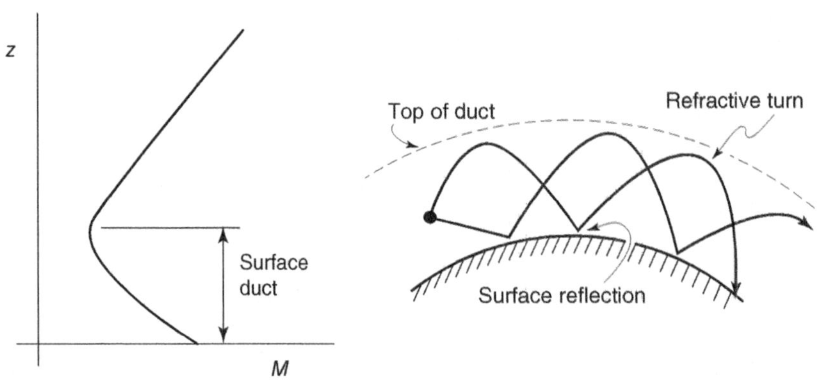

Figure C.14 Conditions in a surface duct. The duct is formed by sub-critical lapse rate above the surface. This causes rays with low elevation angles to be returned to the surface by refraction, where they are reflected at the angle of incidence. Such rays are trapped in the duct. (See text.)

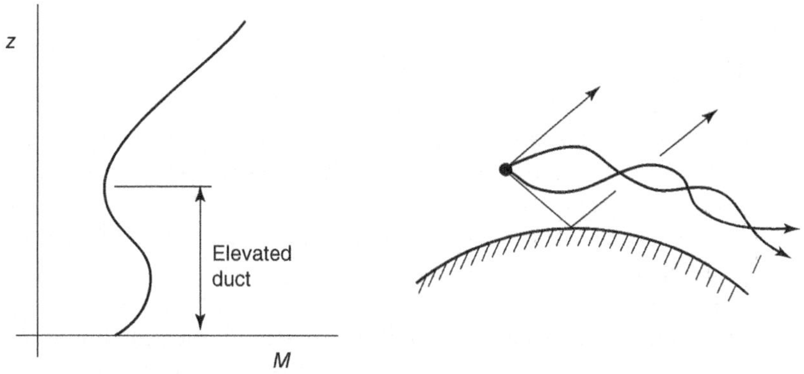

Figure C.15 Conditions in an elevated duct. An elevated duct is similar to the surface duct, Figure C.14 except that the surface is replaced by a lower region of positive refractivity lapse rate. (See text.)

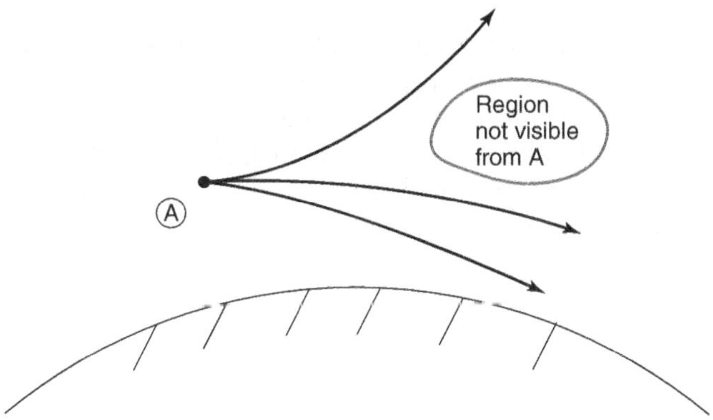

Figure C.16 Ray paths for an excluded region. A lateral variation in refractivity structure can result in regions from which rays are excluded (for a particular source point). Scattering objects in such a region cannot be observed along direct propagation paths by a radar at point A.

Another possibility is the *elevated duct*, illustrated in Figure C.15. In this instance, the duct is located in the altitude region surrounding the local maximum in M. In ideal cases "ducted rays" would form uniform patterns, corresponding to perfect dielectric waveguide modes. This seldom occurs in real atmospheres because of irregularities in the atmospheric structure with lateral position. Nevertheless, ducting is a very real phenomenon that occurs regularly in certain parts of the world. Elevated ducts are regularly present along the West Coast of the U.S., providing reliable communications at VHF north and south along the coast, and with aircraft for a few thousand kilometers out to sea in the direction of Hawaii. In another part of the world there are published reports from WWII of radars located on the west coast of India routinely observing the entire Persian Gulf.

Just as refraction can connect geometrically obscured regions, it can also exclude some regions from observation. An important feature of ducts is that rays propagating in ducts, defined as rays that are at some point horizontal within the duct, can not be observed exterior to the duct (Figure C.16).

C.6.2 Distortion of Antenna Beams

C.6.2.1 Magellan Example at Venus

Refraction acting over the solid angle of an antenna beam distorts the beam shape by the bending of the rays. Differential bending within the beam can significantly affect the gain, as well. These effects can be significant in terms of performance. More commonly, the effect of beam distortion is to degrade the accuracy of the measurements. Angular information derived from pointing of narrow beam antennas is the most seriously affected, as the true direction of a target differs from the direction of arrival of its echo signals.

One example is the case of the NASA *Magellan* radar mapper of Venus. As it happens, the atmosphere of Venus is uniquely deep and dense among the terrestrial planets. Surface conditions are appalling, with very high temperatures and pressures in a gas dominated by carbon dioxide (CO_2),

$$T_{\text{surf}} = 750\ \text{K} \tag{C.40}$$

$$P_{\text{surf}} = 90\ \text{atms} \qquad (\approx 9 \times 10^6\ \text{Pa}) \tag{C.41}$$

Magellan mapped the surface by means of a "synthetic aperture radar" (SAR) system—a type of side looking radar—using a narrow beam antenna looking down through the atmosphere at angles in the range of 45°. The mapping geometry is similar to the example system described in Section 2.4. Refraction in the atmosphere bent the beam downward by as much as 2°, which would have led to a significant mapping error had the refraction not been accurately estimated and the effect corrected. The Magellan situation is depicted in Figure C.17.

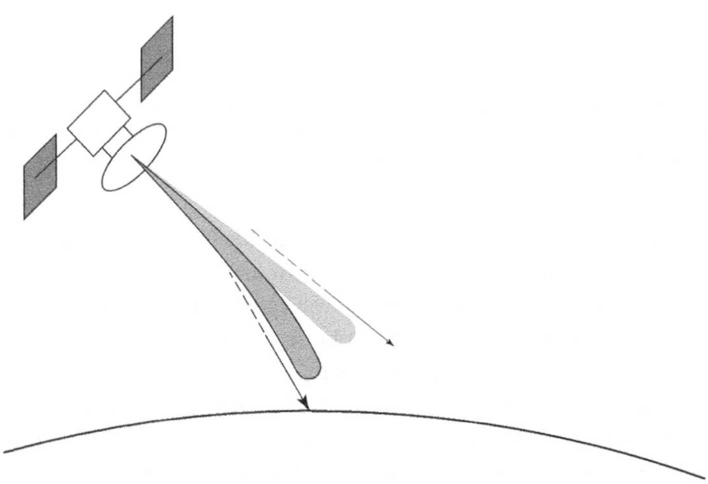

Figure C.17 Refraction of the Magellan beam at Venus. Extreme refraction in the atmosphere of Venus has a profound effect on radio propagation. Magellan Venus Mapping Radar antenna beam direction (dark grey) was deflected significantly with respect to a beam in free space (light grey). The Magellan Radar Mapper of Venus operated at a typical viewing depression angle of about 45°; the antenna beam was deflected downward about 2° under these conditions. Bending in figure is exaggerated by about 10:1.

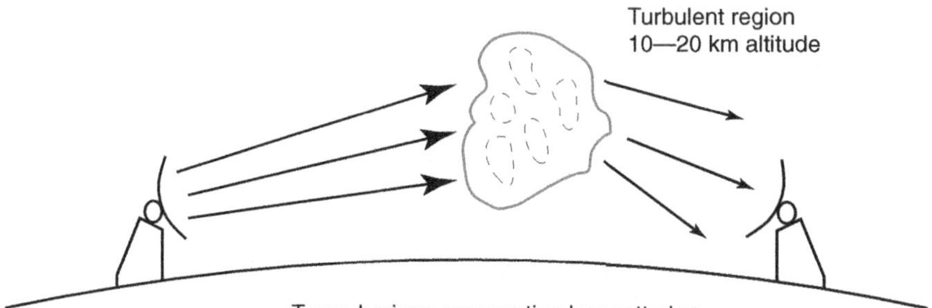

Figure C.18 Tropospheric scattering geometry. Scattering by turbulence in the atmosphere provides reliable over the horizon communications. Considerable transmitter power and large antennas are required.

C.7 Tropospheric Scatter

Tropospheric scattering systems are based on scattering of radar signals by irregularities in the refractive index associated with turbulence in the lower atmosphere. This is an important propagation mechanism at meter and shorter wavelengths. Usually the scattering geometry is arranged for observation of *forward scatter*, as forward scattering is much stronger than in other directions. Figure C.18 indicates the geometry. Tropospheric scatter provides "over-the-horizon" links, typically between high-power, high-gain antenna sites set up to view a common volume of the atmosphere. Tropospheric scatter links are not commonly used as a sensing element of radar systems, although they are sometimes used to study small scale atmospheric and ionospheric processes, and they are sensitive to the presence of scattering objects within the volume of the intersecting antenna beams. These systems are, in fact, a twist on radar itself.

We note that similar systems and geometries are employed in meteor-scatter communications systems. These are true hybrids between bistatic radars and telecommunication systems. As a radar, each terminal of the system "listens" for the presence of an obliquely scattered echo from the other site. Echoes are present only when an ionized meteor trail is created in the common volume and with the correct orientation to produce strong scattering toward the receiving site. Instantiation of the path is detected and high-speed transmissions of accumulated messages initiated.

Epilogue

It has been our goal here to follow the development of radar from its scientific origins, through its rapid growth supporting military and defense needs, to today's systems devoted to better understanding our Earth and managing our interactions with the physical environment. Our approach has been to stress the underlying mathematical and physical foundations upon which the technology is built, with insight into the engineering constraints that govern real-world systems and applications. Our hope is that students of the subject will be able to apply the accumulated insights from radar development to be able to address new situations and problems that arise as our technologies and needs change.

Thus the emphasis here is on seemingly esoteric topics such as radar scattering and detection theory. It is easy to get lost in the math in these areas and lose sight of just what we want to accomplish: the ability to interpret what information is contained in the recorded echoes. Our aim has been to give the engineers and scientists who are using these data means for optimizing and extracting this information.

It is a matter of pride among many in the radar field that this subject has driven so many of the technologies that undergird modern society. The needs for fast computing to be able to react in real time to threats detected by radar, for miniaturization of components that can be deployed in aircraft and spacecraft radar systems, and for measurement and detection theory so valuable in modern communication systems all have roots in radar. And future systems now monitoring many aspects of Earth from the vantage point of space are enabled by the continued development and application of radar remote sensing technologies.

As the authors of these notes, resulting from a series of courses offered at Stanford from 1995 to 2020, we hope we have provided a suitable introduction to the subject, beginning with the basic characteristics of radars from first principles, with enough follow-through on modern system applications, to be suitable for all readers with technical and scientific training who are interested in a broad range of applications. It is this set of readers who will promote thinking about new applications, particularly in remote sensing and scientific uses of radars.

Index

Wiley The manufacturer's authorized representative according to the EU General Product Safety Regulation is Wiley-VCH GmbH, Boschstr. 12, 69469 Weinheim, Germany, e-mail: Product_Safety@wiley.com.

Printed and bound by CPI Group (UK) Ltd, Croydon, CR0 4YY
15/09/2026
02217457-0006